W9-CSB-010

Radar Cross Section Handbook

Volume 2

Radar Cross Section Handbook

by

George T. Ruck
Donald E. Barrick
William D. Stuart
Clarence K. Krichbaum

Battelle Memorial Institute
Columbus Laboratories, Columbus, Ohio

Editor:

George T. Ruck

Volume 2

PLENUM PRESS • NEW YORK - LONDON • 1970

This handbook was supported in part by the Advanced Research Projects Agency of the United States Department of Defense

Library of Congress Catalog Card Number 68-26774
SBN 306-30343-4

A Division of Plenum Publishing Corporation
227 West 17th Street, New York, N. Y. 10011

United Kingdom edition published by Plenum Press, London
A Division of Plenum Publishing Company, Ltd.
Donington House, 30 Norfolk Street, London W.C.2, England

Printed in the United States of America

CONTENTS

Volume 1

CHAPTER 3

Spheres

CHAPTER 4

Cylinders

CHAPTER 5

Ellipsoids and Ogives

CHAPTER 6

Cones, Rings, and Wedges

Volume 2

CHAPTER 7

Planar Surfaces

CHAPTER 8

Complex Bodies

CHAPTER 9

Rough Surfaces

CHAPTER 10

Ionized regions

CHAPTER 11

Radar Cross-Section Measurements

Radar Cross Section Handbook

Volume 2

Chapter 7

PLANAR SURFACES

G. T. Ruck

7.1. LAYERED MEDIA

For a plane wave incident on an infinite medium composed of homogeneous planar layers with various dielectric and magnetic parameters, the primary characteristics of interest are the reflected and transmitted fields. These are specified in terms of reflection and transmission coefficients, which can be defined in several ways. One definition often used is

$$\tilde{\tilde{R}} = \frac{\mathbf{E}^r}{\mathbf{E}^i}, \qquad \tilde{\tilde{T}} = \frac{\mathbf{E}^t}{\mathbf{E}^i} \tag{7.1-1}$$

where $\mathbf{E}^i$ is the incident electric field, and $\mathbf{E}^r$ and $\mathbf{E}^t$ are the reflected and transmitted electric fields, respectively. In Eq. (7.1-1), $\tilde{\tilde{R}}$ and $\tilde{\tilde{T}}$ are, in general, three-dimensional symmetric dyads, which for planar media can always be reduced to two-dimensional symmetric dyads when the coordinates are properly chosen. For isotropic media, which is the only situation considered here, the dyads can be diagonalized by separating the fields into components with the electric field respectively perpendicular to the plane of incidence, and lying in the plane of incidence. Thus only the coefficients $R_\perp$ and $R_\parallel$, describing the perpendicular and parallel polarization cases, respectively, are necessary to describe completely the reflected fields from an isotropic planar layered media. Similarly, $T_\perp$ and $T_\parallel$ allow a complete description of the transmitted fields.

7.1.1. Semi-infinite Media

Considering the simplest problem initially, that of a plane wave of frequency ω, incident on the interface between two semi-infinite media of different material constants, the coordinates of the problem are shown in Figure 7-1, where θ_1 specifies the angle of incidence, ϵ_1, μ_1 the material properties of medium 1, and ϵ_2, μ_2 the material properties of medium 2. Throughout this chapter, the permittivity ϵ and permeability μ will be assumed to be complex, and unless otherwise indicated,

$$\epsilon = \epsilon' + i\epsilon'', \qquad \mu = \mu' + i\mu''$$

thus accounting for both dielectric and magnetic loss in the medium.

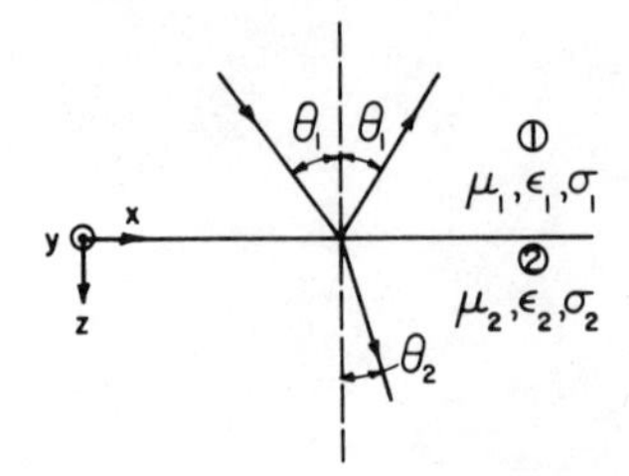

Fig. 7-1. Coordinates for incidence on an interface between two semi-infinite media of different material constants.

For a plane wave incident on an interface between two planar media, application of the electromagnetic boundary conditions at the interface demonstrates that the angle of reflection equals the angle of incidence, this being a consequence of Snell's law,[1] the formal statement of which is

$$k_1 \sin \theta_1 = k_i \sin \theta_i \tag{7.1-2}$$

where

$$k_i = \omega \sqrt{\mu_i \epsilon_i}$$

Applying the boundary conditions at the interface along with Snell's law, the reflection coefficients, $R_\perp$ and $R_\parallel$, defined as

$$R_\perp = E_y{}^r / E_y{}^i, \qquad R_\parallel = E_x{}^r / E_x{}^i \tag{7.1-3}$$

can be expressed in terms of equivalent complex impedances as

$$R_{\parallel,\perp} = -(z_1 - z_2)/(z_1 + z_2) \tag{7.1-4}$$

while the transmission coefficient components are†

$$T_{\perp,\parallel} = 2z_2/(z_1 + z_2) \tag{7.1-5}$$

The z_i are equivalent complex impedances, defined as

$$z_i{}^\perp = \frac{1}{\cos \theta_i} \sqrt{\frac{\mu_i}{\epsilon_i}} = \frac{\eta_i}{\cos \theta_i} = \frac{E_{iy}}{H_{ix}} \tag{7.1-6}$$

† The statement is occasionally seen that, as a consequence of energy conservation, the sum of the squares of the reflection and transmission coefficients must be unity. As these coefficients are defined above, this is not necessarily true. The actual relationship between R and T at any interface is $1 + R_{\perp,\parallel} = T_{\perp,\parallel}$, as can be easily verified from Eqs. (7.1-4) and (7.1-5). Energy conservation, however, does require that the rate at which energy flows into the interface equal the rate at which it leaves, and this condition is indeed satisfied by the reflection and transmission coefficients given above. Only in the case of a lossless dielectric layer with the same medium on both sides does

$$|R|^2 + |T|^2 = 1$$

where R describes the reflected field on the incident side of the layer and T describes the field transmitted entirely through the layer.

for the incident electric field vector perpendicular to the plane of incidence, and

$$z_i^{\parallel} = \cos\theta_i \sqrt{\mu_i/\epsilon_i} = \eta_i \cos\theta_i = E_{ix}/H_{iy} \tag{7.1-7}$$

for the incident electric field vector lying in the plane of incidence. Using these definitions and [from Eq. (7.1-2)]

$$\cos\theta_i = [1 - (k_1/k_i)^2 \sin^2\theta_1]^{1/2} \tag{7.1-8}$$

where k_i is the complex wave number of medium (i), then the reflection coefficients become

$$R_\perp = \frac{\eta_2/\eta_1 \cos\theta_1 - [1 - (k_1/k_2)^2 \sin^2\theta_1]^{1/2}}{\eta_2/\eta_1 \cos\theta_1 + [1 - (k_1/k_2)^2 \sin^2\theta_1]^{1/2}} \tag{7.1-9}$$

and

$$R_\parallel = -\frac{\eta_1/\eta_2 \cos\theta_1 - [1 - (k_1/k_2)^2 \sin^2\theta_1]^{1/2}}{\eta_1/\eta_2 \cos\theta_1 + [1 - (k_1/k_2)^2 \sin^2\theta_1]^{1/2}} \tag{7.1-10}$$

These are known as the Fresnel coefficients,[2,3] and the reflection properties of any arbitrarily polarized plane wave can be determined in terms of $R_\perp$ and $R_\parallel$, since the incident field can always be expressed as the linear combination of a component polarized in the plane of incidence and a component polarized perpendicular to the plane of incidence.

Definitions of the reflection coefficients differing slightly from that given by Eq. (7.1-3) are often used in books on electromagnetic theory. These are

$$R'_\perp = E_y^r/E_y^i \tag{7.1-11}$$

and

$$R'_\parallel = H_y^r/H_y^i \tag{7.1-12}$$

The relationships between these reflection coefficients and those defined by Eq. (7.1-3) are quite simple, i.e., $R_\perp = R'_\perp$, and $R_\parallel = -R'_\parallel$. Relationships analogous to Eq. (7.1-4) hold for the primed reflection coefficients, where

$$R'_\perp = (y'_1 - y'_2)/(y'_1 + y'_2) \tag{7.1-13}$$

and

$$R'_\parallel = (z'_1 - z'_2)/(z'_1 + z'_2) \tag{7.1-14}$$

The parameters, z'_i and y'_i, are equivalent complex impedances and admittances defined by

$$z'_i = (\mu_i/\epsilon_i)^{1/2} \cos\theta_i = E_{xi}/H_{yi} \tag{7.1-15}$$

and

$$y'_i = (\epsilon_i/\mu_i)^{1/2} \cos\theta_i = -H_{xi}/E_{yi} \tag{7.1-16}$$

During the remainder of this section the unprimed reflection coefficients, $R_\perp$ and $R_\parallel$, will be used; however, in Section 7.2 extensive use will be made of the reflection coefficients $R'_\perp$ and $R'_\parallel$ as defined by Eqs. (7.1-11) and (7.1-12).

To illustrate the influence of the angle of incidence, polarization, and material properties on the reflection coefficient, several special cases will be considered:

(1) $k_1 < k_2$; ϵ_1, ϵ_2 *real; and* $\mu_1 = \mu_2 = \mu_0$

Under these conditions R is real and given by

$$R = \frac{\sqrt{\epsilon_1/\epsilon_2}\cos\theta_1 - \sqrt{1-(\epsilon_1/\epsilon_2)\sin^2\theta_1}}{\sqrt{\epsilon_1/\epsilon_2}\cos\theta_1 + \sqrt{1-(\epsilon_1/\epsilon_2)\sin^2\theta_1}} \tag{7.1-17}$$

$$R_\parallel = -\frac{\sqrt{\epsilon_2/\epsilon_1}\cos\theta_1 - \sqrt{1-(\epsilon_1/\epsilon_2)\sin^2\theta_1}}{\sqrt{\epsilon_2/\epsilon_1}\cos\theta_1 + \sqrt{1-(\epsilon_1/\epsilon_2)\sin^2\theta_1}} \tag{7.1-18}$$

Curves of R are given in Figure 7-2 for ϵ_2/ϵ_1 equal to 10, 4, and 2.

From Figure 7-2, it is apparent that $R_\parallel$ is zero for a particular angle of incidence which depends on ϵ_2/ϵ_1. This angle is known as the Brewster angle, and is given by

$$\tan\theta_B = k_2/k_1 = \sqrt{\epsilon_2/\epsilon_1} \tag{7.1-19}$$

For the incident field polarized perpendicular to the plane of incidence, there is, in general, no angle at which zero reflection occurs.

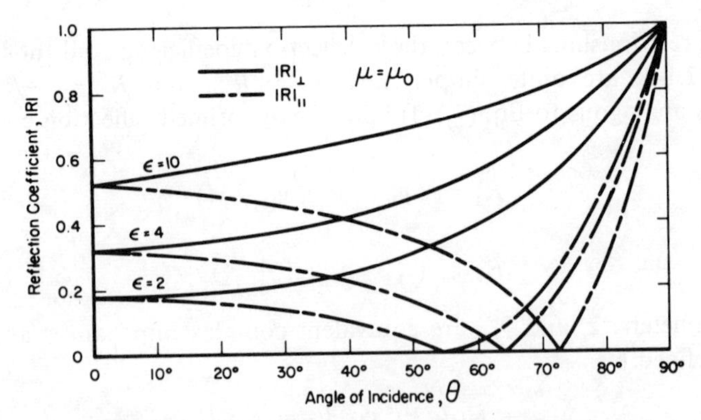

Fig. 7-2. Reflection coefficients versus the angle of incidence for a plane wave incident on a semi-infinite dielectric slab with $\mu_r = 1$, $\epsilon_r = 2$, 4, and 10.

(2) $k_1 < k_2$; $\mu_1 = \mu_2$; *and* ϵ_1 , ϵ_2 *complex*

This case corresponds to reflection from the interface between two dielectric media of which one or both are lossy and, in general, R is complex. When medium (2) is lossy, there is no longer zero reflection at the Brewster angle; although for small values of loss, a reflection minimum occurs at this angle.

Computation of the magnitude and phase of R for various dielectric parameters and angles of incidence is tedious when the medium is lossy; however, using Figures 7-3a, b, c, this can be done by a simple graphical method developed by Panchenko.[4] These figures are valid for an interface between free space and a lossy dielectric with a relative dielectric constant of $\epsilon_2 = | \epsilon_2 | e^{-i\delta}$, and values of $| \epsilon_2 |$ from 4 to 1000. The graphs are used by determining, from Figure 7-3a, the value of the parameter ρ for a particular ϵ_2 and angle of incidence θ_1 .

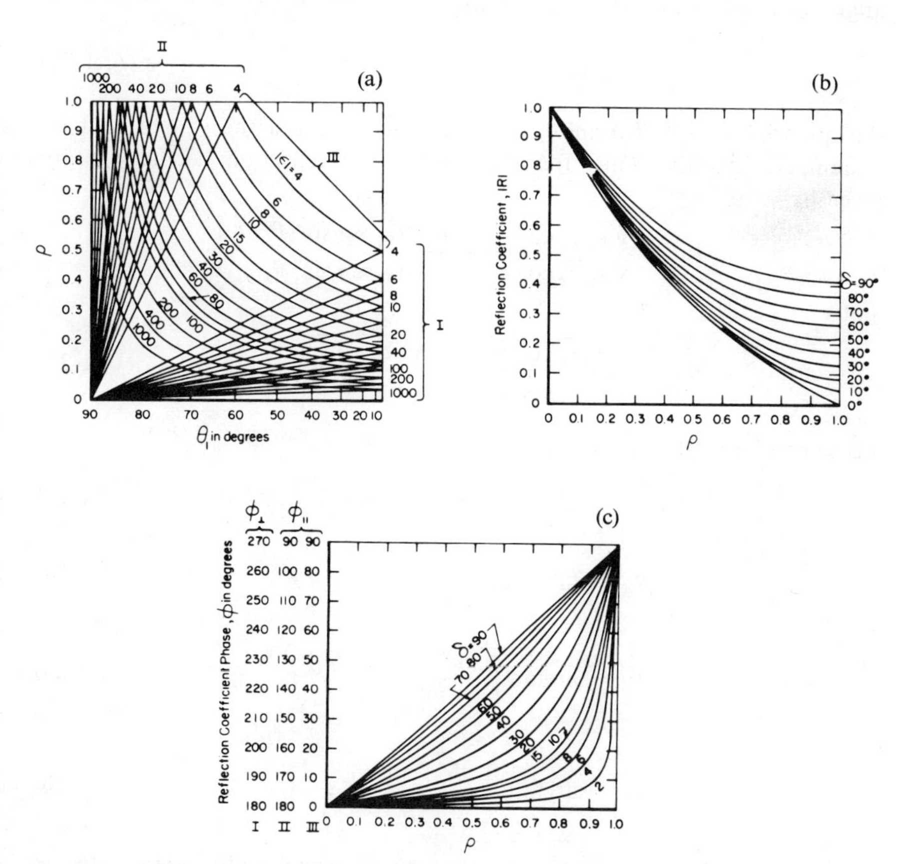

Fig. 7-3. Curves for graphically determining the reflection coefficient for any angle of incidence on a lossy dielectric interface with $4 \leqslant | \epsilon_r V | \leqslant 1000$ [after Panchenko[4]].

Then entering Figures 7-3b and 7-3c with the value of ρ obtained from Figure 7-3a along with the loss angle δ, the magnitude and phase of the reflection coefficient can be found, respectively. The value of ρ obtained from curve set I in Figure 7-3a is used in Figures 7-3b and 7-3c for perpendicular polarization, while the value obtained from sets II and III is used for parallel polarization. Some approximations were made in obtaining these curves, and the maximum error occurs with small $|\epsilon_2|$, and θ_1 near the Brewster angle when δ equals zero. For example, with perpendicular polarization, the maximum error in the reflection coefficient occurs when θ_1 is 35° and $\delta = 0$. For $|\epsilon_2| = 4$, this amounts to 4.5% and falls rapidly as $|\epsilon_2|$ increases.

(3) $k_2 < k_1$; $\mu_1 = \mu_2$; *and* ϵ_1 , ϵ_2 *real*

Under these conditions an examination of Eq. (7.1-8) shows that for angles of incidence greater than a value given by

$$\sin \theta_\tau = \sqrt{\epsilon_2/\epsilon_1} = k_2/k_1 \tag{7.1-20}$$

the quantity $1 - (k_1/k_2) \sin^2 \theta_1$ becomes negative, and the square root an imaginary quantity. Thus for $|\theta_1| \geqslant \theta_\tau$, the reflection coefficients are given by

$$R_\perp = \frac{\sqrt{\epsilon_1/\epsilon_2} \cos \theta_1 - i \sqrt{(\epsilon_1/\epsilon_2) \sin^2 \theta_1 - 1}}{\sqrt{\epsilon_1/\epsilon_2} \cos \theta_1 + i \sqrt{(\epsilon_1/\epsilon_2) \sin^2 \theta_1 - 1}} \tag{7.1-21}$$

and

$$R_\parallel = \frac{\sqrt{\epsilon_2/\epsilon_1} \cos \theta_1 - i \sqrt{(\epsilon_1/\epsilon_2) \sin^2 \theta_1 - 1}}{\sqrt{\epsilon_2/\epsilon_1} \cos \theta_1 + i \sqrt{(\epsilon_1/\epsilon_2) \sin^2 \theta_1 - 1}} \tag{7.1-22}$$

These can be written as

$$R_\perp = (a - ib)/(a + ib) = e^{-2i\phi_\perp} \tag{7.1-23}$$

and

$$R_\parallel = (a' - ib')/(a' + ib') = e^{-2i\phi_\parallel} \tag{7.1-24}$$

with

$$\phi_\perp = \arctan \left[\frac{\sin^2 \theta_1 - (\epsilon_2/\epsilon_1)}{\cos^2 \theta_1}\right]^{1/2} \tag{7.1-25}$$

and

$$\phi_\parallel = \arctan \left[\frac{(\epsilon_1/\epsilon_2) \sin^2 \theta_1 - 1}{(\epsilon_2/\epsilon_1) \cos^2 \theta_1}\right]^{1/2} \tag{7.1-26}$$

Equations (7.1-23) and (7.1-24) show that for angles of incidence greater than or equal to θ_τ , total reflection occurs, and only the phase of the reflected field varies as a function of the angle of incidence. For this case, detailed examination of the field in medium (2) shows that an energy flow parallel to the interface exists even though total reflection results for the incident field. The question of where the energy comes from to sustain this flow can be resolved, when the fact that the incident field is actually finite both in space and time is considered. When a field of finite beamwidth is incident on an interface of this type, it is found that energy enters the second medium, travels laterally along the interface for some distance, and then leaves. The result of this is that the reflected beam is laterally displaced from the incident beam. Expressions for the lateral displacement of an incident ray have been obtained and are as follows:[5]

$$\Delta_{\perp} = \frac{\lambda_0}{\pi} \frac{\tan \theta_1}{\sqrt{\sin^2 \theta_1 - (\epsilon_2/\epsilon_1)}} \tag{7.1-27}$$

$$\Delta_{\parallel} = \frac{\lambda_0}{\pi} \left(\frac{\epsilon_1}{\epsilon_2}\right) \frac{\tan \theta_1}{\sqrt{\sin^2 \theta_1 - (\epsilon_2/\epsilon_1)}} \tag{7.1-28}$$

Any loss in medium (2) prevents total reflection from occurring, and the reflection coefficients can be obtained from the general equations (7.1-9) and (7.1-10). The case where $k_2 < k_1$ is encountered in practice primarily for a wave incident on an air–ionized medium boundary. This will be considered further in Chapter 10.

When $(k_1/k_2)^2 \sin^2 \theta_1 \ll 1$, then this term can be neglected in Eqs. (7.1-9) and (7.1-10) resulting in considerable simplification in the reflection coefficients.

$$R_{\perp} \approx \frac{z_2 \cos \theta_1 - z_1}{z_2 \cos \theta_1 + z_2} \tag{7.1-29}$$

and

$$R_{\parallel} \approx \frac{z_2 - z_1 \cos \theta_1}{z_2 + z_1 \cos \theta_1} \tag{7.1-30}$$

This condition occurs for θ_1 near zero, or for $|k_2| \gg |k_1|$; that is medium (2), relatively lossy, or having a high dielectric constant. When these conditions hold, the components of the fields in medium (1) tangential to the boundary are related by $\mathbf{E}_t \approx \sqrt{\mu_2/\epsilon_2}\,(\hat{\mathbf{n}} \times \mathbf{H}_t)$; this is a special case of the Leontovich boundary condition discussed in Section 2.2.5.

7.1.2. Arbitrary Layered Media

A considerably more complex problem results when a plane wave is incident at an arbitrary angle on a medium which consists of an arbitrary number of plane layers, each of which is homogeneous and has specific

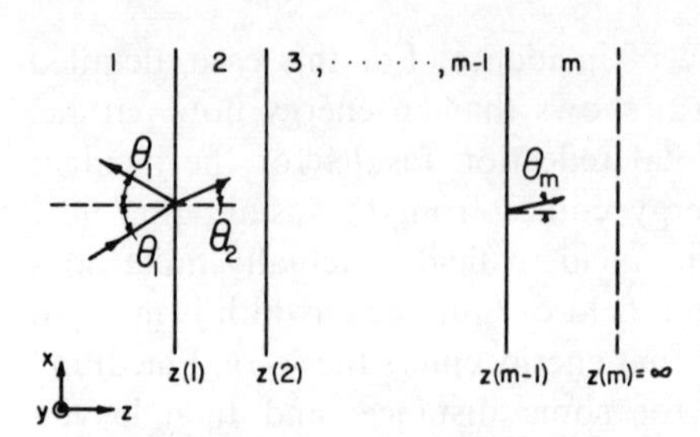

Fig. 7-4. Coordinates for a plane wave incident on a semi-infinite medium consisting of m layers.

dielectric properties. This situation is shown in Figure 7-4, where the fields and parameters pertaining to each layer are labeled by the subscript m. The tangential field components on each layer can be related by the matrix recursion relations,

$$\begin{bmatrix} E_{m-1} \\ H_{m-1} \end{bmatrix} = [A_m] \begin{bmatrix} E_m \\ H_m \end{bmatrix} \tag{7.1-31}$$

or $f_{m-1} = A_m f_m$. In this expression, E_m and H_m represent the y and x components of E and H, respectively, for perpendicular polarization, or the x and y components, respectively, for parallel polarization. The transformation matrix, A_m, is given by

$$[A_m] = \begin{bmatrix} \cos \alpha_m & iz_m \sin \alpha_m \\ \dfrac{i}{z_m} \sin \alpha_m & \cos \alpha_m \end{bmatrix} \tag{7.1-32}$$

where

$$\begin{aligned} \alpha_m &= k_m[z(m) - z(m-1)] \cos \theta_m \\ &= k_m[z(m) - z(m-1)] \sqrt{1 - (k_1/k_m)^2 \sin^2 \theta_1} \end{aligned} \tag{7.1-33}$$

and

$$z_m{}^{\perp} = \frac{1}{\cos \theta_m} \sqrt{\frac{\mu_m}{\epsilon_m}} = \frac{\sqrt{\mu_m/\epsilon_m}}{\sqrt{1 - (k_1/k_m)^2 \sin^2 \theta_1}} \tag{7.1-34}$$

$$z_m{}^{\parallel} = \cos \theta_m \sqrt{\frac{\mu_m}{\epsilon_m}} = \sqrt{1 - \left(\frac{k_1}{k_m}\right)^2 \sin^2 \theta_1} \sqrt{\frac{\mu_m}{\epsilon_m}} \tag{7.1-34}$$

By using these recursion relations then,

$$f_1 = (A_2 A_3 \cdots A_{m-1} A_m) f_m, \qquad \text{or} \quad f_1 = A f_m$$

where

$$A = A_2 A_3 \cdots A_{m-1} A_m$$

Since $z(m)$ is infinite, $A_m = 1$, and A can be written in matrix form as

$$A = \begin{bmatrix} a_{11} & a_{12} \\ a_{21} & a_{22} \end{bmatrix} = A_2 A_3 \cdots A_{m-1} \tag{7.1-35}$$

An effective input impedance for the layer combination can be defined as

$$Z = \frac{a_{11} z_m - a_{12}}{a_{22} - a_{21} z_m} \tag{7.1-36}$$

Using this, the reflection coefficient in region 1 is found to be

$$R = \frac{Z - z_1}{Z + z_1} \tag{7.1-37}$$

and the transmission coefficient for the field in region m is

$$T = \frac{2(Z a_{22} + a_{12})}{Z + z_1} \tag{7.1-38}$$

The above relations provide a conceptually simple and organized procedure for obtaining the reflection coefficient for a plane wave of any polarization and angle of incidence on an arbitrary layered medium. The effects of the wave polarization and angle of incidence are carried in the coefficients α_m and z_m, and do not influence the forms of the general relations (7.1-36) to (7.1-38). Even though the technique for arriving at R and T, as expressed above, is straightforward, the final forms, when expressed in terms of the dielectric parameters, frequency, and angle of incidence, are very complex if many layers are present. However, this formulation is exceptionally well suited for the use of a digital computer to determine the A_m matrices and the resulting A matrix.

The case where $m = 3$, corresponding to a single layer, will be considered in more detail. For $m = 3$, Eq. (7.1-35) gives $A = A_2$, or

$$A = \begin{bmatrix} \cos \alpha_2 & i z_2 \sin \alpha_2 \\ (i/z_2) \sin \alpha_2 & \cos \alpha_2 \end{bmatrix} \tag{7.1-39}$$

Thus from Eq. (7.1-36),

$$Z = \frac{z_3 \cos \alpha_2 - i z_2 \sin \alpha_2}{\cos \alpha_2 - i(z_3/z_2) \sin \alpha_2}$$

or

$$Z = z_2 \frac{z_3 - i z_2 \tan \alpha_2}{z_2 - i z_3 \tan \alpha_2} \tag{7.1-40}$$

where $\alpha_2 = k_2[z(2) - z(1)] \cos \theta_2 = k_2 d \cos \theta_2$

$$= k_2 d \sqrt{1 - (k_1/k_2)^2 \sin^2 \theta_1} \qquad (7.1\text{-}41)$$

and the appropriate values of z_1, z_2, z_3 are used depending on the polarization. The reflection coefficient is then given by Eq. (7.1-37), with Z given by Eq. (7.1-40). Thus the reflection coefficient is

$$R = \frac{z_2(z_3 - z_1) - i(z_2^2 - z_1 z_3) \tan \alpha_2}{z_2(z_3 + z_1) - i(z_2^2 + z_1 z_3) \tan \alpha_2} \qquad (7.1\text{-}42)$$

Similarly, from Eqs. (7.1-38) and (7.1-40), the transmission coefficient for the layer is

$$T = \frac{2 z_2 z_3}{z_2(z_1 + z_3) \cos \alpha_2 - i(z_2^2 + z_1 z_3) \sin \alpha_2} \qquad (7.1\text{-}43)$$

Examining Eq. (7.1-40) for Z, two special cases occur; when $\tan \alpha_2$ is zero, and when $\tan \alpha_2$ is infinite. In both these cases, Z is real for lossless media, and the thickness of the layer corresponds to some integer multiple of one-quarter of the wavelength in medium (2). For example, with $\tan \alpha_2 =$ zero and the wave normally incident, then $\alpha_2 = n\pi$, and the thickness of the layer is some multiple of a half-wave in the layer or $d = (n/2) \lambda_2$. In this case, $Z = z_3$, and the presence of the layer is not apparent since

$$R = (z_3 - z_1)/(z_3 + z_1)$$

is the same as if medium (2) were not present.

If $\tan \alpha_2$ is infinite and the wave normally incident, then $\alpha_2 = (n + \frac{1}{2}) \pi$, and the thickness of the layer is some odd multiple of a quarter wavelength in the layer or $d = [(2n + 1)/4] \lambda_2$. For this case, $Z = z_2^2/z_3$, and

$$R = (z_2^2 - z_1 z_3)/(z_2^2 + z_1 z_3)$$

R becomes zero when

$$z_2^2 = z_1 z_3 \qquad (7.1\text{-}44)$$

and the presence of the layer matches medium (1) to medium (3) without creating a reflection.

For a layer with an electrical thickness much less than a wavelength (a "thin" layer), the effective input impedance becomes

$$Z \approx z_3 \left[1 - i\alpha_2 \left(\frac{z_2}{z_3} - \frac{z_3}{z_2}\right)\right] \qquad (7.1\text{-}45)$$

and the reflection coefficient is

$$R \approx \frac{z_3 - z_1}{z_3 + z_1} + i\,\frac{2\alpha_2 z_1}{z_2(z_3 + z_1)^2}\,(z_3{}^2 - z_2{}^2) \tag{7.1-46}$$

while the transmission coefficient is

$$T \approx \frac{2z_3}{z_1 + z_3} - \frac{i\alpha_2}{(z_1 + z_3)^2}\left[(z_1 + z_2)\left(z_2 - \frac{2z_3{}^2}{z_2}\right) - 2\left(z_2 - \frac{z_3{}^2}{z_2}\right)\right] \tag{7.1-47}$$

The effect of the slab, medium (2), is contained in the imaginary term, and this is a linear function of the slab thickness.

When medium (3) is the same as medium (1), the reflection coefficient is obtained from Eq. (7.1-42) with $z_3 = z_1$, or

$$R = \frac{i(z_1{}^2 - z_2{}^2)\tan\alpha_2}{2z_1 z_1 - i(z_1{}^2 + z_2{}^2)\tan\alpha_2} \tag{7.1-48}$$

For a thin layer, $\alpha_2 \ll 1$, and the reflection coefficient becomes

$$R \approx i\alpha_2\left(\frac{z_1{}^2 - z_2{}^2}{2z_1 z_2}\right) - \alpha_2{}^2\left(\frac{z_1{}^4 - z_2{}^4}{4z_1{}^2 z_2{}^2}\right) \tag{7.1-49}$$

Curves of $|R|^2$ versus d/λ_0 are given in Figure 7-5 for a lossless dielectric slab with $\epsilon/\epsilon_0 = 2, 4, 8$; and given in Figure 7-6 are curves of $|R|^2$ and $|T|^2$ for a lossy dielectric slab with $\epsilon'/\epsilon_0 = 4$ and various values of $\epsilon''/\epsilon_0 = \tan\delta$. For both figures, medium (3) is the same as medium (1). From Figure 7-6, it is apparent that as the loss in the layer increases, the oscillations of the reflection coefficient with layer thickness rapidly damp out, and the reflection coefficient approaches unity as d/λ_0 becomes large.

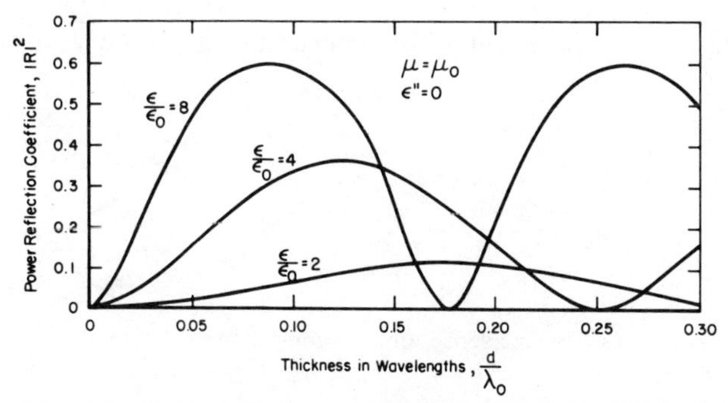

Fig. 7-5. Power-reflection coefficient for normal incidence on a lossless homogeneous dielectric slab versus the slab thickness, $\epsilon_r = 2$, 4, and 8.

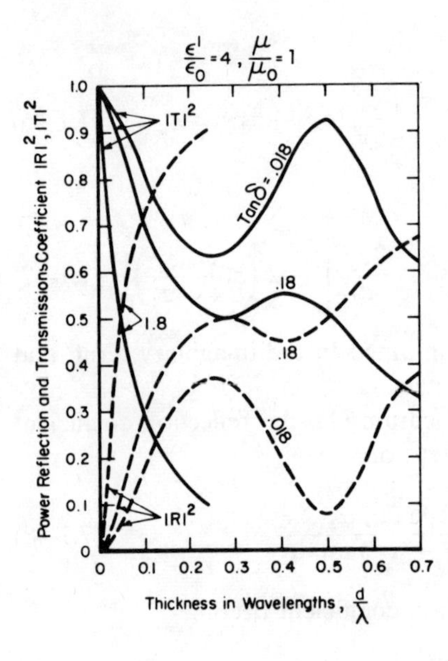

Fig. 7-6. Power reflection and transmission coefficients for normal incidence on a lossy homogeneous dielectric slab versus the slab thickness, with $\mu_r = 1$, $\epsilon_r = 4$ and $\tan\delta = 1.8$, 0.18, and 0.018.

7.2. INHOMOGENEOUS ISOTROPIC MEDIA

Consider the region of space where $z > 0$ to be filled with a material for which ϵ, μ, σ are functions of z only; allow a plane wave to be incident at an arbitrary angle θ_0, and polarized either perpendicular to or parallel to the plane of incidence. An illustration of this is shown in Figure 7-7. As for the homogeneous case, reflection coefficients can be defined which relate the amplitudes of the incident and reflected waves, or†

$$R'_\perp = E_y{}^r/E_y{}^i, \qquad R'_\parallel = H_y{}^r/H_y{}^i = -E_x{}^r/E_x{}^i \tag{7.2-1}$$

A transmission coefficient is not defined quite as simply as in Section 7-1,

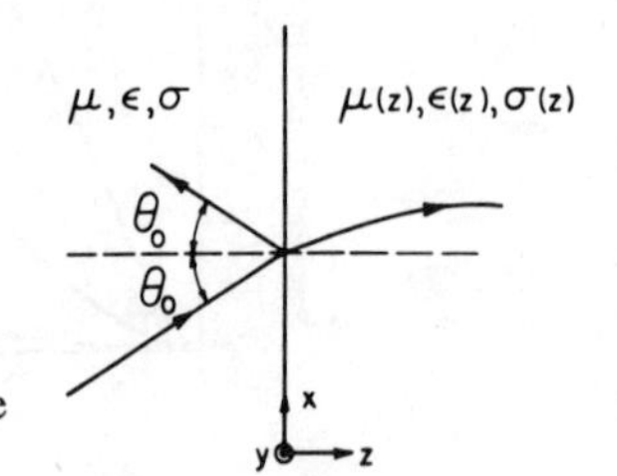

Fig. 7-7. Coordinates for incidence on a semi-infinite inhomogeneous medium.

† Throughout this section the definition of Eq. (7.2-1) for the reflection coefficients will be used; however, the primes will be omitted for the remainder of the section.

since within the inhomogeneous region the field amplitudes are functions of z.†

For an inhomogeneous medium with material constants varying only in the z direction, the reflection coefficients can be obtained by determining the fields which satisfy Maxwell's equations within the medium, and then applying the proper boundary conditions at the interface and infinity. For arbitrary forms of dielectric variation, assuming that this variation is only in the z direction and that the medium is isotropic, field solutions satisfying Maxwell's equations can be obtained from the solutions of a second-order differential equation:

$$d^2\phi/dz^2 + F(z)\,\phi = 0 \tag{7.2-2}$$

where for perpendicular polarization

$$\phi_\perp = E_y/\sqrt{\mu}$$

and

$$F_\perp(z) = k^2(z) - \mu^{1/2}(d^2/dz^2)(\mu^{-1/2}) - k_0^2 \sin^2\theta_0 \tag{7.2-3}$$

For parallel polarization,

$$\phi_\parallel = H_y/\sqrt{\epsilon}$$

and

$$F_\parallel = k^2(z) - \epsilon^{1/2}(d^2/dz^2)(\epsilon^{-1/2}) - k_0^2 \sin^2\theta_0 \tag{7.2-4}$$

In general, analytical solutions to Eq. (7.2-2) have been obtained for only a few specific forms of dielectric variation, and various approximation methods must be used to obtain solutions for arbitrary dielectric variation. Comprehensive discussions of the various approximation techniques in use are given in References 5 and 6.

7.2.1. Inhomogeneous Half-Space

7.2.1.1. *Linear Variation*

An examination of one of the few forms of dielectric variation for which an exact solution exists will serve to illustrate some of the general properties of the reflection coefficients. Choosing k as $k^2(z) = k_0^2(1 + \alpha z)$ for $z > 0$ and $k^2(z) = k_0^2$ for $z < 0$, then in the case of perpendicular polarization and α positive with $\mu_r = 1$ throughout the medium, solutions to Eq. (7.2-2) are given by[5]

$$\phi(z) = A\xi^{1/3}H^{(1)}_{1/3}(\xi)$$

† When the lower case symbol z is used in this section without a subscript, it refers to the coordinate.

where $\xi = (2k_0/3\alpha)(\cos^2\theta_0 + \alpha z)^{3/2}$, $H_{1/3}^{(1)}(\xi)$ is a Hankel function of the first kind of order $\frac{1}{3}$, and A is an undetermined constant. Applying the boundary conditions and solving for $R_\perp$ results in

$$R_\perp = \frac{iH_{1/3}^{(1)}(\xi_0) - H_{-2/3}^{(1)}(\xi_0)}{iH_{1/3}^{(1)}(\xi_0) + H_{-2/3}^{(1)}(\xi_0)} \tag{7.2-5}$$

where $\xi_0 = (2k_0/3\alpha)\cos^3\theta_0$.

In order to examine $R_\perp$ in more detail, the two limiting cases of $\xi_0 \gg 1$ and $\xi_0 \ll 1$ will be considered. In the event $\xi_0 \gg 1$, which corresponds to a slowly varying index of refraction, and θ_0 less than $\pi/2$, the Hankel functions can be expanded asymptotically yielding

$$R_\perp \approx \frac{-i\alpha}{8k_0\cos^3\theta_0} \tag{7.2-6}$$

Since $\alpha = (2/k_0)(dk/dz)|_{z=0}$, $R_\perp$ becomes

$$|R_\perp| \approx \frac{dk/dz\,|_{z=0}}{4k_0^2\cos^3\theta_0} \tag{7.2-7}$$

In the event $\xi_0 \ll 1$, which corresponds to grazing incidence with $\theta_0 \approx \pi/2$ or dk/dz very large, a power series expansion can be used for the Hankel functions giving

$$R_\perp \approx -[1 - 2\gamma(k_0/3\alpha)^{1/3}(\sqrt{3} - i)\cos\theta_0] \tag{7.2-8}$$

with $\gamma = 1.978$. As $\theta_0 \to \pi/2$, then $R_\perp \to -1$.

For the case where α is negative, if we write $k^2(z)$ as $k_0^2(1 - \alpha z)$ for $z > 0$, a general solution to Eq. (7.2-2) for perpendicular polarization and $\mu_r = 1$ is

$$\phi(z) = a\mathrm{Ai}(\zeta) + b\mathrm{Bi}(\zeta)$$

where

$$\zeta = (k_0^2\alpha)^{1/3}[z - (1/\alpha)\cos^2\theta_0]$$

and $\mathrm{Ai}(\zeta)$, $\mathrm{Bi}(\zeta)$ are Airy functions.[7] Applying the boundary conditions at $z \to +\infty$ results in only $\mathrm{Ai}(\zeta)$, giving a valid solution. Thus, the reflection coefficient can be obtained by applying the boundary conditions at $z = 0$, and is

$$R_\perp = \frac{(\zeta_0)^{1/2}\,\mathrm{Ai}(-\zeta_0) + i\mathrm{Ai}'(-\zeta_0)}{(\zeta_0)^{1/2}\,\mathrm{Ai}(-\zeta_0) - i\mathrm{Ai}'(-\zeta_0)} \tag{7.2-9}$$

where

$$\zeta_0 = -(k_0/\alpha)^{2/3}\cos^2\theta_0$$

and Ai' is the derivative of Ai with respect to ζ.

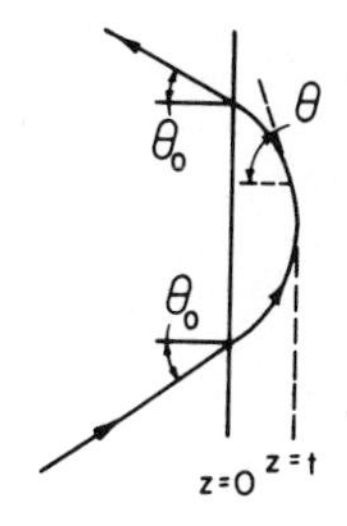

Fig. 7-8. A turning point in an inhomogeneous medium.

When α is real, ζ_0 is a negative real number, and $|R_\perp| = 1$, indicating perfect reflection. In the event $|\zeta_0|$ is large, an asymptotic expansion of the Airy function yields

$$R_\perp \approx -i \exp[i\tfrac{4}{3}(k_0/\alpha) \cos^3 \theta_0] \tag{7.2-10}$$

This same expression, with the exception of the factor i, can be obtained by a ray-tracing procedure which provides some insight to the physical phenomenon involved. Considering a ray incident on the interface $z = 0$ at an angle θ_0, then at any point z within the inhomogeneous medium, the orientation of the ray with respect to the normal to the interface is given by the angle $\theta(z)$, where $k(z) \sin \theta(z) = k_0 \sin \theta_0$. At the point $z = t$, where $\sin \theta(t) = 1$ or $k(t) = k_0 \sin \theta_0$, the ray is parallel to the interface, and this represents a turning point as illustrated in Figure 7-8.† The total phase change along the ray from the $z = 0$ plane to the turning point and back is

$$\phi = 2 \int_0^t k(z) \cos \theta(z)\, dz$$

or

$$\phi = 2k_0 \int_0^t \sqrt{\cos^2 \theta_0 - \alpha z}\, dz$$

giving

$$\phi \approx (4k_0/3\alpha) \cos^3 \theta_0 \tag{7.2-11}$$

which agrees with Eq. (7.2-10) with the exception of the factor i.

When $k(z)$ is complex, corresponding to lossy media, total reflection does not generally occur as is the case with lossless media. Similarly, if the inhomogeneous medium is bounded at some point z_B beyond t, or $k(z) =$ constant for $z > z_B$, then the reflection is not complete, and some energy is transmitted through the layer. This phenomena is analogous to the quantum mechanical "tunneling" of a particle through a potential barrier.

† Turning points of this type are often encountered in ionized media.

7.2.1.2. *Other Forms of Variation*

Exact solutions to Eq. (7.2-2) have also been obtained in the case of perpendicular polarization and oblique incidence for several forms of $k(z)$, other than that just discussed. When $\mu_r = 1$ independently of z, some of these are

(1) $k^2(z) = k_0^2 \left[a^2 - \dfrac{b^2}{(z + z_0)^2}\right]$

(2) $k^2(z) = (k_1^2 - k_2^2)\, e^{-\beta z} + k_2^2$

(3) $k^2(z) = k_1^2\, e^{\beta z}$

(4) $k^2(z) = k_0^2[1 - be^{\alpha z}], \qquad -\infty \leqslant z \leqslant \infty$

(5) $k^2(z) = (a + bz)^2$

For parallel polarization and oblique incidence, again with $\mu_r = 1$ for all z, exact solutions are available in only a very few cases. Among these are (3) and (5), the exponential and quadratic variation of $k^2(z)$; and (1), the hyperbolic variation for $a = 0$.[8]

7.2.2. Inhomogeneous Layer

7.2.2.1. *Transition Layer*

To examine the effects of a smooth transition from one value of wave number k_1 to another value k_2, consider a perpendicularly polarized wave, incident from $z = -\infty$, at an angle θ_0, on a medium of the form

$$k^2(z) = k_0^2 \left[1 - \alpha \frac{\exp[(2k_0/d)\, z]}{1 + \exp[(2k_0/d)\, z]}\right] \tag{7.2-12}$$

where $\mu_r = 1$, and d is a constant which determines the effective thickness of the layer. Figure 7-9 shows $k(z)$ as a function of z. Thus at $-\infty$, $k(z) = k_0$; while at $+\infty$, $k(z) = k_0 \sqrt{1 - \alpha}$. For this form of $k(z)$, solutions to

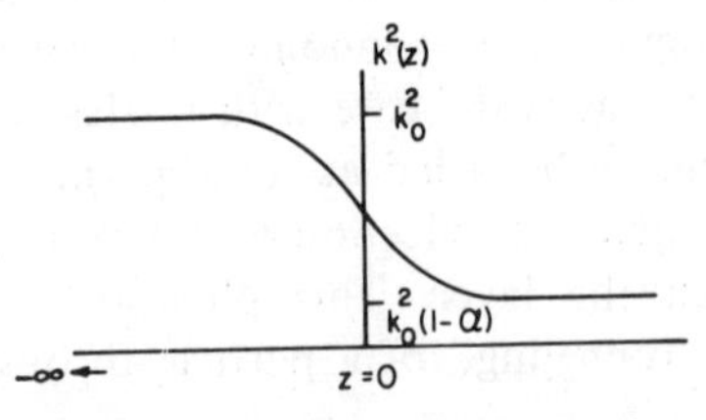

Fig. 7-9. Wave-number profile for a transition layer.

Eq. (7.2-2) are hypergeometric functions, and an exact expression for the reflection coefficient can be obtained, and is given by[5]

$$R = \frac{\left[\begin{array}{l}\Gamma(id\cos\theta_0)\,\Gamma[-i(d/2)(\cos\theta_0+\sqrt{\cos^2\theta_0-\alpha})] \\ \qquad\times\,\Gamma[1-i(d/2)(\cos\theta_0+\sqrt{\cos^2\theta_0-\alpha})]\end{array}\right]}{\left[\begin{array}{l}\Gamma(-id\cos\theta_0)\,\Gamma[i(d/2)(\cos\theta_0-\sqrt{\cos^2\theta_0-\alpha})] \\ \qquad\times\,\Gamma[1+i(d/2)(\cos\theta_0-\sqrt{\cos^2\theta_0-\alpha})]\end{array}\right]} \tag{7.2-13}$$

where $\Gamma(x)$ is the gamma function,[7] and α equals $[k_0^2 - k^2(\infty)]/k_0^2$.

In the case where $\alpha \geqslant \cos^2\theta_0$, Eq. (7.2-13) can be simplified, yielding for the magnitude of the reflection coefficient,

$$|R_\perp| = 1 \tag{7.2-14}$$

This means that at some point $z = t$, the function $F(z)$, defined in Eq. (7.2-3), has a zero which corresponds to a turning point and thus results in perfect reflection.

For $\alpha < \cos^2\theta_0$, Eq. (7.2-13) simplifies to

$$|R_\perp| = \frac{\sinh[(\pi d/2)(\cos\theta_0 - \sqrt{\cos^2\theta_0-\alpha})]}{\sinh[(\pi d/2)(\cos\theta_0 + \sqrt{\cos^2\theta_0-\alpha})]} \tag{7.2-15}$$

In the limit as $d \to 0$, the reflection coefficient becomes

$$|R_\perp| = \frac{k_0\cos\theta_0 - k\sqrt{1-(k_0^2/k)\sin^2\theta_0}}{k_0\cos\theta_0 + k\sqrt{1-(k_0^2/k)\sin^2\theta_0}} \tag{7.2-16}$$

where $k = k_0(1-\alpha)^{1/2}$. This, of course, is the Fresnel coefficient for reflection from an interface between a medium of wave number k_0 and one of wave number k.

For a transition layer which is not smooth, i.e., discontinuities exist in $dk(z)/dz$, the reflection coefficient is not a monotonic function of d as is the case with a smooth transition. In Figure 7-10 is shown the normal-incidence reflection coefficient as a function of l, the layer thickness for a smooth layer described by Eq. (7.2-12) with $\alpha = 0.36$ and $d = 2.09l/\lambda_0$, and a layer described by

$$k(z) = k_0\frac{a/(1-a)}{[a/(1-a)] + z/l}, \qquad 0 \leqslant z/l \leqslant 1 \tag{7.2-17}$$

$$k(z) = k_0, \qquad z/l < 0$$

$$k(z) = ak_0 = 8k_0, \qquad z/l > 1$$

which is illustrated, along with the layer described by Eq. (7.2-12), in Figure 7-11.

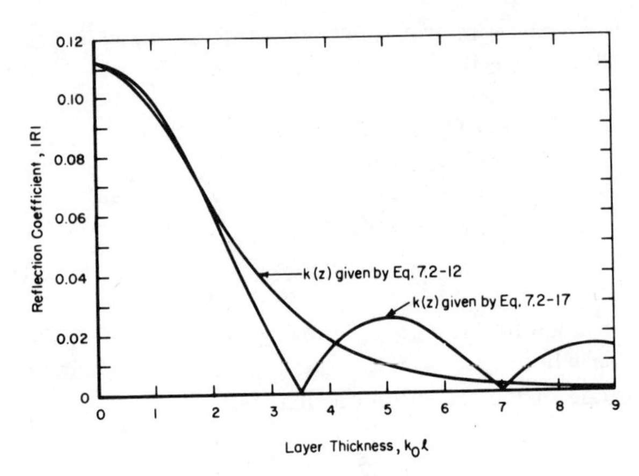

Fig. 7-10. Reflection coefficient versus the layer thickness for a smoothly varying transition layer and one with derivative discontinuities in the wave number at the layer interfaces.

From Figure 7-10, it is apparent that the discontinuities of dk/dz at $z = 0$ and l for the layer given by Eq. (7.2-17) destroy the monotonic variation of the reflection coefficient and cause zeros and maxima to occur in the reflection coefficient at specific values of l.

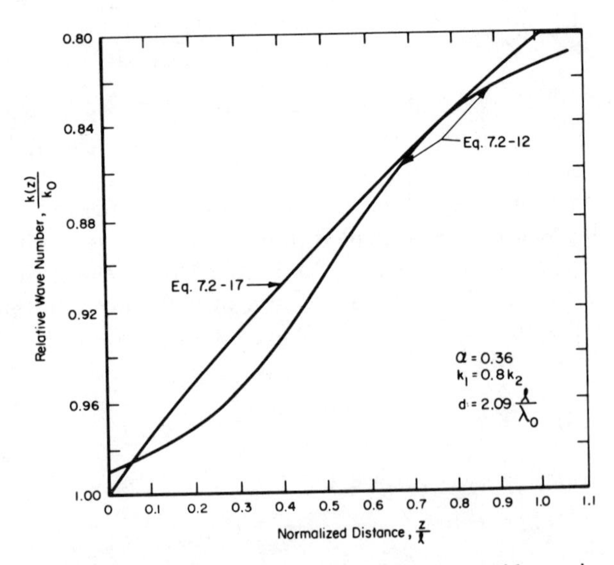

Fig. 7-11. Wave-number profiles for the smoothly varying and discontinuous transition layers.

7.2.2.2. *Symmetrical Layer*

For a medium with a wave number variation as shown in Figure 7-12, which is called a symmetrical layer,

$$k^2(z) = k_0^2 \left[1 - 4\alpha \frac{\exp[(2k_0/d)\, z]}{(1 + \exp[(2k_0/d)\, z])^2}\right] \tag{7.2-18}$$

with $\mu_r = 1$. At $z \to \pm\infty$, $k^2(z) \to k_0^2$, while at $z = 0$, $k^2(z)$ equals $k_0^2(1 - \alpha)$. Using this form of $k^2(z)$ in Eq. (7.2-2), for perpendicular polarization, an exact solution is again available in the form of hypergeometric functions. Solving for the reflection coefficient results in[5]

$$R_\perp = \left[\frac{\Gamma(id\cos\theta_0)}{\Gamma(-id\cos\theta_0)}\, \Gamma\left(\frac{1}{2} - \frac{1}{2}\sqrt{1 - 4d^2\alpha} - id\cos\theta_0\right)\right.$$
$$\left. \times\, \Gamma\left(\frac{1}{2} + \frac{1}{2}\sqrt{1 - 4d^2\alpha} - id\cos\theta_0\right) \frac{1}{\pi} \cos\left(\frac{\pi}{2}\sqrt{1 - 4d^2\alpha}\right)\right] \tag{7.2-19}$$

Again there are two cases of interest; when $4d^2\alpha < 1$,

$$|R_\perp|^2 = \frac{\cos^2(\pi/2)\sqrt{1 - 4d^2\alpha}}{\cos \pi(\frac{1}{2}\sqrt{1 - 4d^2\alpha} + id\cos\theta_0)\cos \pi(\frac{1}{2}\sqrt{1 - 4d^2\alpha} - id\cos\theta_0)} \tag{7.2-20}$$

and when $4d^2\alpha > 1$,

$$|R_\perp|^2 = \frac{\cosh^2(\pi/2)\sqrt{4d^2\alpha - 1}}{\cosh \pi(\frac{1}{2}\sqrt{4d^2\alpha - 1} + d\cos\theta_0)\cosh \pi(\frac{1}{2}\sqrt{4d^2\alpha - 1} - d\cos\theta_0)} \tag{7.2-21}$$

In the first case, when α is negative, corresponding to the layer having a maximum index of refraction greater than free space, the reflection coefficient, $R_\perp$, goes to zero as $d \to \infty$. This is to be expected since, then, $dk(z)/dz$ goes to zero. When α is positive and $4d^2\alpha > 1$, there are still two possible conditions:

(a) $\frac{1}{2}\sqrt{4d^2\alpha - 1} < d\cos\theta_0$

(b) $\frac{1}{2}\sqrt{4d^2\alpha - 1} > d\cos\theta_0$

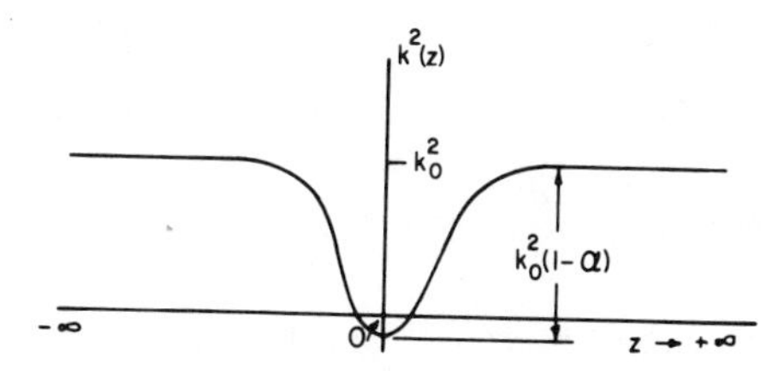

Fig. 7-12. Wave-number profile for a symmetrical layer.

For condition (a), although α is positive, it is small and no turning point occurs in the layer so the reflection coefficient $|R_\perp|$ goes to zero as $d \to \infty$. In case (b), α is positive and large, thus a turning point exists and as $d \to \infty$, $|R_\perp| \to 1$.

7.2.3. Approximation Techniques

7.2.3.1. *Geometrical Optics*

One of simplest approximation techniques, valid for a medium which is slowly varying with respect to the wavelength of the field, is geometrical optics. The geometrical-optics approximation can be shown to be the first term in a perturbation expansion of an exact solution. Consider a plane wave incident on an inhomogeneous medium from $-\infty$ at an arbitrary angle θ_0. In the event of perpendicular polarization, the total electric field at any point in the medium can be written as[9]

$$\mathbf{E} = \hat{\mathbf{y}}E_y = \hat{\mathbf{y}}\left\{\frac{P(z)}{\sqrt{\mathscr{y}}}\exp\left[i\int_{z_0}^{z}\delta(\tau)\,d\tau\right] + \frac{Q(z)}{\sqrt{\mathscr{y}}}\exp\left[-i\int_{z_0}^{z}\delta(\tau)\,d\tau\right]\right\}\exp[ik_0x\sin\theta_0] \quad (7.2\text{-}22)$$

where the time dependence $e^{-i\omega t}$ has been suppressed, and

$$\delta = k(z)\cos\theta(z) = k(z)[1-(k_0/k)^2\sin^2\theta_0]^{1/2} \quad (7.2\text{-}23)$$

$$\mathscr{y}(z) = \sqrt{\frac{\epsilon(z)}{\mu(z)}}\cos\theta(z) = \frac{\cos\theta(z)}{\eta(z)} \quad (7.2\text{-}24)$$

where $\mathscr{y}$ is the effective intrinsic admittance of the medium.

Similarly, for parallel polarization, the total magnetic field in the medium can be expressed as[9]

$$\mathbf{H} = \hat{\mathbf{y}}H_y = \hat{\mathbf{y}}\left\{\frac{P(z)}{\sqrt{\mathscr{z}}}\exp\left[i\int_{z_0}^{z}\delta(\tau)\,d\tau\right] + \frac{Q(z)}{\sqrt{\mathscr{z}}}\exp\left[-i\int_{z_0}^{z}\delta(\tau)\,d\tau\right]\right\}\exp[ik_0x\sin\theta_0] \quad (7.2\text{-}25)$$

where

$$\mathscr{z}(z) = \sqrt{\frac{\mu(z)}{\epsilon(z)}}\cos\theta(z) = \eta(z)\cos\theta(z) \quad (7.2\text{-}26)$$

is the effective intrinsic impedance of the medium.

Writing the field components in the forms shown above is for mathematical convenience, and it should be noted that only in special circumstances can it be said that the two terms represent forward and backward traveling waves. Inserting Eqs. (7.2-22) and (7.2-25) into Maxwell's equations (2.2-1) and (2.2-2) yields a pair of differential equations for the coefficients P and Q;

$$\frac{dP}{dz} = \gamma Q \exp\left[-2i \int_{z_0}^{z} \delta(\tau)\, d\tau\right] \tag{7.2-27}$$

$$\frac{dQ}{dz} = \gamma P \exp\left[2i \int_{z_0}^{z} \delta(\tau)\, d\tau\right] \tag{7.2-28}$$

where

$$\gamma = \frac{1}{2\mathscr{Y}} \frac{d}{dz}(\mathscr{Y}) \quad \text{for } \perp \text{ polarization}$$

and

$$\gamma = \frac{1}{2\mathscr{Z}} \frac{d}{dz}(\mathscr{Z}) \quad \text{for } \parallel \text{ polarization}$$

In these equations, as well as Eqs. (7.2-22) and (7.2-25), the lower limit z_0 of the integral in the exponential is arbitrary, and merely specifies the phase reference point.

Equations (7.2-27) and (7.2-28) are exact, and if solvable for P and Q with the appropriate boundary conditions, would provide exact solutions for the fields in the medium. Should, however, the right-hand sides of Eqs. (7.2-27) and (7.2-28) be small, then approximate solutions can be obtained by a perturbation technique. Thus, the coefficients P and Q are expanded in terms of the small parameter, γ, or

$$\begin{aligned} P(z) &= P_0(z) + \gamma P_1(z) + \gamma^2 P_2(z) + \cdots \\ Q(z) &= Q_0(z) + \gamma Q_1(z) + \gamma^2 Q_2(z) + \cdots \end{aligned} \tag{7.2-29}$$

and substituting into the differential Eqs. (7.2-27) and (7.2-28) gives a pair of differential equations for the components of P and Q:

$$\begin{aligned} \frac{dP_i}{dz} &= \gamma \exp\left[-2i \int_{z_0}^{z} \delta(\tau)\, d\tau\right] Q_{i-1} \\ \frac{dQ_i}{dz} &= \gamma \exp\left[2i \int_{z_0}^{z} \delta(\tau)\, d\tau\right] P_{i-1} \end{aligned} \tag{7.2-30}$$

Integrating these leads to recursion equations for the components:

$$P_i = \int_{-\infty}^{z} \gamma \exp\left[-2i \int_{z_0}^{z'} \delta(\tau)\, d\tau\right] Q_{i-1}(z')\, dz'$$

$$Q_i = \int_{-\infty}^{z} \gamma \exp\left[2i \int_{z_0}^{z'} \delta(\tau)\, d\tau\right] P_{i-1}(z')\, dz' \tag{7.2-31}$$

where P_0 and Q_0 are constants.

The geometrical-optics approximation corresponds to using only the first terms of the expansions for P and Q, Eq. (7.2-29), or $P(z) = P_0$ and $Q(z) = Q_0$. The conditions under which this assumption is valid can be obtained by examining the magnitude of the terms for P_1 and Q_1, which are neglected.

Using (7.2-31),

$$Q_1 = \int_{-\infty}^{z} \gamma \exp\left[2i \int_{z_0}^{z'} \delta(\tau)\, d\tau\right] P_0\, dz'$$

or

$$\frac{Q_1}{P_0} = \int_{-\infty}^{z} \gamma \exp\left[2i \int_{z_0}^{z'} \delta(\tau)\, d\tau\right] dz' \tag{7.2-32}$$

The exponential in Eq. (7.2-32) is a rapidly oscillating function with a period approximately equal to

$$\frac{\pi}{k \cos \theta}$$

and over any half-period with the same sign for the exponential, it can be replaced by one. If this is done and the medium is slowly varying, then γ can be taken outside the integral giving

$$\frac{Q_1}{P_0} \approx \frac{\pi\gamma}{4k \cos \theta} \tag{7.2-33}$$

and the condition for the validity of the geometrical-optics approximation is then given by

$$\frac{\pi}{\mathcal{Y} 8k \cos \theta} \frac{d\mathcal{Y}}{dz} \ll 1 \tag{7.2-34}$$

for perpendicular polarization, and

$$\frac{\pi}{\mathcal{Z} 8k \cos \theta} \frac{d\mathcal{Z}}{dz} \ll 1 \tag{7.2-35}$$

for parallel polarization.

It is apparent that these conditions are violated if μ, ϵ, or $\cos \theta$ goes to zero at any point in the medium, or if the properties of the medium should vary rapidly with respect to the wavelength at any point.

The geometrical-optics approximation then corresponds to the physical picture of a wave moving through the inhomogeneous medium with a constant intensity and a changing direction, which at any point is given with respect to the z axis by $\cos \theta$.

If μ, ϵ, or $\cos \theta$ goes to zero at any point, this represents a turning point in the medium, and geometrical optics cannot provide a valid description near this point. For perpendicular polarization, however, a solution which is valid at a turning point, and which reduces to geometrical optics far from the turning point, is given by[10]

$$E_y = E_0\text{Ai}(-\xi) \tag{7.2-36}$$

where Ai is the Airy function previously used, and ξ is real and given by

$$\xi = \left[\frac{3}{2}\int_z^t k\cos\theta\, dz\right]^{2/3} \tag{7.2-37}$$

The upper limit of the integral is the turning point t, defined by

$$k(t)\cos\theta(t) = 0.$$

By asymptotically expanding Eq. (7.2-36) for $\xi \gg 1$, the reflection coefficient is found to be

$$R_\perp = -i\exp\left[2i\int_0^t k\cos\theta\, dz\right]$$

where the phase reference is the plane $z = 0$.

The above expressions are valid in the event the medium is slowly varying and there are no other turning points or singularities of $k\cos\theta$ near the point t.

For parallel polarization, the situation is far more complex, and it is perhaps best to obtain approximate expressions for the reflection coefficient in the presence of turning points by some other technique, such as that which will now be discussed.

7.2.3.2. *Another Approximation*

An approximate expression for the reflection coefficients has been obtained by Presnyakov and Sobel'man from Eqs. (7.2-22) and (7.2-25), which in the high-frequency limit corresponds to geometrical optics and in the low-frequency limit to the Born approximation.[11] For an arbitrary dielectric variation and frequency, this approximation can be considered to be an interpolation formula.

For perpendicular polarization, the magnitude of the reflection coefficient from an inhomogeneous layer extending from a to b, with homogeneous media elsewhere, is given approximately by

$$|R_\perp| \approx \tanh\left|\int_a^b \frac{1}{2y}\frac{dy}{dz'}\exp\left[2i\int_{z_0}^{z'}\delta(\tau)\,d\tau\right]dz'\right| \tag{7.2-38}$$

For parallel polarization, the reflection coefficient is given approximately by

$$| R_{\parallel} | \approx \tanh \left| \int_a^b \frac{1}{2z} \frac{dz}{dz} \exp \left[2i \int_{z_0}^{z'} \delta(\tau) \, d\tau \right] dz' \right| \qquad (7.2\text{-}39)$$

and $\delta(\tau) = k(\tau) \cos \theta(\tau)$.

The above expressions appear to be valid in the event that the medium is discontinuous or has turning points. As an example of a discontinuity, consider a layer such that for $z \leqslant a$, $z = z_0$, and for $z \geqslant b$, $z = z_1$, while for $a < z < b$, μ and ϵ are arbitrary bounded functions such that

$$[k \cos \theta]_{\max}(b - a) \ll 1$$

Under these conditions the exponential in Eq. (7.2-39) is approximately equal to one giving

$$| R_{\parallel} | = \tanh \left| \ln \sqrt{\frac{z_1}{z_0}} \right|$$

or

$$| R_{\parallel} | = \left| \frac{z_1 - z_0}{z_1 + z_0} \right|$$

the Fresnel coefficient for an interface between two homogeneous media.

At a turning point, $k \cos \theta$, the integral in the exponential of Eqs. (7.2-38) and (7.2-39) is zero. Thus, in the high-frequency limit as $k_0 \to \infty$, the integral over z can be evaluated by stationary phase resulting in

$$\left| \int_a^b \frac{1}{2y} \frac{dy}{dz} \exp \left[2i \int_{z_0}^{z} k \cos \theta \, d\tau \right] dz' \right| \to \infty$$

and

$$| R | \to 1$$

as it should if the medium has a turning point.

As a further example of the use of Eqs. (7.2-38) and (7.2-39), consider the smooth transition layer considered previously which was specified by Eq. (7.2-12), or

$$k^2(z) = \left(1 - \alpha \frac{\exp[(2k_0/d) \, z]}{1 + \exp[(2k_0/d) \, z]} \right) \qquad (7.2\text{-}40)$$

Substituting this into Eq. (7.2-38), the integrations can be approximated for $k_0 = 1$ to give

$$| R_{\perp} | \approx \tanh \left| \exp[-\pi d \sqrt{\cos^2 \theta_0 - \alpha}] \ln \left[\frac{\sqrt{\cos \theta_0}}{\sqrt{\cos^2 \theta_0 - \alpha}} \right] \right| \qquad (7.2\text{-}41)$$

In the limit as $d \to 0$, this approaches

$$| R_\perp | \approx \tanh \left| \ln \left[\frac{\sqrt{\cos \theta_0}}{\sqrt{\cos^2 \theta_0 - \alpha}} \right] \right|$$

or

$$| R_\perp | \approx \frac{\cos \theta_0 - \sqrt{\cos^2 \theta_0 - \alpha}}{\cos \theta_0 + \sqrt{\cos^2 \theta_0 - \alpha}}$$

which is the same as Eq. (7.2-16) obtained from the exact solution if k_0 is set equal to one.

To obtain a feeling for how well Eq. (7.2-41) approximates $| R |$, the exact solution [Eq. (7.2-13)] for the transition will be compared in Figure 7-13

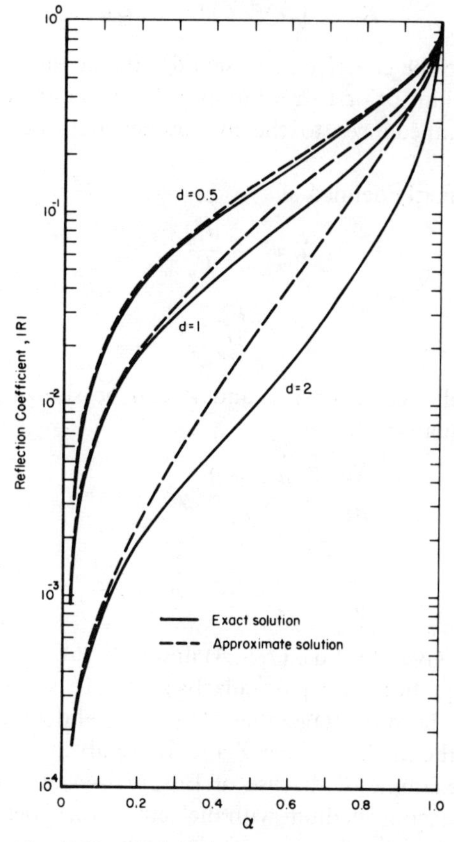

Fig. 7-13. Comparison of the reflection-coefficient magnitude obtained using the approximation technique of Presnyakov and Sobel'man with exact results [after Presnyakov and Sobel'man[11]].

with the results of the approximate formulation [Eq. (7.2-42)] for normal incidence and several values of α and d.

7.2.4. Numerical Solutions

For an inhomogeneous medium with properties varying in the z direction, which begins at $z = a$ and ends at $z = b$, where b may be finite or $+\infty$, exact solutions for the reflection coefficients may be obtained by formulating the problem in terms of circuit concepts. As previously discussed, the reflection coefficients can be given by

$$R_{\perp} = (y_1 - Y)/(y_1 + Y) \tag{7.2-42}$$

and

$$R_{\parallel} = (z_1 - Z)/(z_1 + Z) \tag{7.2-43}$$

where y_1 and z_1 are the effective intrinsic admittance and impedances of the homogeneous medium prior to $z = a$, and Y and Z are the effective input admittances and impedances to the inhomogeneous medium at the point $z = a$.

These are explicitly defined as

$$Y = -\left.\frac{H_x}{E_y}\right|_a$$

$$Z = \left.\frac{E_x}{H_y}\right|_a$$

Using Maxwell's equations, Y and Z can be shown to satisfy Ricatti-type differential equations,[12] or

$$\frac{dY}{dz} = \frac{ik\cos\theta}{\mathscr{y}}(\mathscr{y}^2 - Y^2) \tag{7.2-44}$$

and

$$\frac{dZ}{dz} = \frac{ik\cos\theta}{\mathscr{z}}(\mathscr{z}^2 - Z^2) \tag{7.2-45}$$

where $\mathscr{y}$ and $\mathscr{z}$ are given by Eqs. (7.2-24) and (7.2-26).

The above equations are particularly suited to solution by numerical means, and have the advantage that they are valid at turning points or discontinuities of the medium since Y and Z are always continuous.

As a simple example of the use of Eqs. (7.2-44) and (7.2-45), consider a stratified homogeneous medium with dielectric properties ϵ_1, μ_1 for $z < 0$; ϵ_2, μ_2 for $0 \leqslant z \leqslant l$; and ϵ_3, μ_3 for $z > l$; $\mathscr{z}$ is a constant, z_2; and Eq. (7.2-45) can be integrated as

$$\int \frac{dZ}{{z_2}^2 - Z^2} = i\frac{k_2\cos\theta_2}{z_2}z + C$$

where C is a constant of integration. Performing the integration on the left-hand side gives

$$\frac{1}{z_2}\tanh^{-1}\frac{Z}{z_2} = \frac{ik_2\cos\theta_2}{z_2}z + C$$

or

$$\frac{Z}{z_2} = \tanh(ik_2\cos\theta_2 z + Cz_2) \tag{7.2-46}$$

Now at $z = l$, $Z = z_3$, so

$$z_3/z_2 = \tanh(ik_2\cos\theta_2 l + Cz_2)$$

or

$$\frac{z_3}{z_2} = \frac{\tanh(ik_2\cos\theta_2 l) + \tanh(Cz_2)}{1 + \tanh(ik_2\cos\theta_2 l)\tanh(Cz_2)}$$

giving

$$\tanh Cz_2 = \frac{\tanh(ik_2\cos\theta_2 l) - (z_3/z_2)}{(z_3/z_2)\tanh(ik_2\cos\theta_2 l) - 1} \tag{7.2-47}$$

$$= \frac{z_3 - iz_2\tan(k_2\cos\theta_2 l)}{z_2 - iz_3\tan(k_2\cos\theta_2 l)}$$

At the interface, $z = 0$, then, from (7.2-46) and (7.2-47),

$$Z = z_2\tanh Cz_2$$

or

$$Z = z_2\left[\frac{z_3 - iz_2\tan(k_2\cos\theta_2 l)}{z_2 - iz_3\tan(k_2\cos\theta_2 l)}\right] \tag{7.2-48}$$

This is identical to Eq. (7.1-40), previously given in Section 7-1, for the effective input impedance of a dielectric slab of effective intrinsic impedance, z_2, separating two semi-infinite media with impedances z_1 and z_3.

Using Eqs. (7.2-42), (7.2-43) and (7.2-44), (7.2-45), Ricatti differential equations for the reflection coefficients themselves can be obtained as[13]

$$\frac{dR_\perp}{dz} = \frac{ik\cos\theta}{2}\left[\frac{y_0}{y}(1-R)^2 - \frac{y}{y_0}(1+R)^2\right] \tag{7.2-49}$$

$$\frac{dR_\parallel}{dz} = \frac{ik\cos\theta}{2}\left[\frac{z_0}{z}(1-R)^2 - \frac{z}{z_0}(1+R)^2\right] \tag{7.2-50}$$

These equations for the reflection coefficients are also valid for media which contain discontinuities or turning points, and are well suited for solution by numerical techniques.

7.3. APERTURES IN INFINITE PLANAR MEDIA

Consider an aperture of arbitrary shape in an infinite plane with a plane wave incident as shown in Figure 7-14. The plane $z = 0$ will be considered to be infinitesimally thin and perfectly conducting. For such an infinite plane surface, the radar cross section as normally defined is usually of little interest since, if no apertures existed in the surface, the backscatter cross section would be zero for $\theta \neq 0$ and ∞ at $\theta = 0$. The presence of the aperture, in general, means that for $\theta \neq 0$ the backscatter cross section would no longer be zero; however, at $\theta = 0$, it would still be infinite.

The fields of primary interest for a hypothetical structure of this nature are the total fields in the region $z < 0$ and the fields in the aperture.

The reason for this interest from the radar standpoint is a result of Babinet's principle, since investigation of the field scattered by an aperture in an infinite plane yields results on the fields scattered by the complementary flat plate.

One of the best and most straightforward statements of Babinet's principle has been given by Bouwkamp.[14] Let $\mathbf{E}^i$, $\mathbf{H}^i$ represent a field incident on a perfectly conducting, flat, infinitesimally thin plate of shape S. Define a complementary incident field as $\mathbf{E}_c^{\,i}$, $\mathbf{H}_c^{\,i}$, where $\mathbf{E}_c^{\,i} = -\eta_0\mathbf{H}^i$ and $\mathbf{H}_c^{\,i} = \mathbf{E}^i/\eta_0$, incident on an aperture A of the same shape as S in an infinitely thin, perfectly conducting, infinite plane screen. For the plate S, the total fields everywhere in space are given by $\mathbf{E}^i + \mathbf{E}^s$ and $\mathbf{H}^i + \mathbf{H}^s$, while for the aperture, the regions $z > 0$ and $z < 0$ must be considered separately. For $z > 0$, the total fields in the absence of the aperture are $\mathbf{E}^0$, $\mathbf{H}^0$, while $\mathbf{E}^0 + \mathbf{E}^d$, $\mathbf{H}^0 + \mathbf{H}^d$ are the total fields with the aperture present; and for $z < 0$, $\mathbf{E}^d$, $\mathbf{H}^d$ are the total fields. Babinet's principle then consists of the following statements:

$$\begin{aligned} \mathbf{E}^d(z) &= \eta_0\mathbf{H}^s(z), \\ \mathbf{H}^d(z) &= -\mathbf{E}^s(z)/\eta_0, \end{aligned} \qquad z > 0 \tag{7.3-1}$$

and

$$\begin{aligned} \mathbf{E}^d(z) &= -\eta_0\mathbf{H}^s(-z), \\ \mathbf{H}^d(z) &= \mathbf{E}^s(-z)/\eta_0, \end{aligned} \qquad z < 0 \tag{7.3-2}$$

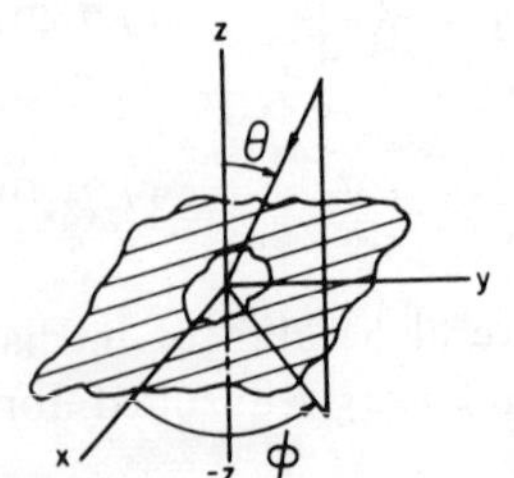

Fig. 7-14. Coordinates for a plane wave incident on an aperture in an infinite perfectly conducting plane.

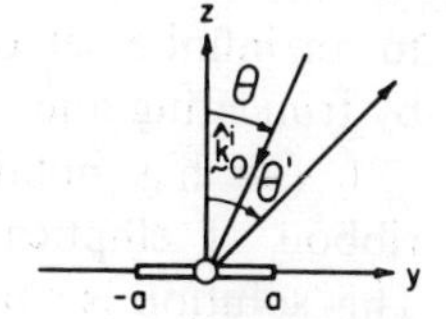

Fig. 7-15. Coordinates for a plane wave incident on an infinite strip.

Thus, if the diffracted field produced by an aperture in an infinite, perfectly conducting plane is known, the scattered fields produced by the complementary plate can be determined from Eqs. (7.3-1) and (7.3-2).

In terms of the radar cross section, Babinet's principle allows the diffraction patterns of an aperture for incident fields polarized parallel and perpendicular to the plane of incidence to be used to determine the radar cross sections of the complimentary plate for incident fields polarized, respectively, perpendicular and parallel to the plane of incidence.

At this point, the subject of the diffracted fields produced by an aperture will be dropped in favor of discussions in the next two sections on the complementary problem. Results for various apertures can be obtained by applying Babinet's principle to the results given there.

7.4. INFINITE STRIPS†

7.4.1. Exact Solutions

For a plane wave incident at an angle θ in the yz-plane, normal to an infinite, perfectly conducting strip of width $2a$, situated along the x-axis as shown in Figure 7-15, several exact analytical solutions for the radar cross section per unit length (scattering width) of the strip have been obtained.[15,16] One was accomplished by obtaining a solution for the scattered field produced by an infinite elliptic cylinder which, in the limiting case of the minor axis going to zero, gives the strip solution (see Section 4.4). This solution is in terms of an infinite series of Mathieu functions, and as discussed in Section 4.4.1.1, is generally not very suitable for numerical computations. Theoretical results for the backscatter cross section obtained by summing the series on a large digital computer are given in Figures 4-53 to 4-59.

In addition, other exact formulations have been obtained in terms of integral equations. One example of this type of solution is that due to Waterman, who obtained numerical results by reducing the integral equation

† Only the two-dimensional solution, with the plane wave incident in the yz-plane will be discussed in this section. The generalization of the results given in this section to the three-dimensional case of a plane wave incident on a perfectly conducting strip at an angle with respect to the x-axis can be accomplished by the technique discussed in Section 2.3.1.1.2.

to an infinite set of linear algebraic equations which were then evaluated by truncating and solving the resultant $n \times n$ set with a computer.[17]

Yeh has obtained solutions for the scattered field from a dielectric ribbon of elliptical cross section, which approximates a dielectric strip. This solution is again in terms of an infinite series of Mathieu functions.[18] This presents the same difficulties that the conducting strip solutions do, in that numerical computations are difficult. Thus, even though exact theoretical solutions are available, various approximations have been obtained and others are being sought which will allow numerical computations to be made readily.

7.4.2. Approximate Solutions

7.4.2.1. *Low-Frequency Region*

For a perfectly conducting strip, when the strip width ($2a$) is small in comparison with the radar wavelength or when $k_0a \ll 1$, expanding the exact result in powers of k_0a and retaining only the first terms gives a solution for the cross section valid in the limit $k_0a \to 0$ (Rayleigh region). Carrying out this process gives, for the backscattered fields,[19]

$$E_z^s = -E_z^i \left[\frac{1 + (k_0^2a^2/4)\cos^2\theta}{\ln(4/\gamma k_0a) + i(\pi/2)}\right]\left(\frac{\pi}{2k_0r}\right)^{1/2} \exp[i(k_0r + \pi/4)] \quad (7.4\text{-}1)$$

$$H_z^s = H_z^i \left[\frac{(2k_0a)^2\cos^2\theta}{8}\right]\left(\frac{\pi}{2k_0r}\right)^{1/2} \exp[i(k_0r + \pi/4)] \quad (7.4\text{-}2)$$

where $\gamma = 1.7811$. The resultant cross sections are†

$$\sigma_\perp^c(\theta, \theta) = \frac{\pi^2}{k_0} \frac{[1 + (k_0^2a^2/4)\cos^2\theta]^2}{[\pi^2/4 + \ln^2(4.48/2k_0a)]} \quad (7.4\text{-}3)$$

$$\sigma_\parallel^c(\theta, \theta) = \frac{\pi^2}{k_0}\left[\frac{(2k_0a)^4\cos^4\theta}{64}\right] \quad (7.4\text{-}4)$$

where $\sigma_\perp^c(\theta, \theta)$ is the backscatter cross section per unit length for the incident field polarized perpendicular to the plane of incidence, and $\sigma_\parallel^c(\theta, \theta)$ is the backscatter cross section for parallel polarization.

Additional terms in the expansion of the scattered fields in powers of k_0a have been obtained by a number of authors, and a good discussion of the status of this problem is given by Bouwkamp.[14]

† The cross section given here is equivalent to the scattering width defined and used in Chapter 4.

For the general bistatic case, the Rayleigh results are

$$\sigma_{\perp}{}^{c}(\theta, \theta') = \frac{\pi^2}{k_0}\left[\frac{(1 + (k_0{}^2a^2/4)\cos\theta\cos\theta')^2}{\pi^2/4 + \ln^2(4.48/2k_0a)}\right] \tag{7.4-5}$$

and

$$\sigma_{\parallel}{}^{c}(\theta, \theta') = \frac{\pi^2}{k_0}\left[\frac{(2k_0a)^4\cos^2\theta\cos^2\theta'}{64}\right] \tag{7.4-6}$$

7.4.2.2. *High-Frequency Region*

7.4.2.2.1. *Monostatic Cross Section.* In the high-frequency case, when $k_0a \gg 1$, approximate solutions have been developed in a variety of ways. The simplest of these is that predicted by physical optics, which gives for the backscatter cross section

$$\sigma^{c}_{\text{P.O.}}(\theta, \theta) = \frac{1}{k_0}\left[\frac{\cos\theta\sin[2k_0a\sin\theta]}{\sin\theta}\right]^2 \tag{7.4-7}$$

For normal incidence this becomes

$$\sigma^{c}_{\text{P.O.}}(0) = 4k_0a^2 \tag{7.4-8}$$

Equations (7.4-7) and (7.4-8), as with all physical-optics solutions, are independent of the polarization.

An improvement over physical optics can be obtained by using the Sommerfeld–Macdonald technique, in which each edge of a wide strip is considered to scatter as if it were a half-plane, and the two contributions are then summed. Additional higher-order terms can be obtained by taking into account the interaction between the edges. Several authors have obtained high-frequency solutions in this fashion.[20,21] Using the version of this technique developed by Ufimtsev,[21] the backscattered fields, taking into account first-order interaction between the edges, can be expressed as

$$E_z{}^s = E_z{}^i\left[B_2{}^2(\theta)\left(\frac{1-\sin\theta}{2\sin\theta}\right)\exp[i2k_0a\sin\theta] - B_1{}^2(\theta)\left(\frac{1+\sin\theta}{2\sin\theta}\right)\exp[-i2k_0a\sin\theta]\right]\frac{\exp[i(k_0r+\pi/4)]}{\sqrt{2\pi k_0r}} \tag{7.4-9}$$

$$H_z{}^s = H_z{}^i\left[A_2{}^2(\theta)\left(\frac{1+\sin\theta}{2\sin\theta}\right)\exp[i2k_0a\sin\theta] - A_1{}^2(\theta)\left(\frac{1-\sin\theta}{2\sin\theta}\right)\exp[-i2k_0a\sin\theta]\right]\frac{\exp[i(k_0r+\pi/4)]}{\sqrt{2\pi k_0r}} \tag{7.4-10}$$

with the resultant cross sections

$$\sigma_{\perp}{}^{c}(\theta, \theta) = \frac{1}{k_0} \left| B_2{}^2(\theta) \left(\frac{1 - \sin\theta}{2\sin\theta}\right) \exp[i2k_0 a \sin\theta] - B_1{}^2(\theta) \left(\frac{1 + \sin\theta}{2\sin\theta}\right) \exp[-i2k_0 a \sin\theta] \right|^2 \quad (7.4\text{-}11)$$

$$\sigma_{\parallel}{}^{c}(\theta, \theta) = \frac{1}{k_0} \left| A_2{}^2(\theta) \left(\frac{1 + \sin\theta}{2\sin\theta}\right) \exp[i2k_0 a \sin\theta] - A_1{}^2(\theta) \left(\frac{1 - \sin\theta}{2\sin\theta}\right) \exp[-i2k_0 a \sin\theta] \right|^2 \quad (7.4\text{-}12)$$

where

$$A_{1,2}(\theta) = \sqrt{2} \exp(-i\pi/4)[C(\gamma_{1,2}) + iS(\gamma_{1,2})] \quad (7.4\text{-}13)$$

$$B_{1,2}(\theta) = A_{1,2}(\theta) + \frac{i}{2\sqrt{\pi k_0 a}} \frac{\exp[i2k_0 a(1 \pm \sin\theta) - i(\pi/4)]}{|\cos[(\pi/4) \mp (\theta/2)]|} \quad (7.4\text{-}14)$$

and

$$\gamma_{1,2} = 2\sqrt{\frac{2k_0 a}{\pi}} \left| \cos\left(\frac{\pi}{4} \mp \frac{\theta}{2}\right) \right| \quad (7.4\text{-}15)$$

where the minus sign goes with γ_1, and the plus with γ_2. The functions $C(\gamma)$ and $S(\gamma)$ of Eq. (7.4-13) are the Fresnel integrals defined by

$$C(\gamma) = \int_0^{\gamma} \cos\left(\frac{\pi}{2} t^2\right) dt$$

and

$$S(\gamma) = \int_0^{\gamma} \sin\left(\frac{\pi}{2} t^2\right) dt$$

These are tabulated in References 22, and 23.

A simplified expression, valid for θ not too far from zero, can be obtained by neglecting the interaction between the edges. Like the physical-optics expression, this is independent of the incident polarization; however, it provides better numerical agreement with experimental and exact theoretical results. This approximation results in

$$\sigma^c \approx \frac{1}{k_0} \left[\frac{\sin^2(2k_0 a \sin\theta)}{\sin^2\theta} + \cos^2(2k_0 a \sin\theta)\right] \quad (7.4\text{-}16)$$

For normal incidence, $\theta = 0$, and Eq. (7.4-16) reduces to

$$\sigma^{c}(0) = \frac{1}{k_0} [1 + (2k_0 a)^2] \quad (7.4\text{-}17)$$

For grazing incidence, Eqs. (7.4-11) and (7.4-12) can be evaluated asymptotically for large k_0a and $\theta = \pi/2$ to give

$$\sigma_\perp{}^c\left(\pm\frac{\pi}{2}\right) \approx \frac{1}{k_0}\left[1 + \frac{\sin(8k_0a)}{16k_0a}\right] \tag{7.4-18}$$

and

$$\sigma_\parallel{}^c(\pm\pi/2) = 0 \tag{7.4-19}$$

For θ near $\pm\pi/2$, the cross section can be expressed as

$$\sigma_\perp{}^c(\theta) \approx \frac{1}{k_0}\left[\frac{1 + |\sin\theta|}{2\sin\theta}\right]^2 \tag{7.4-20}$$

and

$$\sigma_\parallel{}^c(\theta) \approx \frac{1}{k_0}\left[\frac{4k_0a}{\pi}\left(\frac{1-\sin^2\theta}{\sin\theta}\right)\right]^2 \tag{7.4-21}$$

7.4.2.2.2. *Bistatic Cross Sections.* The bistatic cross sections at high frequencies can also be approximated by using the Sommerfeld–Macdonald technique, including the first interaction term. For $0 \leqslant \theta \leqslant \pi/2$, these are

$$\begin{aligned}\sigma_\parallel{}^c(\theta,\theta') = \frac{1}{k_0}\Bigg| & A_2(\theta)\,A_2(\theta')\left[\frac{\sin[(\theta+\theta')/2] + \cos[(\theta-\theta')/2]}{\sin\theta + \sin\theta'}\right] \\ & \times \exp[ik_0a(\sin\theta + \sin\theta')] \mp A_1(\theta)\,A_1(\theta') \\ & \times \left[\frac{\sin[(\theta+\theta')/2] + \cos[(\theta-\theta')/2]}{\sin\theta + \sin\theta'}\right] \exp[-ik_0a(\sin\theta + \sin\theta')]\Bigg|^2\end{aligned} \tag{7.4-22}$$

and

$$\begin{aligned}\sigma_\perp{}^c(\theta,\theta') = \frac{1}{k_0}\Bigg| & B_2(\theta)\,B_2(\theta')\left[\frac{\sin[(\theta+\theta')/2] - \cos[(\theta-\theta')/2]}{\sin\theta + \sin\theta'}\right] \\ & \times \exp[ik_0a(\sin\theta + \sin\theta')] \mp B_1(\theta)\,B_1(\theta') \\ & \times \left[\frac{\sin[(\theta+\theta')/2] + \cos[(\theta-\theta')/2]}{\sin\theta + \sin\theta}\right] \exp[-ik_0a(\sin\theta + \sin\theta')]\Bigg|^2\end{aligned} \tag{7.4-23}$$

with $A_{1,2}$, $B_{1,2}$ given by Eqs. (7.4-13) and (7.4-14), and the upper sign used for $|\theta'| \geqslant \pi/2$, the lower for $|\theta'| \leqslant \pi/2$.

For scattering in the forward direction a simplified expression can be obtained by neglecting the interaction between edges in the Sommerfeld–Macdonald approximation. For $\theta \neq \pm\pi/2$ and $|\theta'| > \pi/2$, this gives

$$\begin{aligned}\sigma_\parallel{}^c(\theta,\theta') = \sigma_\perp{}^c(\theta,\theta') = \frac{1}{k_0}\Bigg[& \frac{\sin^2[k_0a(\sin\theta + \sin\theta')]}{\cos^2[(\theta-\theta')/2]} \\ & + \frac{\cos^2[k_0a(\sin\theta + \sin\theta')]}{\sin^2[(\theta+\theta')/2]}\Bigg]\end{aligned} \tag{7.4-24}$$

The true forward scatter cross section when $\theta' = \theta + \pi$, and $\theta \neq \pm\pi/2$, is then approximately

$$\sigma_{\parallel}{}^{c}(\theta, \theta + \pi) = \sigma_{\perp}{}^{c}(\theta, \theta + \pi) = \frac{1}{k_0}\left[(2k_0a)^2 \cos^2\theta + \frac{1}{\cos^2\theta}\right] \tag{7.4-25}$$

Thus,

$$\sigma^{c}_{\parallel,\perp}(0, \pi) = \frac{1}{k_0}[1 + (2k_0a)^2] \tag{7.4-26}$$

the same as Eq. (7.4-17) for normal incidence backscatter, as expected.

For scattering in the backward hemisphere, where $|\theta'| < \pi/2$ and $\theta \neq \pm\pi/2$, neglecting the interaction between the edges results in

$$\sigma_{\parallel}{}^{c}(\theta, \theta') = \sigma_{\perp}{}^{c}(\theta, \theta') = \frac{1}{k_0}\left[\frac{\sin^2[k_0a(\sin\theta + \sin\theta')]}{\sin^2[(\theta + \theta')/2]} + \frac{\cos^2[k_0a(\sin\theta + \sin\theta')]}{\cos^2[(\theta - \theta')/2]}\right] \tag{7.4-27}$$

7.4.3. Strips of Finite Thickness and Dielectric Strips

Very little information exists regarding the effect of finite thickness on the cross section of a perfectly conducting strip, although some experimental evidence indicates that for near normal angles of incidence, and for strips wide in comparison to the radar wavelength, the effects of strip thickness and rear shape are negligible for thickness less than approximately $\frac{1}{8}$ wavelength. In a series of experiments on the effect of finite thickness on the diffracted fields produced by a plane wave normally incident on a half-plane, Row[24] determined that the deviation from the results, predicted theoretically for an infinitesimally thin, perfectly conducting half-plane, was less than 10% for thicknesses as large as $\lambda_0/4$. An example of these experimental results is shown in Figure 7-16. On the other hand, it is obvious that at incident angles near grazing, the effect of finite-strip thickness could be appreciable, particularly for incident fields polarized in the plane of incidence where the backscatter cross section goes to zero at grazing incidence for an infinitely thin conducting strip.

For an infinite dielectric strip of rectangular cross section, very little information exists on the radar cross section. If the strip is wide in terms of wavelength and for angles of incidence near normal, the radar cross section is approximately given by

$$\sigma^c \approx |R|^2 \sigma^c_{\text{p.c.}} \tag{7.4-28}$$

where $\sigma^c_{\text{p.c.}}$ is the backscatter cross section of a perfectly conducting strip

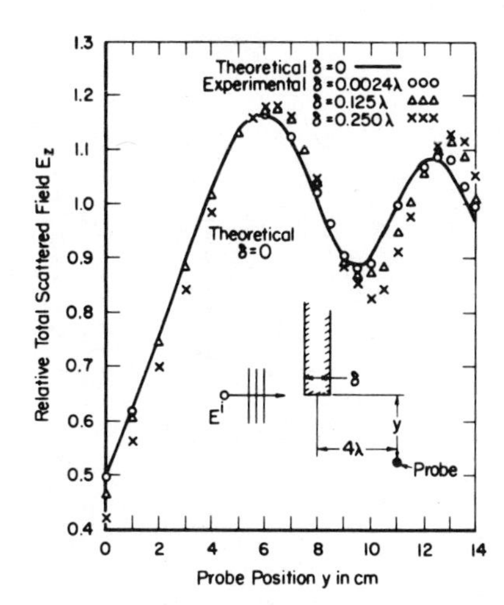

Fig. 7-16. Scattered field intensity versus distance from the edge of an infinite perfectly conducting half-plane of finite thickness [from Row[24]].

and R is the reflection coefficient of an infinite slab of dielectric material of thickness corresponding to the strip thickness, and is given by Eq. (7.1-42).

7.4.4. Experimental Results

The backscatter cross section per unit length of an infinite conducting strip at normal incidence, with the incident field polarized perpendicular to the plane of incidence, has been measured as a function of k_0a by Macrakis.[25] The measurements were made on a parallel plate scattering range, and the results are given in Figure 7-17 and compared with exact theoretical values computed from the eigenfunction series. Measurements were also made for several angles of incidence other than normal, and these are compared in Figure 7-18 with computed values using Eq. (7.4-16).

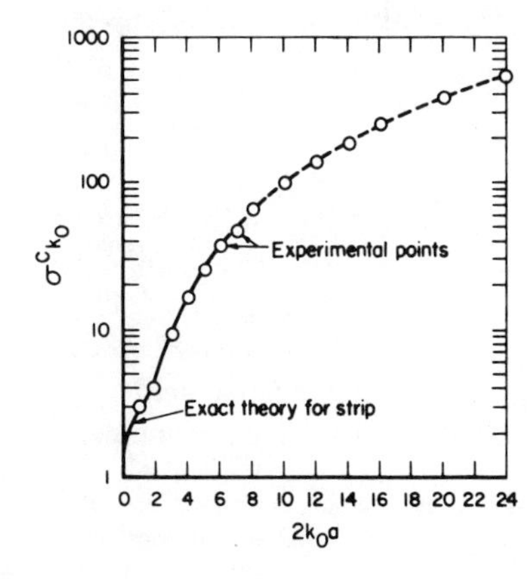

Fig. 7-17. Exact normal-incidence backscatter cross section for a perfectly conducting strip with some measured points, perpendicular polarization [after Macrakis[25]].

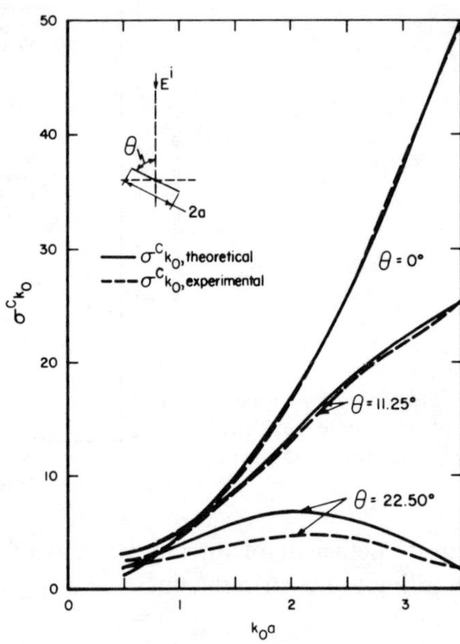

Fig. 7-18. Backscatter cross section of a perfectly conducting strip for several angles of incidence, comparison of experimental and approximate theoretical values [after Macrakis[25]].

A comparison of the various approximations is shown in Figures 7-19 and 7-20, where the backscatter cross sections, $\sigma_{\parallel}{}^{c}$ and $\sigma_{\perp}{}^{c}$, versus the angle of incidence are shown for $k_0 a = 5.29$ and 8.94, using physical optics, Eq. (7.4-7), and the Sommerfeld–Macdonald approximations with and without interaction effects [Eqs. (7.4-11), (7.4-12) and (7.4-16)].† In addition, the exact edge-on backscatter cross section for perpendicular polarization is compared with that given by Eq. (7.4-18) in Figure 7-21, along with the Rayleigh result [Eq. (7.4-3)].

7.5. FLAT PLATES

In this section, the radar cross sections presented by flat plates will be discussed. In all of the following discussion, unless otherwise indicated, the plate will be considered to lie in the xy-plane with a plane wave incident at an angle θ with respect to the z-axis, as shown in Figure 7-22. Again, as previously, two polarizations will be considered; the electric field vector perpendicular to the plane of incidence or parallel to the plane of incidence.

† In Figures 7-19 and 7-20, the curves for the cross sections given by Eqs. (7.4-11) and (7.4-12) agree with the exact results to within graphical accuracy; thus they are not shown.

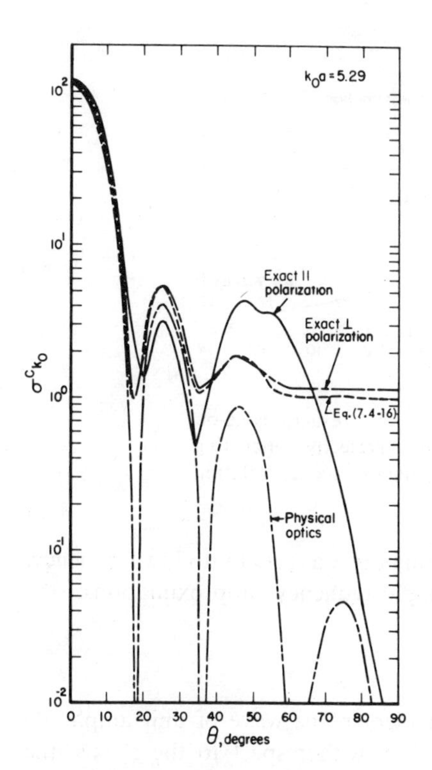

Fig. 7-19. Comparison of the backscatter cross sections for a perfectly conducting strip obtained by various approximation techniques with the exact solution, $k_0a = 5.29$.

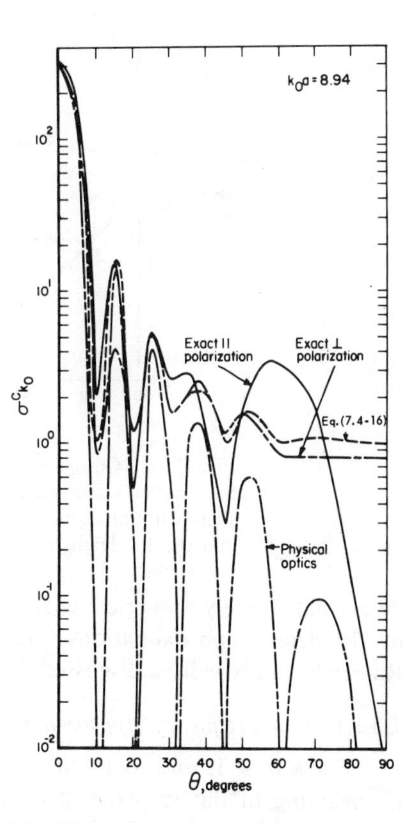

Fig. 7-20. Comparison of the backscatter cross sections for a perfectly conducting strip obtained by various approximation techniques with the exact solution, $k_0a = 8.94$.

7.5.1. Circular Plates

The simplest and best known flat-plate scatterer is a perfectly conducting circular disc of radius a. An exact solution to the disc problem has been obtained by Meixner and Andrejewski, using oblate spheroidal wave functions, which in the limiting case of the minor axis going to zero, are appropriate for the disc.[26] The normal-incidence backscatter cross section has been computed from the exact solution for values of k_0a up to ten. However, complete scattering patterns have been computed for only a few values of k_0a, and the spheroidal wave-function expansions employed in the exact solution of Andrejewski cannot be used easily for numerical computations.

Thus, the requirement arises for approximate expressions which can

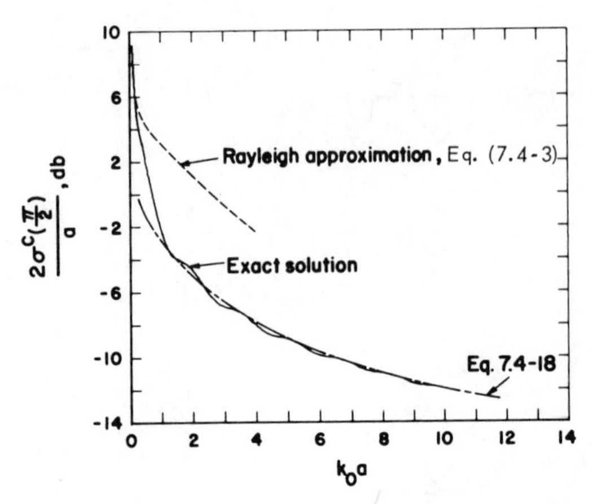

Fig. 7-21. Comparison of the exact, edge-on backscatter cross section of a perfectly conducting strip with an approximate high-frequency solution and the Rayleigh result.

be evaluated easily numerically. Separating these as usual into low frequency, or Rayleigh, approximations and high-frequency approximations, the Rayleigh results will be discussed first.

7.5.1.1. *Low-Frequency Approximations*

For $k_0a \ll 1$, and assuming an incident plane wave of unit amplitude propagating in the xz-plane at an angle θ with respect to the z axis, the low-frequency scattered fields to first order in k_0 are as follows:(27,28)†

For perpendicular polarization ($\mathbf{E}^i = -E_y{}^i\hat{\mathbf{y}}$)

$$E_\theta{}^s(\theta;\theta',\phi') = -k_0{}^2a^3\left(\frac{4}{3\pi}\right)\cos\theta'\sin\phi'\,\frac{\exp(ik_0r)}{r}$$

$$E_\phi{}^s(\theta;\theta',\phi') = -k_0{}^2a^3\left(\frac{2}{3\pi}\sin\theta\sin\theta' + \frac{4}{3\pi}\cos\phi'\right)\frac{\exp(ik_0r)}{r} \qquad (7.5\text{-}1)$$

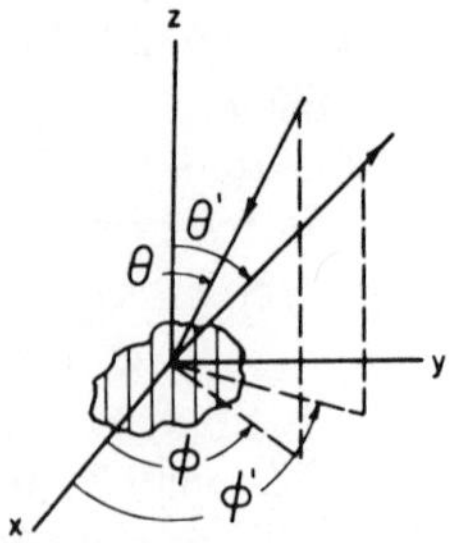

Fig. 7-22. Coordinates for scattering from an arbitrarily shaped flat plate.

† These can also be obtained from the low-frequency perfectly conducting oblate spheroid results of Section 5.1.3 by letting the minor axis, c, go to zero.

For parallel polarization ($\mathbf{H}^i = H_y{}^i\hat{\mathbf{y}}$)

$$E_\theta{}^s(\theta;\, \theta', \phi') = k_0{}^2a^3 \left(\frac{4}{3\pi}\right) \cos\theta \cos\theta' \cos\phi' \,\frac{\exp(ik_0r)}{r}$$

$$E_\phi{}^s(\theta;\, \theta', \phi') = -k_0{}^2a^3 \left(\frac{4}{3\pi}\right) \cos\theta \sin\phi' \,\frac{\exp(ik_0r)}{r} \tag{7.5-2}$$

The backscattered fields can be obtained from the above expressions by setting $\theta' = \theta$ and $\phi' = 0$, giving for the backscatter cross sections

$$\sigma_\perp(\theta) = \frac{16}{9\pi} k_0{}^4a^6[2 + \sin^2\theta]^2 \tag{7.5-3}$$

and

$$\sigma_\parallel(\theta) = \frac{64}{9\pi} k_0{}^4a^6 \cos^4\theta \tag{7.5-4}$$

At normal incidence, the backscatter cross section reduces to

$$\sigma_\parallel(0) = \sigma_\perp(0) = \frac{64}{9\pi} k_0{}^4a^6 \tag{7.5-5}$$

while for edge-on incidence

$$\sigma_\perp\left(\frac{\pi}{2}\right) = \frac{16}{\pi} k_0{}^4a^6 \tag{7.5-6}$$

and

$$\sigma_\parallel\left(\frac{\pi}{2}\right) = 0 \tag{7.5-7}$$

The low-frequency forward scatter cross sections can be obtained from Eqs. (7.5-1) and (7.5-2) by setting $\theta' = \pi - \theta$, $\phi' = \pi$, resulting in

$$\sigma_\perp(\theta, \theta + \pi) = \frac{16}{9\pi} k_0{}^4a^6[2 + \sin^2\theta]^2 \tag{7.5-8}$$

and

$$\sigma_\parallel(\theta, \theta + \pi) = \frac{64}{9\pi} k_0{}^4a^6 \cos^4\theta \tag{7.5-9}$$

Examining how closely these expressions approximate the exact solutions at normal incidence, the Rayleigh expression for the backscatter cross section [Eq. (7.5-5)] gives a cross section approximately 1.5 db lower than the correct result at $k_0a = 0.5$. By $k_0a = 1$, this error has increased to approximately 4.5 db. On the other hand, for the disc observed edge-on,

Eq. (7.5-6) is in error by less than 0.4 db at $k_0a = 1$, although the error increases very rapidly for larger values of k_0a. A considerable improvement in the normal-incidence case can be brought about by including the next higher-order term in the low-frequency expansion.[28] If this is done, the following expression is obtained for the normally incidence backscatter cross section:

$$\sigma_{\perp}(0) = \sigma_{\parallel}(0) = \frac{4}{\pi} k_0^4 a^6 \left[\frac{4}{3} + \frac{32}{45}(k_0a)^2\right]^2 \tag{7.5-10}$$

Using this expression instead of Eq. (7.5-5), the error at $k_0a = 1$ reduces to approximately 0.7 db. At edge-on incidence, introducing the next higher-order term improves slightly the already excellent approximation provided by Eq. (7.5-6) for values of $k_0a < 1$; however, it does not provide significant improvement for $k_0a > 1$ where the Rayleigh expressions are of little use.

In Figure 7-23, the backscatter cross section for normal incidence is plotted for $0 \leqslant k_0a \leqslant 6$ as computed by the exact solution of Meixner and Andrejewski. For comparison, the two Rayleigh expressions, Eqs. (7.5-5) and (7.5-10), are plotted in the same figure along with some measured points.[29]

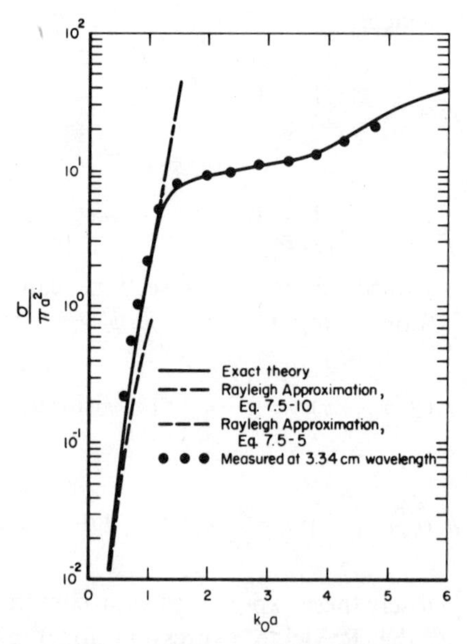

Fig. 7-23. Exact, Rayleigh, higher-order Rayleigh, and experimental backscatter cross sections for normal incidence on a perfectly conducting disc.

7.5.1.2. *High-Frequency Approximations*

7.5.1.2.1. *Physical Optics.* In the high-frequency limit, $k_0 a \gg 1$, the simplest approximation capable of illustrating the angular dependence of the scattered fields is the physical-optics solution. Using physical optics, the backscatter cross section of a circular perfectly conducting disc is

$$\sigma_{\text{p.o.}}(\theta) = \frac{4\pi(\pi a^2)^2 \cos^2 \theta}{\lambda_0^2} \left[\frac{J_1(2k_0 a \sin \theta)}{k_0 a \sin \theta} \right]^2 \tag{7.5-11}$$

which for normal incidence reduces to

$$\sigma_{\text{p.o.}}(0) = \frac{4\pi}{\lambda_0^2} (\pi a^2)^2 = \frac{4\pi A^2}{\lambda_0^2} \tag{7.5-12}$$

or

$$\sigma_{\text{p.o.}}/\pi a^2 = (k_0 a)^2$$

The physical-optics cross section is independent of the polarization, and reduces to zero at grazing incidence, where $\theta = \pm\pi/2$. For angles of incidence near zero, the physical-optics result provides a good approximation to the backscatter cross section for large $k_0 a$. At normal incidence the physical-optics result deviates from the exact solution by less than 1 db for values of $k_0 a$ equal to three or greater, as can be seen from Figure 7-24, where the normal incidence exact solution for $1 \leqslant k_0 a \leqslant 10$ is compared with the physical-optics solution and some measured points.[29] However, below $k_0 a = 3$, the error rapidly increases, and for $k_0 a = 2$, amounts to greater than 3.5 db. For angles of incidence greater than about 45°, the physical-optics result does not give the correct angular dependence for the cross section, and is of no use for grazing or near-grazing angles of incidence.

7.5.1.2.2. *Geometrical Diffraction.* Results for the high-frequency backscatter cross section for a perfectly conducting disc have also been obtained using Keller's "Geometrical Theory of Diffraction."[30] The backscatter cross section predicted by geometrical diffraction for $0 < |\theta| < \pi/2$, using only singly diffracted rays, is

$$\frac{\sigma(\theta)}{\lambda_0^2} = \frac{k_0 a}{4\pi^2 \sin \theta} \left[\cos^2(2k_0 a \sin \theta) + \frac{\sin^2(2k_0 a \sin \theta)}{\sin^2 \theta} \right] \tag{7.5-13}$$

This expression is not valid at $\theta = 0$, where it goes to infinity; in addition, although finite at $\theta = \pm\pi/2$, it is not valid at that point. It is also obviously independent of the polarization.

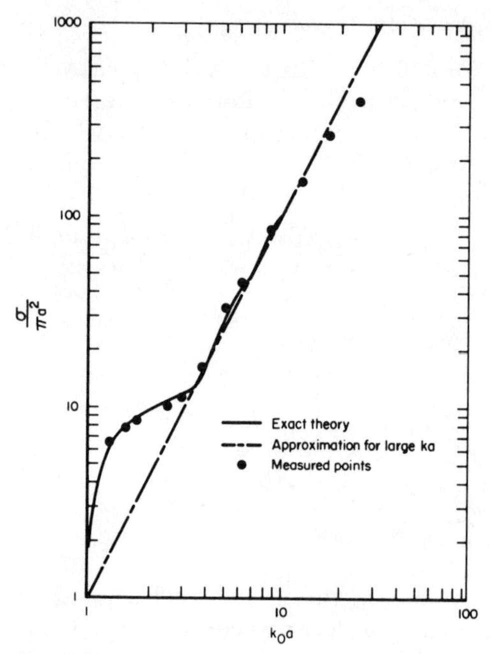

Fig. 7-24. Exact and physical-optics backscatter cross sections for a perfectly conducting disc at normal incidence.

7.5.1.2.3. *Sommerfeld–Macdonald Technique.* An improvement over the physical-optics and geometrical-diffraction results for angles of incidence other than zero can be obtained by a modification, due to Ufimtsev, of the Sommerfeld–Macdonald technique discussed earlier.[31,32] In essence, the scattered field produced by a point on the disc edge is assumed to be the same as would be produced by a half-plane oriented along the tangent to the edge of the disc at the point. The entire scattered field is then obtained by integrating around the edge of the disc. Using an approach similar to this, the backscatter fields for $\theta \neq 0$ are approximated by

$$E_\phi{}^s(\theta) \approx \frac{ia}{2}\Bigg[\left[(B_2(\theta))^2\left(\frac{1-\sin\theta}{2\sin\theta}\right) + (B_1(\theta))^2\left(\frac{1+\sin\theta}{2\sin\theta}\right)\right] J_1(2k_0 a\sin\theta)$$
$$+ i\left[(B_2(\theta))^2\left(\frac{1-\sin\theta}{2\sin\theta}\right)\right.$$
$$\left. - (B_1(\theta))^2\left(\frac{1+\sin\theta}{2\sin\theta}\right)\right] J_2(2k_0 a\sin\theta)\Bigg]\frac{\exp(ik_0 r)}{r} \qquad (7.5\text{-}14)$$

for perpendicular polarization, and

$$H_\phi{}^s(\theta) \approx -\frac{ia}{2}\Big[\Big[(A_1(\theta))^2\left(\frac{1-\sin\theta}{2\sin\theta}\right) + (A_2(\theta))^2\left(\frac{1+\sin\theta}{2\sin\theta}\right)\Big] J_1(2k_0a\sin\theta)$$
$$- i\Big[(A_1(\theta))^2\left(\frac{1-\sin\theta}{2\sin\theta}\right)$$
$$- (A_2(\theta))^2\left(\frac{1+\sin\theta}{2\sin\theta}\right)\Big] J_2(2k_0a\sin\theta)\Big]\frac{\exp(ik_0r)}{r} \quad (7.5\text{-}15)$$

for parallel polarization. The functions, $A_{1,2}(\theta)$, $B_{1,2}(\theta)$, were defined in Eqs. (7.4-13) and (7.4-14) as

$$A_{1,2}(\theta) = \sqrt{2}\exp[-i(\pi/4)]\Big\{C\Big[2\sqrt{\frac{2k_0a}{\pi}}\,\Big|\cos\left(\frac{\pi}{4}\mp\frac{\theta}{2}\right)\Big|\Big]$$
$$+ iS\Big[2\sqrt{\frac{2k_0a}{\pi}}\,\Big|\cos\left(\frac{\pi}{4}\mp\frac{\theta}{2}\right)\Big|\Big]\Big\}$$

and

$$B_{1,2}(\theta) = A_{1,2}(\theta) + \frac{i}{2\sqrt{\pi k_0a}}\,\frac{\exp[i2k_0a(1\pm\sin\theta) - i(\pi/4)]}{\left|\cos\left(\frac{\pi}{4}\mp\frac{\theta}{2}\right)\right|}$$

where $C(\gamma)$ and $S(\gamma)$ are Fresnel integrals. The backscatter cross sections are then

$$\sigma_\perp = 4\pi r^2 \mid E_\phi{}^s \mid^2$$

and

$$\sigma_\parallel = 4\pi r^2 \mid H_\phi{}^s \mid^2$$

In Figure 7-25, the backscatter cross sections, calculated from Eqs. (7.5-14) and (7.5-15) above, are plotted for $k_0a = 5$.

Simpler expressions than (7.5-14) and (7.5-15) can be obtained by expanding the Fresnel integrals in inverse powers of k_0a, and retaining only the first-order terms. If this is done, then the expressions for the scattered fields become

$$E_\phi{}^s = \frac{ia}{2}\left[\frac{J_1(2k_0a\sin\theta)}{\sin\theta} - iJ_2(2k_0a\sin\theta)\right]\frac{\exp(ik_0r)}{r} \quad (7.5\text{-}16)$$

and

$$H_\phi{}^s = \frac{-ia}{2}\left[\frac{J_1(2k_0a\sin\theta)}{\sin\theta} + iJ_2(2k_0a\sin\theta)\right]\frac{\exp(ik_0r)}{r} \quad (7.5\text{-}17)$$

valid for k_0a large and $\mid\theta\mid < \pi/2$. Thus, the backscatter cross section is given by

$$\sigma = \pi a^2\left[\frac{J_1{}^2(2k_0a\sin\theta)}{\sin^2\theta} + J_2{}^2(2k_0a\sin\theta)\right] \quad (7.5\text{-}18)$$

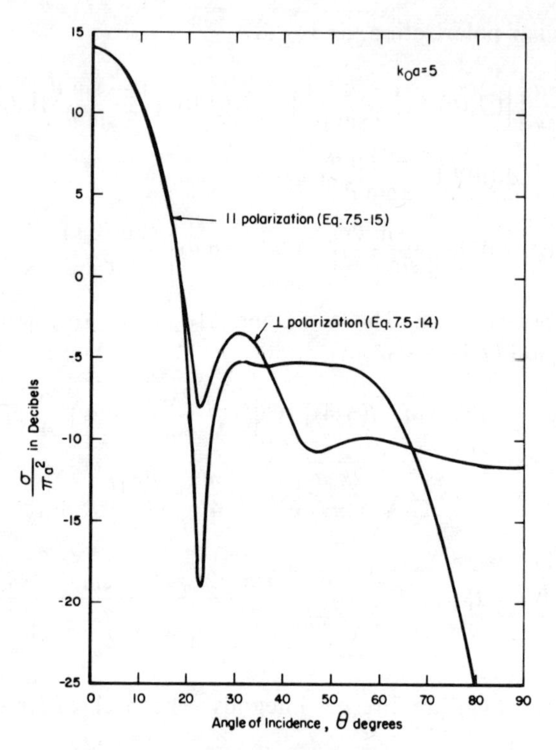

Fig. 7-25. Sommerfeld–Macdonald approximation backscatter cross sections for a perfectly conducting disc versus the angle of incidence, $k_0a = 5.0$.

which for $\theta = 0$ reduces to $4\pi(A^2/\lambda_0^2)$, the result previously obtained by physical optics.

For $\phi = \pi/2$ or grazing incidence, the backscattered field for perpendicular polarization becomes, from Eq. (7.5-14),

$$E_\phi^s\left(\frac{\pi}{2}\right) \approx \left(\frac{ia}{2}\right) B_1^2\left(\frac{\pi}{2}\right)[J_1(2k_0a) - iJ_2(2k_0a)]\,\frac{\exp(ik_0r)}{r}$$

giving for the cross section

$$\sigma_\perp = \pi a^2 \left| B_1^2\left(\frac{\pi}{2}\right)\right|^2 [J_1^2(2k_0a) + J_2^2(2k_0a)] \qquad (7.5\text{-}19)$$

This can be simplified still further by expanding the Bessel functions and $B_1(\pi/2)$ asymptotically, giving to the first order in k_0^{-1}

$$\sigma_\perp\left(\frac{\pi}{2}\right) \approx \pi a^2\left(\frac{1}{\pi k_0 a}\right) \approx \frac{a}{k_0} \qquad (7.5\text{-}20)$$

For parallel polarization, the backscattered field for $\theta = \pi/2$, given by Eq. (7.5-15), goes to zero as it should.

In Figure 7-26, the backscatter cross section for edge-on incidence and perpendicular polarization, as given by Eq. (7.5-20), is compared with experimental results obtained by Hey *et al.*[33] From this figure, it is apparent that Eq. (7.5-20) does not correctly predict the observed cross-section fluctuations, although it does represent approximately an average cross section. This might have been expected, since at grazing incidence, Eqs. (7.5-14) and (7.5-15) are interpolation formulas, and the half-plane results from which they are derived are not capable of correctly predicting the effects of the rear edge.

Examining Figure 7-26, the experimental data can be fitted by assuming the disc to consist of two interacting scatterers. If these are assumed to be the front and rear edges, the cross section should be of the form

$$\sigma_\perp \approx 2\sigma_1[1 + \cos(4k_0a + \phi)] \tag{7.5-21}$$

where σ_1 is the cross section of the individual contributors, assumed to be

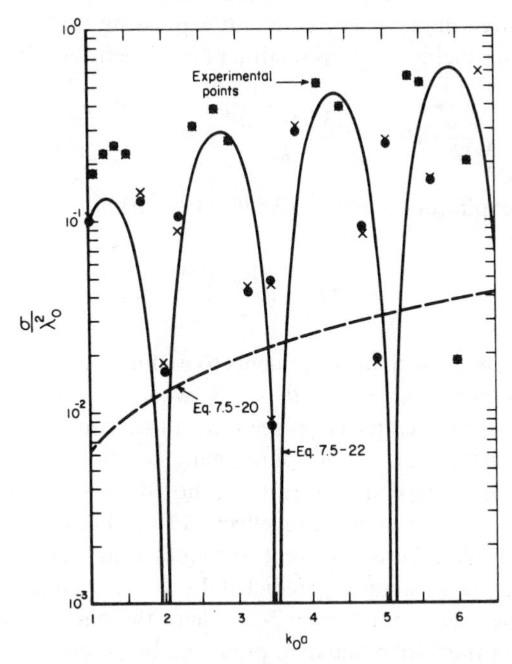

Fig. 7-26. Measured edge-on backscatter cross section of a perfectly conducting disc for perpendicular polarization, compared with two approximate solutions [after Hey *et al.*[33]].

equal. If σ_1 is assumed to be a/k_0, as given in Eq. (7.5-20), then $\sigma_\perp$ can be expressed as

$$\sigma_\perp \approx \frac{4a}{k_0} \cos^2 \left(2k_0 a + \frac{\phi}{2}\right)$$

Estimating $\phi/2$ to be approximately $-3\pi/4$, then

$$\sigma_\perp \approx \frac{4a}{k_0} \cos^2 \left(2k_0 a - \frac{3\pi}{4}\right) \tag{7.5-22}$$

This is also plotted in Figure 7-26, and provides only a fair fit with the experimental data, since the periodicity is incorrect.

A much better fit can be provided by assuming that the two contributors are not equal. The first, as before, is assumed to be the forward edge with a magnitude and phase of

$$\sqrt{\sigma_1} = -\sqrt{a/k_0}\, e^{i\pi/4}$$

The second is assumed to resemble a creeping wave launched at the shadow boundary which travels around the rear rim of the disc. Using this assumption, Senior has obtained an empirical expression for this second contributor by fitting measured data.[34] The resultant expression for the perpendicularly polarized, grazing-incidence cross section of a perfectly conducting thin disc is

$$\sigma = \left| -\sqrt{\frac{a}{k_0}}\, e^{i\pi/4} + \frac{35\sqrt{\pi}\, m^4}{k_0}\, e^{i\pi/4} \left[1 + \frac{i\pi}{20}\left(1 - \frac{2}{k_0 a}\right)\right] \times \exp[ik_0 a(4 + \pi) - i0.528\pi(1 - 2i)\, m] \right|^2 \tag{7.5-23}$$

where

$$m = \left(\frac{k_0 a}{2}\right)^{1/3}$$

This expression agrees with measurements to within one db for $1 \leqslant k_0 a \leqslant 30$.

In the angular region, $\pi/2 \geqslant \theta \geqslant \pi/4$, where Eqs. (7.5-16) and (7.5-17) are not valid, the backscatter cross section can be computed using the field Eqs. (7.5-14) and (7.5-15), or an approximate solution can be obtained for parallel polarization by considering the echo in this angular region to be due primarily to a traveling wave effect. From Figure 7-25, for parallel polarization, a large lobe is apparent at angles near edge-on. The position of this lobe occurs at a point predicted by the traveling wave theory of Peters, as can be seen from Figure 7-27 where the angular positions of the first, second, and third-lobe maxima, predicted by Peters' theory, are plotted versus l/λ_0, where l is the length of the long thin scatterer. From Eq. (2.3-100) of Chapter 2, the backscatter cross section due to a traveling wave is given by

$$\sigma \approx (\lambda_0^2 \gamma^2/\pi)\, F \tag{7.5-24}$$

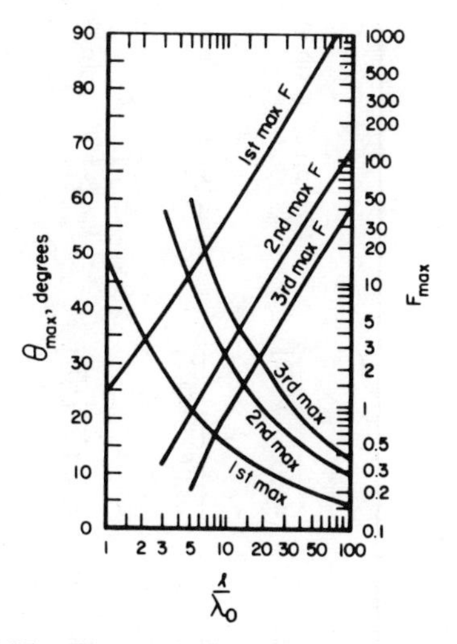

Fig. 7-27. The value of F at the lobe maximum, and the angular position of the lobe maximum for the first three traveling wave lobes versus the length of the body in wavelengths.

where γ is a reflection coefficient depending on the shape of the rear edge of the body and the angle of incidence, and F is given by [Eq. (2.3-100)]

$$F = \left[\frac{\cot^2(\theta/2)\sin^2[k_0 l \sin(\theta/2)]}{\ln 2k_0 l - 0.4228}\right]^2$$

The values of F at the lobe maxima are plotted versus l/λ_0 in Figure 7-27. Experimental results on a number of flat plates tend to indicate that γ is proportional to $\cos\theta$, thus Eq. (7.5-24) can be written as

$$\frac{\sigma}{\lambda_0^2} \approx \frac{F}{\pi}\cos^2\theta(\gamma')^2 \tag{7.5-25}$$

where γ' is then determined by the shape of the rear edge. Using Eq. (7.5-25), the values of F obtained from Figure 7-27, and the cross sections at the peak of the last lobe for parallel polarization given in Figure 7-25, the effective value of $(\gamma')^2$ for discs of this size ($k_0 a \approx$ 3–6) is found to be approximately 1.5. The fact that the effective value of γ is greater than one can be attributed to the generation of parallel traveling waves along disc elements parallel to the plane of incidence.

Bistatic scattering can be computed by the same techniques used in

Table 7-1

θ'	ϕ'	g_1	g_2	f_1	f_2	ζ
$0 \leqslant \theta' \leqslant \pi/2$	0	$\dfrac{1}{\sin[(\theta+\theta')/2]}$	$\dfrac{-1}{\cos[(\theta-\theta')/2]}$	$\dfrac{-1}{\sin[(\theta+\theta')/2]}$	$\dfrac{-1}{\cos[(\theta-\theta')/2]}$	$k_0 a(\sin\theta' + \sin\theta)$
	π	$\dfrac{1}{\sin[(\theta-\theta')/2]}$	$\dfrac{1}{\cos[(\theta+\theta')/2]}$	$\dfrac{1}{\sin[(\theta-\theta')/2]}$	$\dfrac{1}{\cos[(\theta+\theta')/2]}$	$k_0 a(\sin\theta' - \sin\theta)$
$\pi/2 \leqslant \theta' \leqslant \pi$	0	$\dfrac{1}{\cos[(\theta-\theta')/2]}$	$\dfrac{-1}{\sin[(\theta+\theta')/2]}$	$\dfrac{1}{\cos[(\theta-\theta')/2]}$	$\dfrac{1}{\sin[(\theta+\theta')/2]}$	$k_0 a(\sin\theta' + \sin\theta)$
	π	$\dfrac{-1}{\cos[(\theta+\theta')/2]}$	$\dfrac{-1}{\sin[(\theta-\theta')/2]}$	$\dfrac{-1}{\cos[(\theta+\theta')/2]}$	$\dfrac{1}{\sin[(\theta-\theta')/2]}$	$k_0 a(\sin\theta' - \sin\theta)$

the monostatic case, and Ufimtsev's technique, neglecting edge interaction, gives, for the scattered fields,[32]

$$E_\phi^s(\theta, \theta') = \frac{ia}{2}\,[g_1(\theta, \theta')\,J_1(\zeta) + ig_2(\theta, \theta')\,J_2(\zeta)]\,\frac{\exp(ik_0 r)}{r} \qquad (7.5\text{-}26)$$

for perpendicular polarization, and

$$H_\phi^s(\theta, \theta') = \frac{ia}{2}\,[f_1(\theta, \theta')\,J_1(\zeta) + if_2(\theta, \theta')\,J_2(\zeta)]\,\frac{\exp(ik_0 r)}{r} \qquad (7.5\text{-}27)$$

for parallel polarization; where g_1, g_2, f_1, f_2, and ζ are given in Table 7-1 for various angular regions, where ϕ is considered to be zero and $0 \leqslant \theta \leqslant \pi/2$.

For forward scatter, the cross sections obtained from the above field expressions become

$$\sigma_{\perp,\parallel}(\theta, \pi + \theta) = \pi a^2 \left[\frac{1 + k_0^2 a^2 \cos^2\theta}{\cos^2\theta}\right] \qquad (7.5\text{-}28)$$

and for $\theta = 0$, this is identical to the normal-incidence backscatter as expected.

Somewhat more accurate expressions for the bistatic scattered field can be obtained by taking into account the interaction between the disc edges; the resultant expressions are complex, however, and not easy to evaluate numerically. For details of this, Ufimtsev should be consulted.[32]

7.5.1.3. *Dielectric Discs*

For dielectric discs, no exact solutions are available, and few approximate results exist. However, for discs which are very small compared with the wavelength, or Rayleigh discs, the backscatter cross sections can be approximated by using the oblate spheroid results [Eqs. (5.1-87) and (5.1-88)]. Thus, for $k_0 a \ll 1$ and $a \gg c$, with $2c$ the disc thickness,

$$\begin{aligned}\sigma_\perp(\theta) &\approx \frac{16}{9\pi}\,k_0^4(4\pi a^2 c)^2 \left[\frac{4(\epsilon_r - 1)}{(\epsilon_r - 1)\,\pi(c/a) + 4}\right.\\ &\quad \left. - \frac{4(\mu_r - 1)\cos^2\theta}{(\mu_r - 1)\,\pi(c/a) + 4} - \frac{2(\mu_r - 1)\sin^2\theta}{2\mu_r - (\mu_r - 1)\,\pi(c/a)}\right]^2 \\ \sigma_\parallel(\theta) &\approx \frac{16}{9\pi}\,k_0^4(4\pi a^2 c)^2 \left[\frac{4(\epsilon_r - 1)\cos^2\theta}{(\epsilon_r - 1)\,\pi(c/a) + 4}\right.\\ &\quad \left. + \frac{2(\epsilon_r - 1)\sin^2\theta}{2\epsilon_r - (\epsilon_r - 1)\,\pi(c/a)} - \frac{4(\mu_r - 1)}{(\mu_r - 1)\,\pi(c/a) + 4}\right]^2\end{aligned} \qquad (7.5\text{-}29)$$

The above expressions were obtained from Eqs. (5.1-87) and (5.1-88) by setting $I_a \approx \pi/2a^3$ and $I_c \approx (2/a^2c) - (\pi/a^3)$ valid for $a/c \gg 1$. Note that the Rayleigh backscatter cross sections for a perfectly conducting disc

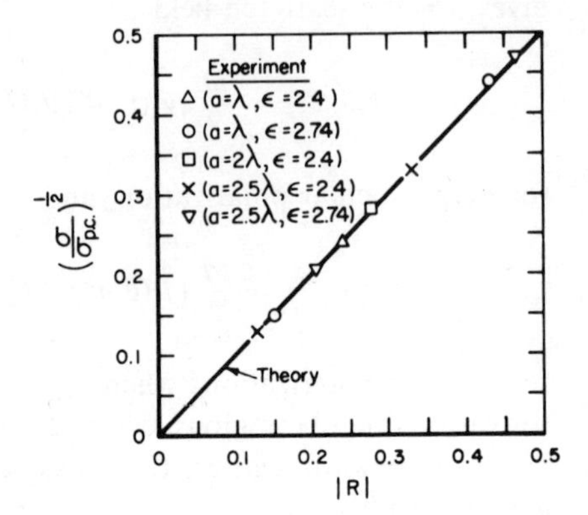

Fig. 7-28. A comparison of the reflection-coefficient magnitude obtained by cross section measurements of dielectric discs and computed from the Fresnel coefficients.

[(Eqs. (7.5-3) and (7.5-4)] can be obtained from the above expressions by setting $\mu_r = 0$, $\epsilon_r \to \infty$, and letting $c \to 0$.

For dielectric discs which are very large compared with the wavelength and angles of incidence near normal, the backscatter cross sections are given by

$$\sigma = | R |^2 \sigma_{\text{p.c.}} \tag{7.5-30}$$

where R is the reflection coefficient of an infinite slab of the same thickness and dielectric material as the disc, and $\sigma_{\text{p.c.}}$ is the backscatter cross section of a perfectly conducting disc of the same size. This result has been verified experimentally at normal incidence by several investigators. In Figure 7-28 is shown $| R |$, as computed from Eq. (7.5-30), using the results of normally incident backscatter cross-section measurements made on several dielectric discs by Severin and von Baeckmann,[35] along with the value of $| R |$ obtained from Eq. (7.1-42), using the dielectric constant and thickness of the disc. The agreement is excellent even for discs as small as $a = \lambda_0$.

Equation (7.5-30) can be obtained theoretically from the results of Neugebauer, who obtained expressions for the scattered fields produced by a plane wave normally incident on an aperture in a dielectric screen.[36] By applying a suitable modification of Babinet's principle, the scattered fields from a dielectric disc can be derived; and for backscatter, the resultant cross section is that given by Eq. (7.5-30).

For arbitrary incidence, Eq. (7.5-30) should give adequate results for large $k_0 a$ over a range of angles of incidence from normal to approximately 45°, provided a reflection coefficient appropriate to the angle of incidence and the polarization is used. For grazing angles of incidence, Eq. (7.5-30) is of no use, and no valid expression is known for this range of incidence.

7.5.2. Rectangular Plates

7.5.2.1. *Low Frequency*

In the low-frequency or Rayleigh region, no analytical results are available for a rectangular plate. An empirical expression based on experimental measurements for the normal-incidence backscatter cross section for a square perfectly conducting plate is

$$\sigma(0) \approx 114 \frac{a^6}{\lambda_0^4} \tag{7.5-31}$$

with a the length of the plate side.

7.5.2.2. *High Frequency*

For plates of other shapes than circular, no exact analytical results are available and few experimental results exist. Physical optics can be utilized to obtain the value of the specular return in the high-frequency limit for an arbitrarily shaped plate giving

$$\sigma(0) = 4\pi A^2/\lambda_0^2 \tag{7.5-32}$$

where A is the total plate area. For angles of incidence other than normal, the physical-optics integrals can be evaluated only for a few specific shapes.

One of the shapes for which the physical optics integral can be evaluated is the rectangular plate. Consider a perfectly conducting plate lying in the xy-plane, with $-a/2 \leqslant x \leqslant a/2$ and $-b/2 \leqslant y \leqslant b/2$, and a plane wave incident, as shown in Figure 7-22. Physical optics yields a backscatter cross section of

$$\sigma_{\text{p.o.}}(\theta) = \frac{4\pi A^2}{\lambda_0^2} \cos^2\theta \left[\frac{\sin(k_0 a \sin\theta\cos\phi)}{k_0 a \sin\theta\cos\phi}\right]^2 \left[\frac{\sin(k_0 b \sin\theta\sin\phi)}{k_0 b \sin\theta\sin\phi}\right]^2 \tag{7.5-33}$$

As usual, this result is independent of the polarization of the incident field, and is valid for angles of incidence near normal and plates large compared with the wavelength.

In Figure 7-29, some measured normal-incidence cross sections for a square plate are shown along with the cross section computed by the variational technique of Kouyoumjian.[37] The measured results and the cross section computed by the variational method agree quite well, while physical optics results in an error $\geqslant 3$ db for $k_0 a \leqslant \pi$.

As a function of the angle of incidence, the physical-optics result agrees relatively well for angles near zero, or normal incidence, but not at all for large angles of incidence. In Figure 7-30 is shown the measured backscatter cross section of a square plate with $k_0 a = 14.7$ as a function of the angle of incidence for both perpendicular and parallel polarization. In addition,

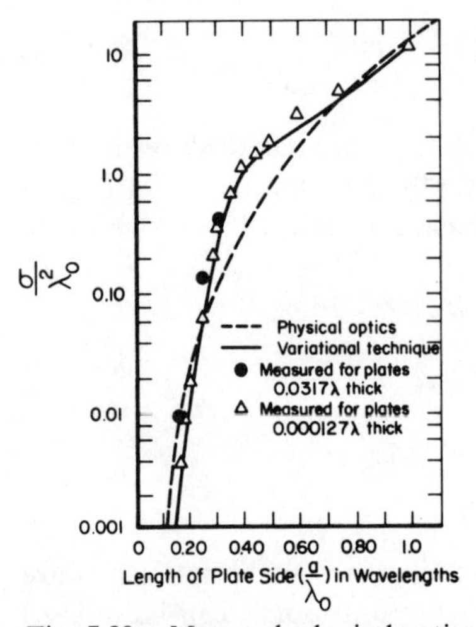

Fig. 7-29. Measured, physical-optics, and Kouyoumjian's variational backscatter cross sections of a perfectly conducting square plate [after Kouyoumjian[37]].

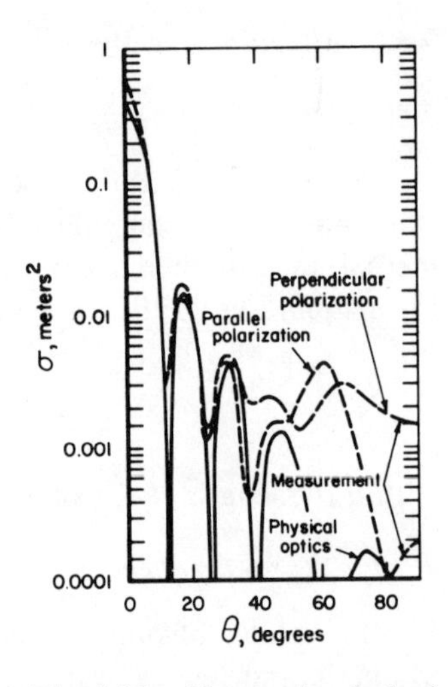

Fig. 7-30. Measured and physical-optics backscatter cross section for a perfectly conducting square plate, $k_0a = 14.7$.

the cross section predicted by physical optics, Eq. (7.5-33), is shown for comparison. It is apparent that for angles of incidence greater than about 40°, the physical-optics results are not valid.

An expression for the high-frequency backscatter cross section for a perfectly conducting rectangular plate has been obtained by Ross[38] using geometrical diffraction. Ross used the geometrical diffraction results for an infinite strip and then considered only a finite section. The results of such a procedure are

$$\sigma_\perp(\theta) = \frac{b^2}{\pi}\left|\left[\cos(k_0a\sin\theta) - \frac{i\sin(k_0a\sin\theta)}{\sin\theta}\right] - \frac{\exp[i(k_0a-\pi/4)]}{\sqrt{2\pi}(k_0a)^{3/2}}\right.$$
$$\times\left[\frac{1}{\cos\theta} + \frac{\exp[i(k_0a-\pi/4)]}{4\sqrt{2\pi}(k_0a)^{3/2}}\left(\frac{(1+\sin\theta)\exp[-ik_0a\sin\theta]}{(1-\sin\theta)^2}\right.\right.$$
$$\left.\left.\left.+\frac{(1-\sin\theta)\exp[ik_0a\sin\theta]}{(1+\sin\theta)^2}\right)\right]\left[1-\frac{\exp[i(2k_0a-\pi/2)]}{8\pi(k_0a)^3}\right]^{-1}\right|^2 \tag{7.5-34}$$

and

$$\sigma_\parallel(\theta) = \frac{b^2}{\pi}\left|\left[\cos(k_0a\sin\theta) + \frac{i\sin(k_0a\sin\theta)}{\sin\theta}\right] - \frac{4\exp[i(k_0a+\pi/4)]}{\sqrt{2\pi}(k_0a)^{1/2}}\right.$$
$$\times\left[\frac{1}{\cos\theta} - \frac{\exp[i(k_0a+\pi/4)]}{2\sqrt{2\pi}(k_0a)^{1/2}}\left(\frac{\exp[-ik_0a\sin\theta]}{1-\sin\theta}\right.\right.$$
$$\left.\left.\left.+\frac{\exp[ik_0a\sin\theta]}{1+\sin\theta}\right)\right]\left[1-\frac{\exp[i(2k_0a+\pi/2)]}{2\pi(k_0a)}\right]^{-1}\right|^2 \tag{7.5-35}$$

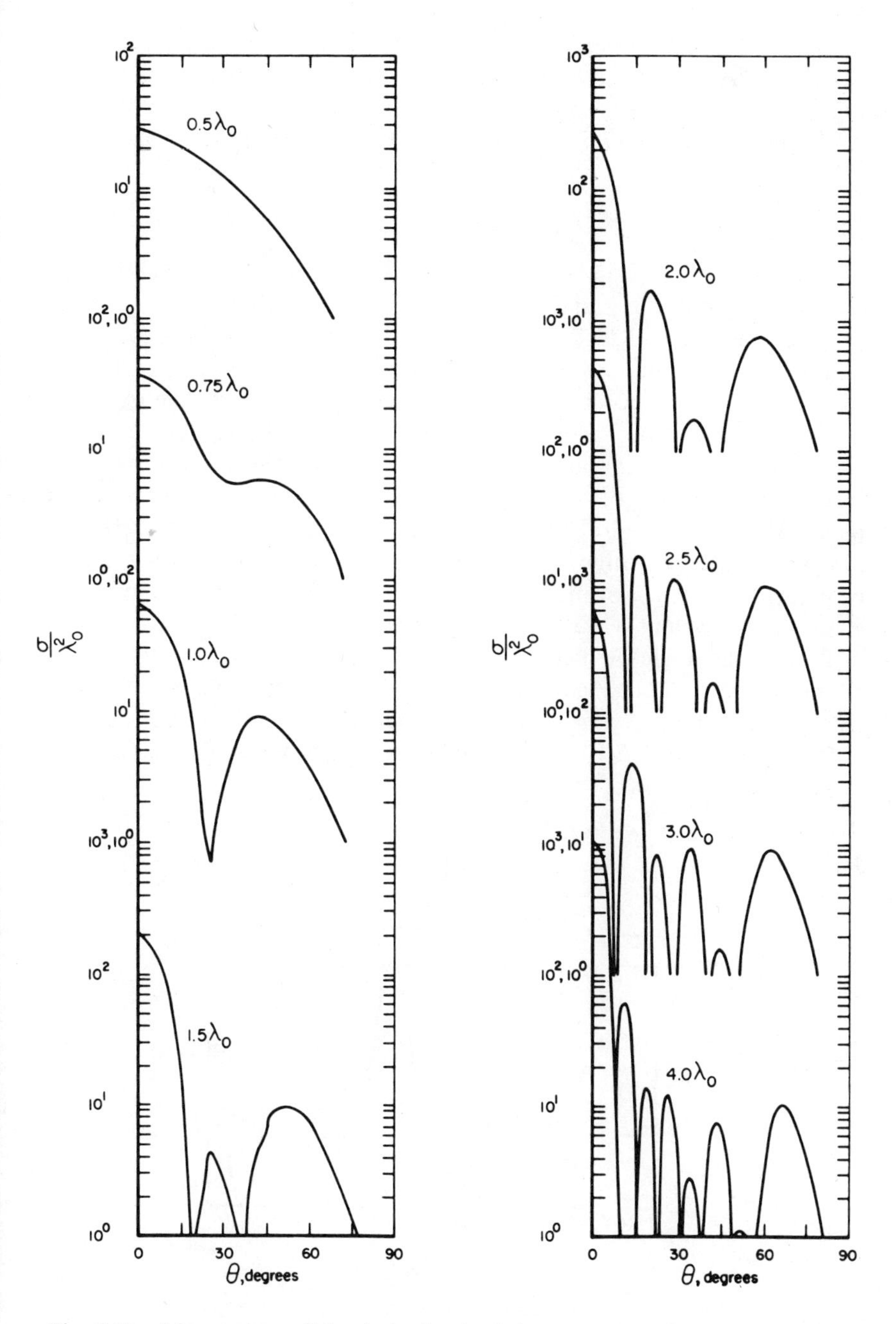

Fig. 7-31. Measured parallel polarization backscatter cross sections versus the angle of incidence for perfectly conducting rectangular plates with heights of $2\lambda_0$ and ranging in length from 0.5 to $4\lambda_0$.

Equations (7.5-34) and (7.5-35) are not valid at $\theta = \pi/2$ where they become infinite. For angles of incidence less than about 70°, however, good agreement has been obtained between measured cross sections and calculations using Eqs. (7.5-34) and (7.5-35) for plates as small as $3.13\lambda_0$ by $3.13\lambda_0$.

7.5.2.3. *Discussion*

For parallel polarization, the cross section shown in Figure 7-30, has a large lobe at about 61°, and appears similar to that observed for circular discs when the incident field has the same polarization. This can be again attributed to a traveling wave echo and, from Figure 7-27, using $l/\lambda_0 = 7.34$, it can be predicted to occur at $\theta \approx 58°$, which is very close to the observed location. The magnitude of the effective reflection coefficient to be used in Eq. (7.5-24) for determination of the contribution of the traveling wave to σ can be empirically estimated by utilizing the measured results shown in Figure 7-31. In this figure, the backscatter cross sections for a series of rectangular plates $2\lambda_0$ high, and ranging in length from 0.5 to $4\lambda_0$, are shown versus the angle of incidence for parallel polarization. It is apparent that the first traveling wave lobe starts to form when $l \approx 0.75\lambda_0$ at about 45°, close to the angle predicted by Figure 7-27, and becomes progressively more pronounced as l increases. The amplitude, however, remains essentially constant at about 9, varying only slightly over the range of lengths used.

For grazing incidence, neither the physical-optics nor the geometrical

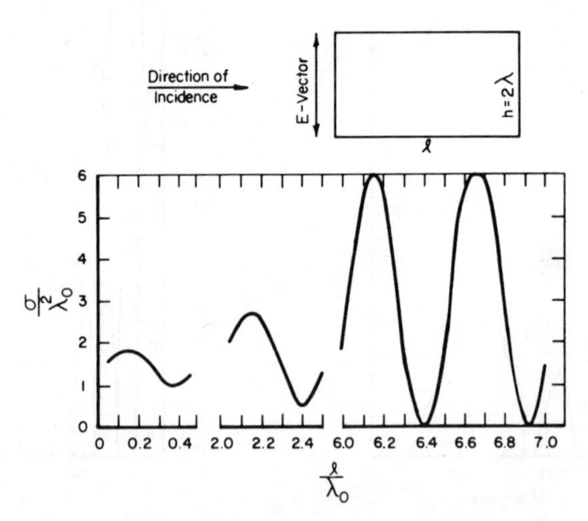

Fig. 7-32. Measured perpendicular polarization edge-on backscatter cross section for a perfectly conducting rectangular plate versus the plate length in wavelengths [from Hey and Senior[39]].

theory of diffraction result is valid. Measurements made by Hey and Senior, shown in Figure 7-32, indicate that for perpendicular polarization, the backscatter cross section is an oscillatory function of the length of the plate parallel to the direction of incidence.[39] The oscillation has a period of $\lambda_0/2$, and increases in magnitude as the length of the plate increases. The mechanism responsible for this behavior appears to be interaction between the front and rear edges of the plate; however, the form of this interaction is not the same as when an infinite strip is viewed edge-on as is apparent by comparing Figures 7-21 and 7-32. In this case, the presence of the top and bottom edges of the plate apparently contributes significantly to the interaction between the front and rear edges.

An expression for the grazing-incidence backscatter cross section has been obtained empirically by Ross,[37] by fitting an extensive set of measured data. Ross's result is

$$\sigma\left(\frac{\pi}{2}\right) = \frac{l}{8\lambda_0} h^2 \left[\left[1 + \frac{\pi}{2}\frac{1}{(l/\lambda_0)^2}\right] + \left[1 - \frac{\pi}{2}\frac{1}{(l/\lambda_0)^2}\right] \cos\left[2k_0 l - \frac{3\pi}{5}\right]\right] \tag{7.5-36}$$

and in Figure 7-33, this is compared with his measured data.

7.5.3. Elliptical Plates

7.5.3.1. *Low Frequency*

For a perfectly conducting elliptical plate lying in the xy-plane with semi-axis a and b as shown in Figure 7-34, the Rayleigh scattered fields,

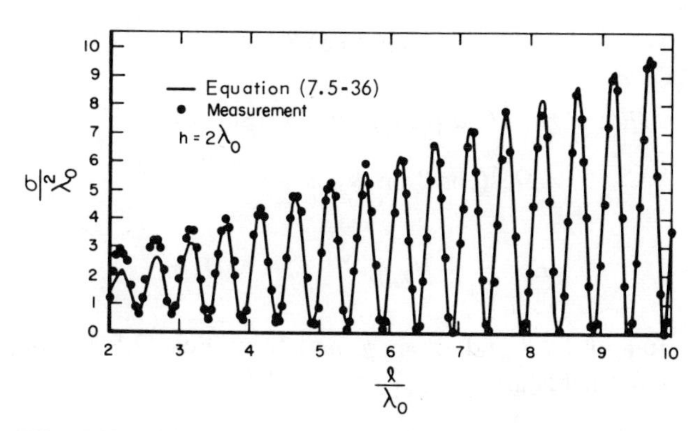

Fig. 7-33. Comparison of Ross' empirical expression for the edge-on backscatter cross section of a perfectly conducting rectangular plate with measured values [from Ross[37]].

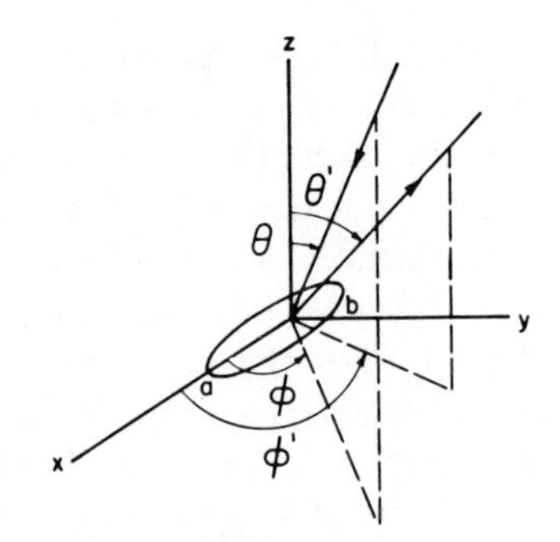

Fig. 7-34. Coordinates for scattering from an elliptical plate.

determined from Eqs. (5.1-30), (5.1-31) and (5.1-34), (5.1-35) by letting $c = 0$, $\mu = 0$, $\epsilon \to \infty$, are

$$E^s_{\phi_\perp} = \tfrac{2}{3}k_0{}^2 E_\perp{}^i a^3[I_a^{-1}(\sin\phi\sin\phi') + I_b^{-1}(\cos\phi\cos\phi') + (I_a + I_b)^{-1}\sin\theta\sin\theta'] \frac{\exp(ik_0 r)}{r} \quad (7.5\text{-}37)$$

$$E^s_{\theta_\perp} = -\tfrac{2}{3}k_0{}^2 E_\perp{}^i a^3[I_a^{-1}(\sin\phi\cos\theta'\cos\phi') - I_b^{-1}(\cos\phi\cos\theta'\sin\phi')] \frac{\exp(ik_0 r)}{r} \quad (7.5\text{-}38)$$

where

$$H^s_{\phi_\perp} = E^s_{\theta_\perp}/\eta, \qquad H^s_{\theta_\perp} = -E^s_{\phi_\perp}/\eta$$

and

$$H^s_{\phi_\parallel} = -\tfrac{2}{3}k_0{}^2 H_\parallel{}^i a^3[I_a^{-1}(\cos\theta\cos\phi\cos\theta'\cos\phi') + I_b^{-1}(\cos\theta\sin\phi\cos\theta'\sin\phi')] \frac{\exp(ik_0 r)}{r} \quad (7.5\text{-}39)$$

$$H^s_{\theta_\parallel} = \tfrac{2}{3}k_0{}^2 H_\parallel{}^i a^3[I_a^{-1}(\cos\theta\cos\phi\sin\phi') - I_b^{-1}(\cos\theta\sin\phi\cos\phi')] \frac{\exp(ik_0 r)}{r} \quad (7.5\text{-}40)$$

where

$$E^s_{\phi_\parallel} = -H^s_{\theta_\parallel}\eta, \qquad E^s_{\theta_\parallel} = H^s_{\phi_\parallel}\eta$$

For backscatter, $\theta' = \theta$ and $\phi' = \phi$, and from Eqs. (7.5-37) and (7.5-39), the backscattered fields are

$$E^s_{\phi_\perp} = \tfrac{2}{3}k_0{}^2 E_\perp{}^i a^3[I_a^{-1}\sin^2\phi + I_b^{-1}\cos^2\phi + (I_a + I_b)^{-1}\sin^2\theta] \frac{\exp(ik_0 r)}{r} \quad (7.5\text{-}41)$$

and

$$H^s_{\phi_\parallel} = -\tfrac{2}{3}k_0{}^2 H_\parallel{}^i a^3[I_a^{-1}\cos^2\theta\cos^2\phi + I_b^{-1}\cos^2\theta\sin^2\phi]\,\frac{\exp(ikr)}{r} \qquad (7.5\text{-}42)$$

Thus, the backscatter cross sections are

$$\sigma_\perp = \frac{16\pi}{9}\,k_0{}^4 a^6[I_a^{-1}\sin^2\phi + I_b^{-1}\cos^2\phi + (I_a + I_b)^{-1}\sin^2\theta]^2 \quad (7.5\text{-}43)$$

and

$$\sigma_\parallel = \frac{16\pi}{9}\,k_0{}^4 a^6[I_a^{-1}\cos^2\theta\cos^2\phi + I_b^{-1}\cos^2\theta\sin^2\phi]^2 \qquad (7.5\text{-}44)$$

The constants, I_a and I_b, in the above equations, are given by

$$I_a = a^3\int_0^\infty \frac{dx}{(a^2+x)\sqrt{(a^2+x)(b^2+x)\,x}} \qquad (7.5\text{-}45)$$

and

$$I_b = a^3\int_0^\infty \frac{dx}{(b^2+x)\sqrt{(a^2+x)(b^2+x)\,x}} \qquad (7.5\text{-}46)$$

where it is assumed that $a \geqslant b$. These integrals can be evaluated in terms of the complete elliptic integrals of the first and second kinds, giving

$$I_a = \frac{2}{[1-(b/a)^2]}\left[\mathrm{K}\left(\cos^{-1}\frac{b}{a}\right) - \mathrm{E}\left(\cos^{-1}\frac{b}{a}\right)\right] \qquad (7.5\text{-}47)$$

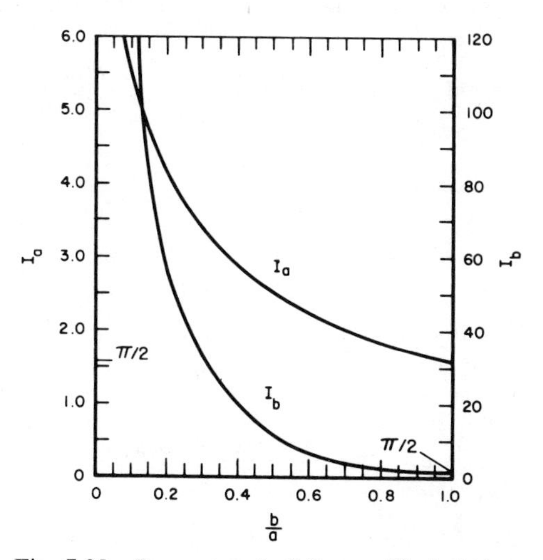

Fig. 7-35. Parameters I_a, I_b for an elliptical plate versus the ratio of the minor to major axis, b/a.

and

$$I_b = \frac{2}{(b/a)^2\,[1-(b/a)^2]}\left[\mathrm{K}\left(\cos^{-1}\frac{b}{a}\right) - \left(\frac{b}{a}\right)^2 \mathrm{E}\left(\cos^{-1}\frac{b}{a}\right)\right] \quad (7.5\text{-}48)$$

These are plotted in Figure 7-35 for values of b/a from zero to one.

At normal incidence, Eqs. (7.5-43) and (7.5-44) reduce to

$$\sigma_{\perp}(0, \phi) = \frac{16\pi}{9}(k_0{}^4a^6)[I_a^{-1}\sin^2\phi + I_b^{-1}\cos^2\phi]^2 \quad (7.5\text{-}49)$$

and

$$\sigma_{\parallel}(0, \phi) = \frac{16\pi}{9}(k_0{}^4a^6)[I_a^{-1}\cos^2\phi + I_b^{-1}\sin^2\phi]^2 \quad (7.5\text{-}50)$$

For forward scattering, the cross sections are obtained from Eqs. (7.5-37) and (7.5-39) by setting $\theta' = \pi - \theta$, and $\phi' = \pi + \phi$, giving

$$\sigma_{\perp} = \frac{16\pi}{9}(k_0{}^4a^6)[I_a^{-1}\sin^2\phi + I_b^{-1}\cos^2\phi - (I_a + I_b)^{-1}\sin^2\theta]^2 \quad (7.5\text{-}51)$$

and

$$\sigma_{\parallel} = \frac{16\pi}{9}(k_0{}^4a^6)[I_a^{-1}(\cos^2\theta\cos^2\phi) + I_b^{-1}(\cos^2\theta\sin^2\phi)]^2 \quad (7.5\text{-}52)$$

7.5.3.2. *High Frequency*

In the high-frequency limit, the physical-optics integral for the elliptical plate can be evaluated asymptotically by the stationary-phase method, giving

$$\sigma_{\text{p.o.}} = \frac{4\pi(\pi ab)^2}{\lambda_0{}^2}\cos^2\theta\left[\frac{J_1{}^2[2k_0(a^2\sin^2\theta\cos^2\phi + b^2\sin^2\theta\sin^2\phi)^{1/2}]}{[k_0{}^2a^2\sin^2\theta\cos^2\phi + k_0{}^2b^2\sin^2\theta\sin^2\phi]}\right] \quad (7.5\text{-}53)$$

which for $\theta = 0$ reduces to

$$\sigma = \frac{4\pi}{\lambda_0{}^2}(\pi^2a^2b^2) = \frac{4\pi}{\lambda_0{}^2}A^2 \quad (7.5\text{-}54)$$

In general, for any convex perfectly conducting flat plate, the high-frequency backscatter cross section in the physical-optics approximation is given by

$$\sigma_{\text{p.o.}} \approx \frac{4\pi A^2\cos^2\theta}{\lambda_0{}^2}[\Lambda_1(x)]^2 \quad (7.5\text{-}55)$$

where

$$\Lambda_1(x) = \frac{2J_1(x)}{x} \quad (7.5\text{-}56)$$

and $\Lambda_1(x)$ is tabulated for values of x from 0 to 10 [page 181 of Jahnke and Emde[22]].

For a circular plate,

$$x = 2k_0 a \sin\theta \tag{7.5-57}$$

while for an elliptical plate,

$$x = 2k_0(a^2 \sin^2\theta \cos^2\phi + b^2 \sin^2\theta \sin^2\phi)^{1/2} \tag{7.5-58}$$

and, in general, x is $k_0 \sin\theta$ times the maximum linear dimension of the plate at the intersection with the plane of incidence.

7.5.4. Triangular Plates

7.5.4.1. *Low Frequency*

In the Rayleigh region, no analytical results are available for a triangular plate.

7.5.4.2. *High Frequency*

The high-frequency backscatter cross section for a triangular plate can be evaluated by physical optics. For an arbitrary isosceles triangle oriented as shown in Figure 7-36, the physical-optics backscatter cross section is

$$\sigma_{\text{p.o.}} = \frac{4\pi A^2}{\lambda_0^2}\cos^2\theta \times \left[\frac{[\sin^2\alpha - \sin^2(\beta/2)]^2 + (1/4\sin^2\phi)[(2a/b)\cos\phi\sin\beta - \sin\phi\sin 2\alpha]^2}{[\alpha^2 - (\beta/2)^2]}\right] \tag{7.5-59}$$

where $\alpha = k_0 a \sin\theta\cos\phi$, $\beta = k_0 b\sin\theta\sin\phi$, and $A = ab/2$. For the wave incident in the plane $\phi = 0$, this reduces to

$$\sigma_{\text{p.o.}} = \frac{4\pi A^2}{\lambda_0^2}\cos^2\theta\left[\frac{\sin^4\alpha}{\alpha^4} + \frac{(\sin 2\alpha - 2\alpha)^2}{4\alpha^4}\right] \tag{7.5-60}$$

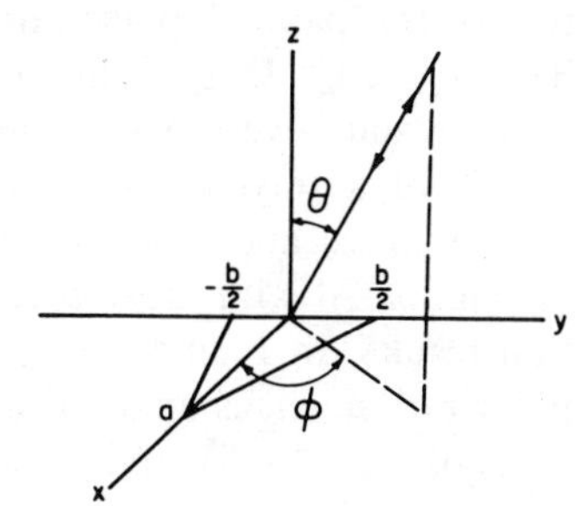

Fig. 7-36. Coordinates for backscatter from a perfectly conducting isosceles triangular plate.

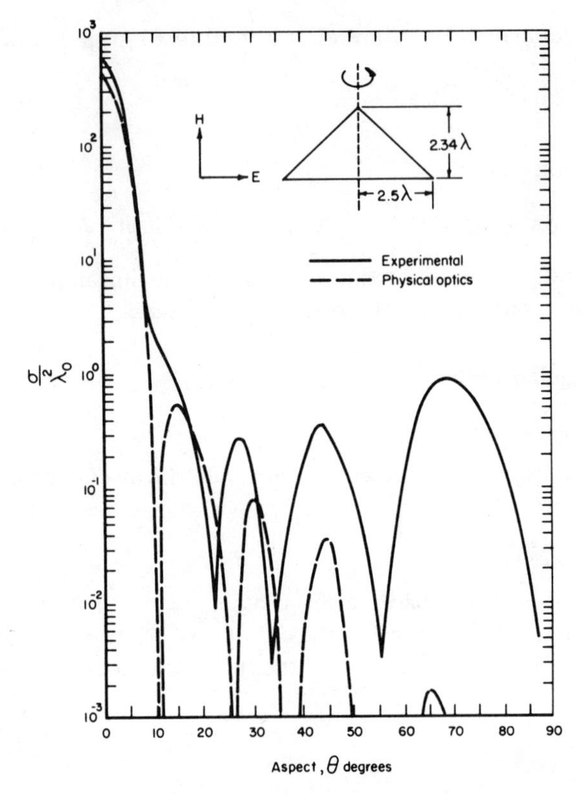

Fig. 7-37. Comparison of experimental and physical-optics backscatter cross sections for a triangular plate versus the angle of incidence, parallel polarization.

and for incidence in the $\phi = \pi/2$ plane

$$\sigma_{\text{p.o.}} = \frac{4\pi A^2}{\lambda_0^2} \cos^2 \theta \left[\frac{\sin^4(\beta/2)}{(\beta/2)^4} \right] \tag{7.5-61}$$

In Figure 7-37, the physical-optics result for an isosceles triangle with a wave incident in the $\phi = \pi/2$ plane is compared with measured results for parallel polarization for a triangle of height $a = 2.34\lambda_0$ and base $b = 5\lambda_0$. It is apparent that far from normal incidence the physical-optics result is very poor; however, in this region the lobe structure characteristic of a traveling wave echo is again observed as in the case of both circular discs and rectangular plates. If the positions of the lobe maxima are compared with those predicted by Figure 7-27, again very good correlation results. In Figure 7-38 measured backscatter cross sections for parallel polarization versus angle of incidence are given for several different orientations of a $2\lambda_0$ by $5\lambda_0$ right triangle.

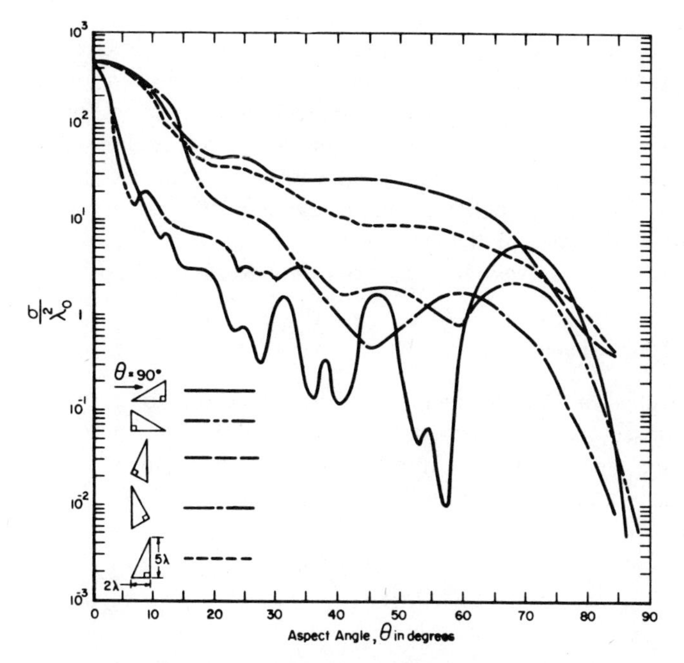

Fig. 7-38. Measured backscatter cross sections for a perfectly conducting right triangular plate versus the angle of incidence for various aspects, parallel polarization.

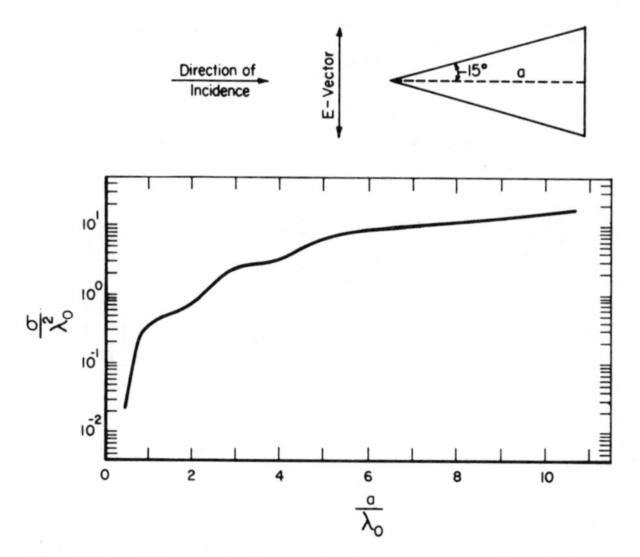

Fig. 7-39. Measured edge-on backscatter cross section for a 15° perfectly conducting isosceles triangular plate versus the plate length in wavelengths, perpendicular polarization [after Hey and Senior[39]].

For an isosceles triangle viewed edge-on along a bisector with perpendicular polarization, the oscillatory behavior with respect to frequency which occurs for the disc and rectangular plate is not observed. In Figure 7-39, measured results for the edge-on backscatter cross section of an isosceles triangle are shown, and the variation with a/λ_0 is almost smooth.[39] This essentially monotonic behavior indicates that the primary source of the scattering is the rear of the triangle. The major contributors are the rear-edge and in-phase traveling waves from the two slant edges.

7.5.5. Other Shapes

The physical-optics integral can also be solved for a plate shaped as a circular sector, as illustrated in Figure 7-40.[40] The radii bounding the sector are a_1 and a_2, and the angular extent of the sector is α. The physical-optics backscatter cross section of such a sector is

$$\sigma_{\text{p.o.}}(\theta, \phi) = \frac{4\pi \cos^2 \theta}{\lambda_0^2} [W^2(\theta, \phi) + V^2(\theta, \phi)] \tag{7.5-62}$$

where

$$W(\theta, \phi) = a_2^2 \left\{ \alpha \left(\frac{J_1(2k_0 a_2 \sin \theta)}{2k_0 a_2 \sin \theta} - \frac{a_1^2}{a_2^2} \frac{J_1(2k_0 a_1 \sin \theta)}{2k_0 a_1 \sin \theta} \right) + 2 \sum_{n=1}^{\infty} \frac{2n+1}{n(n+1)} \left[\sum_{m=1}^{n} (-1)^m \sin m\alpha \cos 2m\phi \left(\frac{J_{2n+1}(2k_0 a_2 \sin \theta)}{2k_0 a_2 \sin \theta} - \frac{a_1^2}{a_2^2} \frac{J_{2n+1}(2k_0 a_1 \sin \theta)}{2k_0 a_1 \sin \theta} \right) \right] \right\} \tag{7.5-63}$$

and

$$V(\theta, \phi) = 16a_2^2 \left\{ \sum_{n=1}^{\infty} \frac{n}{(2n-1)(2n+1)} \left[\sum_{m=1}^{n} (-1)^n \sin(2m-1) \frac{\alpha}{2} \cos(2m-1)\phi \times \left(\frac{J_{2n}(2k_0 a_2 \sin \theta)}{2k_0 a_2 \sin \theta} - \frac{a_1^2}{a_2^2} \frac{J_{2n}(2k_0 a_1 \sin \theta)}{2k_0 a_1 \sin \theta} \right) \right] \right\} \tag{7.5-64}$$

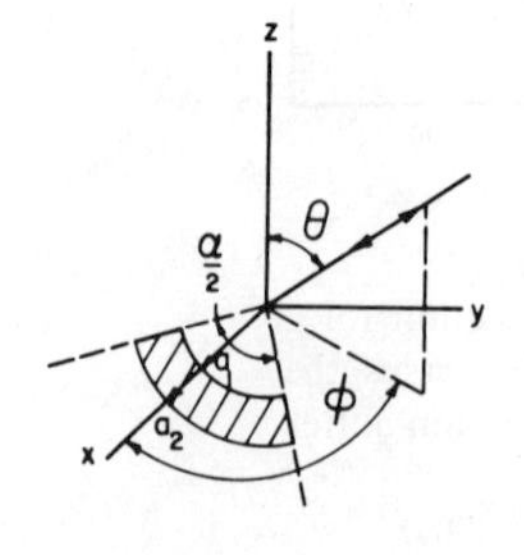

Fig. 7-40. Coordinates for backscattering from a circular sector.

For an annular ring, $\alpha = 2\pi$; so $V = 0$, and

$$\sigma_{\text{p.o.}} = \frac{4\pi}{\lambda_0^2} \cos^2 \theta (\pi a_2^2)^2 \left[\frac{2J_1(2k_0 a_2 \sin \theta)}{2k_0 a_2 \sin \theta} - \frac{2a_1^2}{a_2^2} \frac{J_1(2k_0 a_1 \sin \theta)}{2k_0 a_1 \sin \theta}\right]^2 \tag{7.5-65}$$

which reduces for normal incidence to

$$\sigma_{\text{p.o.}}(0) = \frac{4\pi}{\lambda_0^2} [\pi(a_2^2 - a_1^2)] \tag{7.5-66}$$

as it should.

For a semicircle, $\alpha = \pi$, and for a plane wave incident in the yz-plane, $\phi = \pi/2$, so

$$\sigma_{\text{p.o.}} = \frac{\pi(\pi a^2)^2 \cos^2 \theta}{\lambda_0^2} \left[\frac{2J_1(2k_0 a \sin \theta)}{2k_0 a \sin \theta}\right]^2 \tag{7.5-67}$$

For the plane of incidence, the xz-plane, $\phi = 0$ and

$$\sigma_{\text{p.o.}} = \frac{\pi \cos^2 \theta}{\lambda_0^2} (\pi a^2)^2 \left[\left(\frac{2J_1(2k_0 a \sin \theta)}{2k_0 a \sin \theta}\right)^2 + \frac{256}{\pi^2} \left(\sum_{n=1}^{\infty} \frac{2n^2}{(2n-1)(2n+1)} \frac{J_{2n}(2k_0 a \sin \theta)}{2k_0 a \sin \theta}\right)^2\right] \tag{7.5-68}$$

For an angular sector, $a_1 = 0$, and for the plane of incidence, the yz-plane, $\phi = \pi/2$, so

$$\sigma_{\text{p.o.}} = \frac{4\pi \cos^2 \theta}{\lambda_0^2} \left[\frac{\alpha a^2 J_1(2k_0 a \sin \theta)}{2k_0 a \sin \theta} + 2 \sum_{n=1}^{\infty} \frac{2n+1}{n(n+1)} \left(\sum_{m=1}^{n} \sin m\alpha \frac{J_{2n+1}(2k_0 a \sin \theta)}{2k_0 a \sin \theta}\right)\right]^2 \tag{7.5-69}$$

For a wave incident in the xz-plane, $\theta = 0$, and

$$\sigma_{\text{p.o.}} = \frac{4\pi \cos^2 \theta}{\lambda_0^2} \left\{\left[\frac{\alpha a^2 J_1(2k_0 a \sin \theta)}{2k_0 a \sin \theta} + 2 \sum_{n=1}^{\infty} \frac{2n+1}{n(n+1)} \left(\sum_{m=1}^{n} (-1)^m \sin m\alpha \frac{J_{2n+1}(2k_0 a \sin \theta)}{2k_0 a \sin \theta}\right)\right]^2 + \left[16a^2 \sum_{n=1}^{\infty} \frac{n}{(2n-1)(2n+1)} \left(\sum_{m=1}^{n} (-1)^m \times \sin(2m-1) \frac{\alpha}{2} \frac{J_{2n}(2k_0 a \sin \theta)}{2k_0 a \sin \theta}\right)\right]^2\right\} \tag{7.5-70}$$

7.6. REFERENCES

1. Born, M. and Wolf, E., *Principles of Optics*, Pergamon, New York (1959), p. 37.
2. Stratton, J. A., *Electromagnetic Theory*, McGraw–Hill, New York (1941), p. 492.
3. Panofsky, W. K. H., and Phillips, M., *Classical Electricity and Magnetism*, Addison–Wesley, Reading, Mass. (1955), p. 178.
4. Panchenko, V. S., Nomograms for the Determination of the Modulus and the Phase of the Reflection Coefficient of a Radio Wave, *Izvestia VUZ Radiotekhnika* **7**:139 (1964).
5. Brekhovski, L. M., *Waves in Layered Media*, Academic Press, New York (1961).
6. Wait, J. R., *Electromagnetic Waves in Stratified Media*, Macmillan, New York (1962).
7. Abramowitz, M., and Stegun, I. A., *Handbook of Mathematical Functions*, National Bureau of Standards of Department of Commerce (June 1964).
8. Ginzberg, V. L., *Propagation of Electromagnetic Waves in Plasma*, Gordon and Breach, New York (1961).
9. Bremmer, H., The Propagation of Electromagnetic Waves Through a Stratified Medium and Its W. K. B. Approximation for Oblique Incidence, *Physica* **15**:593 (1949).
10. Jones, D. S., *The Theory of Electromagnetism*, Macmillan, New York (1964), p. 364.
11. Presnyakov, L. P., and Sobel'man, I. I., On the Propagation of Electromagnetic Waves in a Medium with a Variable Refractive Index, *Izvestia VUZ Radiofizika* **8**:57 (1965).
12. Schelkunoff, S. A., Remarks Concerning Wave Propagation in Stratified Media. *In Theory of Electromagnetic Waves*, John Wiley (Interscience Div.), New York (1951), p. 117.
13. Redheffer, R. M., Reflection and Transmission Equivalence of Dielectric Media, *Proc. IRE* **39**:503 (1951).
14. Bouwkamp, C. J., Diffraction Theory, *Rept. Progr. Phys.* **17**:35 (1954).
15. Morse, P. M., and Rubenstein, P. J., The Diffraction of Waves by Ribbons and by Slits, *Phys. Rev.* **54**:895 (1938).
16. Skavlem, S., On the Diffraction of Scalar Plane Waves by a Slit of Infinite Length, *Arch. Math. Naturvidenskab* **51**:61 (1951).
17. Waterman, P. C., Exact Theory of Scattering by Conducting Strips, AVCO Corp., RAD-TM-63-78 (December 1963), AF 04(694)-239.
18. Yeh, C., The Diffraction of Waves by a Penetrable Ribbon, *J. Math. Phys.* **4**:64 (1963).
19. Morse, P. M., and Feshbach, H., *Methods of Theoretical Physics, Pt. II*, McGraw–Hill, New York (1953), p. 1428.
20. Karp, S. N., and Russek, A., Diffraction by a Wide Slit, *J. Appl. Phys.* **27**:886 (1956).
21. Ufimtsev, P. Ya., Secondary Diffraction of Electromagnetic Waves by a strip, *Zh. Tekh. Fiz. (USSR)* **28**:569 (1958).
22. Jahnke, E., and Emde, F., *Tables of Functions*, 4th ed., Dover, New York (1945).
23. Pearcey, T., *Table of the Fresnel Integral*, Cambridge Univ. Press, London (1956).
24. Row, R. V., Microwave Diffraction Measurements in a Parallel Plate Region, *J. Appl. Phys.* **24**:1448 (1953).
25. Macrakis, M. S., Theoretical and Experimental Study of the Backscattering Cross Section of an Infinite Ribbon, *J. Appl. Phys.* **31**:2261 (1960).
26. Meixner, J., and Andrejewski, W., Strenge Theorie der Beugung ebener elektromagnetischer Wellen an der vollkommen leitenden Kreisscheibe und an der kreisförmigen Öffnung im leitenden ebenen Schirm, *Ann. Physik* **7**:157 (1950).
27. Bethe, H. A., Theory of Diffraction by Small Holes, *Phys. Rev.* **66**:163 (1944).
28. Bouwkamp, C. J., On the Diffraction of Electromagnetic Waves by Small Circular Discs and Holes, *Philips Res. Rept.* **5**:401 (1950).
29. Schmitt, H. J., Backscattering Measurements with a Space-Separation Method, Harvard Univ., Cruft Lab. Report No. 14(November 1957).
30. Case, K. M., An Approximate Method for Diffraction Problems, *Rev. Modern Phys.* **36**:669 (1964).
31. Ufimtsev, P. Ya., Approximate Calculation of the Diffraction of Plane Waves by Certain Metal Objects, II, *Zh. Tekh. Fiz.* **28**:2604 (1958).

32. Ufimtsev, P. Ya., Secondary Diffraction of Electromagnetic Waves by Disk, *Zh. Tekh. Fiz.* **28**:583 (1958).
33. Hey, J. S., Stewart, G. S., Pinson, J. T., and Prince, P. E. V., The Scattering of Electromagnetic Waves by Conducting Spheres and Discs, *Proc. Phys. Soc.* **69B**:1038 (1956).
34. Senior, T. B. A. (unpublished work).
35. Severin, H., and von Baeckmann, W., Beugung elektromagnetischer Zentimeterwellen an metallischen und dielektrischen Scheiben, *Z. Angew. Phys.* **3**:22 (1951).
36. Neugebauer, H. E. J., Extension of Babinet's Principle to Absorbing and Transparent Materials and Approximate Theory of Backscattering by Plane Absorbing Disks, *J. Appl. Phys.* **28**:302 (1957).
37. Kouyoumjian, R. G., The Calculation of the Echo Areas of Perfectly Conducting Objects by the Variational Method, The Ohio State University Antenna Lab. Report No. 444-13 (November 1953), AD 48214.
38. Ross, R. A., Radar Cross Section of Rectangular Flat Plates as a Function of Aspect Angle, *IEEE Trans.* **AP-14**:329 (1966).
39. Hey, J. S., and Senior, T. B. A., Electromagnetic Scattering by Thin Conducting Plates at Glancing Incidence, *Proc. Phys. Soc. (GB)* **72**:981 (1958).
40. Mahan, A. I., Bitterli, C. V., and Cannon, S. M., Far-Field Diffraction Patterns of Single and Multiple Apertures Bounded by Arcs and Radii of Concentric Circles, *J. Opt. Soc. A*m. **54**:721 (1964).

Chapter 8

COMPLEX BODIES

G. T. Ruck

8.1. ANALYSIS

Methods for obtaining analytical estimates of the radar cross sections of complex bodies will be considered in this section. The difficulty of obtaining such an estimate depends upon several factors. One is the relative size of the body with respect to the radar wavelength, thus determining whether high- or low-frequency approximations can be used. Another, perhaps the most important, is the complexity of the body. This can range from very complex shapes to shapes which are only slight perturbations from a regular shape for which exact results or good analytical approximations are available. For bodies of the latter type, no difficulties are usually encountered in making engineering estimates of the radar cross section, and they will not be considered here.

For a large class of radar targets which can be considered combinations of various simple geometric shapes, the situation is quite different, and such targets will be discussed in the following paragraphs.

8.1.1. Combinations of Simple Shapes

The analytical estimation of the radar cross sections of bodies which are composed of geometrical combinations of simple shapes, such as have been considered in the previous chapters, will now be discussed. Exact analytical solutions are not available for any objects of this type with the exception of the spherically-capped cone discussed in Chapter 6. The general class of objects considered are those for which only a few basic geometrical shapes are involved, and for which it is expected that reasonable analytical estimates of the backscatter cross section can be obtained over at least restricted ranges of frequency and aspect angle. Thus, quite complex bodies such as aircraft, some satellites, etc., for which often only average cross sections or cross section envelopes can be estimated, will be considered in a later section. In addition, only perfectly conducting bodies or bodies for which a Leontovich-type boundary condition holds will be considered in this section. This specifically excludes dielectric bodies where the effects of internal reflections or refraction must be considered.

The techniques available for estimating the radar cross section of a shape, which is not a coordinate surface of one of the coordinate systems for which the vector or scalar wave equations are separable, are quite few. In addition, the ability of the cross-section analyst to apply these techniques properly is critically important in obtaining results which agree reasonably well with measurements. Often the proper application of the available techniques is as much art as science, and is due to a mixture of physical intuition and long experience on the part of the analyst. This situation will be alleviated only as more information becomes available on the detailed physical phenomena involved in the scattering process. It is hoped that the recent use of digital computers to solve numerically for the surface currents and the scattered fields for a variety of simple shapes, as discussed in Section 2.2.2.8, when coupled with more and better measurements and improved analytical techniques, will be of considerable value in accomplishing this.

8.1.1.1. *Analytical Techniques*

8.1.1.1.1. *High-Frequency Techniques.* When the body dimensions are large in terms of the radar wavelength, the general technique for estimating the radar cross section of a body (which is not simply a small perturbation of a regular shape for which exact or good approximate solutions exist) is essentially the same as for obtaining approximate solutions for regular bodies for which no exact results are available. This consists of establishing a model in which the cross section is considered to be due to a number of component scatterers. These components can be both geometric and analytic in nature. That is, the body is separated into a number of geometric shapes, and the scattered fields from each geometric shape are, in turn, attributed to a combination of various analytic components, such as the specular return, surface diffraction effects, edge effects, etc. These analytic components have often been encountered in previous sections of this book, some typical examples are creeping wave contributions, traveling wave contributions, tip scattering, etc. Considerable care must be taken when establishing a model for estimating the cross section of a complex shape, since these analytic components of the cross section, of a particular geometric shape, may not be the same when the shape is a geometric component of a more complex body as they are when the shape is an isolated scatterer.

Once a model has been established for the various component scatterers, the relative magnitudes of the scattered fields from these components are determined, and the major contributions are then added with the proper phase to provide an estimate of the cross section of the entire body. In general, the complexity of the model established for a particular body will be determined by how accurate an estimate of the true cross section is desired, although this is not always the case. For some bodies, very simple models may produce good results; while for others, even very complex models may produce relatively poor results.

To illustrate the general techniques, the various analytic components, and methods for estimating their magnitudes, will be reviewed; then several specific simple complex bodies will be discussed. With few exceptions, the high-frequency radar cross section of a geometric shape can be attributed to combinations of the following analytic components:

(1) Specular scattering points
(2) Scattering from surface discontinuities
 (a) Edges
 (b) Corners
 (c) Tips
(3) Scattering from surface derivative discontinuities
(4) Creeping waves or shadow-boundary scattering
(5) Traveling-wave scattering
(6) Scattering from concave regions
(7) Multiple scattering points

8.1.1.1.1.1. *Specular Points.* Discussing specular scattering initially, the specular contribution to the cross section of a complex body can be computed by either geometrical optics, including the Luneberg–Kline series technique, or physical optics. Of these techniques, the simplest to apply is geometrical optics. It must be noted, however, that geometrical optics is not valid for flat surfaces or surfaces which have an infinite or zero radius of curvature. In Section 2.2, it was shown that the geometrical-optics cross section for a perfectly conducting surface is given by [Eq. (2.2-40)]

$$\sigma = \pi \rho_1 \rho_2 \tag{8.1-1}$$

where ρ_1, ρ_2 are the principal radii of curvature at the specular points. If the shape of the surface at the specular point can be described by an analytical expression, such as $F(x, y, z) = 0$, or $f(x, y) = z$, then expressions for σ are given by

$$\sigma = \pi \frac{1 + (\partial f/\partial x)^2 + (\partial f/\partial y)^2}{(\partial^2 f/\partial x^2)(\partial^2 f/\partial y^2) - (\partial^2 f/\partial x\, \partial y)} \tag{8.1-2}$$

or

$$\sigma = \pi \frac{(\partial F/\partial x)^2 + (\partial F/\partial y)^2 + (\partial F/\partial z)^2}{D} \tag{8.1-3}$$

where

$$D = -\begin{vmatrix} \dfrac{\partial^2 F}{\partial x^2} & \dfrac{\partial^2 F}{\partial x\, \partial y} & \dfrac{\partial^2 F}{\partial x\, \partial z} & \dfrac{\partial F}{\partial x} \\ \dfrac{\partial^2 F}{\partial x\, \partial y} & \dfrac{\partial^2 F}{\partial y^2} & \dfrac{\partial^2 F}{\partial z\, \partial y} & \dfrac{\partial F}{\partial y} \\ \dfrac{\partial^2 F}{\partial x\, \partial z} & \dfrac{\partial^2 F}{\partial y\, \partial z} & \dfrac{\partial^2 F}{\partial z^2} & \dfrac{\partial F}{\partial z} \\ \dfrac{\partial F}{\partial x} & \dfrac{\partial F}{\partial y} & \dfrac{\partial F}{\partial z} & 0 \end{vmatrix} \tag{8.1-4}$$

In the above expressions, all the partial derivatives are to be evaluated at the specular point.

For dielectric or imperfectly conducting surfaces, geometrical optics can still be used to determine the specular contribution to the cross section. The process is considerably more complex than for perfectly conducting bodies since, in general, internal reflection and refractive effects must be accounted for. General techniques for determining the geometrical-optics scattering from dielectric bodies are discussed in References 1 and 2; the results of such a technique are given for dielectric spheres in Section 3.3.2, and for dielectric cylinders in Section 4.1.3.4. In the event that internal reflection and refractive effects can be neglected, then the geometrical cross section of a dielectric body can be estimated by

$$\sigma = \pi\rho_1\rho_2 \mid R \mid^2 \tag{8.1-5}$$

where R is the reflection coefficient discussed in Section 7.1.

The specular contributions to the cross section of a body can also be computed using physical optics as discussed in Section 2.2.2. In general, the use of physical optics requires evaluation of an integral over the illuminated portion of the body surface and, in many cases, this integral cannot be integrated analytically. In this event, one can often perform a stationary phase evaluation, which however gives results identical to geometrical optics, as is shown in Section 2.2.2.3.4. Another technique which can be used when the physical-optics integral cannot be integrated exactly has as its starting point Eq. (2.2-86), which in slightly different notation is

$$\mid H^s \mid = \left| \frac{-ik_0 \exp[ik_0 r]}{2\pi r} \int_0^l \exp[i2k_0 l] \frac{\partial A}{\partial l}\, dl \right| \tag{8.1-6}$$

where $A(l)$ is the area of the body projected on a plane normal to $\hat{\mathbf{k}}_0{}^i$. Integrating this by parts gives

$$\mid H^s \mid = \left| \frac{-ik_0 \exp[ik_0 r]}{2\pi r} I \right| \tag{8.1-7}$$

or

$$\sigma = \frac{4\pi}{\lambda_0{}^2} \mid I \mid^2 = \frac{k_0{}^2}{\pi} \mid I \mid^2 \tag{8.1-8}$$

where

$$I = \left[\exp[2ik_0 l] \left(\frac{-1}{2k_0 i} \frac{\partial A}{\partial l} - \frac{1}{4k_0{}^2} \frac{\partial^2 A}{\partial l^2} + \frac{1}{i8k_0{}^3} \frac{\partial^3 A}{\partial l^3} + \cdots \right)\right]_0^L \tag{8.1-9}$$

The integral is taken over the illuminated region and can be evaluated as

$$I = \sum_{j=1}^{N} \exp[2ik_0 l_j] \left\{ \frac{-1}{2k_0 i} \left[\frac{\partial A}{\partial l_j}\bigg|_{l_{j+}} - \frac{\partial A}{\partial l_j}\bigg|_{l_{j-}} \right] - \frac{1}{4k_0^2} \left[\frac{\partial^2 A}{\partial l_j^2}\bigg|_{l_{j+}} - \frac{\partial^2 A}{\partial l_j^2}\bigg|_{l_{j-}} \right] + \cdots \right\} \tag{8.1-10}$$

where the l_j are points at which discontinuities in the projected area exist. A general expression for I then is

$$I = \sum_{j=1}^{N} \left[\sum_{n=1}^{\infty} \frac{(-1)^n}{(2k_0 i)^n} \exp[2ik_0 l_j] \left(\frac{\partial^n A}{\partial l^n_{j+}} - \frac{\partial^n A}{\partial l^n_{j-}} \right) \right] \tag{8.1-11}$$

As an example of the use of this, consider the axial-incidence cross section of a prolate spheroid. For the spheroid shown in Figure 8-1, the projected area $A(l)$ is given by

$$A(l) = \pi b^2/a^2[a^2 - (l - a)^2] \tag{8.1-12}$$

Thus,

$$\partial A/\partial l = -2\pi b^2/a^2(l - a) \tag{8.1-13}$$

and

$$\partial^2 A/\partial l^2 = -2\pi b^2/a^2 \tag{8.1-14}$$

Now,

$$\frac{\partial A}{\partial l}\bigg|_{0-} = 0, \qquad \frac{\partial A}{\partial l}\bigg|_{0+} = 2\pi b^2/a \tag{8.1-15}$$

and

$$\frac{\partial^2 A}{\partial l^2}\bigg|_{0-} = 0, \qquad \frac{\partial^2 A}{\partial l^2}\bigg|_{0+} = -2\pi b^2/a^2 \tag{8.1-16}$$

so

$$\sigma = \frac{k_0^2}{\pi} \left[\frac{1}{2k_0 i} \left(\frac{2\pi b^2}{a} \right) - \frac{1}{4k_0^2} (2\pi b^2/a^2) \right]^2 \tag{8.1-17}$$

or

$$\sigma = \pi \frac{b^4}{a^2} \left[1 + \frac{1}{4k_0^2 a^2} \right] \tag{8.1-18}$$

The first term of this is the same as predicted by geometrical optics, and agrees with the first term of the cross section given in Eq. (5.1-69), obtained

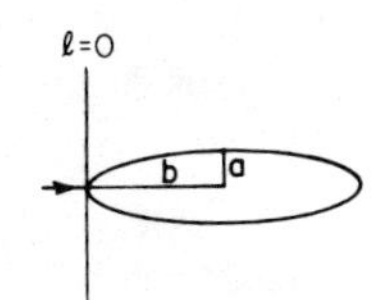

Fig. 8-1. Prolate-spheroid dimensions.

by an exact integration of the physical-optics integral. In deriving Eq. (8.1-18), only the contribution of the specular point to the cross section was considered, and no discontinuity in $A(l)$ or its derivatives was assumed at the shadow boundary. If $A(l)$ and all derivatives are assumed to be zero for $l > a$, then the result obtained by this technique for the prolate spheroid agrees exactly with that given by Eq. (5.1-69).

Unlike the stationary-phase approximation to the physical-optics integral, Eq. (8.1-11) is valid for flat surfaces as well as for curved surfaces, and in the limit of normal incidence on a flat surface, I simply becomes the total area A. However, for a flat surface, and angles of incidence other than normal, the edge is the dominant contributor to the scattered field, and techniques other than physical optics can provide better estimates of the edge contribution.

8.1.1.1.1.2. *Surface Discontinuities.* The contributions from surface discontinuities can be estimated using geometrical diffraction or the Sommerfeld–Macdonald technique discussed in Section 2.3.1. For most bodies, equivalent results will be obtained from either approach. The diffraction coefficient to be used in Keller's geometrical diffraction theory is known, in general, only for perfectly conducting edges and for certain directions of incidence on perfectly conducting tips. For perfectly conducting corners or dielectric edges or tips, no specific diffraction coefficient is available, as discussed in Section 2.2.2.2. Similarly, the Sommerfeld–Macdonald technique can, in general, be applied only to bodies with perfectly conducting edges, owing to the lack of information concerning the edge currents at corners or on dielectric or absorbing bodies.

Reviewing briefly the geometrical diffraction results for scattering from an edge, the singly diffracted far-zone backscattered field from a perfectly conducting convex edge is given by (see Section 2.2.2.2)

$$E^s = D_\perp \sqrt{\rho'} \frac{\exp[i(k_0 r + \psi)]}{r} \tag{8.1-19}$$

and

$$H^s = D_\parallel \sqrt{\rho'} \frac{\exp[i(k_0 r + \psi)]}{r} \tag{8.1-20}$$

where the incident E and H fields have been normalized to unity, and the designation $\perp$, $\parallel$ refers to the magnetic field perpendicular and parallel to the edge, respectively. The functions, D, ρ', and ψ, are the diffraction coefficients, the radius of curvature of the diffracted wavefront, and the phase of the diffracted ray with respect to the phase-reference point, respectively.†

† ψ is positive for points behind the phase-reference plane, and negative for points in front.

For a general perfectly conducting convex edge, the backscatter diffraction coefficients are

$$D_{\perp \atop \parallel} = \frac{\exp(i(\pi/4))\sin(\pi/n)}{n\sqrt{2\pi k_0}}\left[\left(\cos\frac{\pi}{n}-1\right)^{-1} \mp \left(\cos\frac{\pi}{n}-\cos\frac{2\phi}{n}\right)^{-1}\right] \tag{8.1-21}$$

with

$$n = \gamma/\pi \tag{8.1-22}$$

The angle ϕ equals the angle between an incident ray and its projection in the nearest face of the wedge; γ is the external wedge angle; and the minus or plus sign goes with $\perp$ or $\parallel$, respectively. The radius of curvature of the diffracted wavefront, ρ', is given in general by

$$\rho' = \frac{\rho}{\cos\beta_i + \cos\beta_s} \tag{8.1-23}$$

where β_i and β_s are the angles between the incident and diffracted rays and the principal normal to the edge, and ρ is the radius of curvature of the edge at the point of incidence. For backscatter, ρ' is positive if the diffracted rays converge behind the edge, or negative if they converge in front of the edge.

In the event that a ray is incident normal to one of the wedge faces forming the edge, the above diffraction coefficients must be replaced by

$$D_{\perp \atop \parallel} = \frac{-\exp[i(\pi/4)]}{2\pi k_0}\left[\frac{1}{n}\cot\frac{\pi}{2n} \mp \frac{1}{2n}\cot\frac{\pi}{n}\right] \tag{8.1-24}$$

If the backscatter direction is an axial caustic, then the scattered field is independent of the polarization and is given by†

$$\frac{E^s}{H^s} = \mp\frac{(D_\parallel - D_\perp)}{2}\, a\sqrt{2\pi k_0}\,\frac{\exp\{i[k_0 r + \psi - (\pi/4)]\}}{r} \tag{8.1-25}$$

with a the radius of curvature of the edge, and $D_\parallel - D_\perp$, for singly diffracted rays, given by

$$D_\parallel - D_\perp = \left[\frac{(2/n)\sin(\pi/n)}{\cos(\pi/n) - \cos(2\phi/n)}\right]\frac{\exp[i(\pi/4)]}{\sqrt{2\pi k_0}} \tag{8.1-26}$$

The parameter n and the angle ϕ are defined the same as in Eq. (8.1-21).

Doubly and higher-order diffracted rays can be taken into account in the same manner as the singly diffracted rays considered above, with one

† E^s and H^s are the components of the scattered fields parallel to the edge.

difference. The field incident upon the second diffraction point is the singly diffracted field from the first point, thus the bistatic diffraction coefficients,

$$D_{\substack{\perp \\ \parallel}} = \frac{\exp[i(\pi/4)]}{\sqrt{2\pi k_0}} \left(\frac{\sin(\pi/n)}{n}\right) \left\{ \left[\cos\frac{\pi}{n} - \cos\frac{(\phi - \phi')}{n}\right]^{-1} \mp \left[\cos\frac{\pi}{n} - \cos\frac{(\phi + \phi')}{n}\right]^{-1} \right\} \tag{8.1-27}$$

must be used. Since the two diffraction points may not be separated sufficiently that the far-field results apply, the field dependence in general has the form

$$U = \sqrt{\frac{\rho'}{r(\rho' + r)}} \tag{8.1-28}$$

given in Eq. (2.2-56), rather than the $\sqrt{\rho'/r}$ far-field dependence.

If the contribution of edge diffracted rays to the bistatic cross section is desired, then Eq. (8.1-27) must be used for either the singly or higher-order diffracted rays.

For corners, the diffraction coefficients are not known in general (see Section 2.2.2.2); while for tips, physical optics gives good results for small tip angles and angles of incidence near axial. Tip scattering is discussed in more detail in Sections 2.3.1.2 and 6.2.1.2.

8.1.1.1.1.3. *Surface Derivative Discontinuities.* The contribution of surface derivative discontinuities on illuminated positions of a surface can be estimated by physical optics, using the method of Section 8.1.1.1.1.1. As an example of how this is done, the contribution of the cone–sphere join to the axial-incidence backscatter from a cone-sphere will be derived using the method of Section 8.1.1.1.1.1. Consider a cone-sphere with at most a second-derivative discontinuity in the profile, such as illustrated in Figure 8-2. From Eq. (8.1-11), the scattering function I is given by

$$I = \sum_{j=1}^{N} \left\{ \sum_{n=1}^{\infty} \frac{(-1)^n}{(2k_0 i)^n} \exp[2ik_0 l_j] \left[\frac{\partial^n A}{\partial l_j^{\,n}} \bigg|_+ - \frac{\partial^n A}{\partial l_j^{\,n}} \bigg|_- \right] \right\} \tag{8.1-29}$$

For the join discontinuity, j in the above expression equals one, and $l_1 = h$. Now for $l \leqslant h$, the expression for A is

$$A = \pi l^2 \tan^2 \alpha \tag{8.1-30}$$

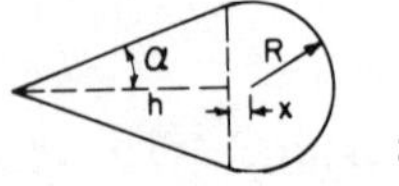

8-2. Cone-sphere dimensions

while for $l > h$, the expression for A is

$$A = \pi[R^2 - (h + x - l)^2] \tag{8.1-31}$$

Using these, then,

$$\left.\frac{\partial A}{\partial l}\right|_+ - \left.\frac{\partial A}{\partial l}\right|_- = 2\pi x - 2\pi h \tan^2 \alpha \tag{8.1-32}$$

and

$$\left.\frac{\partial^2 A}{\partial l^2}\right|_+ - \left.\frac{\partial^2 A}{\partial l^2}\right|_- = -2\pi - 2\pi \tan^2 \alpha \tag{8.1-33}$$

Substituting these in Eq. (8.1-29) gives

$$I = -\exp[2ik_0h]\left[\frac{(2\pi x - 2\pi h \tan^2 \alpha)}{2k_0 i} - \frac{(2\pi + 2\pi \tan^2 \alpha)}{4k_0^2}\right]$$

or

$$I = \frac{\pi \exp[2ik_0h]}{2k_0^2}[\sec^2 \alpha] \tag{8.1-34}$$

This is related to the scattered field by

$$H^s = (ik_0/2\pi)\, I \tag{8.1-35}$$

and

$$\sigma = (k_0^2/\pi)|\, I\,|^2 \tag{8.1-36}$$

This technique can be used to approximate the contribution to the total cross section of surface derivative discontinuities for arbitrary angles of incidence, provided the discontinuity is in the illuminated region, although the calculations in general would be more complex than for the above example.

For surface derivative discontinuities at the shadow boundary or in the shadow region, other techniques must be used. In general, surface derivative discontinuities far in the shadow region can be ignored. Derivative discontinuities near or on the shadow boundary have been examined by Weston,[3] who obtained expressions for the fields in the shadow region due to several specific discontinuity models. Weston gives results which can be used to determine the creeping-wave launch coefficients due to these discontinuities.

8.1.1.1.1.4. *Creeping Waves.* Shadow-boundary or creeping waves are often major contributors to the high-frequency scattered fields from a complex object, and can be estimated by using geometrical diffraction, Fock theory, or asymptotic expansions of an exact solution.

For example, by using geometrical diffraction, the creeping-wave contribution to the bistatic scattered field from a general convex body of revolution for which a Leontovich boundary condition is satisfied at the body surface is given by[4]

$$\mathbf{E}^s(P) = E_n^i(Q_1)\, V(P_1, Q_1)\, \hat{\mathbf{n}}(P_1) + E_b^i(Q_1)\, W(P_1, Q_1)\, \hat{\mathbf{b}}(P_1) \quad (8.1\text{-}37)$$

where the points P, Q, P_1, Q_1 are defined in Figure 8-3, and $\hat{\mathbf{n}}$, $\hat{\mathbf{b}}$ are unit vectors defined by

$$\hat{\mathbf{b}} = \hat{\mathbf{n}} \times \hat{\boldsymbol{\tau}} \quad (8.1\text{-}38)$$

where $\hat{\boldsymbol{\tau}}$ is a unit vector tangent to a surface diffracted ray, $\hat{\mathbf{n}}$ an outward unit vector normal to the surface, and $\hat{\mathbf{b}}$ the unit binormal to the ray. $E_n^i(Q_1)$ is the incident electric field component normal to the surface at the point Q_1, while $E_b^i(Q_1)$ is the incident field component parallel to the surface at the point Q_1. At the surface these two components are required to satisfy the following boundary conditions:

$$\partial E_n/\partial n = -ik_0 \eta E_n \quad (8.1\text{-}39)$$

$$\partial E_b/\partial n = (-ik_0/\eta)\, E_b \quad (8.1\text{-}40)$$

where η is the surface impedance, or $\eta = \sqrt{\mu_r/\epsilon_r}$ for a homogeneous lossy body. The functions $V(P_1, Q_1)$ and $W(P_1, Q_1)$ are given by

$$V(P_1, Q_1) = \exp[ik_0(r+s)] \left(\frac{d\sigma(Q_1)}{d\sigma(P_1)}\right)^{1/2} \left(\frac{\rho_1}{r(\rho_1 + r)}\right)^{1/2} \times \sum_{m=1}^{\infty} \bar{D}_m(P_1)\, \bar{D}_m(Q_1) \exp\left[-\int_0^s \bar{\alpha}_m(x)\, dx\right] \quad (8.1\text{-}41)$$

and

$$W(P_1,Q_1) = \exp[ik_0(r+s)] \left(\frac{d\sigma(Q_1)}{d\sigma(P_1)}\right)^{1/2} \left(\frac{\rho_1}{r(\rho_1 + r)}\right)^{1/2} \times \sum_{m=1}^{\infty} D_m(P_1)\, D_m(Q_1) \exp\left[-\int_0^s \alpha_m(x)\, dx\right] \quad (8.1\text{-}42)$$

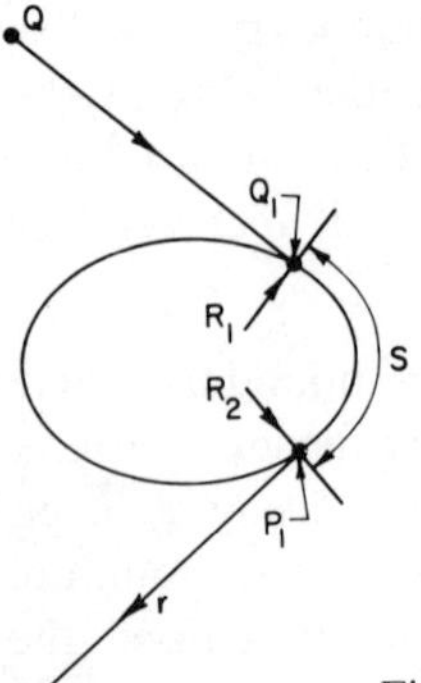

Fig. 8-3. A surface-diffracted ray.

The parameter ρ_1 is the radius of curvature (in the plane tangent to the ray) of the diffracted wavefront at P_1. The ratio, $d\sigma(Q_1)/d\sigma(P_1)$, is the limit of the ratio of the strip width between two adjacent rays at Q_1 and the strip width between the same two rays at P_1, as the strip width goes to zero. The quantities $\bar{\alpha}_m$, α_m are given by

$$\bar{\alpha}_m = \exp[-i(\pi/6)] \left(\frac{k_0}{6R^2}\right)^{1/3} \bar{q}_m \tag{8.1-43}$$

and

$$\alpha_m = \exp[-i(\pi/6)] \left(\frac{k_0}{6R^2}\right)^{1/3} q_m \tag{8.1-44}$$

with $\bar{q}_m$ and q_m the roots of

$$\frac{A'(\bar{q}_m)}{A(\bar{q}_m)} = \exp[i5(\pi/6)] \left(\frac{k_0 R}{6}\right)^{1/6} \eta \tag{8.1-45}$$

and

$$\frac{A'(q_m)}{A(q_m)} = \exp[i5(\pi/6)] \left(\frac{k_0 R}{6}\right)^{1/6} \frac{1}{\eta} \tag{8.1-46}$$

The function $A(q)$ is defined by

$$A(q) = \int_0^\infty \cos(x^3 - qx)\, dx \tag{8.1-47}$$

and is related to the Airy functions used in previous chapters (see Sections 2.2.2.5, and 2.3) by

$$A(q) = \frac{\pi}{3^{1/3}} \operatorname{Ai}\left(-\frac{q}{3^{1/3}}\right) \tag{8.1-48}$$

In the above equations, R is the radius of curvature of the surface in the plane tangent to the ray. The coefficients, $\bar{D}_m$, D_m, are obtained from

$$\exp[-i5(\pi/4)] \left(\frac{k_0}{2\pi}\right)^{1/2} \bar{D}_m{}^2 = \frac{\pi \exp[i5(\pi/6)]}{6} \left(\frac{k_0 R}{6}\right)^{1/3} \Big[[A'(\bar{q}_m)]^2 + \bar{q}_m A^2(\bar{q}_m)/3\Big]^{-1} \tag{8.1-49}$$

and

$$\exp[-i5(\pi/4)] \left(\frac{k_0}{2\pi}\right)^{1/2} D_m{}^2 = \frac{\pi \exp[i5(\pi/6)]}{6} \left(\frac{k_0 R}{6}\right)^{1/3} \Big[[A'(q_m)]^2 + q_m A^2(q_m)/3\Big]^{-1} \tag{8.1-50}$$

The total scattered field at the point P is the sum of the fields of all possible rays from Q to P taken along geodesics of the body surface.

For an absorbing or dielectric surface ($\eta \neq 0$), the parameters α_m and the diffraction coefficients D_m are in general not known, since $\bar{q}_m$ and q_m are very difficult to evaluate as functions of the complex parameter η.

If $\eta \to 0$, representing a perfectly conducting surface, then $\bar{q}_m$ goes to $\bar{q}_m{}^0$ defined by

$$A'(\bar{q}_m{}^0) = 0 \tag{8.1-51}$$

and q_m goes to $q_m{}^0$ defined by

$$A(q_m{}^0) = 0 \tag{8.1-52}$$

In Table 8-1 (from Reference 5), the first five roots, $q_m{}^0$, $\bar{q}_m{}^0$, and the turning points are given.†

Table 8-1

Roots and Turning Points for $A(q)$, $A'(q)$

m	$q_m{}^0$	$A'(q_m{}^0)$	$\bar{q}_m{}^0$	$A(\bar{q}_m{}^0)$
1	3.372134	− 1.059053	1.469354	1.16680
2	5.895843	1.212955	4.684712	− 0.91272
3	7.962025	− 1.306735	6.951786	0.82862
4	9.788127	1.375676	8.889027	− 9.77962
5	11.457423	− 1.430780	10.632519	0.74562

For a body of revolution, the axis represents a caustic of the back-scattered field, and Eq. (8.1-37) for the scattered field cannot be used. The correct form for the scattered field will be derived to illustrate one manner in which the effect of the caustic can be taken into account. Consider a unit incident field to be propagating in the negative z direction, with the electric field polarized along the positive y axis. The normal component $E_n{}^i$ at the point Q_1 on the shadow boundary is then given by

$$E_n{}^i(Q_1) = \sin \phi \tag{8.1-53}$$

while the parallel component is given by

$$E_n{}^i(Q_1) = \cos \phi \tag{8.1-54}$$

The unit vectors, $\hat{\mathbf{n}}$ and $\hat{\mathbf{b}}$, at the point P_1 are

$$\begin{aligned} \hat{\mathbf{n}} &= -\hat{\mathbf{x}} \cos \phi - \hat{\mathbf{y}} \sin \phi, \\ \hat{\mathbf{b}} &= -\hat{\mathbf{x}} \sin \phi + \hat{\mathbf{y}} \cos \phi \end{aligned} \tag{8.1-55}$$

When a ray traveling around the rear of the body passes through the rear caustic, the amplitude changes by a phase factor of $-(\pi/2)$, and this must be

† Logan's notation is opposite to that used here. He defines q_m as the roots of $A'(q) = 0$, and $\bar{q}_m$ as the roots of $A(\bar{q}) = 0$.

included in the expression for the diffracted field produced by that ray; thus, the scattered field due to a ray incident at ϕ is, from Eq. (8.1-37),

$$\mathbf{E}^s = -i[\sin\phi V(P_1, Q_1)(-\hat{\mathbf{x}}\cos\phi - \hat{\mathbf{y}}\sin\phi) + \cos\phi W(P_1, Q_1)(-\hat{\mathbf{x}}\sin\phi + \hat{\mathbf{y}}\cos\phi)] \tag{8.1-56}$$

This has the form of a cylindrical wave emanating from an infinitesimal segment of the shadow boundary at the point $\pi + \phi$. To convert it to a spherical wave coming from the infinitesimal line segment, $dl = a\,d\phi$, the above field expression must be multiplied by

$$\frac{a\,d\phi \exp[-i(\pi/4)]}{(r\lambda_0)^{1/2}} \tag{8.1-57}$$

where a is the radius of the circle formed by the shadow boundary. Multiplying (8.1-56) by this, and integrating over ϕ, gives the total backscattered field as

$$\mathbf{E}^s = \frac{-ia \exp[-i(\pi/4)]}{(r\lambda_0)^{1/2}} \Big[V(P_1, Q_1) \int_0^{2\pi} (-\hat{\mathbf{x}}\cos\phi\sin\phi - \hat{\mathbf{y}}\sin^2\phi)\,d\phi + W(P_1, Q_1) \int_0^{2\pi} (-\hat{\mathbf{x}}\cos\phi\sin\phi + \hat{\mathbf{y}}\cos^2\phi)\,d\phi\Big] \tag{8.1-58}$$

or

$$\mathbf{E}^s = \frac{-i\pi a \exp[-i(\pi/4)]}{(r\lambda_0)^{1/2}} [-V(P_1, Q_1) + W(P_1, Q_1)]\,\hat{\mathbf{y}} \tag{8.1-59}$$

In the functions $V(P_1, Q_1)$ and $W(P_1, Q_1)$, the parameter ρ_1 is infinite, the ratio$[d\sigma(Q_1)/d\sigma(P_1)]$ is unity, and $D_m(P_1) = D_m(Q_1)$, so

$$V(P_1, Q_1) = \frac{\exp[ik_0(r+s)]}{\sqrt{r}} \sum_{m=1}^{\infty} \bar{D}_m^2(P_1) \exp\Big[-\int_0^s \bar{\alpha}_m(x)\,dx\Big] \tag{8.1-60}$$

and

$$W(P_1, Q_1) = \frac{\exp[ik_0(r+s)]}{\sqrt{r}} \sum_{m=1}^{\infty} D_m^2(P_1) \exp\Big[-\int_0^s \alpha_m(x)\,dx\Big] \tag{8.1-61}$$

Various methods other than that given above have been used to obtain the scattered field on the axial caustic;[(6)] however, they all give results equivalent to those of Eq. (8.1-59).

8.1.1.1.1.5. *Traveling Waves.* For long thin bodies with a polarization such that the incident electric field has a component parallel to the body axis, the near nose-on incidence radar cross section can consist of a large traveling-wave component. The cross section due to such a traveling-wave component is discussed in more detail in Section 2.3.3.

8.1.1.1.1.6. *Modified Physical Optics* (*Adachi's Method*). For long, thin, axially symmetric bodies having sharp apices, such as ogives, double-cones, etc., Adachi[7] has obtained a modified physical-optics expression for the axial-incidence backscatter cross section which appears to agree well with experiment. Using a variational expression for the backscatter cross section with the physical-optics current as a trial function results in an expression for the backscatter cross section which is equivalent to assuming the physical-optics current over the entire surface of the scatterer, that is, both in the illuminated and shadow regions.

Adachi's expression for the axial-incidence backscatter cross section, then, is identical to Eq. (8.1-8), if l is considered to be the length of the entire scatterer, or

$$\sigma(0) \approx \frac{k_0^2}{\pi} \left| \int_0^l \frac{\partial A}{\partial l} \exp[2ik_0 l] \, dl \right|^2 \tag{8.1-62}$$

with l the length of the scatterer. For example, for a double-cone of half-angle α and length l, Adachi obtains, for the cross section,

$$\sigma(0) = \pi \tan^4 \alpha \left| l \exp[ik_0 l] - \frac{i}{2k_0} (1 - \exp[i2k_0 l]) \right|^2 \tag{8.1-63}$$

This is plotted in Figure 8-4 (from Reference 7) for double-cones of lengths up to $k_0 l = 20$. The normalization used in Figure 8-4 is $\sigma(0)/(\pi a^2 \tan^2\alpha)$, so this figure is universally valid for any cone angle, subject only to the restriction that the body be relatively long and thin.

8.1.1.1.1.7. *Concave Regions.* For bodies which have regions where the body shape is concave, the determination of the contributions to the

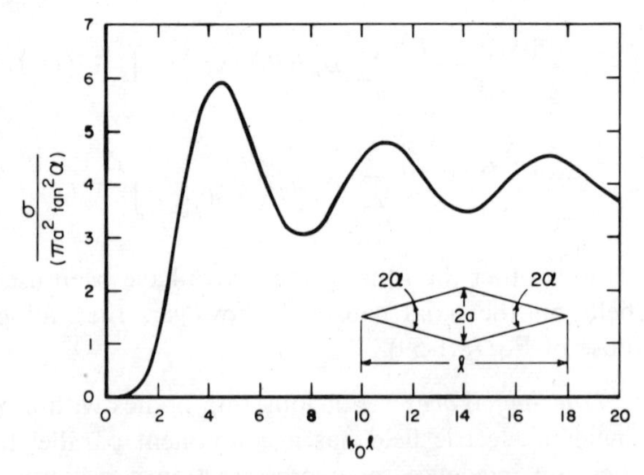

Fig. 8-4. Approximate axial-incidence backscatter cross section for a perfectly conducting double-cone, obtained by Adachi [from Adachi[7]].

cross section is in general much more difficult than for convex bodies, since multiple-scattering effects must often be taken into consideration. To the extent that multiple scattering can be neglected, physical optics can be used to obtain high-frequency estimates of the contribution to the total cross section from such regions, and geometrical diffraction can in some cases be used to determine the contributions from edges which are formed by convex wedges (i.e., wedges with interior angles greater than π). Weil and Raymond[8] have derived the diffraction coefficients for an edge which has an exterior angle γ such that $\pi > \gamma > \pi/2$. Using these results, the backscatter diffraction coefficients to be used in Eqs. (8.1-19) and (8.1-20) are

$$D_{\substack{\perp\\ \parallel}} = \frac{\exp[i(\pi/4)]}{\sqrt{2\pi k_0}} \frac{\sin[(\pi-\gamma)/n]}{n} \left\{ \left[\cos\frac{\pi-\gamma}{n} + 1 \right]^{-1} \mp \left[\cos\frac{\pi-\gamma}{n} + \cos\frac{2\phi}{n} \right]^{-1} \right\} \tag{8.1-64}$$

where γ is the exterior wedge angle, and

$$n = \gamma/\pi$$

For an axial caustic, then, using Eq. (8.1-25),

$$E^s = -\left(\frac{D_\parallel - D_\perp}{2}\right) a\sqrt{2\pi k_0}\, \frac{\exp\{i[k_0 r - (\pi/4) + \psi]\}}{r} \tag{8.1-65}$$

or

$$E^s = -\frac{a}{2r} \exp[i(k_0 r + \psi)] \left[\frac{(2/n)\sin[(\pi-\gamma)/n]}{\cos[(\pi-\gamma)/n] + \cos(2\phi/n)} \right] \tag{8.1-66}$$

For bistatic scattering from a concave wedge region, the diffraction coefficients are

$$D_{\substack{\perp\\ \parallel}} = \frac{\exp[i(\pi/4)]}{\sqrt{2\pi k_0}} \frac{\sin[(\pi-\gamma)/n]}{n} \left\{ \left[\cos\frac{\pi-\gamma}{n} + \cos\frac{\phi-\phi'}{n} \right]^{-1} \mp \left[\cos\frac{\pi-\gamma}{n} + \cos\frac{\phi+\phi'}{n} \right]^{-1} \right\} \tag{8.1-67}$$

The above results were obtained from the work of Weil and Raymond, and no experimental verification is known to have been made; thus, their general validity is unknown.

In the event that multiple scattering occurs, geometrical optics can be used to account partially for any multiple scattering effects. This is discussed in Section 8.1.1.1.1.8.

8.1.1.1.1.8. *Multiple Scattering*. For concave bodies or concave regions of a complex body, a high-frequency analysis indicates that multiply scattered rays are often major contributors to the cross section. The contribution of such rays to the cross section can be determined using a geometrical-optics approach. What is required is the geometrically reflected fields from an arbitrary surface having finite principal radii of curvature when the incident field is not necessarily a plane wave. Fock[9] has obtained expressions for these reflected fields as follows

$$\mathbf{E}^r = \mathbf{A}\sqrt{\frac{D(0)}{D(r)}}\exp[ik_0 r] \tag{8.1-68}$$

$$\mathbf{H}^r = \mathbf{B}\sqrt{\frac{D(0)}{D(r)}}\exp[ik_0 r] \tag{8.1-69}$$

The parameters **A**, **B**, D, in the above equations, are relatively complex, and depend upon the geometrical properties of the reflecting surface, the geometrical properties of the incident wavefront, and the material properties of the surface. The vectors **A** and **B** will be considered first, and for a dielectric surface are given by†

$$\mathbf{A}(s) = \frac{1}{\sin^2\theta}[\hat{\mathbf{n}}\cdot\mathbf{E}^i(s)][\hat{\mathbf{n}}\cos 2\theta + \hat{\mathbf{k}}_0{}^i\cos\theta]\,R_{\parallel}(s) + \eta[\hat{\mathbf{n}}\cdot\mathbf{H}^i(s)][\hat{\mathbf{n}}\times\hat{\mathbf{k}}_0{}^i]\,R_{\perp}(s) \tag{8.1-70}$$

$$\mathbf{B}(s) = \frac{-1}{\sin^2\theta}[\hat{\mathbf{n}}\cdot\mathbf{H}^i(s)][\hat{\mathbf{n}}\cos 2\theta + \hat{\mathbf{k}}_0{}^i\cos\theta]\,R_{\perp}(s) - (1/\eta)[\hat{\mathbf{n}}\cdot\mathbf{E}^i(s)][\hat{\mathbf{n}}\times\hat{\mathbf{k}}_0{}^i]\,R_{\parallel}(s) \tag{8.1-71}$$

In the above equations, the angle θ is the angle between the incident ray and the normal to the surface or $\cos\theta = -(\hat{\mathbf{k}}_0{}^i\cdot\hat{\mathbf{n}})$, and these equations are to be evaluated at the reflection point, s, on the surface.

For a perfectly conducting surface, expressions for **A** and **B** are

$$\mathbf{A}(s) = \mathbf{E}^i(s) - 2\hat{\mathbf{n}}\times[\mathbf{E}^i(s)\times\hat{\mathbf{n}}] \tag{8.1-72}$$

$$\mathbf{B}(s) = \mathbf{H}^i(s) - 2[\hat{\mathbf{n}}\cdot\mathbf{H}^i(s)]\,\hat{\mathbf{n}} \tag{8.1-73}$$

Having expressions for the vectors **A** and **B**, the function $D(r)$ remains to be evaluated. In order to express D in the simplest manner, tensor notation will be used without the summation convention holding. Consider the reflecting surface to be described parametrically by $x = x(u^1, u^2)$, $y = y(u^1, u^2)$, and $z = z(u^1, u^2)$, where u^1, u^2 are curvilinear coordinates on the reflecting surface. A covariant metric tensor, g_{ij}, for the surface is defined as

$$g_{ij} = \frac{\partial x}{\partial u^i}\frac{\partial x}{\partial u^j} + \frac{\partial y}{\partial u^i}\frac{\partial y}{\partial u^j} + \frac{\partial z}{\partial u^i}\frac{\partial z}{\partial u^j} \tag{8.1-74}$$

† Rays which enter the dielectric medium, are refracted or reflected, and exit, are not considered here; the expressions given are valid only for an externally reflected ray.

and we define g as

$$g = g_{11}g_{22} - (g_{12})^2 \tag{8.1-75}$$

The square of the differential arc length on the surface is given in terms of the metric tensor by

$$dl^2 = \sum_{i,j=1}^{2} g_{ij}\, du^i\, du^j \tag{8.1-76}$$

The function $D(r)$ is defined in terms of a symmetric tensor, T_{ij}, by

$$D(r) = \frac{1}{g \cos \theta} \begin{vmatrix} T_{11} & T_{12} \\ T_{12} & T_{22} \end{vmatrix} \tag{8.1-77}$$

with T_{ij} defined by

$$T_{ij} = g_{ij} - \Omega_i\Omega_j + r(\Omega_{ij} - \cos \theta G_{ij}) \tag{8.1-78}$$

The functions, Ω_i, Ω_j, Ω_{ij}, in the above equation are defined in terms of derivatives of the phase of the incident field on the surface. Let the incident field have a phase $\psi = k_0\Omega(u^1, u^2)$ on the surface of the reflecting object, then

$$\Omega_i = \partial\Omega/\partial u^i \tag{8.1-79}$$

and

$$\Omega_{ij} = \frac{\partial^2\Omega}{\partial u^i\, \partial u^j} - \begin{Bmatrix} i \\ ij \end{Bmatrix} \Omega_i - \begin{Bmatrix} j \\ ij \end{Bmatrix} \Omega_j \tag{8.1-80}$$

The bracketed quantities are the Christoffel symbols of the second kind,[10] and are defined as

$$\begin{Bmatrix} k \\ ij \end{Bmatrix} = \sum_{n=1}^{2} g^{kn}[ij, n] \tag{8.1-81}$$

where $[ij, n]$ are the Christoffel symbols of the first kind, or

$$[ij, n] = \frac{1}{2}\left(\frac{\partial g_{in}}{\partial u^j} + \frac{\partial g_{jn}}{\partial u^i} + \frac{\partial g_{ij}}{\partial u^n}\right) \tag{8.1-82}$$

and g^{kn} is the contravariant metric tensor given by

$$g^{11} = \frac{g_{22}}{g}, \qquad g^{22} = \frac{g_{11}}{g} \tag{8.1-83}$$

$$g^{12} = \frac{-g_{12}}{g} = g^{21}$$

The only remaining undefined quantity of Eq. (8.1-78) for the tensor T_{ij} is the tensor G_{ij}. This is defined as follows:

$$G_{ij} = -\left(\frac{\partial n_x}{\partial u^i}\frac{\partial x}{\partial u^j} + \frac{\partial n_y}{\partial u^i}\frac{\partial y}{\partial u^j} + \frac{\partial n_z}{\partial u^i}\frac{\partial z}{\partial u^j}\right) \tag{8.1-84}$$

where n_x, n_y, n_z are the components of the normal to the surface, given by

$$n_x = \frac{1}{\sqrt{g}}\left(\frac{\partial y}{\partial u^1}\frac{\partial z}{\partial u^2} - \frac{\partial y}{\partial u^2}\frac{\partial z}{\partial u^1}\right)$$

$$n_y = \frac{1}{\sqrt{g}}\left(\frac{\partial z}{\partial u^1}\frac{\partial x}{\partial u^2} - \frac{\partial z}{\partial u^2}\frac{\partial x}{\partial u^1}\right) \qquad (8.1\text{-}85)$$

$$n_z = \frac{1}{\sqrt{g}}\left(\frac{\partial x}{\partial u^1}\frac{\partial y}{\partial u^2} - \frac{\partial x}{\partial u^2}\frac{\partial y}{\partial u^1}\right)$$

Although the above formalism appears to be quite complex, in actuality, only simple differentiation and algebraic manipulations are required to obtain expressions for the reflected fields. Once the fields resulting from one reflection have been obtained, they can be considered as the incident fields for determining the effects of a second reflection, and this process can be continued as long as desired.

An example of the above method to calculate the doubly scattered rays from two spheres lying on a line normal to the incident-field direction has been given by Crispin *et al.*[11] In addition, they give a discussion of the use of the stationary phase method to determine the multiple-scattering contribution to the cross section.

8.1.1.1.2. *Low-Frequency Techniques.* The low-frequency or Rayleigh cross section of a body is that cross section which results when all the dimensions of the body are small in terms of the radar wavelength. In this event, often the radar cross section can be expanded in a Taylor series in powers of $k_0 a$, where a is some characteristic dimension, with coefficients which depend on the body size, shape, material, etc. For sufficiently small bodies, with respect to the radar wavelength, only the first terms of this series should contribute significantly.

The first term of the expansion of the far-zone scattered field can be interpreted as being due to the radiation of electric and magnetic dipoles located at the scatterer and induced by the incident field. To evaluate the dipole moments, the static fields induced on the body by a constant applied field must be determined. In other words, solutions to Laplace's equation appropriate to the body geometry must be found, as discussed in Section 2.2.2.7. Although this is easier than solving the wave equation, solutions are available for only a few regular geometrical shapes. Thus, in general, approximate techniques are necessary to estimate the Rayleigh cross section of most bodies.

The most characteristic feature of Rayleigh scattering is the inverse fourth-power dependence on wavelength and, with the exception of flat or wirelike objects, the dependence on the square of the volume. On this basis, an approximate solution for an arbitrary body can be postulated to be of the form

$$\sigma \approx A k_0^4 V^2 \tag{8.1-86}$$

where A depends on the body shape, material, incident-field polarization, etc. Starting with this postulate, Siegel obtained an estimate of the form of A for the axial-backscatter cross section of perfectly conducting bodies of revolution.[12] His general expression is

$$\sigma \approx \frac{4}{\pi} k_0^4 V^2 \left[1 + \frac{e^{-\tau}}{\pi\tau}\right]^2 \tag{8.1-87}$$

where τ is characteristic of the body length-to-width ratio, determined for each body shape by allowing the axial dimension of the body to go to zero, and choosing τ so as to obtain the correct disk results. Using this technique expressions can be obtained for the backscatter cross sections for axial incidence for a variety of bodies of revolution. Table 8-2 lists the results for a number of configurations.

To estimate the cross section for angles of incidence other than along the body axis, the cross section of the equivalent spheroid with the same volume and length-to-width ratio can be used. There is, of course, some question as to precisely what the proper length-to-width ratio is for most bodies; however, an average length should give relatively good results since the cross section is in general not highly shape-dependent in the Rayleigh region. Recalling from Chapter 5, Eqs. (5.1-62) and (5.1-63), the Rayleigh cross sections of a perfectly conducting spheroid can be expressed as

$$\sigma_{\perp} = \frac{1}{\pi} k_0^4 \left(\frac{4\pi}{3} ab^2\right)^2 [A_1 + A_2 \sin^2 \phi + A_3 \cos^2 \phi]^2 \tag{8.1-88}$$

and

$$\sigma_{\parallel} = \frac{1}{\pi} k_0^4 \left(\frac{4\pi}{3} ab^2\right)^2 [A_1' + A_2' \sin^2 \phi + A_3' \cos^2 \phi]^2 \tag{8.1-89}$$

where the coefficients A, A' for a prolate spheroid are as follows:

$$A_1 = [ab^2(I_a + I_b)]^{-1}, \qquad A_2 = [ab^2 I_a]^{-1}, \qquad A_3 = [ab^2 I_b]^{-1} \tag{8.1-90}$$

and

$$A_1' = [ab^2 I_b]^{-1}, \qquad A_2' = [2ab^2 I_b]^{-1}, \qquad A_3' = [ab^2(I_a + I_b)]^{-1} \tag{8.1-91}$$

Curves of I_a, I_b are given in Figures 5-2 and 5-3 as a function of the axial ratio of the spheroid, a/b.

For bodies which do not possess rotational symmetry about some axis, the Rayleigh cross section can be estimated by replacing the body with

Table 8-2

Approximate Rayleigh-Region Axial Backscatter Cross Section for Perfectly Conducting Bodies of Revolution

Body	Approximate Rayleigh Cross Section	
(1) *Flat-base cone:* 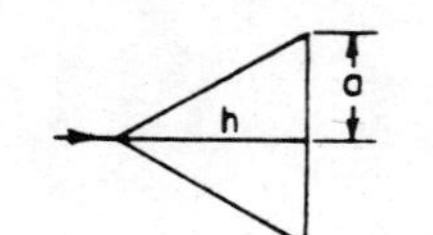	$\sigma \approx \frac{4}{\pi} k_0{}^4 \left[\frac{\pi a^2 b}{3}\left(1 + \frac{4a}{\pi h} \exp\left[-h/4a\right]\right)\right]^2$	(8.1-87a)
(2) *Cone-hemispheroid:*	$\sigma \approx \frac{4}{\pi} k_0{}^4 \left[\left(\frac{2\pi a^2 b}{3} + \frac{\pi a^2 h}{3}\right)\left(1 + \frac{4a \exp\left[-(h+2b)/4a\right]}{\pi(h+2b)}\right)\right]^2$	(8.1-87b)
(3) *Cone-cylinder:*	$\sigma \approx \frac{4}{\pi} k_0{}^4 \left[\frac{\pi a^2}{3}(h + 3l)\left(1 + \frac{4a \exp\left[-(h+3l)/4a\right]}{\pi(h+3l)}\right)\right]^2$	(8.1-87c)
(4) *Cone-cylinder-flare:*	$\sigma \approx \frac{4}{\pi} k_0{}^4 \left[\frac{\pi}{3}\left(a^2 h + 3a^2 l_1 + l_2 \frac{b^3 - a^3}{b - a}\right)\right]^2 \left[1 + 4a \exp\left\{-\left(h + 3l_1 + \frac{l_2}{a^2}\frac{b^3 - a^3}{b - a}\right)\right\}\right]^2$	(8.1-87d)

(5) *Double-cone:*

$$\sigma \approx \frac{4}{\pi} k_0^4 \left[\frac{\pi a^2}{3}(h_1 + h_2)\left(1 + \frac{4a \exp[-(h_1 + h_2)/4a]}{\pi(h_1 + h_2)}\right)\right]^2 \tag{8.1-87e}$$

(6) *Prolate spheroid:*

$$\sigma \approx \frac{4}{\pi} k_0^4 \left[\frac{4}{3}\pi a^2 b\left(1 + \frac{a \exp[-b/a]}{\pi b}\right)\right]^2 \tag{8.1-87f}$$

(7) *Oblate spheroid:*

$$\sigma \approx \frac{4}{\pi} k_0^4 \left[\frac{4}{3}\pi a b^2\left(1 + \frac{b \exp[-a/b]}{\pi a}\right)\right]^2 \tag{8.1-87g}$$

(8) *Spherical segment:*

$$\sigma \approx \frac{4}{\pi} k_0^4 \left[\pi a^2 d - \frac{\pi}{3}(d - a)^3 - \frac{\pi a^3}{3}\right]^2 \times \left[1 + \frac{\exp\{3[a^2 d - (d - a)^3/3 - a^3/3]/[4(2ad - d^2)^{3/2}]\}}{3\pi[a^2 d - (d - a)^3/3 - a^3/3]/[4(2ad - d^2)^{3/2}]}\right]^2 \tag{8.1-87h}$$

Table 8.2 (*Continued*)

Body	Approximate Rayleigh Cross Section	
(9) *Semi-spheroid:*	$\sigma \approx \frac{4}{\pi} k_0^4 \left[\pi a^2 h - \frac{\pi a^2}{3b^2}(h-b)^3 - \frac{\pi a^2 b}{3}\right]^2 \times \left[1 + \frac{\exp\{-3[a^2h - (a^2/3b^2)(h-b)^3 - a^2b/3]/[(4a^3/b^3)(2hb-b^2)^{3/2}]\}}{3\pi[a^2h - (a^2/3b^2)(h-b)^3 - a^2b/3]/[(4a^3/b^3)(2hb-b^2)^{3/2}]}\right]^2$	(8.1-87i)
(10) *Finite paraboloid:* 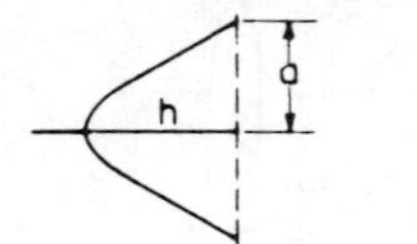	$\sigma \approx \frac{4}{\pi} k_0^4 \left[\frac{\pi a^2 h}{2}\right]^2 \left[1 + \frac{8a \exp[-3h/8a]}{3\pi h}\right]^2$	(8.1-87j)
(11) *Circular ogive:* 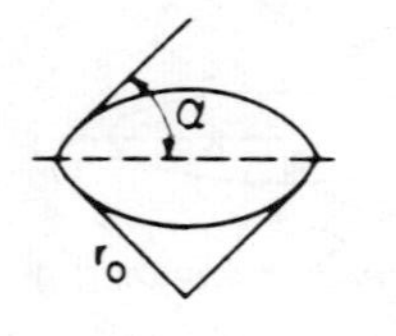	$\sigma \approx \frac{4}{\pi} k_0^4 V^2 \left[1 + \frac{e^{-\tau}}{\pi\tau}\right]^2$ $\tau = 3/2 \frac{\sin\alpha - \alpha\cos\alpha - \frac{1}{3}\sin^3\alpha}{(1-\cos\alpha)^3}$ $V = 2\pi r_0^2 (\sin\alpha - \alpha\cos\alpha - \frac{1}{3}\sin^3\alpha)$	(8.1-87k)
(12) *Parabolic ogive:* 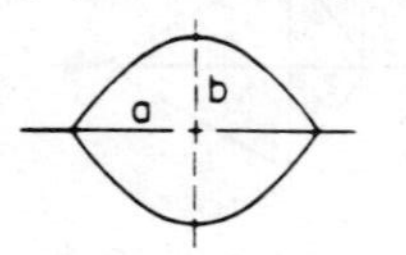	$\sigma \approx \frac{4}{\pi} k_0^4 \left[\left(\frac{16}{15}\pi a b^2\right)\left(1 + \frac{5b}{4\pi a}\exp[-4a/5b]\right)\right]^2$	(8.1-87l)

(13) *Finite cylinder:*

$$\sigma \approx \frac{4}{\pi} k_0{}^4 \left[(\pi a^2 l)\left(1 + \frac{4a}{3\pi l} \exp\left[-\,3l/4a\right]\right)\right]^2 \qquad (8.1\text{-}87m)$$

(14) *Lens:*

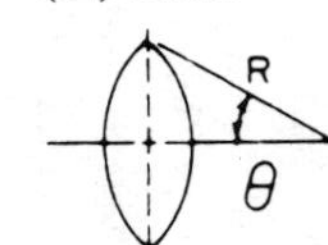

$$\sigma \approx \frac{16\pi}{9} k_0{}^4 V^2 \left[1 + \frac{4d^3}{3V} \exp\left[-\,3V/4\pi d^3\right]\right]^2$$

$$V = R^3(1 - \cos\theta)(1 - \cos\theta + \sin^2\theta) \qquad (8.1\text{-}87n)$$

(15) *Hemisphere-cylinder:*

$$\sigma \approx \frac{4}{\pi} k_0{}^4 \left[\pi a^2(2/3a + l)\right]^2 \left[1 + \frac{4a \exp\left[-(2a + 3l)/4a\right]}{(2a + 3l)\pi}\right]^2 \qquad (8.1\text{-}87o)$$

(16) *Paraboloid-cylinder:*

$$\sigma = \frac{4}{\pi} k_0{}^4 \left[\pi a^2 \left(\frac{h}{2} + l\right)\right]^2 \left[1 + \frac{4a \exp\left[-\,(3h/2 + 3l)/4a\right]}{3\pi(h/2 + l)}\right]^2 \qquad (8.1\text{-}87p)$$

an equivalent ellipsoid having the same volume and axial ratios. The results given in Eqs. (5.1-52) and (5.1-53) of Chapter 5 for the Rayleigh cross section of an ellipsoid can be used for this purpose.

For dielectric bodies, techniques similar to those used for perfectly conducting bodies can be used. That is, the body can be approximated by an equivalent spheroid or ellipsoid having the same dielectric parameters, and axial ratios equivalent to some average length-to-width ratios for the body. Then Eqs. (5.1-40) and (5.1-41) can be used to predict the low-frequency cross section.

8.1.1.1.3. *Resonance-Region Techniques.* In the resonance region, few techniques are available for analytically estimating the radar cross section of even simple shapes. One of the most useful is the extension of high-frequency techniques down into the resonance region. This is accomplished in general by including higher-order terms in $1/k_0x$ in the expressions for the contributions of the various analytic components to the cross section. For some shapes at least, such an approach is capable of giving good cross-section estimates through the entire resonance region. An example of this is the tip-on backscatter cross section of a cone-sphere, where good estimates of the cross section through the entire resonance region can be obtained by improving the accuracy of a high-frequency approximation. In particular, the entire backscatter cross section is assumed to be due to a join contribution plus a creeping-wave contribution, as discussed in Section 6.3.1.2.2, and if a sufficient number of terms (terms to order $k_0a^{-4/3}$) are taken in the expansion for the creeping-wave contribution, the analytical expression agrees well with experimental data for values of k_0a down to one. Senior[13] discusses the extension of high-frequency techniques into the resonance region. He indicates, for example, that if a sphere is considered to be a complex body with its backscatter cross section attributable to the sum of a specular contribution and a creeping-wave contribution, then if the first two terms of the specular contribution are taken along with sufficient terms for the creeping-wave expansion, the deviation from the exact result becomes significant only for $k_0a < 0.5$. At $k_0a = 1$, the error in the approximation for the scattered field is only 3% in amplitude and 2.7° in phase. In fact, Senior indicates that for $0.7 < k_0a < 1$, the high-frequency approximation for the cross section is better than the Rayleigh result.

For some shapes, however, there may exist a frequency range where resonance effects associated with the entire body become dominant and an extension of the high-frequency analysis would be of no use. In such cases at the present time, the most promising technique for obtaining valid analytical resonance-region cross-section estimates for a conducting body is the impulse-response approach discussed in Section 2.2.2.4. Although this technique has been applied to only a few bodies, namely the sphere, cone-sphere, and the prolate spheroid, it has been very successful in those cases.

8.1.1.2. *Examples*

Some simple examples of how the techniques discussed in Section 8.1.1.1 can be applied to the determination of the backscatter cross section at various aspect angles will now be given for several simple perfectly conducting bodies.

8.1.1.2.1. *Finite Cone with Rounded Tip*

8.1.1.2.1.1. *High-Frequency Approximations.* Consider a flat cone of half angle α, height h, base radius a, with a tip which has been smoothly rounded off, and with a radius b as illustrated in Figure 8-5. For axial incidence on the tip, there are three major contributors to the backscatter cross section. These are the specular contribution from the tip, the contribution from the join, and the contribution from the rear edge of the cone. The specular and join contribution can be estimated using physical optics as discussed in Sections 8.1.1.1.1.1 and 8.1.1.1.1.3. The cross section produced by these is given by Eq. (6.2-31). The contribution from the base can be estimated using the methods of Section 8.1.1.1.1.2, and the cross section due to this including doubly diffracted rays is given by Eq. (6.3-49).

The observed cross section at any aspect angle is given by summing, with the proper phases, the contributions from those analytic components which are the major contributors at the specified aspect angle. Thus,

$$\sigma \approx \left| \sum_{n=1}^{N} \sqrt{\sigma_n}\, e^{i\Psi_n} \right|^2 \tag{8.1-92}$$

The total nose-on backscatter cross section would then be the sum of the square roots of the above cross sections added with the proper phase factors. Retaining only the first-order terms in k_0, then

$$\sigma(0) \approx \left| \pi^{1/2} b \left[1 - \frac{\sin[2k_0 b(1 - \sin\alpha)]}{k_0 b \cos^2\alpha}\right]^{1/2} + 2\pi^{1/2} a \left[\frac{\sin(\pi/n)}{n}\right]\left[\cos\frac{\pi}{n} - \cos\frac{2\alpha}{n}\right]^{-1} e^{2ik_0 l} \right|^2 \tag{8.1-93}$$

with

$$n = 3/2 + \alpha/\pi$$

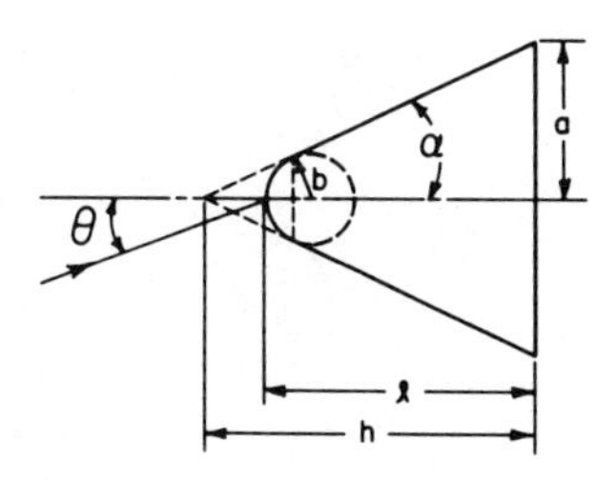

Fig. 8-5. Rounded-tip cone dimensions.

The only terms which are included in the above expression are the specular contribution from the tip, the first-order tip–cone join contribution, and singly diffracted rays from the base. The contributions from higher-order derivative discontinuities at the tip-cone join and doubly diffracted rays from the base are of second order in $1/k_0$, and are not included in Eq. (8.1-93). In order to approximate the nose-on backscatter cross section for lower frequencies, or further down into the resonance region for the body, however, these second-order terms should be included.

If the angle of incidence is such that $0 < |\theta| < \alpha$, then the major contributors are still the specular return from the tip, and the singly diffracted rays from the base. In this case, the base contribution comes from points on opposite sides of the base, and Eqs. (8.1-19) and (8.1-20) should be used since the scattered-field direction is not a caustic of the diffracted rays. Thus, the backscatter cross section for $0 < |\theta| < \alpha$ is

$$\begin{aligned}\sigma_{\perp \atop \parallel}(\theta) \approx \Big| \, & \pi^{1/2} b \exp(i2k_0[(h - l + b)\cos\theta - b]) \\ & \mp \frac{\sin(\pi/n)}{n\sqrt{k_0/2}} \sqrt{(a/2)\csc\theta}\exp[i\pi/4 + 2ik_0(h - a\tan\theta)\cos\theta] \\ & \times \left[\left(\cos\frac{\pi}{n} - 1\right)^{-1} \mp \left(\cos\frac{\pi}{n} - \cos\frac{2(\alpha+\theta)}{n}\right)^{-1}\right] \\ & \mp \frac{\sin(\pi/n)}{n\sqrt{k_0/2}} \sqrt{-(a/2)\csc\theta}\exp[i\pi/4 + 2ik_0(h + a\tan\theta)\cos\theta] \\ & \times \left[\left(\cos\frac{\pi}{n} - 1\right)^{-1} \mp \left(\cos\frac{\pi}{n} - \cos\frac{2(\alpha-\theta)}{n}\right)^{-1}\right]\Big|^2 \end{aligned} \tag{8.1-94}$$

with

$$n = 3/2 + \alpha/\pi$$

In this and all subsequent equations, the upper and lower signs go with the upper and lower subscripts on σ.

For angles of incidence, $\alpha < |\theta| < \pi/2 - \alpha$, the contributors to the cross section are the specular return from the tip and a singly diffracted ray from one point on the base; thus

$$\begin{aligned}\sigma_{\perp \atop \parallel}(\theta) \approx \Big| \, & \sqrt{\pi}\, b \exp(i2k_0[(h - l + b)\cos\theta - b]) \\ & \mp \frac{\sin(\pi/n)}{n\sqrt{k_0/2}} \sqrt{\frac{a\csc\theta}{2}} \exp\left[i\frac{\pi}{4} + 2ik_0(h - a\tan\theta)\cos\theta\right] \\ & \times \left[\left(\cos\frac{\pi}{n} - 1\right)^{-1} \mp \left(\cos\frac{\pi}{n} - \cos\frac{2\alpha + 2\theta}{n}\right)^{-1}\right]\Big|^2 \end{aligned} \tag{8.1-95}$$

with

$$n = 3/2 + \alpha/\pi$$

For $|\theta| \approx \pi/2 - \alpha$ then, the primary contributor to the backscatter cross section is the specular flare from the conical side. This can be estimated using physical optics and considering the shape to be that of a frustum. The backscatter cross section of such a conical frustum is given by Eq. (6.3-45), or for $|\theta| = \pi/2 - \alpha$,

$$\sigma_{\perp \atop \parallel}(\theta) \approx \frac{4}{9} k_0 \cos\alpha \cot^2\alpha [a^{3/2} - (b\cos\alpha)^{3/2}]^2 \tag{8.1-96}$$

This expression was obtained by the cylindrical current method of Dawson;[14] physical optics gives an expression which essentially agrees with Eq. (8.1-96) for small cone angles α.

For $\pi/2 - \alpha < |\theta| < \pi/2$, the only major contributor to the backscatter cross section is the singly diffracted ray from the base, or

$$\sigma_{\perp \atop \parallel}(\theta) \approx \left[\frac{\sin(\pi/n)}{n\sqrt{k_0/2}} \sqrt{(a/2)\csc\theta} \left[\left(\cos\frac{\pi}{n} - 1\right)^{-1} \mp \left(\cos\frac{\pi}{n} - \cos\frac{2(\alpha+\theta)}{n}\right)^{-1}\right]\right]^2 \tag{8.1-97}$$

with

$$n = 3/2 + \alpha/\pi$$

In the region where θ is greater than $\pi/2$, the flat base of the cone is observed, and the backscatter cross section is very similar to that for a disc. For $\pi/2 < |\theta| < \pi$, the cross section consists of diffracted rays from opposite points on the edge of the base. Taking only singly diffracted rays, then the cross section is approximately given by

$$\begin{aligned}\sigma_{\perp \atop \parallel}(\theta) \approx \frac{2\sin^2(\pi/n)}{n^2 k_0} \Bigg| & \sqrt{(a/2)\csc\theta} \exp\left[i\frac{\pi}{4} + 2ik_0(h - a\tan\theta)\cos\theta\right] \\ & \times \left[\left(\cos\frac{\pi}{n} - 1\right)^{-1} \mp \left(\cos\frac{\pi}{n} - \cos\frac{2(3\pi/2 - \theta)}{n}\right)^{-1}\right] \\ & + \sqrt{-(a/2)\csc\theta} \exp\left[i\frac{\pi}{4} + 2ik_0(h + a\tan\theta)\cos\theta\right] \\ & \times \left[\left(\cos\frac{\pi}{n} - 1\right)^{-1} \mp \left(\cos\frac{\pi}{n} - \cos\frac{2(\pi/2 - \theta)}{n}\right)^{-1}\right]\Bigg|^2\end{aligned} \tag{8.1-98}$$

with

$$n = 3/2 + \alpha/\pi$$

At $\theta = \pi$, the flat base is viewed normally, and the physical-optics result for a disc can be used, or

$$\sigma(\pi) \approx \pi k_0^2 a^4 \tag{8.1-99}$$

In the previous paragraphs, approximate backscatter cross sections were given over the entire range of aspect angles for a flat-based cone with a rounded tip. As would be expected, these results differ from those for a pointed cone only for aspects near nose-on.

8.1.1.2.1.2. *Resonance-Region Approximations.* In the resonance region, the high-frequency results of the above section should be carried as far as possible. That is, for example, doubly diffracted terms should be used for the base contributions, any surface derivative discontinuities should be included, etc.

8.1.1.2.1.3. *Low-Frequency Approximations.* In the Rayleigh region, the axial incidence backscatter cross section of the rounded-tip cone can be estimated using Siegel's technique, discussed in Section 8.1.1.1.2. Applying this approach, the backscatter cross section is

$$\sigma(0) = \sigma(\pi) \approx \frac{4}{\pi} k_0^4 \left(\frac{\pi^2}{9}\right) [b^3(2 - 3 \sin\alpha + \sin^3\alpha) + \cot\alpha(a - b\cos\alpha)(a^2 + b^2\cos^2\alpha + ab\cos\alpha)]^2 \times \left[1 + \frac{4\exp[-(\cot\alpha)/4]}{\pi\cot\alpha}\right]^2 \quad (8.1\text{-}100)$$

For other angles of incidence, the Rayleigh results for the spheroid can be used by choosing an equivalent spheroid with the same volume and an axial ratio equal to the average axial ratio of the cone. Thus, the equivalent spheroid will have an axial ratio of $a_e/b_e = \cot\alpha$ or $(\cot\alpha)/4$, depending upon whether α is small or large, and a volume of

$$V = (\pi/3)[b^3(2 - 3\sin\alpha + \sin^3\alpha) + \cot\alpha(a - b\cos\alpha)(a^2 + b^2\cos^2\alpha + ab\cos\alpha)] \quad (8.1\text{-}101)$$

The size of the equivalent spheroid having been specified, then Eqs. (8.1-88) and (8.1-89) can be used to determine the cross section for any aspect angle.

A somewhat simpler approximation would be to ignore the rounded tip and consider the body to be a finite cone. In this event, the equivalent spheroid volume to be used would be

$$V = (\pi/3)\, a^3 \cot\alpha \quad (8.1\text{-}102)$$

The Rayleigh cross sections would then be

$$\sigma_\perp = (k_0^4/\pi)\, V^2[A_1 + A_2\sin^2\theta + A_3\cos^2\theta]^2 \quad (8.1\text{-}103)$$

$$\sigma_\parallel = (k_0^4/\pi)\, V^2[A_1' + A_2'\sin^2\theta + A_3'\cos^2\theta]^2 \quad (8.1\text{-}104)$$

where the coefficients A_i', A_i are given by Eqs. (8.1-90) and (8.1-91). These can be obtained from curves 5-2 and 5-3 where the axial ratio of $b_e/a_e = 4 \tan \alpha$ is used as the ordinate.

8.1.1.2.2. *Cone-Cylinder.* Another example of a simple body which consists of the combination of several geometric shapes is a cone-cylinder, such as illustrated in Figure 8-6. Approximate analytical expressions for the backscatter cross section for various frequency regions and angles of incidence will be given in the following sections.

8.1.1.2.2.1. *High-Frequency Approximations.* For incidence axially on the conical tip, the primary contributions to the cross section will come from the conical tip and the surface discontinuity at the cone–cylinder join. In the event that the cone has a sharp tip such as the one illustrated in Figure 8-6, then the tip contribution can, in general, be neglected. If the tip is rounded like the body discussed in Section 8.1.1.2.1, then a specular contribution from the tip must be included.

Since the axis is a caustic of the diffracted field from the join discontinuity, the singly diffracted join contribution can be determined using Eqs. (8.1-25) and (8.1-26); thus the nose-on backscatter cross section is

$$\sigma(0) \approx \pi a^2 \left[\frac{(2/n)\sin(\pi/n)}{\cos(\pi/n) - \cos(2\alpha/n)}\right]^2 \qquad (8.1\text{-}105)$$

where

$$n = 1 + \alpha/\pi$$

The tip contribution has been neglected in this expression.

Very little experimental data exists for the nose-on cross section of a cone-cylinder, however, there is some evidence (see Figure 6-21) that Eq. (8.1-105) does not describe adequately the cross section. Apparently, even for cylinders as long as $k_0 l = 30$, there is a significant contribution from the rear of the cylinder, although the rear discontinuity is shadowed by the cone-cylinder join. Conventional geometrical diffraction theory is unable to account for the effect of the rear edge, since the diffraction coefficients are not valid for grazing incidence. In order to obtain a valid estimate of the nose-on cross section, a technique such as discussed in Section 2.3.1.1.3 should be used, where the edge currents are to be integrated over the body surface in order to obtain the contribution of the edges to the scattered far fields. A similar approach was utilized by Morse[15] to obtain the forward scattered fields from a rectangular cylinder.

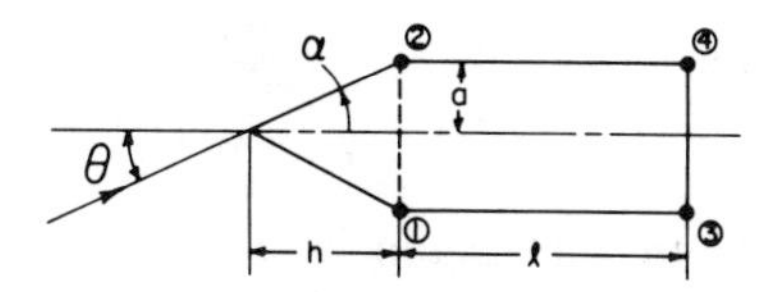

Fig. 8-6. Cone-cylinder dimensions.

An empirical expression which seems to agree reasonably well with the results of Figure 6-21 can be obtained by assuming the incident field at the rear discontinuity to be the same as at the cone–cylinder join, except that it would be shifted in phase by twice the cylinder length in radians plus $\pi/4$. The axial, singly diffracted contributions from the join and the cylinder rear are then added, giving

$$\sigma(0) \approx 4\pi a^2 \left|\left[\frac{(1/n_1)\sin(\pi/n_1)}{\cos(\pi/n_1) - \cos(2\alpha/n_1)}\right] + \exp\left[i\left(2k_0 l + \frac{\pi}{4}\right)\right] \left[\frac{(1/n_2)\sin(\pi/n_2)}{\cos(\pi/n_2) - 1}\right]\right|^2 \quad (8.1\text{-}106)$$

where

$$n_1 = 1 + \alpha/\pi, \qquad n_2 = 3/2$$

For $0 < |\theta| < \alpha$, the scattering comes from two opposite points; (1) and (2) on the join and from a point, and (3) on the rear of the cylinder. Equations (8.1-19) to (8.1-21) can be used to estimate the singly diffracted scattered field from these points, and summing these with the proper phases gives the total backscattered field. The resulting backscatter cross section is, for $0 < |\theta| < \alpha$,

$$\begin{aligned}
\sigma_{\substack{\perp \\ \parallel}}(\theta) \approx \frac{2}{k_0} \Bigg| &\frac{\sin(\pi/n_1)}{n_1} \sqrt{(a/2)\csc\theta}\, \exp\left[i\frac{\pi}{4} + 2ik_0(h - a\tan\theta)\cos\theta\right] \\
&\times \left[\left(\cos\frac{\pi}{n_1} - 1\right)^{-1} \mp \left(\cos\frac{\pi}{n_1} - \cos\frac{2(\alpha+\theta)}{n_1}\right)^{-1}\right] \\
&+ \sqrt{-(a/2)\csc\theta}\, \exp\left[i\frac{\pi}{4} + 2ik_0(h + a\tan\theta)\cos\theta\right] \\
&\times \left[\left(\cos\frac{\pi}{n_1} - 1\right)^{-1} \mp \left(\cos\frac{\pi}{n_1} - \cos\frac{2(\alpha-\theta)}{n_1}\right)^{-1}\right] \\
&+ \frac{\sin(\pi/n_2)}{n_2} \sqrt{(a/2)\csc\theta}\, \exp\left[i\frac{\pi}{4} + 2ik_0(l + h - a\tan\theta)\cos\theta\right] \\
&\times \left[\left(\cos\frac{\pi}{n_2} - 1\right)^{-1} \mp \left(\cos\frac{\pi}{n_2} - \cos\frac{2\theta}{n_2}\right)^{-1}\right]\Bigg|^2
\end{aligned} \quad (8.1\text{-}107)$$

with

$$n_1 = 1 + \alpha/\pi, \qquad n_2 = 3/2$$

In the region $\alpha < |\theta| < \pi/2$, excluding the point $|\theta| = \pi/2 - \alpha$, only points (1) and (3) contribute to the backscatter cross section. Thus, the singly diffracted backscatter cross section is

$$\sigma(\theta)_{\substack{\perp \\ \parallel}} \approx \frac{2}{k_0} \left| \frac{\sin(\pi/n_1)}{n_1} \sqrt{(a/2)\csc\theta}\, \exp\left[i\frac{\pi}{4} + 2ik_0(h - a\tan\theta)\cos\theta\right] \right.$$
$$\times \left[\left(\cos\frac{\pi}{n_1} - 1\right)^{-1} \mp \left(\cos\frac{\pi}{n_1} - \cos\frac{2(\alpha+\theta)}{n_1}\right)^{-1}\right]$$
$$+ \frac{\sin(\pi/n_2)}{n_2} \sqrt{(a/2)\csc\theta}\, \exp\left[i\frac{\pi}{4} + 2ik_0(l + h - a\tan\theta)\cos\theta\right]$$
$$\left. \times \left[\left(\cos\frac{\pi}{n_2} - 1\right)^{-1} \mp \left(\cos\frac{\pi}{n_2} - \cos\frac{2\theta}{n_2}\right)^{-1}\right] \right|^2 \qquad (8.1\text{-}108)$$

with

$$n_1 = 1 + \alpha/\pi, \qquad n_2 = 3/2$$

Near the point $|\theta| \approx \pi/2 - \alpha$, the major contribution to the backscatter cross section is the specular flare from the cone side. This can be estimated using Eq. (6.3-46) with $a_1 = 0$; or, for $|\theta| \approx \pi/2 - \alpha$,

$$\sigma(\theta) \approx \tfrac{4}{9} k_0 a^3 \cos\alpha \cot^2\alpha \qquad (8.1\text{-}109)$$

For $|\theta| \approx \pi/2$, the major contribution is the specular flare from the cylinder side, and this can be estimated using physical optics. An improvement on the physical-optics result can be obtained by using Eq. (4.3-43), where

$$\sigma(\theta) \approx \frac{k_0 l^2}{\pi} \sin^2\theta\, \sigma^c \left[\frac{\sin(k_0 l\cos\theta)}{k_0 l\cos\theta}\right]^2 \qquad (8.1\text{-}110)$$

with σ^c the normal incidence scattering width of an infinite cylinder. Using the first two terms of the Luneberg series result for σ^c gives

$$\sigma_{\substack{\perp \\ \parallel}}(\theta) \approx k_0 a l^2 \left| 1 + \frac{i}{2k_0 a}\left(-\frac{3}{8} \mp 1\right)\right|^2 \sin^2\theta \left[\frac{\sin(k_0 l\cos\theta)}{k_0 l\cos\theta}\right]^2 \qquad (8.1\text{-}111)$$

and, for $|\theta| = \pi/2$,

$$\sigma_{\substack{\perp \\ \parallel}}\left(\frac{\pi}{2}\right) \approx k_0 a l^2 \left| 1 + \frac{i}{2k_0 a}\left(-\frac{3}{8} \mp 1\right)\right|^2 \qquad (8.1\text{-}112)$$

For $\pi/2 < |\theta| < \pi$, points (1), (3), and (4) contribute, so the backscatter cross section is

$$\sigma_{\substack{\perp \\ \parallel}}(\theta) \approx \frac{2}{k_0}\left| \frac{\sin(\pi/n_1)}{n_1} \sqrt{(a/2)\csc\theta}\, \exp\left[i\frac{\pi}{4} + 2ik_0(h - a\tan\theta)\cos\theta\right] \right.$$
$$\times \left[\left(\cos\frac{\pi}{n_1} - 1\right)^{-1} \mp \left(\cos\frac{\pi}{n_1} - \cos\frac{2(\alpha+\theta)}{n_1}\right)^{-1}\right]$$
$$+ \frac{\sin(\pi/n_2)}{n_2}\sqrt{(a/2)\csc\theta}\, \exp\left[i\frac{\pi}{4} + 2ik_0(l + h - a\tan\theta)\cos\theta\right]$$
$$\times \left[\left(\cos\frac{\pi}{n_2} - 1\right)^{-1} \mp \left(\cos\frac{\pi}{n_2} - \cos\frac{2\theta}{n_2}\right)^{-1}\right]$$
$$+ \frac{\sin(\pi/n_2)}{n_2}\sqrt{-(a/2)\csc\theta}\, \exp\left[i\frac{\pi}{4} + 2ik_0(l + h + a\tan\theta)\cos\theta\right]$$
$$\left. \times \left[\left(\cos\frac{\pi}{n_2} - 1\right)^{-1} \mp \left(\cos\frac{\pi}{n_2} - \cos\frac{2[\theta - (\pi/2)]}{n_2}\right)^{-1}\right]\right|^2 \qquad (8.1\text{-}113)$$

with

$$n_1 = 1 + \alpha/\pi, \qquad n_2 = 3/2$$

For $\theta \approx \pi$, the primary contribution to the backscatter cross section arises from the specular return from the cylinder base. This is the same as the backscatter cross section for normal incidence on a circular disc, and can be approximated by physical optics. Thus, for $\theta = \pi$,

$$\sigma(\pi) \approx \pi k_0^2 a^4 \qquad (8.1\text{-}114)$$

8.1.1.2.2.2. *Resonance-Region Approximations.* Results which are approximately valid in the resonance region can be obtained by utilizing the high-frequency techniques discussed above, and including as many higher-order terms as practicable.

8.1.1.2.2.3. *Low-Frequency Approximations.* In the Rayleigh region, the axial-incidence backscatter cross section obtained by Siegel's technique is given in Table 8-2 by Eq. (8.1-87c), or

$$\sigma(0) \approx \frac{4}{\pi} k_0^4 \left[\frac{\pi a^2}{3}(h + 3l)\, 1 + \frac{4a \exp - [(h + 3l)/4a]}{\pi(h + 3l)}\right]^2 \qquad (8.1\text{-}115)$$

For angles of incidence other than axial, the backscatter cross section can be approximated by using the Rayleigh results for a spheroid of the same volume and an approximate axial ratio. The volume of the equivalent spheroid is

$$V = \pi a^2(l + h/3) \qquad (8.1\text{-}116)$$

and an equivalent axial ratio based on the average length and width of the body is

$$\frac{a_e}{b_e} = \frac{(l + h)^2}{a(2l + h)} \tag{8.1-117}$$

8.1.1.2.3. *Cone-Spheroid.* All the shapes previously considered had surface discontinuities (edges) which were large, often dominant, contributors to the cross section. It is instructive to now consider a perfectly conducting shape which has no surface discontinuities, except a conical tip. The shape to be considered will be formed by joining a spheroid smoothly with a cone, as illustrated in Figure 8-7.

8.1.1.2.3.1. *High-Frequency Approximations.* For a plane wave incident axially upon the conical tip, there will be three primary contributors to the backscatter cross section. These will be the tip contribution, either conical or rounded, which is discussed in Section 8.1.1.2.1.1; the contribution from the cone–spheroid join; and the shadow-boundary or creeping-wave contribution. The join contribution can be estimated using physical optics as discussed in Section 8.1.1.1.1.3, and is given by

$$\sqrt{\sigma_j} \approx \frac{i\pi}{k_0}(b^2/c^2 - \tan^2 \alpha) \tag{8.1-118}$$

$$\Psi_j = 2k_0 h \tag{8.1-119}$$

The contribution from the shadow boundary can be obtained by using the geometrical theory of diffraction as discussed in Section 8.1.1.1.1.4, or Fock theory as discussed in Section 2.2.2.5. Since the axis is a caustic of the diffracted field, the backscattered field is given by Eq. (8.1-59), and is independent of the polarization. By an analysis similar to that used to obtain Eq. (8.1-59), the creeping-wave contribution to the backscattered field can be obtained in terms of the Fock functions $\hat{p}$, $\hat{q}$ as

$$E^{cw} \approx -ib\left(\frac{k_0 R_0}{2}\right)^{1/3} [\hat{q}(\xi) - \hat{p}(\xi)] \frac{\exp[ik_0(r + s)]}{r} \tag{8.1-120}$$

where

$$\xi = \int_0^s \left(\frac{k_0}{2R^2(l)}\right)^{1/3} dl \tag{8.1-121}$$

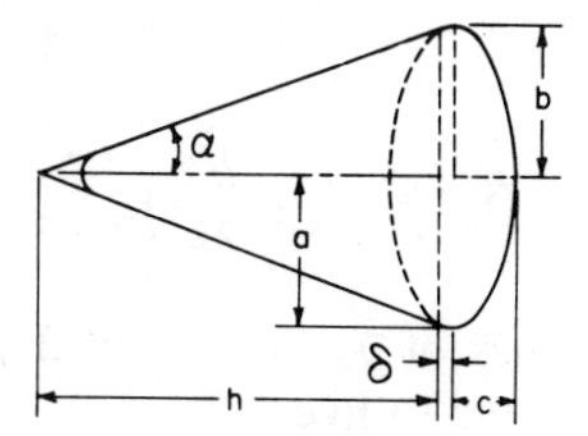

Fig. 8-7. Cone-spheroid dimensions.

and s equals the arc length around the shadowed portion of the body. The parameter R_0 is the radius of curvature at the shadow boundary in the plane of the axis. For the spheroid base, these parameters are

$$\xi = 2\left(\frac{k_0 c^2}{2b}\right)^{1/3} K(\epsilon^2) \tag{8.1-122}$$

$$s = 2bE(\epsilon^2) \tag{8.1-123}$$

$$R_0 = c^2/b \tag{8.1-124}$$

and

$$\epsilon^2 = 1 - c^2/b^2 \tag{8.1-125}$$

Thus the shadow-boundary contribution is

$$\sqrt{\sigma_s} \approx -i2\sqrt{\pi}\, b\left(\frac{k_0 c^2}{2b}\right)^{1/3} [\hat{q}(\xi) - \hat{p}(\xi)] \exp[i2k_0 bE(\epsilon^2)] \tag{8.1-126}$$

$$\Psi_s = 2k_0(h + \delta) \tag{8.1-127}$$

The functions $K(\epsilon^2)$ and $E(\epsilon^2)$ in the above equations are complete elliptic integrals of the first and second kinds, respectively.[16]

An essentially equivalent expression for the shadow-boundary contribution can be obtained using the geometrical diffraction results of Section 8.1.1.1.1.4. Thus, from Eqs. (8.1-59) to (8.1-61),

$$\sqrt{\sigma_s} \approx -i\pi\sqrt{2k_0}\, b \exp[i(k_0 s - \pi/4)] \times\left[-\sum_{m=1}^{\infty} \bar{D}_m{}^2 \exp(\bar{q}_m{}^0 \xi \exp[-i(\pi/3)]/3^{1/3}) + \sum_{m=1}^{\infty} D_m{}^2 \exp(q_m{}^0 \xi \exp[-i(\pi/3)]/3^{1/3})\right] \tag{8.1-128}$$

$$\Psi_s = 2k_0(h + \delta) \tag{8.1-129}$$

In these equations, s and ξ are the same as defined above (Eqs. 8.1-121, 122), and the coefficients, $D_m{}^2$, $\bar{D}_m{}^2$, are

$$D_m{}^2 = \frac{\pi}{6}\left(\frac{k_0}{2\pi}\right)^{-1/2}\left(\frac{k_0 c^2}{6b}\right)^{1/3} [A'(\bar{q}_m{}^0)]^{-2} \exp(i\pi/12) \tag{8.1-130}$$

$$\bar{D}_m{}^2 = \frac{\pi}{2}\left(\frac{k_0}{2\pi}\right)^{-1/2}\left(\frac{k_0 c^2}{6b}\right)^{1/3} [q_m{}^0 A^2(q_m{}^0)]^{-1} \exp(i\pi/12) \tag{8.1-131}$$

The values of $\bar{q}_m{}^0$, $q_m{}^0$, $A(\bar{q}_m{}^0)$, $A'(q_m{}^0)$ are given in Table 8-1 for values of m from one to five.

The total axial-incidence backscatter cross section, then, is the sum of all the contributions, or

$$\sigma \approx | \sqrt{\sigma_t} + \sqrt{\sigma_j}\, e^{i\Psi_j} + \sqrt{\sigma_s}\, e^{i\Psi_s} |^2 \tag{8.1-132}$$

For $b = c$, this, of course, gives the backscatter cross section for a cone-sphere.

8.1.1.2.3.2. *Low-Frequency Approximations.* In the low-frequency case, the axial-incidence backscatter cross section can be approximated using Siegel's technique. Equation (8.1-87b) in Table 8-2 gives the result for a cone-hemispheroid. This should be a good estimate for the cone-spheroid when δ is small in comparison with h and c.

8.1.1.3. *Experimental Results*

In this section some experimental cross sections will be given for a variety of perfectly conducting shapes formed by simple geometric combinations, along with discussions of these in terms of the analytic components which constitute the total observed cross section.

8.1.1.3.1. *Capped Cones.* One example of a capped cone, which was discussed previously in Section 6.3.1, is the cone-sphere. Examples of several others are given in Figure 8-8. The backscatter cross sections for both parallel and perpendicular polarizations as a function of aspect angle are given for these shapes in Figures 8-9 to 8-13. Examining the forms of these backscatter patterns, some characteristic features can be observed. Starting with axial incidence on the conical tip, the observed cross section is due

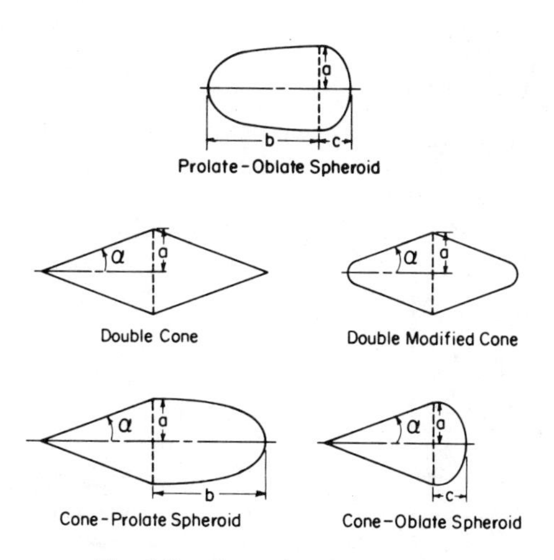

Fig. 8-8. Capped-cone dimensions.

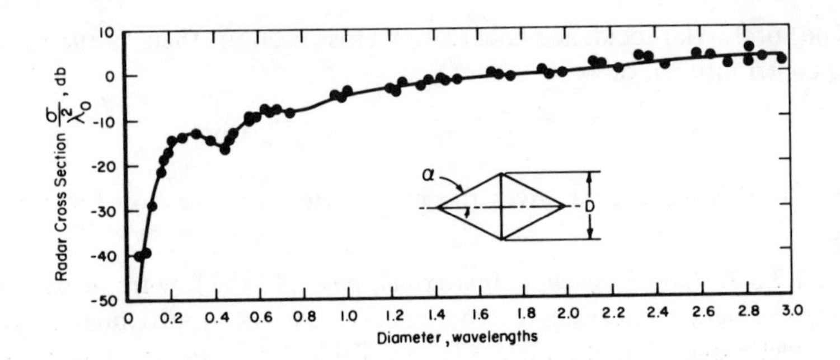

Fig. 8-9. Measured axial-incidence backscatter cross section of a 20° perfectly conducting double-cone versus the cone diameter in wavelengths [from Blore and Royer[17]].

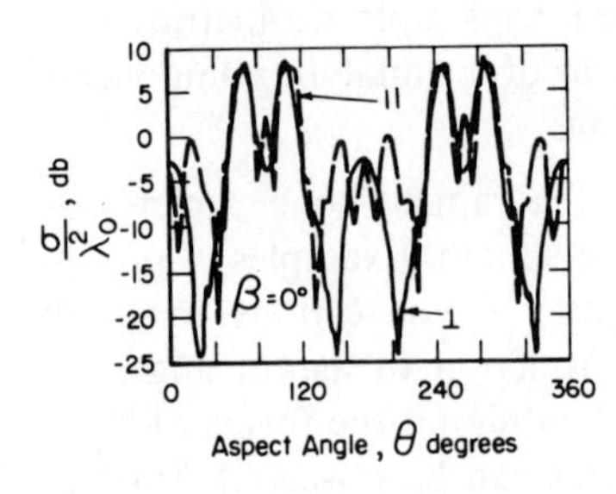

Fig. 8-10. Measured backscatter cross section of a 20° perfectly conducting double-cone versus the angle of incidence, $k_0a = 4.2$ [from Eberle and St. Clair[18]].

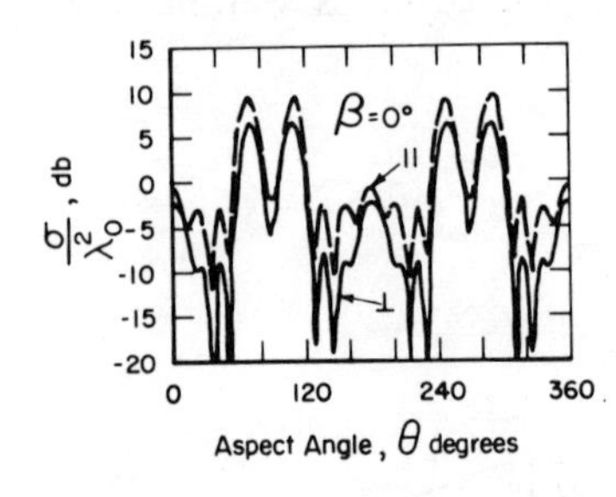

Fig. 8-11. Measured backscatter cross section of a 20° perfectly conducting modified double-cone versus the angle of incidence, $k_0a = 4.2$ [from Eberle and St. Clair[18]].

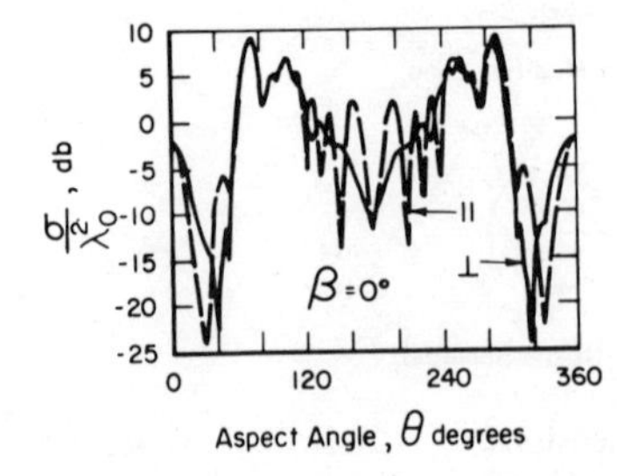

Fig. 8-12. Measured backscatter cross section of a 20° perfectly conducting cone-prolate spheroid versus the angle of incidence, $k_0a = 4.2$ [from Eberle and St. Clair[18]].

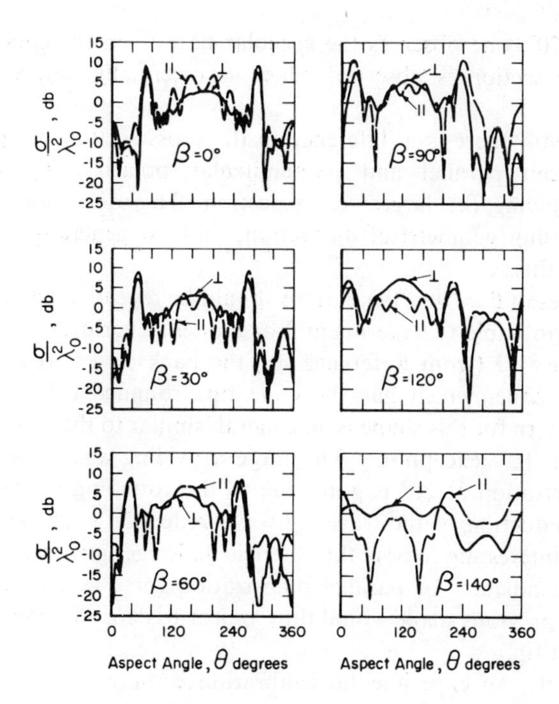

Fig. 8-13. Measured bistatic cross sections of a 20° perfectly conducting cone-oblate spheroid versus the angle of incidence for various bistatic angles, $k_0 a = 4.2$ [from Eberle and St. Clair[18]].

largely to the join, since a finite surface discontinuity (edge) exists at this point. This is verified by Figure 8-9, which presents experimental results for the axial-incidence backscatter from a 20° double-cone.[17] From this figure, one observes that for $d/\lambda_0 > 0.5$, the cross section increases monotonically with the cone diameter; and no resonance effects are exhibited, indicating that there is only one major contribution to the cross section in this frequency region, which is the edge at the join. Thus, the nose-on cross section for this shape should be readily predictable using geometrical diffraction.

Figure 8-10 (from Reference 18) shows the backscatter cross section versus the angle of incidence for a 20° double-cone. As the aspect changes from nose-on, the first interesting feature is the peak in the cross section occurring for parallel polarization at about 20°. At this point, the parallel polarization cross section exceeds that for perpendicular polarization by about 15 db. This peak can be attributed to a traveling-wave echo, such as discussed in Section 2.3.3, and is typical of bodies which are somewhat longer than they are wide, and which have a relatively sharp point at the rear.

At $\theta = 70°$, one observes the specular flare from the conical side, and a large cross section is observed which is essentially independent of the polarization.

At $\theta = 90°$, there is a difference in the cross section of approximately 4–5 db between parallel and perpendicular polarization, with parallel polarization giving the largest cross section. This does not appear to be explainable using geometrical diffraction, and no generally valid estimate is known for this.

For values of $\theta > 90°$, the pattern should be repeated, and any deviation would be attributable to experimental or modeling errors.

In Figure 8-11 (from Reference 18), the backscatter cross section from a 20° double-cone, which has the cone tips rounded off, is shown. The scattering pattern for this shape is, in general, similar to that for the preceding shape, with a few exceptions. The large traveling-wave contribution for parallel polarization at 20° is gone due to the rounding of the tip and the consequent reduction of the traveling-wave reflection coefficient at the rear tip. It is also interesting to note that there is no longer any essential difference between perpendicular or parallel polarization for $\theta \approx 90°$, and the effect noted for the previous shape would thus appear to depend upon the sharpness of the conical tip.

Apparently, an experimental calibration error was made in obtaining the curves of Figure 8-11, since they should agree for axial incidence ($\theta = 0{,}180°$) and at the specular points, while there is a consistant 2 to 3 db difference.

Figure 8-12 (from Reference 18) gives the backscatter cross section for a 20° cone with a prolate spheroidal base versus the angle of incidence. For incidence on the conical tip, the backscatter cross section is about the same as for the double-cones, and is apparently due primarily to the discontinuity at the join. This should be predicted relatively well by geometrical diffraction methods. As the aspect angle moves away from the axis, the cross section drops for both polarizations as would be expected, and no appreciable traveling-wave component appears.

For θ near 70°, the specular flare from the conical side appears. For aspects beyond 70°, the specular contribution from the spheroid dominates for perpendicular polarization, while for parallel polarization the cross section is apparently the sum of the specular contribution and a large traveling-wave component. The pattern, of course, repeats for $\theta > 180°$.

Figure 8-13 (from Reference 18) gives several bistatic scattering patterns for a 20° cone with an oblate-spheroidal base. For backscatter ($\beta = 0$), the scattering pattern is very similar to that for the cone-prolate spheroid, except that the specular return from the oblate base in the rear hemisphere ($\pi/2 < \theta < 3\pi/2$) is much greater.

From the bistatic patterns one sees that for small bistatic angles

($\beta < 60°$), the scattering patterns do not change appreciably from the monostatic result, indicating that for this shape and size of object the monostatic–bistatic theorem (see Section 2.1.2) holds relatively well. That is the bistatic cross section at aspect angle θ is given approximately by the monostatic cross section at an aspect of $\theta + \beta/2$.

8.1.1.3.2. *Other Shapes.* In Figure 8-14 (from Reference 17), the axial-incidence backscatter cross section is shown for a body consisting of a double cone which has had the join smoothly rounded off. Curves are given for cone angles of 7.5, 20, and 37.5°. Comparing these curves with that of

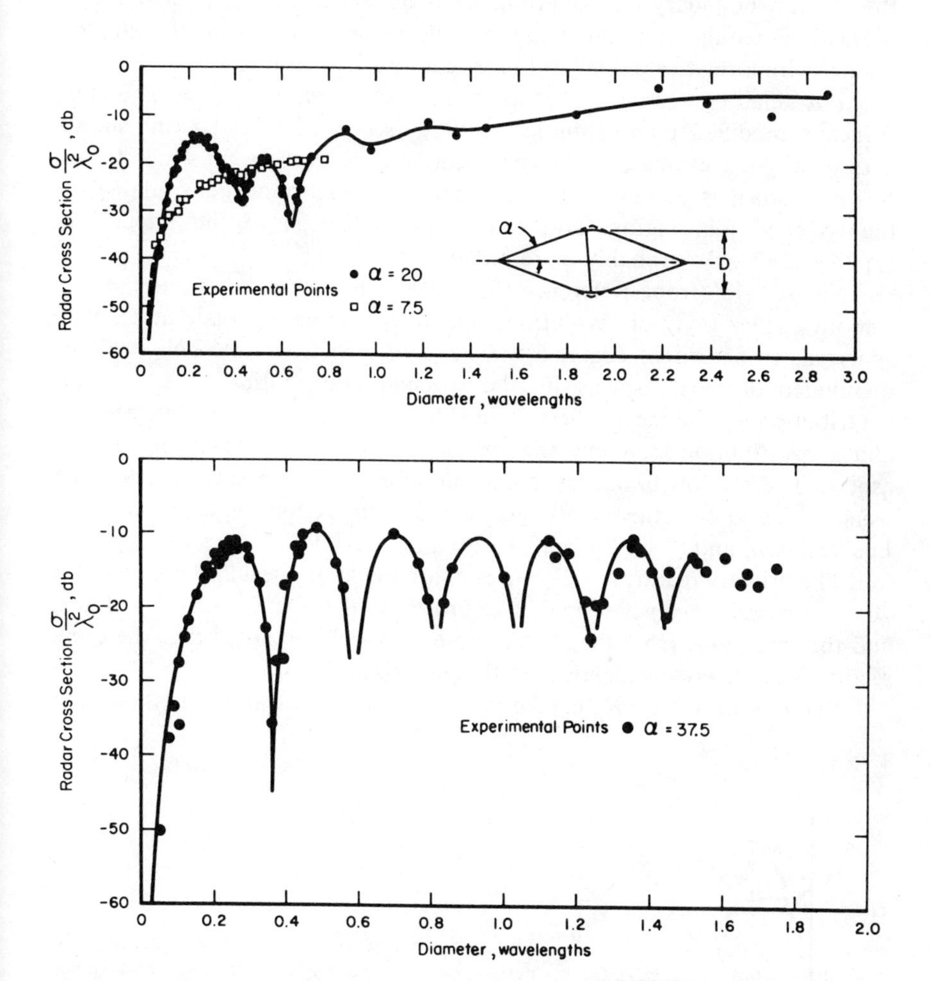

Fig. 8-14. Measured axial-incidence backscatter cross sections for 7.5, 20, and 37.5° double-modified cones versus the cone diameter in wavelengths [from Blore and Royer[17]].

Figure 8-10, where the cone join has not been smoothed, one observes that the cross section is considerably smaller and resonance effects more prominent. Instead of the single large contributor (the join), as is the case with the double-cone, the effect of rounding the join is to reduce the join contribution to the extent that other sources contribute with the same order of magnitude. For this particular shape then, the axial-incidence cross section can most likely be attributed to the sum of field contributions from the surface derivative discontinuities at the join and from the shadow boundary. The shadow-boundary contribution is difficult to predict in this case due to the sharp tip at the rear of the body. All of the techniques for estimating the shadow-boundary contribution, such as Fock theory or geometrical diffraction, require that the body be smooth and convex in the shadow region, with a relatively large radius of curvature at all points.

For small cone angles, such that the body is relatively long and thin, Adachi's modified physical-optics method (see Section 8.1.1.1.1.6) should provide a good estimate of the backscatter cross section. In order to use this method, it is necessary that an analytical description be available for the rounded join region, and numerical evaluation of the integral in Eq. (8.1-62) would probably be required.

Figure 8-15 (from Reference 18) shows the backscatter pattern for a smooth convex body of revolution formed by joining a prolate and oblate spheroid, as shown in Figure 8-8. The scattering from this body will be dominated at most aspects by the specular contribution. The specular contribution will be the smallest for incidence axially on the prolate section, and a contribution from the shadow boundary may be significant at this aspect. The shadow-boundary contribution in this case can be estimated using either Fock theory or geometrical diffraction, and is given by Eqs. (8.1-126) and (8.1-127) or (8.1-128) and (8.1-129).

The primary features of Figure 8-15 are the large specular lobe at about 70°, the broad, somewhat smaller return from the oblate base ($\theta \approx 180°$), and the small lobe from the prolate nose. It should be noted that the cross section is relatively independent of the polarization.

Figure 8-16 (from Reference 19) gives the backscatter cross section

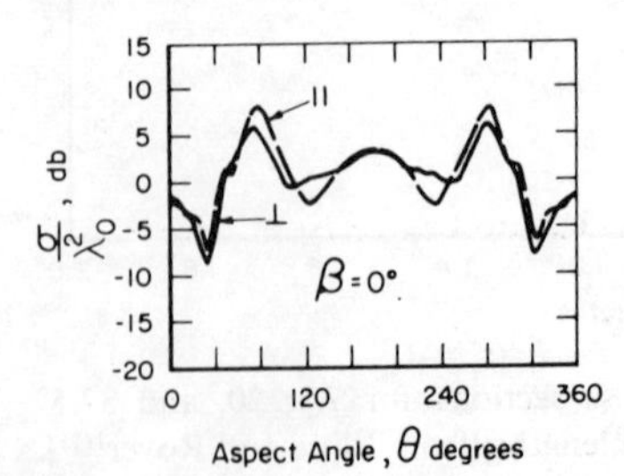

Fig. 8-15. Measured backscatter cross section of a perfectly conducting prolate-oblate spheroid versus the angle of incidence, $k_0a = 4.2$ [from Eberle and St. Clair[18]].

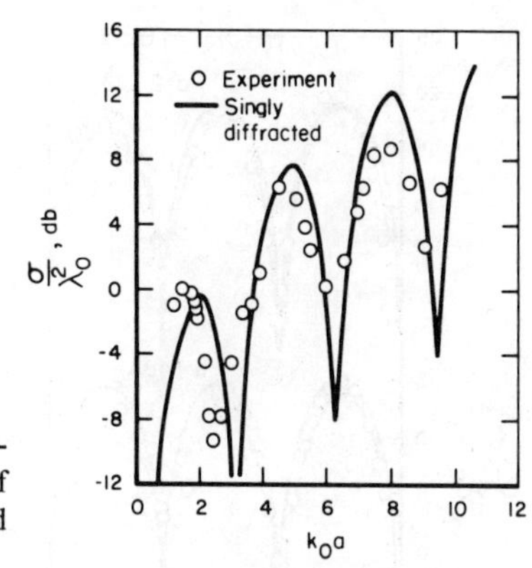

Fig. 8-16. Comparison of experimental and theoretical values of the backscatter cross section of metal hemispheres versus k_0a [from Blore and Musal[19]].

of a perfectly conducting hemisphere versus k_0a with a the radius of the hemisphere. The field is incident upon the spherical portion of the hemisphere. The figure gives experimental results, along with a theoretical estimate obtained by adding a singly diffracted contribution from the rear edge to the physical-optics result for a sphere. The agreement is seen to be only fair. The authors of Reference 19 indicate that including doubly diffracted terms in the analysis did not improve the agreement. One possible reason for the poor agreement was the authors' use of the complete physical-optics result for the sphere, including the terms introduced by the assumed current discontinuity at the shadow boundary. Thus, in adding the sphere physical-optics result and the edge contribution from the geometrical theory of diffraction, the edge is being accounted for twice. It is important to remember that when physical optics is being used to obtain an estimate for the specular contribution to the cross section, the terms arising from the shadow boundary are spurious and should be discarded. In fact, for moderate values of k_0a, geometrical optics (Luneberg–Kline series) should be used instead of physical optics, since physical optics will correctly generate only the first term of an asymptotic series in inverse powers of k_0.

Figure 8-17 (from Reference 20) gives the backscatter cross section of perfectly conducting spherical shell segments versus the depth of the segment for incidence axially on the spherical face. Included in this figure are the physical-optics result and an analytical estimate obtained by Raybin, using a version of the Sommerfeld–Macdonald technique. Raybin's expression for the backscatter cross section is

$$\sigma \approx 4\pi a^2 \sin^2 k_0a(1 - \cos \gamma) \tag{8.1-133}$$

where a is the radius of the sphere, and γ is the angle at the center of the sphere subtended by the segment. It is apparent from Figure 8-17 that this agrees well with the measured data.

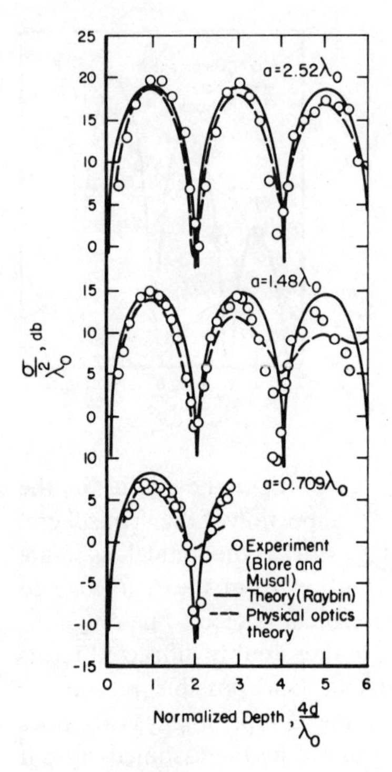

Fig. 8-17. Comparison of experimental and theoretically predicted backscatter cross sections of conducting spherical shell segments versus the shell depth in wavelengths [from Raybin[20]].

8.1.2. Arbitrarily Shaped Finite Bodies

8.1.2.1. *Analytical Techniques*

Analytically estimating the radar cross section of an arbitrarily shaped finite body is a difficult task; often only average values or maximum bounds can be obtained. For a complex target, such as an aircraft or some satellites, there are for some aspect angles, a large number of widely separated contributors to the high-frequency cross section. As a result, the high-frequency backscatter cross section generally scintillates very rapidly as the body is viewed over a range of aspect angles. This is illustrated, for example, in Figure 2-4, where the X-band cross section of the Telstar satellite is shown as a function of aspect angle. For some complex bodies, the only clearly recognizable features of the high-frequency cross section are a few large specular returns. For others there may be certain aspect angles where the techniques discussed in the previous sections can be used, while for the remaining aspects these techniques must be modified.

For any body, whether relatively simple or complex, there are several steps that must be carried out in obtaining an analytical estimate of the cross section. The first step is to resolve the shape into a number of simple geometric segments or components. The scattered-field contribution from each of these geometric components is in turn attributed to the combination of various analytic components. At this point, decisions must be made as to the relative magnitudes of the contributions from the various analytic and geometric components, and only the major contributors retained. Having decided upon the major contributors, one has established a model for the body. This model will in general differ for different aspects and frequencies.

The above steps must be carried out whether the body is simple or complex, and at this point there is no difference in the approach used. Once models have been established for the particular aspects and frequencies of interest, the cross section can be computed by combining the contributions from the various analytic components. In the previous section, where simple bodies were discussed, these contributions were combined by taking into account the relative phases between the analytic components. This results in an estimate of the actual cross section variation as a function of aspect angle and frequency.

For a complex body, at some aspects there may be a large number of components of the same relative magnitude which contribute to the cross section. In this event, small errors in the estimate of the relative phase between these contributors can lead to large errors in the estimated cross section patterns. If this occurs, an average cross section can be obtained by assuming the relative phases between the contributors to be randomly distributed, and directly adding the cross sections produced by the various analytic components. Thus,

$$\langle\sigma(\theta, \phi; \theta', \phi')\rangle \approx \sum_{n=1}^{N} \sigma_n(\theta, \phi; \theta', \phi') \tag{8.1-134}$$

An estimate of the deviation from the average cross section is given by

$$S^2 = \left[\sum_{n=1}^{N} \sigma_n\right]^2 - \sum_{n=1}^{N} \sigma_n^2 \tag{8.1-135}$$

and the probable value of the cross section then lies within a range of values given by[†]

$$\langle\sigma\rangle \pm S \tag{8.1-136}$$

The maximum possible value of σ is given by

$$\left|\sum_{n=1}^{N} \sqrt{\sigma_n}\right|^2 \tag{8.1-137}$$

[†] If the relative phases are random with a uniform probability density, then the scattered field would have a Rayleigh distribution asymptotically, and the standard deviation S would equal $\langle\sigma\rangle$. Thus the cross section could be expected to fall in the range $0 \leqslant \sigma \leqslant 2\langle\sigma\rangle$ in the limit of very large N.

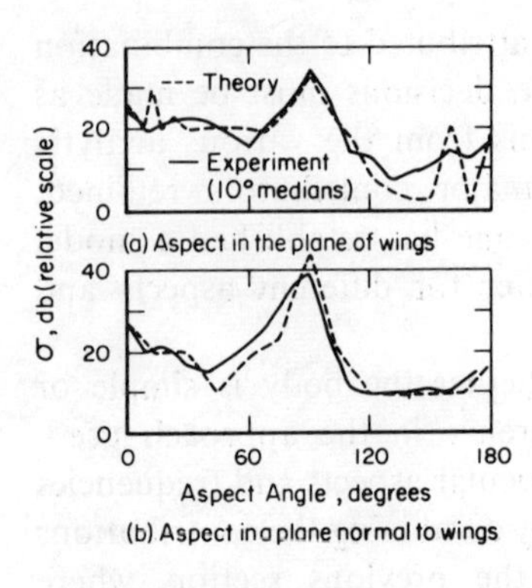

Fig. 8-18. Comparison of 10° average measured and calculated backscatter cross sections for a large aircraft versus the aspect angle in planes through and normal to the wings [from Crispin and Maffett[22]].

The approach outlined above was first applied by researchers at the University of Michigan.[21] Some results of this approach, obtained in the period 1955-1958, are shown in Figures 8-18 and 8-19. Figure 8-18 (from Reference 22) compares measured and theoretical average high-frequency radar cross sections for a large manned aircraft for aspects in, and normal to, the plane of the wings. The experimental results shown in this figure are based on median values over 10° aspect intervals, and the experimental errors were relatively large for the lower cross section values. Figure 8-19 (from Reference 22) also compares the theoretical average high-frequency radar cross section of a manned aircraft with experimental results. In this figure the experimental results are given for both sides of the aircraft.

These figures prove that the technique described above can provide high-frequency, average, cross-section values for complex bodies which agree well with experiment. In fact, analyses performed today on the same aircraft used to obtain the results of Figures 8-18 and 8-19 should yield even better agreement, since improved models can be devised and better estimates of the contributions of the various analytic components are now available.

In the Rayleigh region, the same techniques used for simple shapes can be applied to estimate the cross section of a complex body. On the

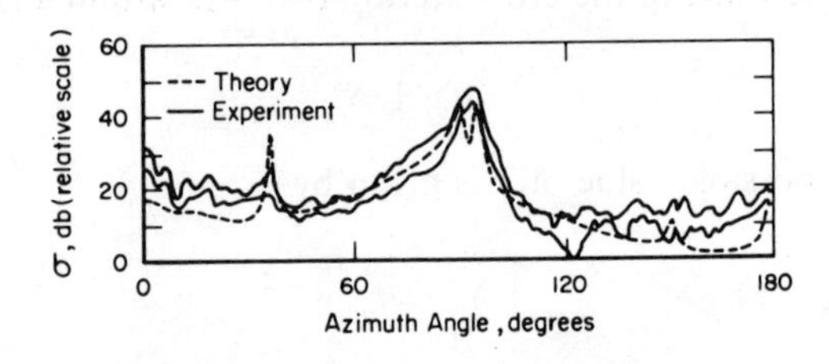

Fig. 8-19. Comparison of average measured and calculated backscatter cross sections for a large aircraft versus aspect angle in a plane through the wings [from Crispin and Maffett[22]].

other hand, for the resonance region, no generally reliable analytical techniques are available for complex shapes, and experimental results must be utilized.

8.1.2.2. *Some Experimental Results*

To illustrate the general nature of radar cross section patterns as a function of the aspect angle, some high-frequency experimental results for typical complex bodies are shown in Figures 8-20 through 8-26. Figures 8-20 to 8-21 (from Reference 23) show the radar cross section for both perpendicular and parallel polarizations of the Project Mercury capsule at frequencies of 440, 933.3, 1184, and 2800 Mcps. The shape and dimensions of

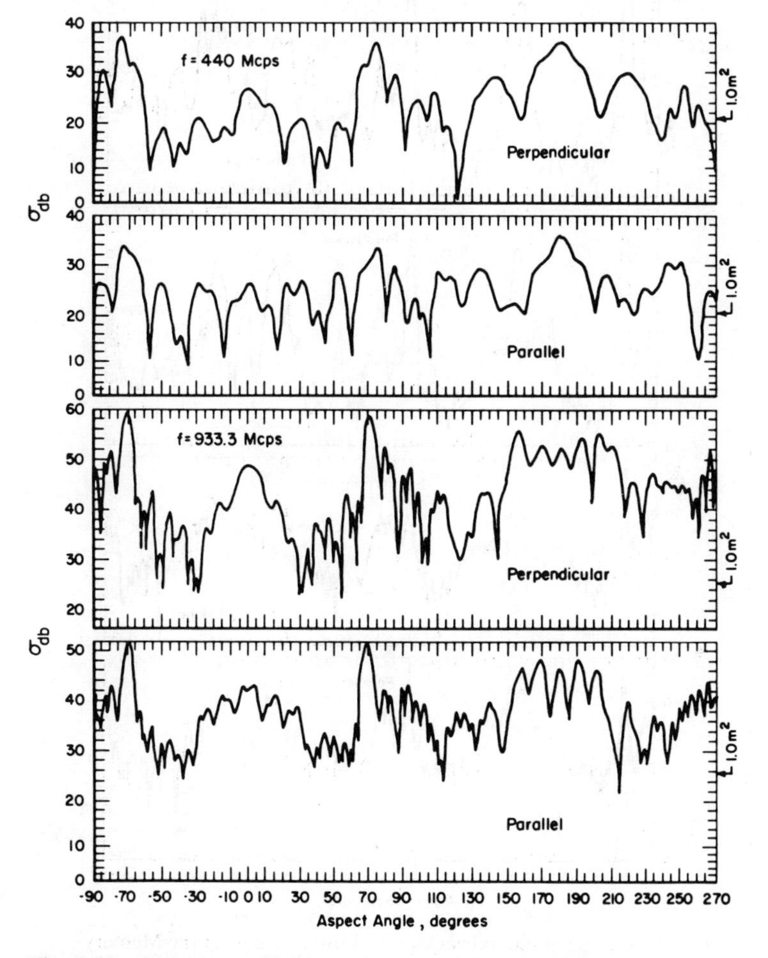

Fig. 8-20. Experimental backscatter cross sections for the Mercury capsule versus the angle of incidence, at frequencies of 440 and 933.3 Mcps.

the capsule are illustrated in Figure 8-22. Zero-degree aspect angle in Figures 8-20 and 8-21 corresponds to incidence on the small diameter end. As one examines these figures, the only obviously recognizable features are the specular flares from the capsule sides at $\pm 70°$ and $\pm 90°$, and the specular flare from the disc-like nose at $0°$.

Figures 8-23 and 8-24 (from Reference 24) illustrate the radar cross section of the BQM-34A target drone for both perpendicular and parallel polarizations at 9150 Mcps. The BQM-34A is a small jet aircraft, and zero-degree aspect corresponds to nose-on incidence, with the direction

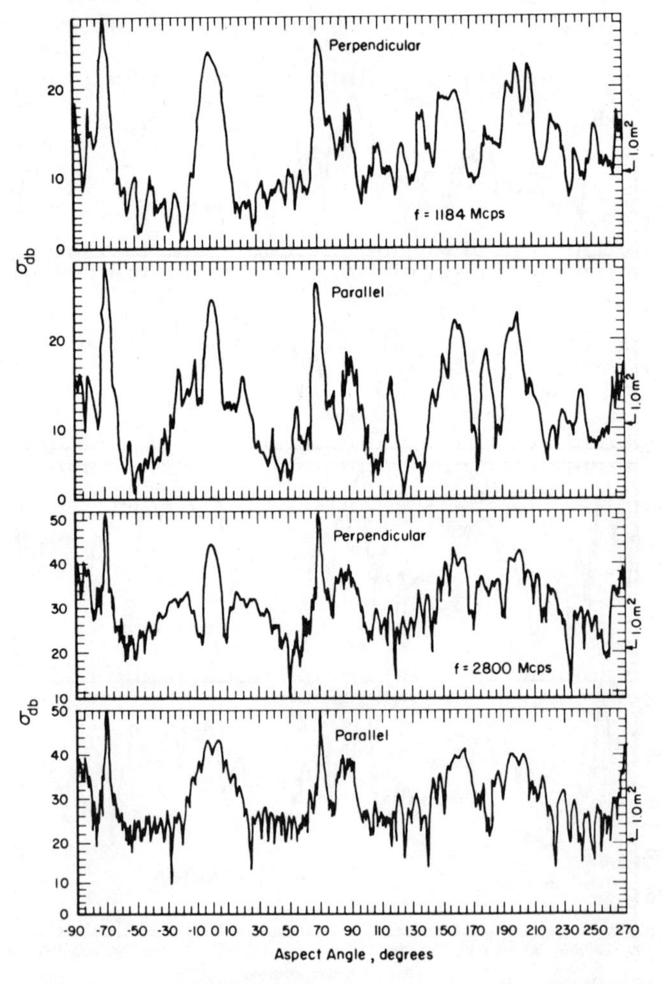

Fig. 8-21. Experimental backscatter cross sections for the Mercury capcule versus the angle of incidence, at frequencies of 1184 and 2800 Mcps.

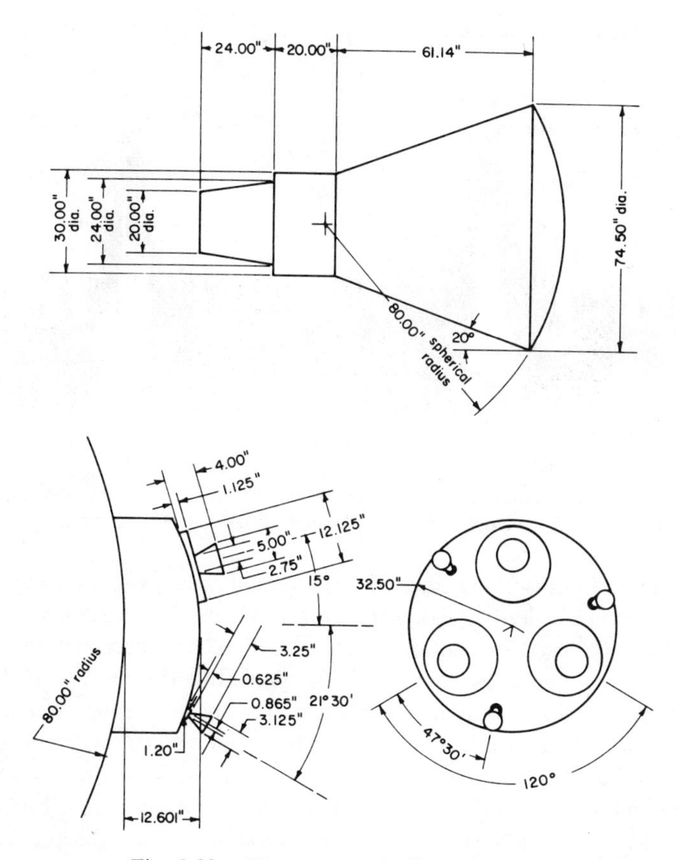

Fig. 8-22. Mercury capsule dimensions.

of incidence being in the plane of the wings. Again, only the specular flare from the sides of the approximately cylindrical body is clearly recognizable.

Figures 8-25 and 8-26 (from Reference 25) show the radar cross section of the Agena vehicle for perpendicular and parallel polarization at 10,000 Mcps. This vehicle is a large, second-stage booster rocket, and again only a few characteristic features are observable in the cross-section patterns. These are the reduction in the average cross section for near nose-on aspects, and the specular flares from the conical and cylindrical sides and the tail.

8.2. RADAR CROSS-SECTION ENHANCEMENT

In certain applications it is desirable to enhance the radar cross section, over specific ranges of aspect angles, above the normal cross section of a body at these angles. This may be the case, for example, when an aircraft

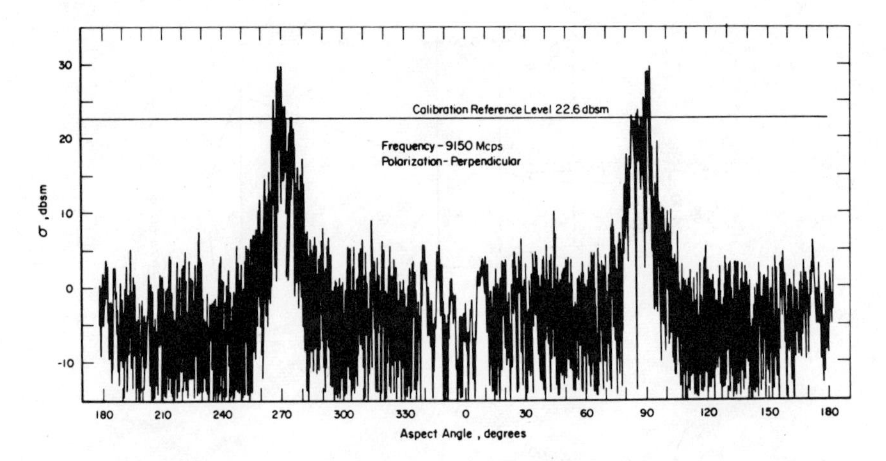

Fig. 8-23. Experimental backscatter cross sections for the BQM-34A target drone versus the angle of incidence, perpendicular polarization at 9150 Mcps.

or missile in flight is being tracked by a ground radar, and a strong, relatively steady return is desired for good tracking accuracy. There are several methods which can be used to obtain a cross-section enhancement at particular aspect angles. The most obvious, and probably the least useful method, is to shape the body so as to present a strong specular echo over the aspects of interest. This method usually cannot be applied since other considerations than the radar cross section normally impose constraints on the body shape. Another method is to disturb the current distribution on the body so as to increase the cross section at the desired aspects by utilizing discrete impedance

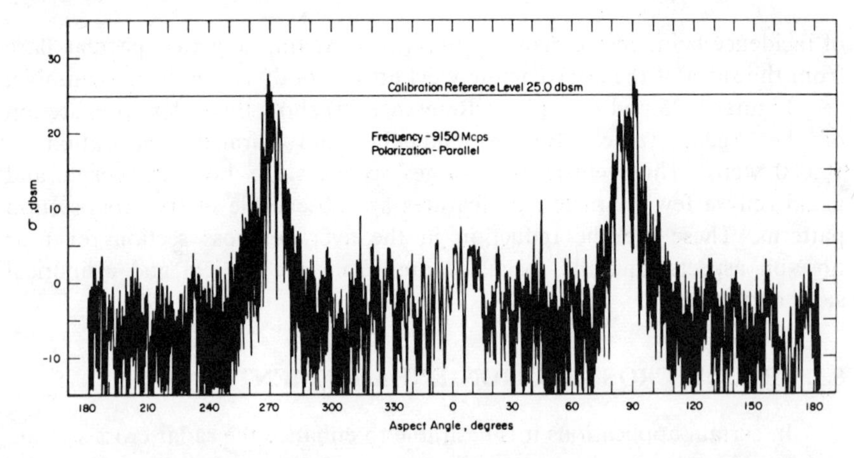

Fig. 8-24. Experimental backscatter cross section for the BQM-34A target drone versus the angle of incidence, parallel polarization at 9150 Mcps.

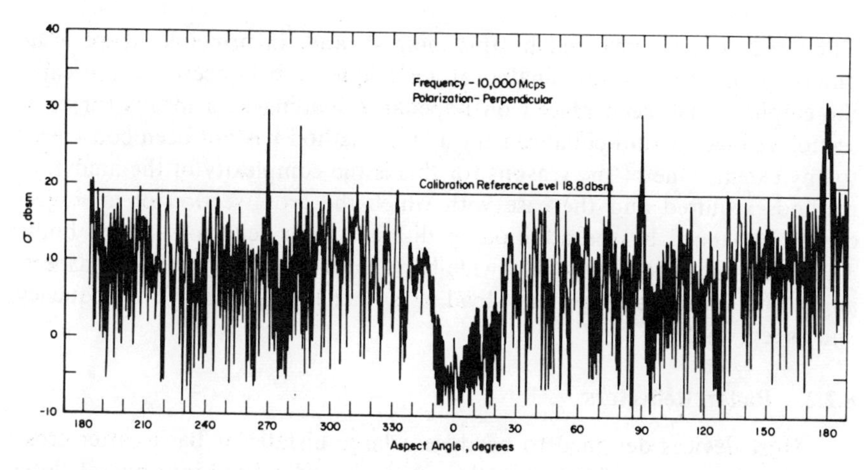

Fig. 8-25. Experimental backscatter cross section for the Agena booster versus the angle of incidence, perpendicular polarization at 10,000 Mcps.

loading at one or more points. A third method is the addition to the body of radar reflectors which have been designed specifically to produce large cross sections over specified angular regions. The last two methods will be discussed in more detail in the following sections.

8.2.1. Impedance Loading

The question of controlling the cross section of a body by adding lumped impedances at various points on the body is discussed in Section 8.3.3. In general, the cross section at a specific aspect angle and frequency can be

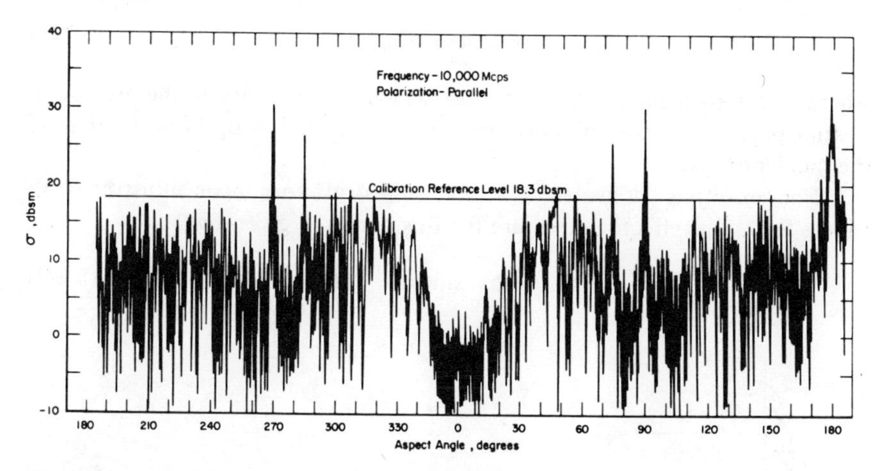

Fig. 8-26. Experimental backscatter cross section for the Agena booster versus the angle of incidence, parallel polarization at 10,000 Mcps.

either reduced or enhanced by this method, and, in principle, there is no difference insofar as the analytical techniques are concerned. Basically, the emphasis has been placed on impedance loading as a means for cross section reduction, and enhancement by this method has not been considered to any extent. One of the reasons for this is the complexity of the analytical methods required and the ease with which the cross section of an object can be enhanced by the addition of discrete radar reflectors to the body. These can, in general, provide wide-band enhancement over large aspect-angle ranges, whereas impedance loading is often highly sensitive to frequency and aspect angle.

8.2.2. Radar Reflectors

Most devices designed to produce a large bistatic or backscatter cross section over a range of aspect angles can be broadly fitted into one of three categories; these are multiple scatterers, dielectric lenses, or retrodirective arrays. Representative devices in each of these categories will be discussed in the following sections.

8.2.2.1. *Multiple Scatterers*

8.2.2.1.1. *Corner Reflectors.* One of the best known radar reflectors is the corner reflector. A corner reflector can be designed to provide a large monostatic or bistatic cross section over a wide range of aspect angles and frequencies. There are a number of different types of corner reflectors; however, the high-frequency cross section of these devices can be estimated using geometrical optics and taking into account multiple scattering. In general, the backscatter cross section of a corner reflector is given by

$$\sigma \approx 4\pi A_e^2/\lambda_0^2 \qquad (8.2\text{-}1)$$

where A_e equals the area of the aperture which has the shape of the projection of that part of the corner reflecting rays in a direction opposite to that of the incident rays.

For example, consider a simple dihedral corner as illustrated in Figure 8-27. The effective aperture for this corner is given by[26]

$$A = 2ab\sin(\pi/4 + \phi) \qquad (8.2\text{-}2)$$

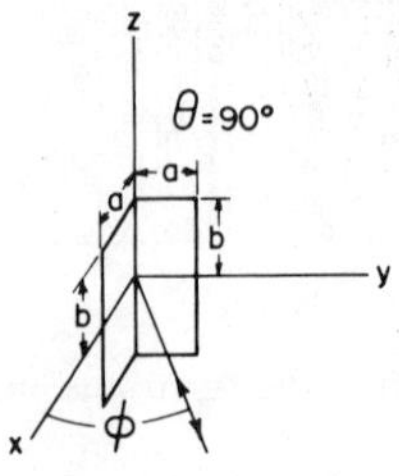

Fig. 8-27. Coordinates for dihedral corner reflector.

and the backscatter cross section is

$$\sigma \approx \frac{16\pi a^2 b^2 \sin^2(\pi/4 + \phi)}{\lambda_0^2} \tag{8.2-3}$$

The total aspect-angle range over which the cross section is within 3 db of the maximum value is 30° for the dihedral corner. It should be noted that for polarization either in the plane of incidence (no z component in terms of the coordinates of Figure 8-27) or perpendicular to the plane of incidence (only a z component), the scattered field is not depolarized. For any other polarization, however, the scattered field will be depolarized to some extent.

The dihedral corner can be used as a bistatic reflector for small bistatic angles. Peters[27] has presented a general theory for the analysis of corner reflectors as bistatic enhancement devices. He assumes that the bistatic cross section of the corner reflector has the form

$$\sigma(\beta) = \sigma(0)\, e^{-2\gamma\beta} \tag{8.2-4}$$

where β is the bistatic angle between the transmitter and receiver directions, $\sigma(0)$ is the monostatic cross section, and γ is an arbitrary design parameter. The design procedure is to choose γ so as to approximate the desired bistatic pattern, and then determine the parameter Φ/a' from Figure 8-28. The angle Φ is the relative phase in radians at the edge of the aperture, and a' is the effective width of the aperture in the plane of bistatic coverage. Knowing the desired monostatic cross section, $\sigma(0)$, the required effective aperture width a' can be determined from Eq. (8.2-1), thus Φ can be determined and related to the necessary error in the angle between the walls to give the desired bistatic coverage.

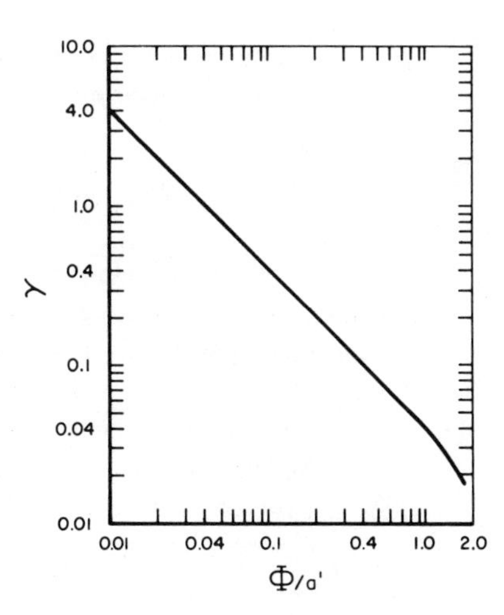

Fig. 8-28. Peters' design parameter γ versus Φ/a' [after Peters[27]].

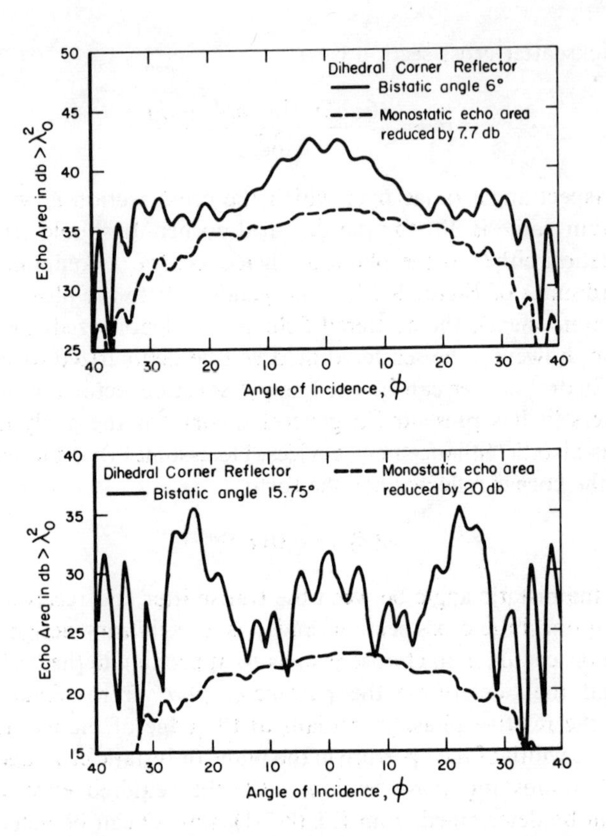

Fig. 8-29. Experimental bistatic cross sections for a dihedral corner reflector versus the angle of incidence at bistatic angles of 6 and 15.75° [from Peters[27]].

For the dihedral corner, the wall angle should be $90° + \epsilon$, where for ϵ small, it is given approximately by

$$\epsilon \approx \frac{\lambda_0 \Phi}{2\sqrt{2}\pi a} \tag{8.2-5}$$

Figure 8-29 (from Reference 27) shows experimental scattering patterns versus the angle of incidence for two different bistatic angles, 6 and 15.75°. The monostatic pattern is also shown in this figure. The corner used in this figure had an angle of 92.6° between the walls, and $a = b = 7.79\lambda_0$. The value of γ for this corner was 0.146.

The dihedral corner reflector has the obvious disadvantage that it provides a large cross section only in the plane perpendicular to the corner walls. This can be alleviated to some extent by the use of a trihedral corner as illustrated in Figure 8-30. The maximum radar cross section, average

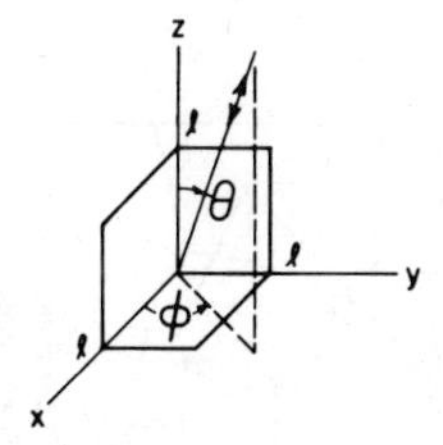

(a) Square Trihedral Corner Reflector

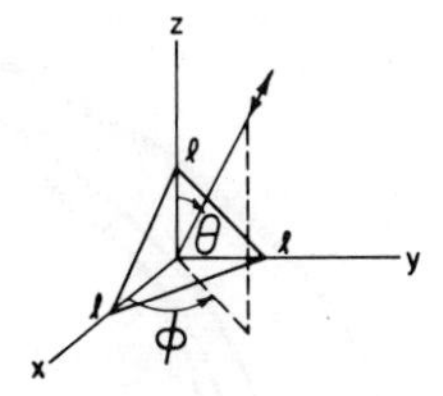

(b) Triangular Trihedral Corner Reflector

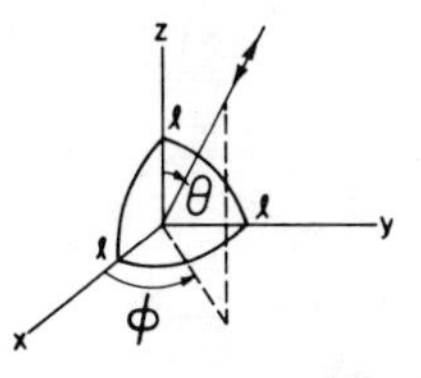

(c) Circular Trihedral Corner Reflector

Fig. 8-30. Trihedral corner-reflector coordinates and dimensions.

cross section over all aspects, and the angular coverage between points at which the backscatter cross section falls by 3 db are shown in Table 8-3 for several types of trihedral corner reflectors.[28]

Table 8-3

Comparison of Trihedral Corner Reflectors

Trihedral Corner Reflector Type	Maximum Radar Cross Section	Average Radar Cross Section	Angular Coverage
Square	$\sigma = \dfrac{12\pi l^4}{\lambda_0^2}$	$\langle\sigma\rangle = \dfrac{0.7 l^4}{\lambda_0^2}$	23° cone about symmetry axis
Triangular	$\sigma = \dfrac{4\pi l^4}{\lambda_0^2}$	$\langle\sigma\rangle = \dfrac{0.17 l^4}{\lambda_0^2}$	40° cone about symmetry axis
Circular	$\sigma = \dfrac{15.6 l^4}{\lambda_0^2}$	$\langle\sigma\rangle = \dfrac{0.47 l^4}{\lambda_0^2}$	32° cone about symmetry axis

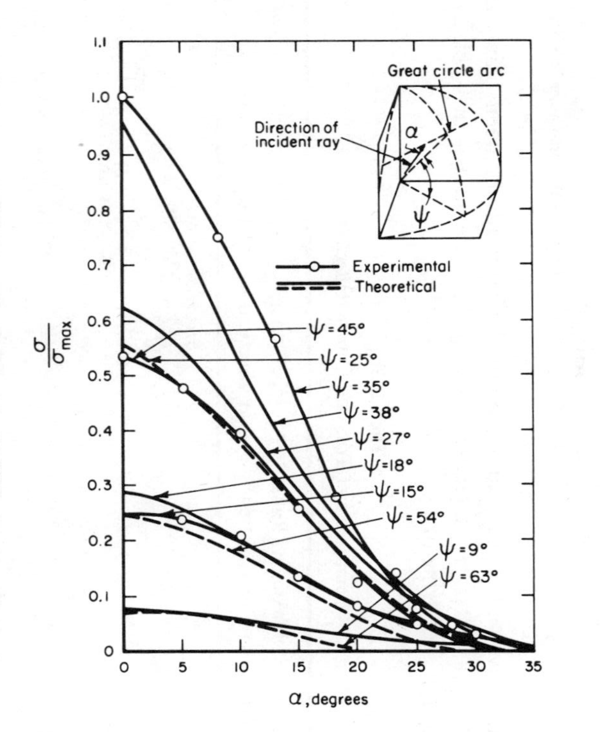

Fig. 8-31. Relative theoretical and measured backscatter cross sections for a square trihedral corner reflector versus the angle α for several angles ψ.

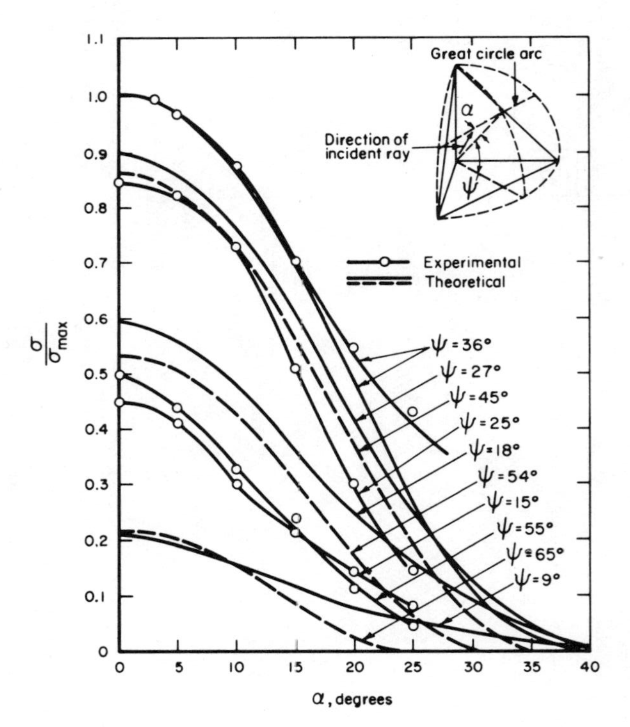

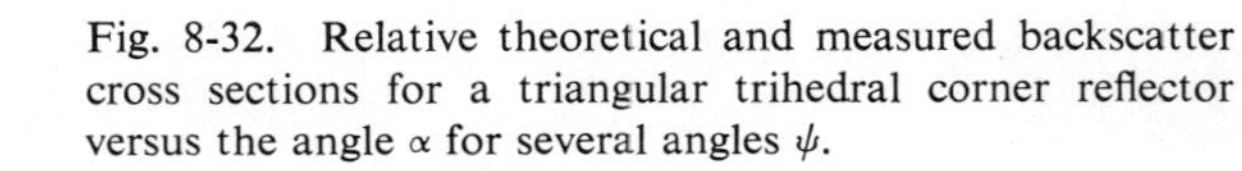

Fig. 8-32. Relative theoretical and measured backscatter cross sections for a triangular trihedral corner reflector versus the angle α for several angles ψ.

Figures 8-31 and 8-32 compare theoretical and measured backscatter cross sections versus the angles of incidence, α, ψ, for square and triangular trihedral corner reflectors, respectively.[29] These angles are illustrated in the figures, and are defined as $\psi = 90° - \theta$ and $\alpha \approx (\phi - 45°) \sin \theta$.

For a square corner reflector, the radar cross section as a function of the spherical angles θ and ϕ, for θ, ϕ near the symmetry axis (or $\theta \approx 54.74°$, $\phi \approx 45°$), is given by

$$\sigma(\theta, \phi) \approx \frac{4\pi}{\lambda_0^2} \cos^2 \theta (4 - \cot \phi)^2 l^4 \tag{8.2-6}$$

For a triangular corner reflector, the radar cross section as a function of the spherical angles θ and ϕ, for angles near the symmetry axis, is

$$\sigma(\theta, \phi) \approx \frac{4\pi}{\lambda_0^2} l^4[\cos \theta + \sin \theta(\sin \phi + \cos \phi) - 2[\cos \theta + \sin \theta(\sin \phi + \cos \phi)]^{-1}]^2 \tag{8.2-7}$$

Figure 8-33 shows experimental backscatter cross sections for a triangular trihedral corner reflector versus the angles of incidence α, δ ($\delta = \theta - 54.74°$).[26] The peaks at the edges of the patterns are due to the corner scattering in a dihedral mode when the incident field direction is parallel to one of the planes comprising the corner. By modifying the corner shape, the pattern can be made flatter as a function of α. Figure 8-34

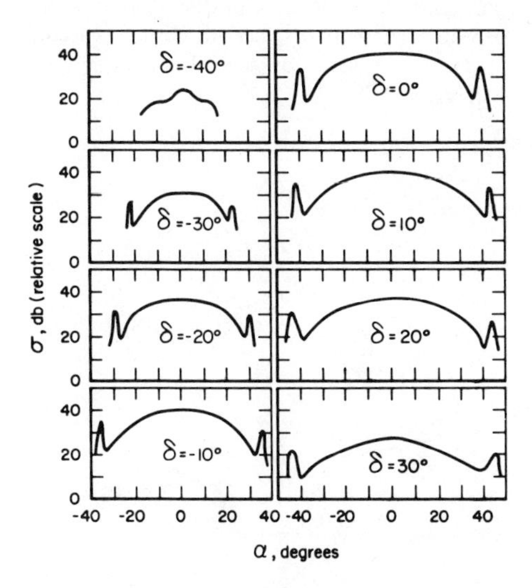

Fig. 8-33. Experimental relative backscatter cross sections of a triangular trihedral corner reflector versus the angle α for several angles δ [after Robertson[26]].

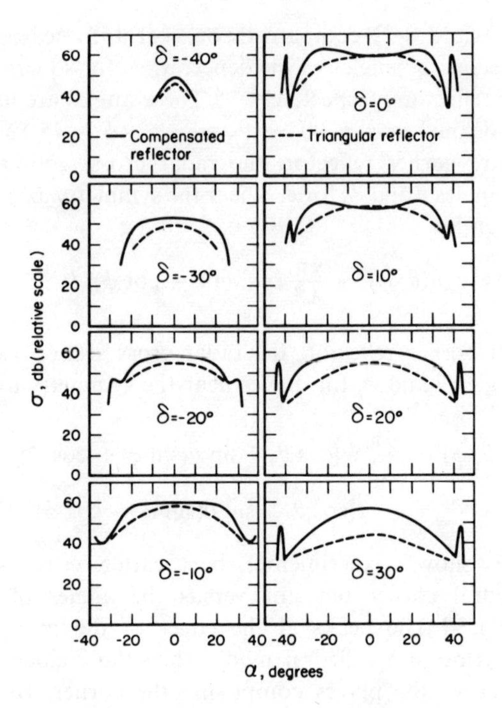

Fig. 8-34. Comparison of experimental relative backscatter cross sections of triangular and compensated triangular trihedral corner reflectors versus the angle α for several angles δ [after Robertson[26]].

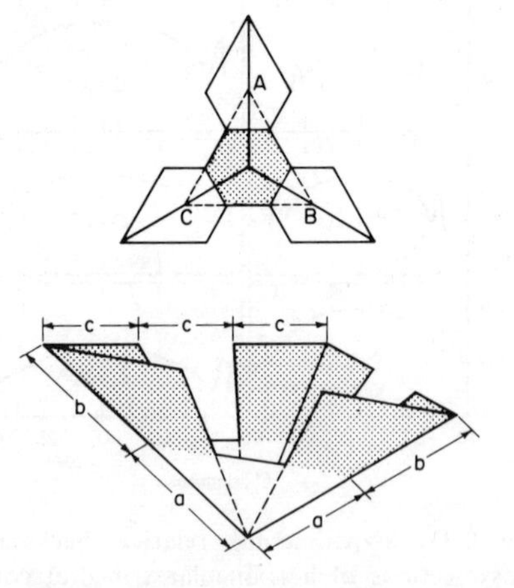

Fig. 8-35. Compensated triangular trihedral corner reflector.

compares the backscatter cross section of the compensated triangular trihedral, shown in Figure 8-35, with that of an uncompensated triangular corner reflector. It is apparent that the compensated reflector can have a relatively flat response over 40 to 60° angular range.

Peters' technique for obtaining a desired bistatic response of the form of Eq. (8.2-4) can also be applied to trihedral corner reflectors, as discussed in References 27 and 28. Figure 8-36 (from Reference 27) shows the response

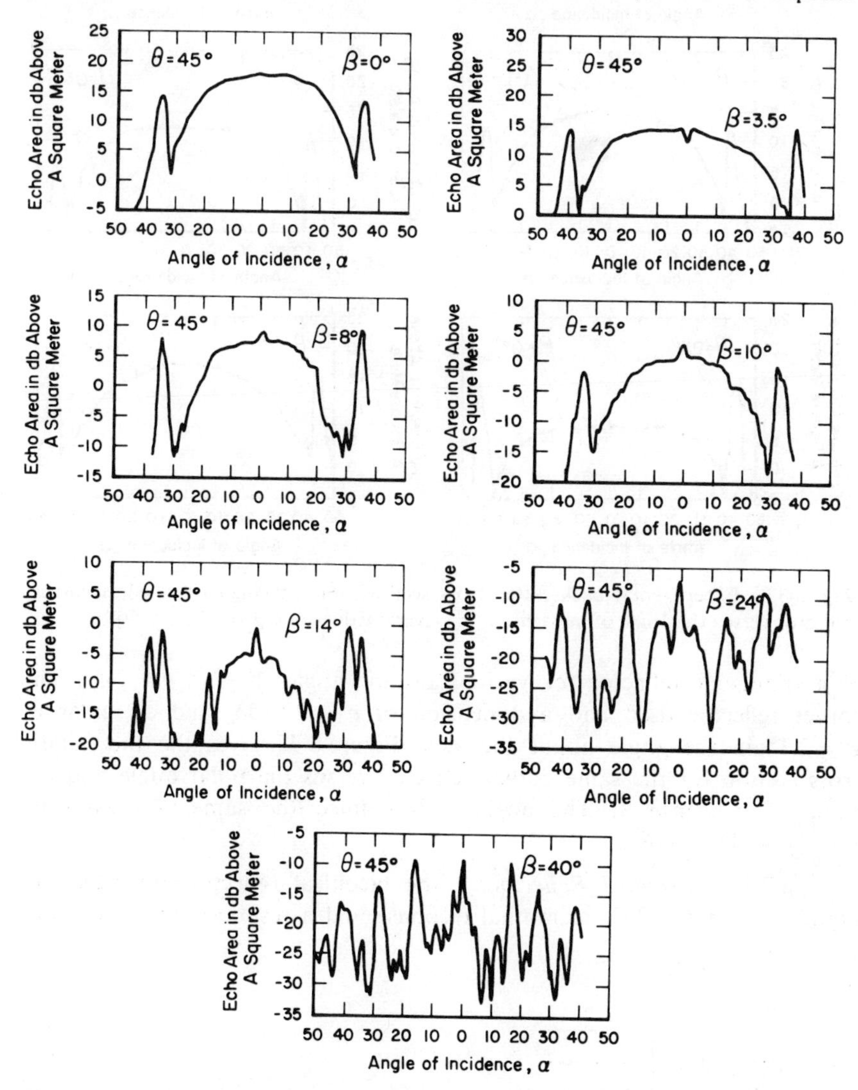

Fig. 8-36. Experimental bistatic cross sections for a triangular trihedral corner reflector versus the angle of incidence α for several bistatic angles, and $\theta = 45°$ [from Peters[27]].

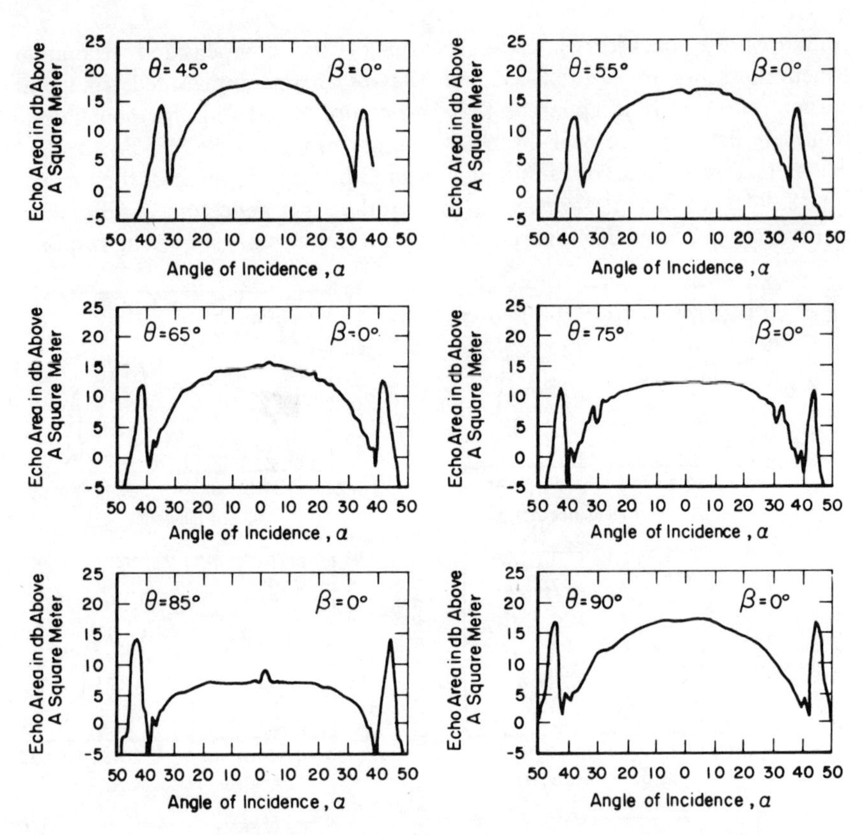

Fig. 8-37. Experimental backscatter cross sections for a triangular trihedral corner reflector versus the angle of incidence α for several values of θ [from Peters[27]].

of a triangular reflector for various bistatic angles β, and $\theta = 45°$. The corner reflector used had a dimension of $a = 12.15\lambda_0$ and an error of $\epsilon = 2.1°$ in the angle about the axis. Figure 8-37 gives the monostatic cross section for the same corner reflector versus the polar angle and the angle of incidence α. The angle α is defined the same as previously, $\alpha = (\phi - 45°) \sin \theta$.

8.2.2.1.2. *Biconical Reflectors.* The biconical reflector, illustrated in Figure 8-38, has a 360° azimuthal response and a response like a dihedral

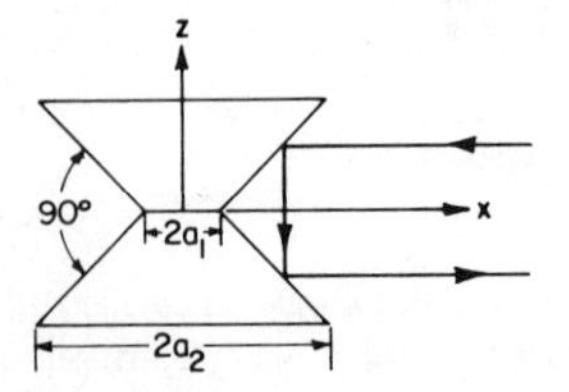

Fig. 8-38. Biconical reflector dimensions.

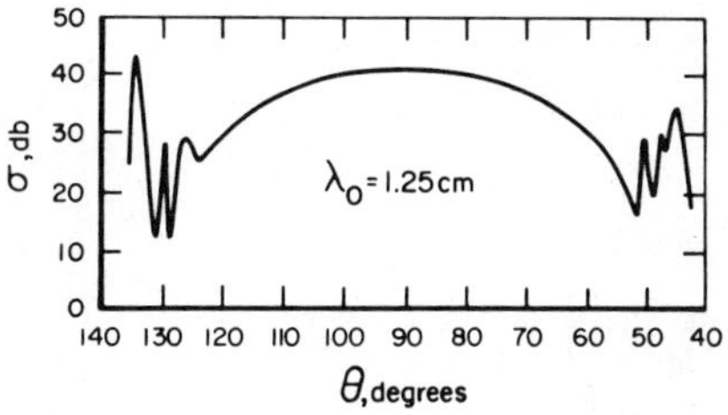

Fig. 8-39. Experimental backscatter cross section for a biconical reflector versus the polar angle θ, σ for $\theta = 90°$ is 0.015 square meters [after Robertson[26]].

corner in θ. The maximum backscatter cross section for $\theta = 0$ is

$$\sigma_{\max} = \frac{32\pi}{9\lambda_0}\left[a_2 - \sqrt{2a_2 - a_1} - a_1^{3/2}\right]^2 \tag{8.2-8}$$

Figure 8-39 illustrates the backscatter cross section of a biconical reflector as a function of the angle of incidence θ.[26] The dimensions of the reflector used were $a_1 = 1.5$ inches, $a_2 = 8.5$ inches, and the wave-length was $\lambda_0 = 1.25$ cm.

8.2.2.2. *Dielectric Lenses*

Monostatic and bistatic enhancement over relatively wide frequency and angular ranges can be obtained with dielectric lenses such as the Luneberg lens, the cylindrical Eaton–Lippman lens, the spherical lenses of Kay,[30] etc. The spherical Luneberg lens is discussed in Section 3.4.3.3, and the cylindrical Eaton–Lipman lens in Section 4.1.8.3.

The cylindrical Eaton–Lippman lens has a normal-incidence high-frequency backscatter cross section of

$$\sigma(0) \approx \frac{4k_0^2a^2l^2}{\pi} \tag{8.2-9}$$

with a the lens radius and l its length. This particular lens type has the advantage that no reflective cap is required and that 360° azimuthal coverage is provided. Among its disadvantages, however, are poor elevation angle coverage and the requirement for an infinite dielectric constant at the lens center which is, of course, not realizable physically.

The spherical Luneberg lens can provide wide-angle coverage in both azimuth and elevation by employing a reflective cap covering most of a hemisphere of the lens. The high-frequency backscatter cross section for the Luneberg lens is given by

$$\sigma = \pi k_0^2a^4 \tag{8.2-10}$$

with a the lens radius. Figure 8-40 shows the measured backscatter cross sections of several commercially available[31] Luneberg lenses as a function

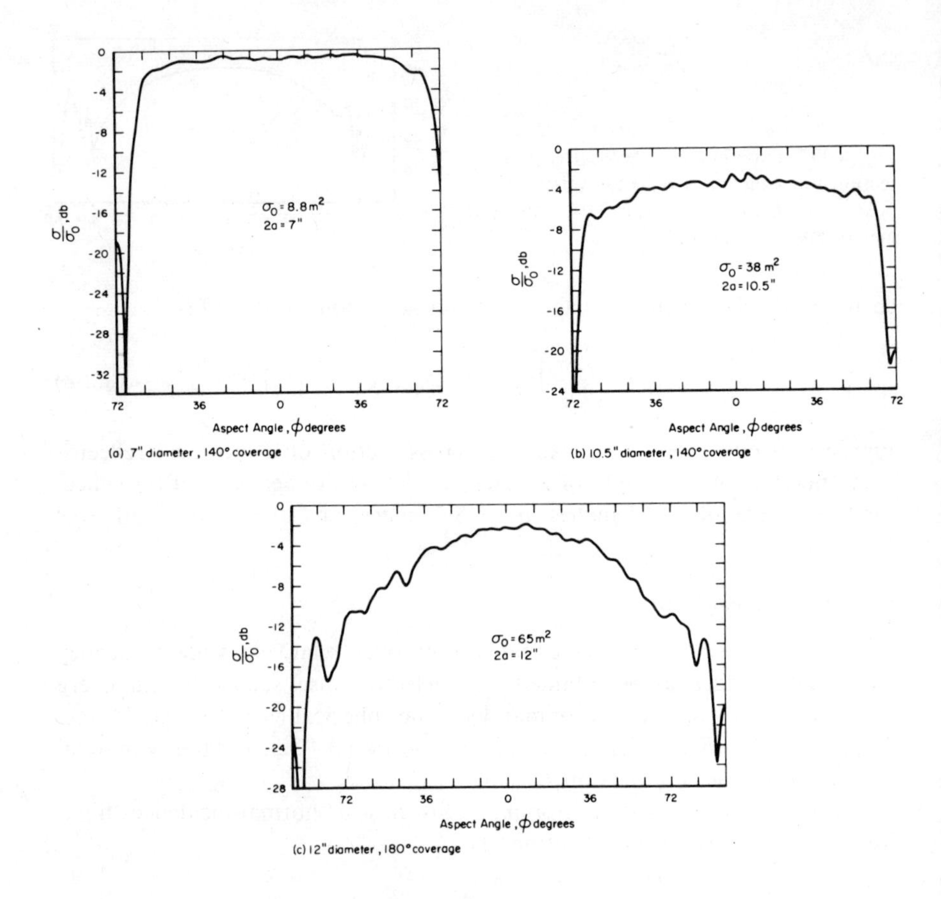

Fig. 8-40. Comparison of backscatter cross sections versus aspect angle for several commercially available Luneberg lenses at 9375 Mcps.

of the angle of incidence. These measurements were made at a frequency of 9375 Mcps. One major disadvantage of the Luneberg lens in the past has been its weight; however, there has been steady improvement in this regard, and the lenses used to obtain the results illustrated in Figure 8-40 weighed 0.6, 1.9, and 2.3 lb, respectively, for the 7-, 10.5-, and 12-inch lenses.

It is interesting to compare the maximum high-frequency backscatter cross sections of various corner reflectors, lenses, and simple perfectly-conducting bodies; this is done in Table 8-4.

Figure 8-41 compares the angular coverage provided by three different Luneberg lenses with a cluster of four rectangular trihedral corner reflectors.[33] It is obvious from this figure that the Luneberg lens provides substantially better angular coverage.

Table 8-4

Maximum High-Frequency Backscatter Cross Section of Various Radar Reflectors

Reflector Type	σ_{max}	σ_{max} (Experimental at 9400 Mcps)† (m²)	Characteristic Dimensions (cm)
Luneberg lens	$\frac{4\pi^3 a^4}{\lambda_0^2}$	31.7	$a = 12.7$
Circular trihedral corner reflector	$\frac{15.6 l^4}{\lambda_0^2}$	7.6	$l = 14.7$
Triangular trihedral corner reflector	$\frac{4\pi l^4}{3\lambda_0^2}$	2.41	$l = 15.5$
Square trihedral corner reflector	$\frac{12\pi l^4}{\lambda_0^2}$	17.1	$l = 14.7$
Cylindrical Eaton–Lippman lens	$\frac{16\pi a^2 l^2}{\lambda_0^2}$	—	—
Dihedral corner reflector	$\frac{8\pi a^2 b^2}{\lambda_0^2}$	12.85	$a = 12.7$ $b = 9$
Circular flat metal plate	$\frac{4\pi^3 a^4}{\lambda_0^2}$	31.7	$a = 12.7$
Metal sphere	πa^2	—	—
Finite metal cylinder	$\frac{2\pi a l^2}{\lambda_0}$	0.62	$a = 7.35$ $l = 23.5$

† Experimental values obtained from Reference 32.

8.2.2.3. *Retrodirective Arrays*

The retrodirective array was first conceived by Van Atta,[34] and in its simplest form (as illustrated in Figure 8-42), the antenna elements are symmetrical pairs connected by transmission lines of equal length. This configuration is suitable for a planar array. In general, the requirement for a retrodirective array is that the phase of the scattered field at each element of the array be the complex conjugate of the incident field phase at that element. The total scattered field will then sum in the far field into a wave propagating toward the incident-field source. This process may be carried out passively as for the Van Atta array, or it may be accomplished by active devices.

The experimental backscatter cross sections obtained from a two-dimensional 4 × 4 Van Atta array of half-wave dipoles (Figure 8-43) are shown in Figures 8-44 and 8-45 (Reference 35). Figure 8-44 shows the cross section as a function of θ in the $\phi = 0$ plane, and Fig. 8-45 shows the cross section in the $\phi = \pi/2$ plane. The cross section was measured at a frequency of 2850 Mcps, and the dipole elements of the array had a separation

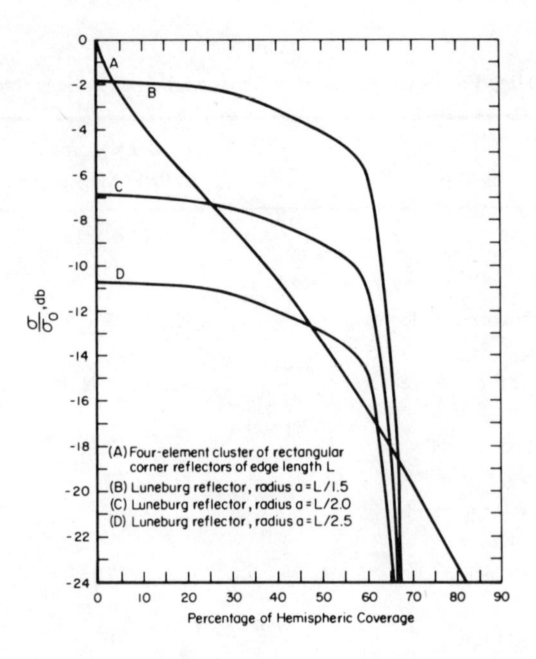

Fig. 8-41. Comparison of the angular performance of a corner-reflector cluster with several Luneberg lenses [after Holt[33]].

of $0.61\lambda_0$ between elements and were spaced a quarter wave above a ground plane. The point σ_0 in these figures is the measured normal-incidence cross section of a conducting flat plate of the same size as the array. A substantially broader scattering pattern was exhibited by the array than by the plate.

In general for an array of this type, the high-frequency backscatter cross section is given by

$$\sigma(\theta, \phi) \approx N^2 \frac{\Gamma^2 G^2(\theta, \phi)\, \lambda_0^2}{\pi} \tag{8.2-11}$$

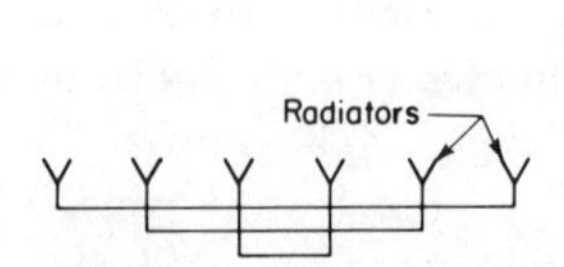

Fig. 8-42. Van Atta array.

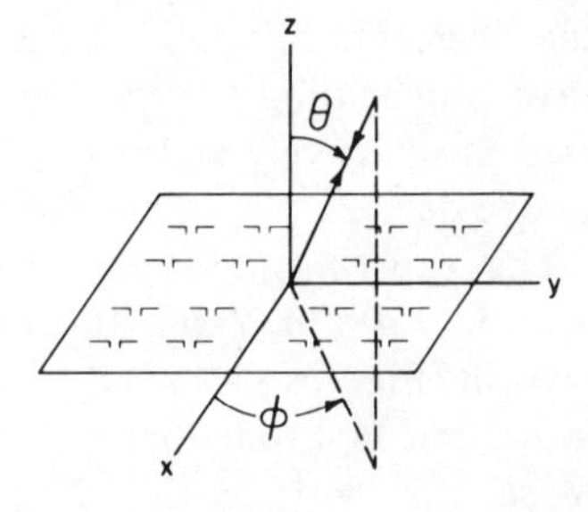

Fig. 8-43. Two-dimensional Van Atta array.

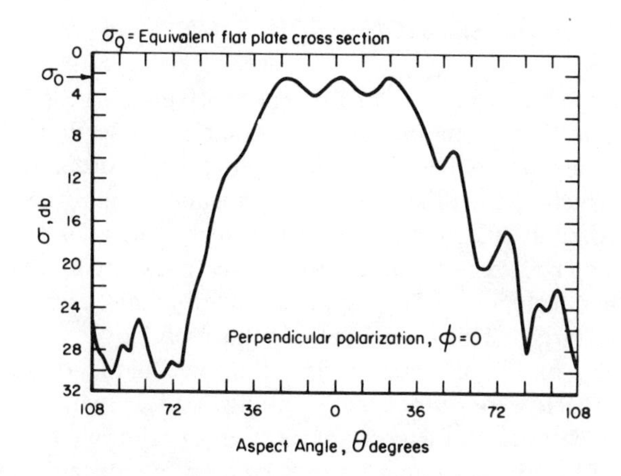

Fig. 8-44. Backscatter cross section in the $\phi = 0$ plane of a 4×4 Van Atta array versus the angle of incidence θ at 2850 Mcps [after Sharp[35]].

where G is the gain pattern of the array, Γ is the voltage-reflection coefficient at the terminals of an array element, and N is the number of elements in the array. This equation assumes identical elements and neglects the effects of mutual coupling between the elements.

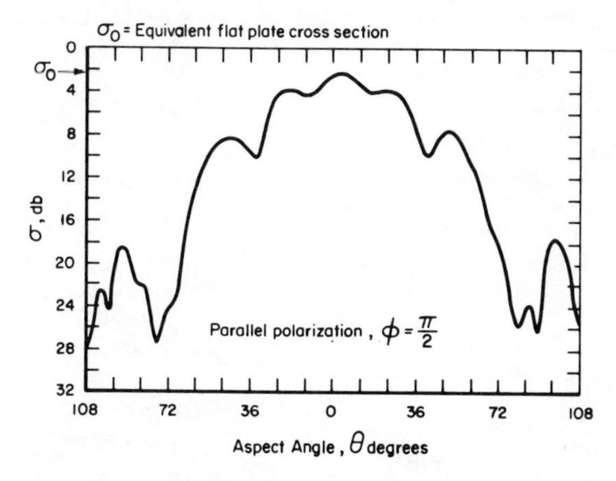

Fig. 8-45. Backscatter cross section in the $\phi = \pi/2$ plane of a 4×4 Van Atta array versus the angle of incidence θ at 2850 Mcps [after Sharp[35]].

8.3. RADAR CROSS SECTION REDUCTION

A problem often occurring in the design of military aircraft, missiles, satellites, etc., is that of minimizing the radar cross section of the body. There are several approaches that one might take in order to accomplish this, any of which, under the proper circumstances, may be of use. The first of these to be discussed is shaping. This technique is useful in reducing the backscatter cross section of a target over a selected range of aspect angles. Another technique is the application of radar-absorbing materials (RAM) to the target. This technique can be useful over a selected range of aspect angles or for all aspects. The third technique to be considered is known as reactive or impedance loading, and in general, is effective only over selected ranges of aspect angles. For an arbitrary body, all or any combination of the above techniques may be used to reduce the cross section at selected aspects and frequencies.

8.3.1. Body Shaping

8.3.1.1. *High-Frequency and Resonance Region*

In considering what shape a specific perfectly conducting body should take to minimize the backscatter cross section at selected aspect angles, it is of interest to examine the dependence of the various analytic components of the cross section with frequency. In Table 8-5 the approximate dependence

Table 8-5

Scattering Component	Approximate Wavelength Dependence
Specular	
Flat surface	$\sigma \propto \lambda_0^{-2}$
Single-curved surface	$\sigma \propto \lambda_0^{-1}$
Double-curved surface	$\sigma \propto$ constant
Edge and Tip	
Single edge	$\sigma \propto \lambda_0$
Multiple edge (Diffracted n times)	$\sigma \propto \lambda_0^n$
Tip	$\sigma \propto \lambda_0^2$
Surface Derivative Discontinuities	
2nd derivative	$\sigma \propto \lambda_0^2$
nth derivative	$\sigma \propto \lambda_0^{2(n-1)}$
Creeping Wave	$\sigma \propto \exp[-\gamma\lambda_0^{-2/3}]$
Traveling Wave	$\sigma \propto$ constant

of σ on λ is given for a number of components when the frequency ranges upward from the upper resonance region. In the limit as the frequency goes to infinity, the creeping-wave dependence becomes exponential and hence smaller than tip or derivative discontinuity terms.

From Table 8-5 it is obvious that in order for the cross section to become smaller with increasing frequency, no specular contribution to the scattered field should exist. Thus, the body should appear as a tip or an edge at the aspect at which the cross section is desired to be minimized. Examining the table further shows that if only tip, multiply diffracted edge, second or higher-order surface derivative discontinuity, or creeping wave contributions exist, then each would contribute as λ_0^2 or higher. Thus, the body should, in general, present a tip toward the radar with no edges capable of scattering a singly diffracted ray visible. This requires an ogival or conical shape with the cone or ogive base smoothly matched, and having no less than a second derivative discontinuity at the matching point. In the case of the ogival shape, the rear tip must be eliminated in order to prevent the large cross-section peak near nose-on caused by traveling waves for parallel polarization.

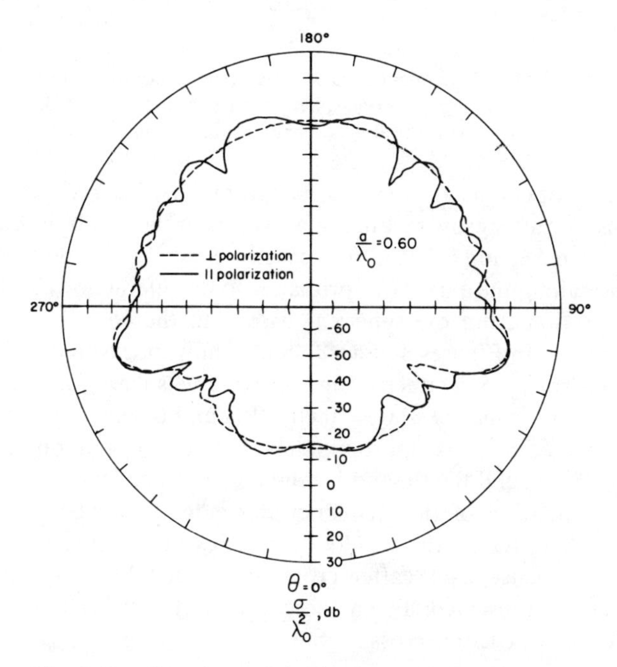

Fig. 8-46. Experimental backscatter cross sections of a perfectly conducting 25° cone-sphere with a base radius of 0.6 wavelengths versus the angle of incidence [after Pannell *et al.*[36]].

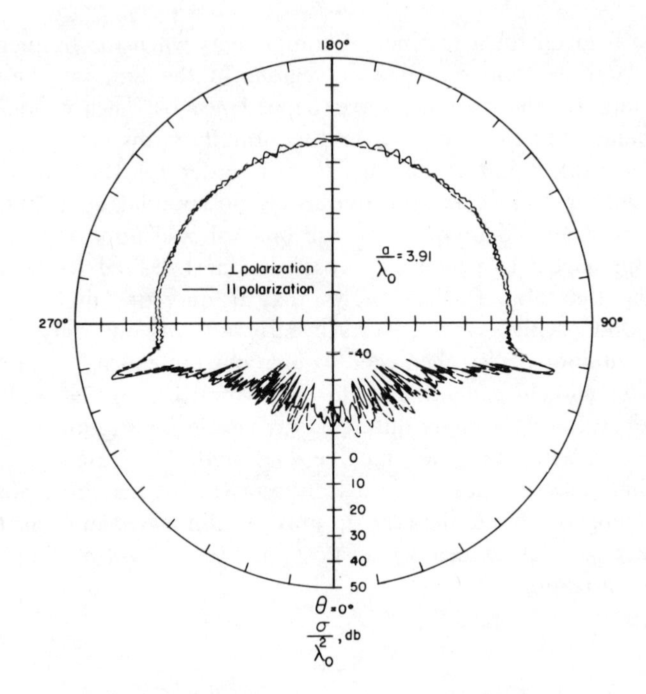

Fig. 8-47. Experimental backscatter cross sections of a perfectly conducting 25° cone-sphere with a base radius of 3.91 wavelengths versus the angle of incidence [after Pannell *et al.*[36]].

A shape that fits the above requirements and has been widely proclaimed for its low backscatter cross sections at aspects near the axis is the cone-sphere. As discussed in Section 6.3, the backscatter cross section of a cone-sphere for incidence along the axis is due primarily to the interaction of the creeping-wave contribution from the spherical base with the contribution from the cone-sphere join. In Figures 8-46 to 8-48 are shown measured[36] backscatter cross sections for a 25° perfectly conducting cone-sphere for both $\perp$ and $\|$ polarizations and values of $a/\lambda_0 = 0.60$, 3.91, and 6.09.† It is apparent from these figures that σ/λ_0^2 is approximately -10 db over a range of aspect angles of $\pm 45°$ around the tip and is relatively independent of the frequency. A closer examination of the frequency dependence can be obtained from Figure 8-49a, b, c, where the results of a series of measurements[17] of the tip-on axial-incidence backscatter cross section for 15, 30, 40, 60, and 75° cone-spheres are presented. From these figures it appears that the average axial-incidence backscatter cross section is

$$\sigma_{\text{av}} \approx 0.2\lambda_0^2 \tag{8.3-1}$$

† The parameter a is the radius of the spherical base.

and is bounded by maximum and minimum values of

$$\sigma_{\text{max}} \approx 0.4\lambda_0^2 \tag{8.3-2}$$

and

$$\sigma_{\text{min}} \approx 0.01\lambda_0^2 \tag{8.3-3}$$

relatively independently of the cone angle.

In principle, the near tip-on cross section can be further reduced by improving the match at the cone–base join, and by reducing the creeping-wave contribution to the cross section. Some experimental measurements were made by MIT of the backscatter cross sections produced by several shapes for which this was attempted.[37] These shapes are shown in Figure 8-50. Three of the models, designated "King," "Foxtrot," and "Golf," were designed to improve the derivative match at the join. The model designated as "King" has no surface discontinuities of any order except at the tip, while "Golf" has, at most, a third derivative discontinuity at points other than the tip and "Foxtrot" has a second derivative discontinuity like a cone-sphere, only of smaller magnitude. The model "Juliet" was designed to reduce the creeping-wave contribution to the cross section. The measured

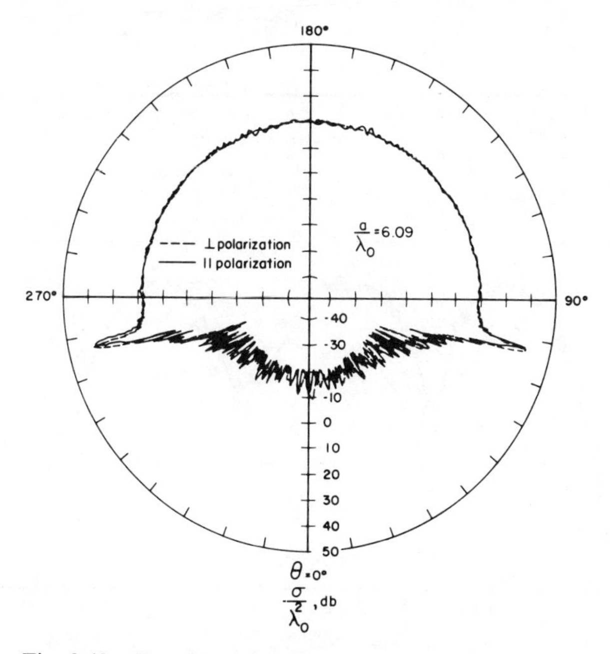

Fig. 8-48. Experimental backscatter cross sections of a perfectly conducting 25° cone sphere with a base radius of 6.09 wavelengths versus the angle of incidence [after Pannell *et al.*[36]].

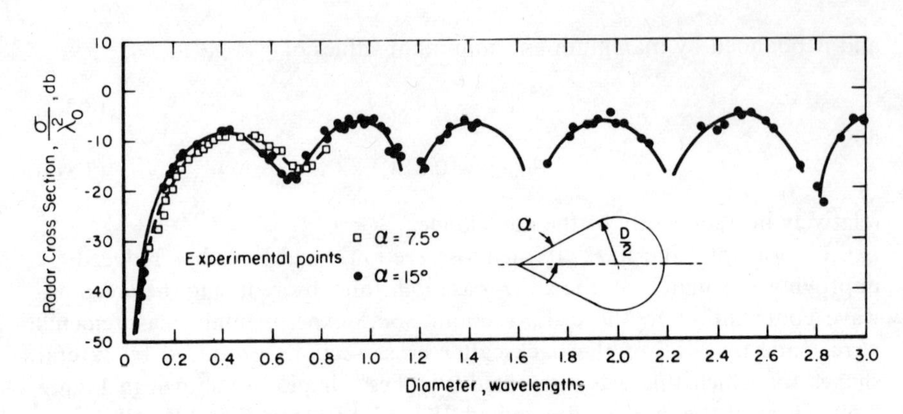

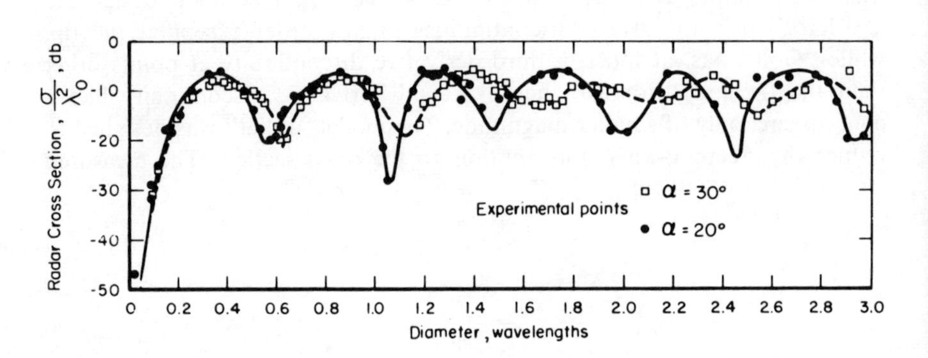

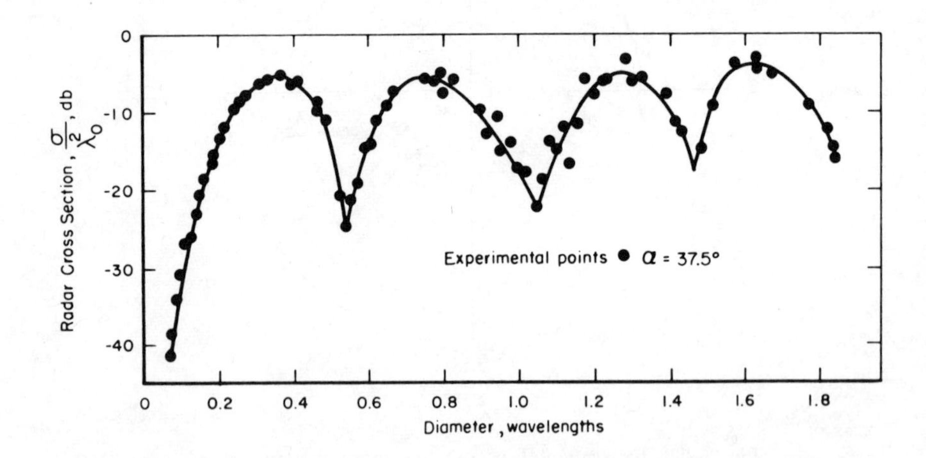

Fig. 8-49. Experimental tip-on backscatter cross sections for 7.5, 15, 20, 30, and 37.5° perfectly conducting cone-spheres versus the base diameter in wavelengths [from Blore and Royer[17]].

cross sections of these four configurations are shown in Figures 8-51 to 8-54. It is apparent from these figures that no major difference in cross section resulted from the difference between these models. In the case of the first three, this is to be expected since for the values of a/λ_0 used, the creeping wave component dominates the join contribution; thus, the effects of differences in the surface derivative discontinuities would be masked by the creeping-wave return. In the case of the "Juliet" model, the results are not as easily explained. Knot and Senior[38] have shown that objects at the rear of a cone-sphere can produce significant differences in the nose-on

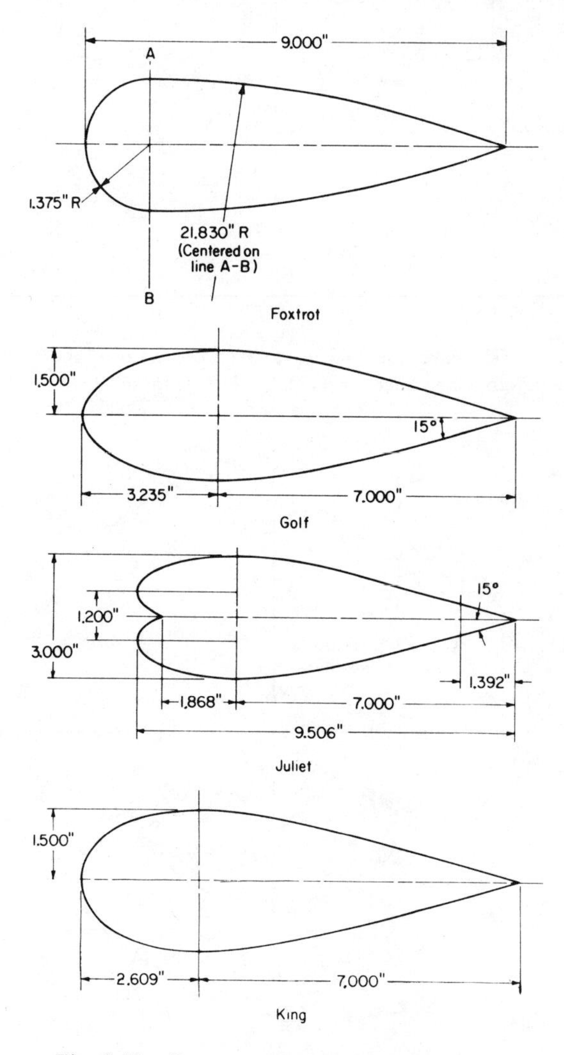

Fig. 8-50. Low-cross-section shape dimensions.

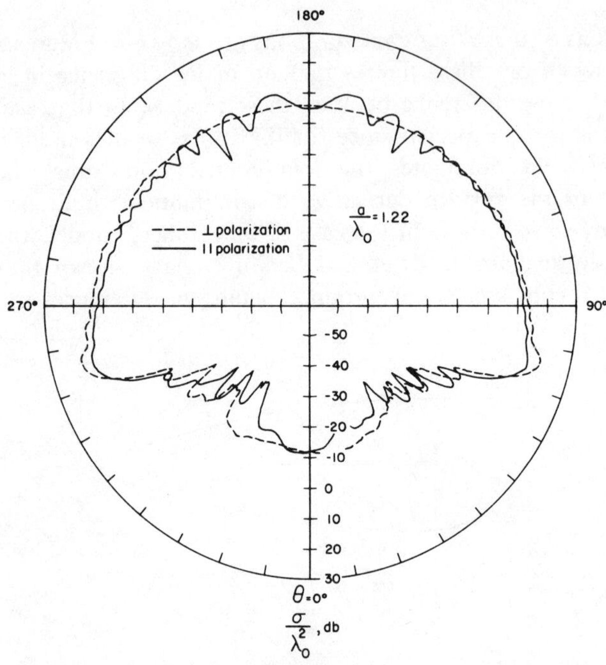

Fig. 8-51. Experimental backscatter cross section of the "King" shape versus the angle of incidence [after Smith[37]].

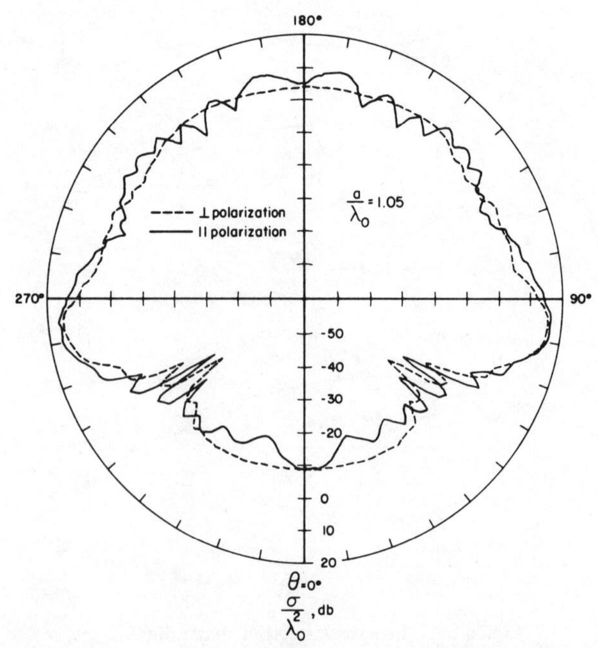

Fig. 8-52. Experimental backscatter cross section of the "Foxtrot" shape versus the angle of incidence [after Smith[37]].

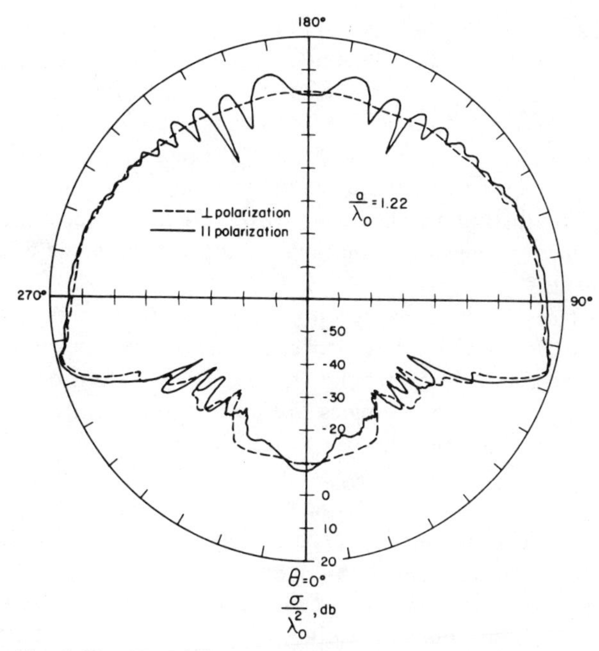

Fig. 8-53. Experimental backscatter cross section of the "Golf" shape versus the angle of incidence [after Smith[37]].

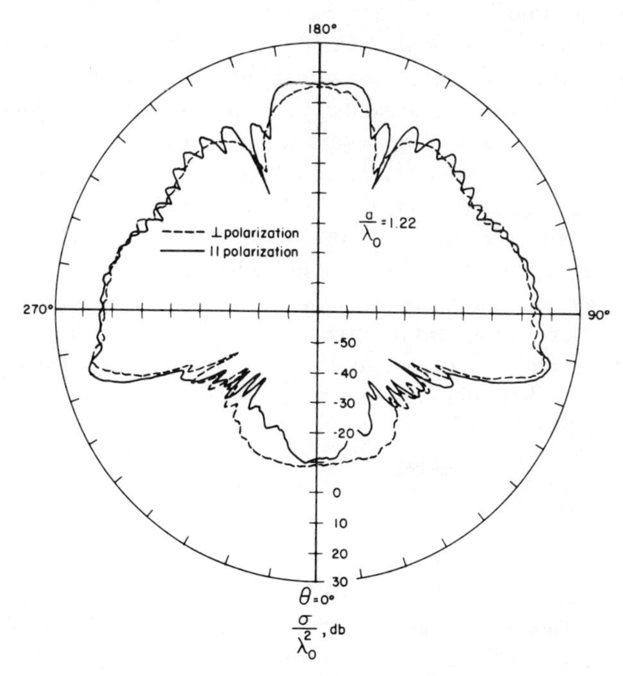

Fig. 8-54. Experimental backscatter cross section of the "Juliet" shape versus the angle of incidence [after Smith[37]].

cross section as would be predicted by creeping-wave theory (see Figures 6-55 to 6-57 of Section 6.3.1.3.4). The shape of the shadowed portion of a low-cross-section object apparently can exert considerable influence on the near-nose-on cross section, although no analytical techniques are available in most cases for predicting the precise effect of the rear shape. Simply lengthening the effective creeping-wave path around the rear of the object or providing a sharp edge to disrupt the creeping waves does not necessarily cause a reduction in the creeping-wave component, and may in some cases produce an enhancement.

The effect of shaping a body into a configuration similar to that of a cone-sphere is to produce a low backscatter cross section over a range of aspect angles of about $\pm 45°$ around the tip. The question occurs as to the cross section at other aspect angles, as well as the bistatic cross section. At other aspect angles, the cone-sphere return is dominated by the specular contributions from the conical sides and the spherical base. For parallel polarization and not too high a frequency, traveling-wave effects cause some fluctuations in the cross section at rear aspects over the angular region $120° \leqslant \theta \leqslant 240°$, as illustrated in Figure 8-46. These fluctuations can be attributed to phase interaction between the specular component, produced by the spherical base, and a traveling-wave component, reflected by the conical tip. This interpretation is validated by the absence of fluctuations in this region for perpendicular polarization.

For the bistatic case, the largest bistatic cross section in the high-frequency limit occurs at forward aspects. When the incident field is nearly axial, the forward scatter cross section is approximately the same as for a sphere of the same radius as the cone-sphere base. The largest high-frequency bistatic cross section occurs for forward scatter when the cone-sphere axis is normal to the incident field direction, and for a 10° cone-sphere is approximately 8 db greater than the axial-incidence forward scatter cross section. For a 30° cone-sphere, this difference reduces to only 2 db. Thus, for example, modifying an essentially spherical body into a cone-sphere shape in order to reduce its backscatter cross section may enhance its bistatic characteristics at certain aspect angles; however, this enhancement would on the average be of less consequence than the backscatter enhancement occurring when the specular return from the cone side is viewed. Some measured[18] bistatic cross sections for a perfectly conducting 20° cone sphere with $a/\lambda_0 = 0.66$ are given in Figure 8-55.†

† The body measured had a surface derivative discontinuity at the cone–sphere join, which coincided with the shadow boundary. A better terminology for this shape is perhaps "cone-hemisphere."

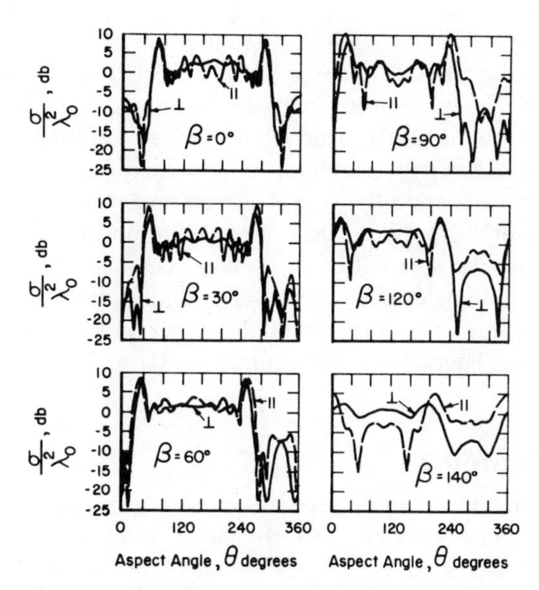

Fig. 8-55. Experimental bistatic cross sections for a 20° perfectly conducting cone-hemisphere versus the angle of incidence for several bistatic angles [from Eberle and St. Clair[18]].

8.3.1.2. *Low Frequency*

In the low-frequency or Rayleigh region, shaping is in general completely ineffective as a means for cross-section reduction, since the cross section depends primarily on the body volume rather than its shape.

8.3.2. Radar-Absorbing Materials

In addition to shaping an object so as to produce a low radar cross section over a specified range of aspect angles, the radar cross section of an object may in some cases be reduced over a specific range of frequencies and aspect angles by applying to the surface of the object a material which "absorbs" the incident energy. The word "absorbs" was set off by quotation marks in the above sentence to emphasize that many so-called absorbing materials, when applied to actual radar targets, achieve a cross-section reduction not entirely by absorption but by a combination of absorption and redirection of the scattered field. The development of radar-absorbing materials, henceforth referred to as RAM, has been pursued vigorously for a number of years. The search for suitable RAM was initiated in the early 1940's both in the United States and Germany. Ideally, the optimum RAM would be a paint-like material effective at all polarizations over a broad range of frequencies and angles of incidence. Unfortunately, such a material currently does not exist and the probability of its being developed is rather remote. Practically, the type of absorber which would be most effective in a given situation is highly dependent upon the radar frequency, target shape and dimensions, bandwidth required, and the physical bounds such as weight, thickness, strength, environment, etc., which are placed on the absorber. Classically, RAM development has been devoted to developing absorbers which when placed on an infinite-planar surface would reduce the Fresnel reflection coefficients, $R_{\parallel}$ and $R_{\perp}$, over a range of frequencies and angles of incidence. The initial discussion in this section will also be confined to absorbers of this type, and RAM for real targets will be discussed later. Thus what might be termed "flat-plate RAM" is the present topic.

8.3.2.1. *Flat-Plate RAM*

RAM of this type can be separated into two broad categories, one of which is designated resonant and the other broadband. The separation into these categories is not completely unique in that some so-called resonant RAM may have, in fact, bandwidths exceeding those of some of the broadband types. The fundamental difference between the two types is for resonant materials, the conditions for reduction of $R_{\parallel}$ and/or $R_{\perp}$ are satisfied in general at one or more discrete frequencies; while for broadband absorbers, the conditions are valid in principle at all frequencies and the absorber becomes ineffective due to certain assumptions inherent in the material design being violated outside a particular band of frequencies.

8.3.2.1.1. *Resonant RAM*

8.3.2.1.1.1. *Salisbury Screen.* One of the oldest and simplest types of resonant absorbers is known as a Salisbury screen.[39] The configuration of such an absorber is shown in Figure 8-56.

The absorber consists of a thin layer of lossy material, shown as medium (1) in Figure 8-56, separated by a lossless layer of thickness l from a backing of infinite conductivity. From Eq. (7.1-37), the reflection coefficient is

$$R = \frac{Z(1) - z_0}{Z(1) + z_0} \tag{8.3-4}$$

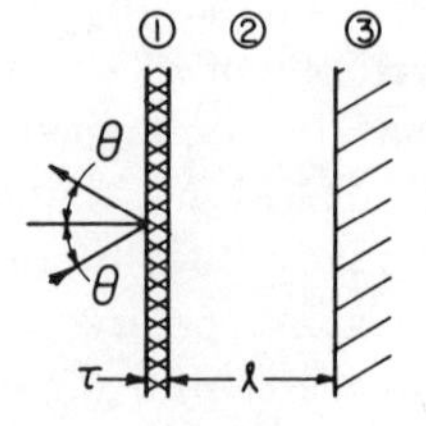

Fig. 8-56. Electric Salisbury screen.

where $Z(1)$ is the effective input impedance at the interface between free space and medium (1), and z_0 is given by Eq. (7.1-6). Now for $Z(1)$, Eq. (7.1-40) gives

$$Z(1) = z_1 \frac{Z(2) - iz_1 \tan \alpha_1}{z_1 - iZ(2) \tan \alpha_1} \tag{8.3-5}$$

where $Z(2)$ is the effective input impedance at the interface between medium (1) and (2), and $\alpha_1 = k_1\tau \cos \theta_1$.

Assuming for the moment that $Z(2) \gg z_1$, then from (8.3-5)

$$Z(1) \approx i \frac{z_1}{\tan \alpha_1} \tag{8.3-6}$$

For τ sufficiently small, $\alpha_1 \ll 1$ and $\tan \alpha_1 \approx \alpha_1$, so that

$$\tan \alpha_1 \approx k_1\tau \left[1 - \left(\frac{k_0}{k_1}\right)^2 \sin^2 \theta_0\right]^{1/2} \tag{8.3-7}$$

and at normal incidence, $\tan \alpha_1 \approx k_1\tau$, or

$$Z(1) \approx i \frac{z_1}{k_1\tau} \approx i \frac{\sqrt{\mu/\epsilon}}{\omega\sqrt{\mu\epsilon}\,\tau} \approx \frac{i}{\omega\epsilon\tau} \tag{8.3-8}$$

If medium (1) has sufficient conductivity and a permittivity such that $\epsilon_1'' \gg \epsilon_1'$, then

$$\epsilon_1 \approx i(\sigma/\omega) \tag{8.3-9}$$

and

$$Z(1) \approx 1/\sigma\tau \tag{8.3-10}$$

For zero reflection, using Eq. (8.3-10) in (8.3-4) indicates that

$$1/\sigma\tau = \eta_0 \tag{8.3-11}$$

This, along with the requirements that $Z(2)$ be large and $\epsilon_1'' \gg \epsilon_1'$, constitute the conditions for an electric Salisbury screen. The requirement that $Z(2)$ be large can be satisfied by making medium (2) a lossless dielectric and $l = \lambda/4$ in the medium. Under these conditions $Z(2) \approx \infty$. Using a circuit analog, the electric Salisbury screen corresponds to a resistive material of 377 ohms per square in parallel with a very high reactance. In fact, the circuit analog of the electric Salisbury screen is a parallel resonant circuit as shown in Figure 8-57. This is due to the fact that near the point $\alpha_2 = \pi/2$, $Z(2) = iz_2 \tan \alpha_2$ behaves in the same manner as a lossless parallel resonant circuit.

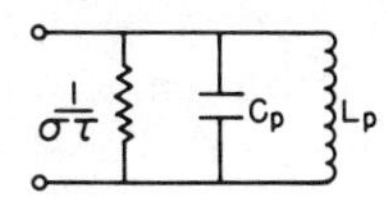

Fig. 8-57. Equivalent circuit for an electric Salisbury screen.

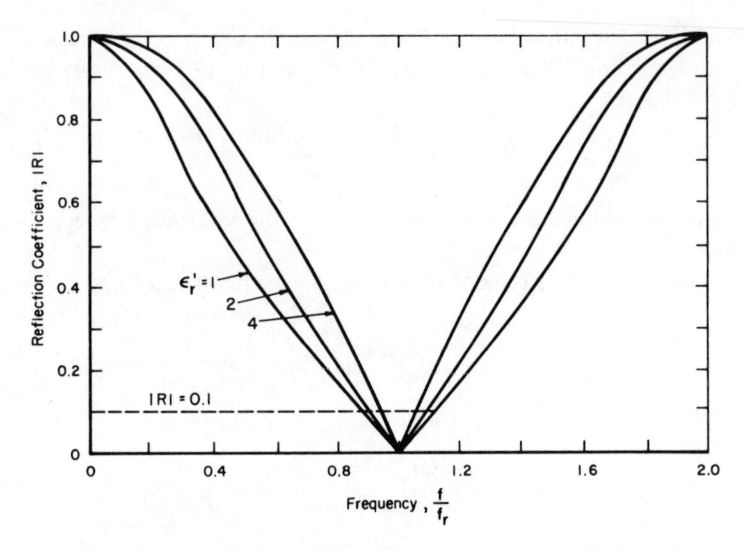

Fig. 8-58. Normal-incidence reflection coefficient of an electric Salisbury screen versus frequency for several values of dielectric constant.

The variation of $Z(2)$ with wavelength determines the frequency dependence of an electric Salisbury screen, and in Figure 8-58, the normal-incidence reflection coefficient of a Salisbury screen is plotted for values of $\epsilon_2 = 1, 2, 4$ with μ_2 assumed to be 1. It is apparent that the bandwidth decreases as ϵ_2 increases. Thus the thinner the overall absorber is made by increasing ϵ_2, the narrower the bandwidth becomes.

An additional factor of interest with such absorbers is the angular dependence of the reflection coefficient. For an electric Salisbury screen, matched for normal incidence with ϵ_2 sufficiently large that the variation of $Z(2)$ with θ can be neglected, the reflection coefficients off normal incidence are approximately given by

$$R_{\parallel} \approx \frac{1 - \cos\theta}{1 + \cos\theta}$$
$$R_{\perp} \approx \frac{\cos\theta - 1}{\cos\theta + 1} \tag{8.3-12}$$

and their magnitudes are plotted in Figure 8-59. From this figure, the range of angles of incidence over which $|R_{\parallel}|, |R_{\perp}| \leqslant 0.1$ is seen to be $|\theta| \leqslant 35°$.

If the absorber is designed for zero reflection at some particular angle of incidence θ_0 for a particular polarization, say perpendicular, then it is of interest to inquire about the angular dependence of R for the other polarization as well as for the design case. This is illustrated in Figure 8-60, where $|R_{\perp}|$ and $|R_{\parallel}|$ are plotted versus θ for an absorber designed for

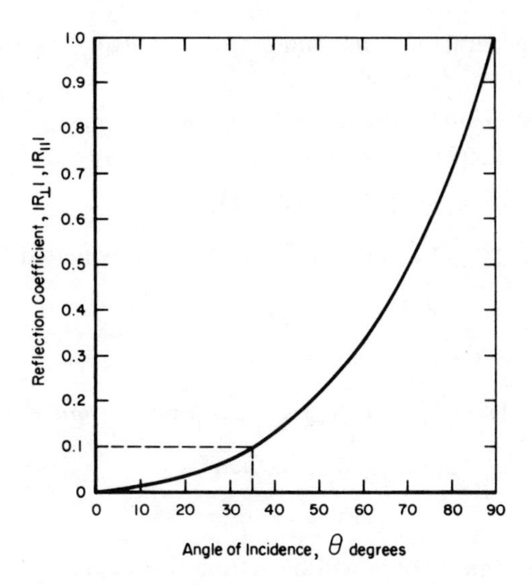

Fig. 8-59. Reflection coefficient of an electric Salisbury screen versus the angle of incidence.

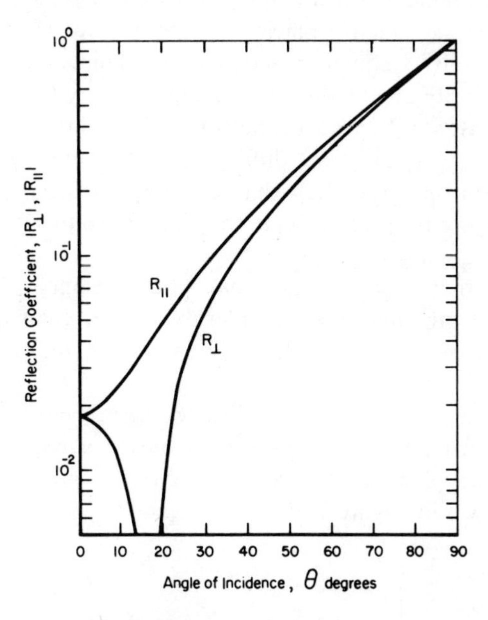

Fig. 8-60. Reflection coefficient versus the angle of incidence for an electric Salisbury screen designed for zero reflection at $\theta = 15°$ for perpendicular polarization.

$R_{\perp}(15°) = 0$. In general, the maximum value of θ_0 for which both $| R_{\perp} |$, $| R_{\parallel} | < 0.1$ for $| \theta | \leqslant \theta_0$ is 25°.

8.3.2.1.1.2. *Magnetic Salisbury Screen.* Returning to Eq. (8.3-5), consider the situation if $Z(2) \ll z_1$; then

$$Z(1) \approx -iz_1 \tan \alpha_1 \tag{8.3-13}$$

For τ small, $\alpha_1 \ll 1$, and again $\tan \alpha_1$ can be approximated by α_1, so that

$$Z(1) \approx -iz_1\alpha_1 \approx -iz_1k_1\tau$$

at normal incidence. Or substituting for k_1 and z_1,

$$Z(1) \approx -i(\sqrt{\mu_1/\epsilon_1})(\omega \sqrt{\mu_1\epsilon_1})\, \tau \approx -i\omega\mu_1\tau \tag{8.3-14}$$

Now, $\mu_1 = \mu_1' + i\mu_1''$, and if $\mu_1'' \gg \mu_1'$, then

$$Z(1) \approx \omega\mu_1''\tau \tag{8.3-15}$$

and zero reflection can be obtained by setting this equal to η_0 or

$$\omega\mu''\tau = \eta_0 \tag{8.3-16}$$

This, along with the conditions that $Z(2) \ll z_1$, $\alpha_1 \ll 1$, and $\mu_1'' \gg \mu_1'$, constitute the requirements for a magnetic Salisbury screen. The condition that $Z(2) \ll z_1$ can be met by placing the thin lossy layer, medium (1), directly on a perfectly conducting plate, or in Figure 8-56, medium (2) is required to be perfectly conducting. In principle, a very thin absorber could be devised in this fashion. In practice, however, the conditions that $\mu_1'' \gg \mu_1'$, $\alpha_1 \ll 1$, and $\omega\mu''\tau = \eta_0$ are very difficult to meet simultaneously. These conditions require a material such that $\mu_1'' \gg \epsilon_1$, μ_1', and $\mu_1'' \propto 1/\omega$. The only materials having properties resembling these are the ferrites, and then only over small frequency ranges.

In the event that a satisfactory magnetic Salisbury screen could be realized, its bandwidth might be considerably better than for the electric Salisbury screen since it would be determined only by the frequency dependence of ϵ_1 and μ_1.

The bandwidth of an electric Salisbury-screen type absorber can be improved by constructing a multiple-screen absorber, which is analogous to a number of coupled parallel resonant circuits. An example of such an absorber is shown in Figure 8-61.

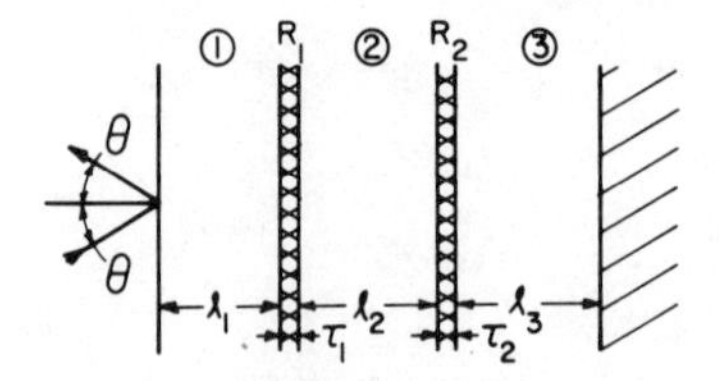

Fig. 8-61. Multiple-layer electric Salisbury screen.

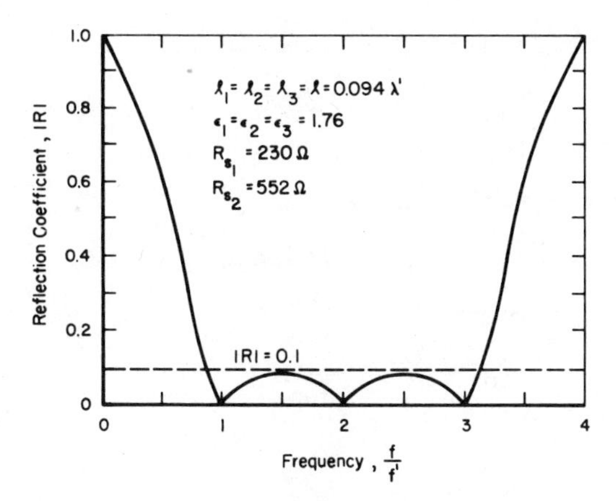

Fig. 8-62. Normal-incidence reflection coefficient of a three-layer Salisbury screen versus frequency.

The frequency dependence of the normally incident reflection coefficients for such an absorber is shown in Figure 8-62. The input impedance of an absorber of this type can be obtained using the theory for arbitrary layered media outlined in Section 7.1. If there are n dielectric layers, the general requirements on $Z(1)$ are $\operatorname{Re} Z(1) = z_0$ at n frequencies and $\operatorname{Im} Z(1) = 0$ at these frequencies. To satisfy these $2n$ equations, assuming $\mu_n = \mu_{n-1} = \cdots = \mu_1 = \mu_0$ and $\tau_1 \approx \tau_2 \approx \cdots \tau_{n-1} \ll 1$ along with media (1), (3),..., $(2n - 1)$ lossless, there are $3n - 1$ parameters, $l_1, l_2, \ldots, l_n$; $\epsilon_1, \epsilon_2, \ldots, \epsilon_n$; and $n - 1$ surface resistivities of the resistive sheets. In the event $l_1 = l_2 = \cdots = l_n = l$, these reduce to $2n$ parameters; sufficient to satisfy the $2n$ equations.

8.3.2.1.1.3. *Homogeneous Layer.* A second type of resonant absorber, sometimes called a Dällenbach layer, can be constructed of a homogeneous lossy layer backed by a metallic plate, such as shown in Figure 8-63.

From Eq. (7.1-40), the input impedance at the interface $z = 0$ with $Z(3) = 0$ is

$$Z(0) = -iz_1 \tan \alpha_1$$

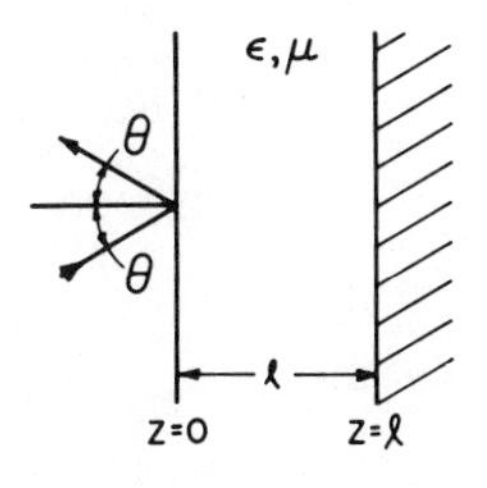

Fig. 8-63. Homogeneous or Dällenbach layer absorber.

where $\alpha_1 = k_1 l \cos\theta$; so the reflection coefficient is

$$R = \frac{-iz_1 \tan\alpha_1 - z_0}{-iz_1 \tan\alpha_1 + z_0} \tag{8.3-17}$$

The condition for zero reflection is

$$z_0 = -iz_1 \tan\alpha_1 \tag{8.3-18}$$

or for perpendicular polarization,

$$1 = -i\sqrt{\frac{\mu_r}{\epsilon_r}} \frac{\cos\theta_0}{[1 - (\sin^2\theta_0)/\mu_r\epsilon_r]^{1/2}} \tan\left\{\frac{2\pi l}{\lambda_0}\left[\sqrt{\mu_r\epsilon_r}\left[1 - \frac{\sin^2\theta_0}{\mu_r\epsilon_r}\right]^{1/2}\right]\right\} \tag{8.3-19}$$

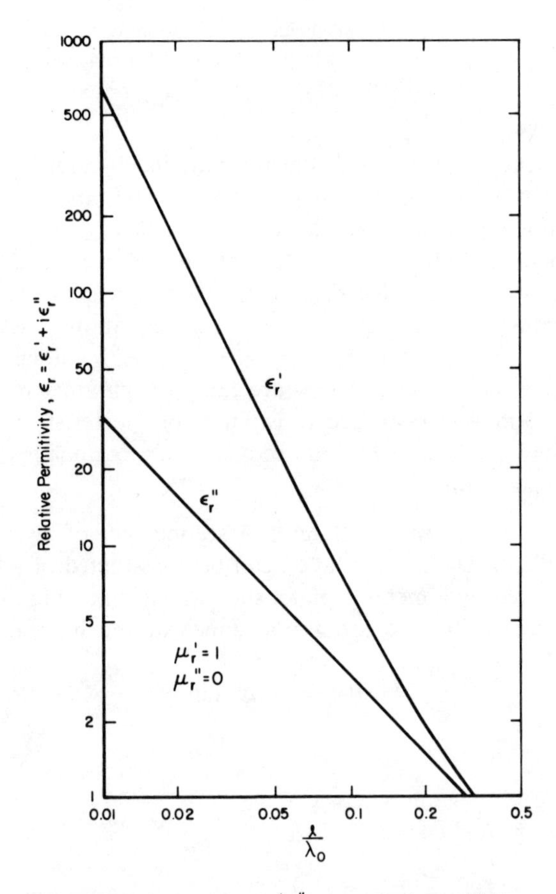

Fig. 8-64. The values of ϵ_r', ϵ_r'' versus the layer thickness giving a zero normal-incidence reflection coefficient for a homogeneous layer with $\mu_r = 1$.

while for parallel polarization,

$$1 = -i\sqrt{\frac{\mu_r}{\epsilon_r}}\,\frac{[1 - (\sin^2\theta_0)/\mu_r\epsilon_r]^{1/2}}{\cos\theta_0}\tan\left\{\frac{2\pi l}{\lambda_0}\left[\sqrt{\mu_r\epsilon_r}\left[1 - \frac{\sin^2\theta_0}{\mu_r\epsilon_r}\right]^{1/2}\right]\right\} \tag{8.3-20}$$

These expressions are quite complex in general, but can be simplified considerably for normal incidence, and if several special cases are examined. In the event that the permeability of the layer is the same as free space, i.e., $\mu = \mu_0$, then the required values of ϵ_r' and ϵ_r'' as a function of l/λ_0 for a zero normal-incidence reflection coefficient are given in Figure 8-64. For $\epsilon_r = \epsilon_r' =$ constant and $\mu_r = \mu_r' + \mu_r''$, then the required values of μ_r' and μ_r'' for zero normal-incidence reflection are shown in Figure 8-65 versus $\epsilon_r l/\lambda_0$.

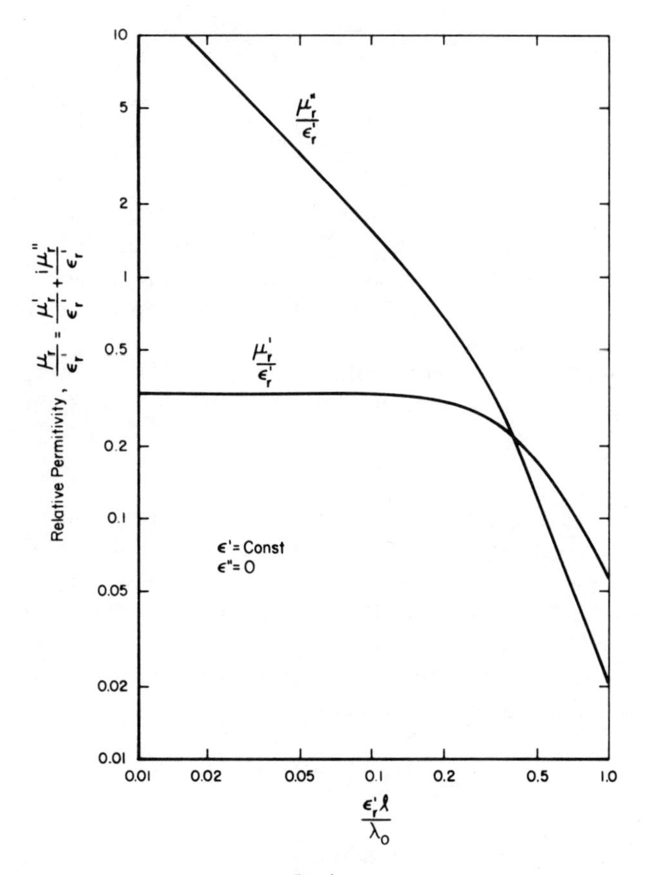

Fig. 8-65. The values of μ_r'', μ_r' versus the layer thickness giving a zero normal-incidence reflection coefficient for a homogeneous layer with $\epsilon_r = \epsilon_r' =$ constant.

The angular dependence of the reflection coefficients for a homogeneous layer can be obtained readily from Eq. (8.3-17), when $\sqrt{\mu\epsilon} \gg 1$. For example, if the layer is designed to give zero reflection at normal incidence, then at angles of incidence other than normal the reflection coefficients are

$$|R_\perp| = |R_\parallel| = \frac{1 - \cos\theta}{1 + \cos\theta} \tag{8.3-21}$$

This is the same as Eq. (8.3-12) for the Salisbury screen, so Figure 8-59 also shows the angular dependence of the homogeneous layer.

Similarly, if the layer is designed for zero reflection at a particular angle of incidence for a specific polarization, say perpendicular, then the reflection coefficients for arbitrary incidence are given by

$$|R_\perp| = \frac{\cos\theta - \cos\theta_0}{\cos\theta + \cos\theta_0} \tag{8.3-22}$$

and

$$|R_\parallel| = \frac{1 - \cos\theta\cos\theta_0}{1 + \cos\theta\cos\theta_0} \tag{8.3-23}$$

where θ_0 is the angle of zero reflection for perpendicular polarization. If the layer is designed for zero reflection at some angle for parallel polarization, the same equations are valid by interchanging $R_\perp$ and $R_\parallel$. These are plotted in Figure 8-60 for $\theta_0 = 15°$, and again the angular dependence is identical to that for the Salisbury screen.

The frequency dependence of the homogeneous layer absorber can be examined readily for the normal incidence case, where (8.3-18) reduces to

$$-i \tan\alpha = \sqrt{\epsilon_r/\mu_r} \tag{8.3-24}$$

and where ϵ_r and μ_r are the relative complex permittivity and permeability of the layer, respectively. Thus at some frequency f' and layer thickness l, the dielectric parameters, ϵ_r and μ_r, required for zero reflection can be determined by solving Eq. (8.3-24) or from Figures 8-64 and 8-65 for the special cases of no magnetic loss or no electric loss. At any frequency, once ϵ_r, μ_r, and l are specified, the reflection coefficient is given by (8.3-17). Using this and Eq. (8.3-21), the reflection coefficient is

$$|R| = \frac{\tan\alpha - \tan\alpha'}{\tan\alpha + \tan\alpha'} \tag{8.3-25}$$

Now examining the case where $\alpha \approx \alpha'$,

$$|R| \approx \frac{\alpha - \alpha'}{\sin 2\alpha'} \tag{8.3-26}$$

and expressing sin $2\alpha'$ in terms of tan α',

$$|R| \approx \frac{\alpha - \alpha'}{2 \tan \alpha'/(1 + \tan^2 \alpha')}.$$

or

$$|R| \approx \frac{[1 - (\epsilon_r/\mu_r)](\alpha - \alpha')}{2i\sqrt{\epsilon_r/\mu_r}} \tag{8.3-27}$$

Expressing α and α' in terms of μ, ϵ, l, and f, then

$$|R| \approx \frac{|\mu_r - \epsilon_r|}{2}\left[\frac{f - f'}{f}\right]\frac{2\pi l}{\lambda'} \tag{8.3-28}$$

Writing this in terms of the bandwidth gives

$$B = 2\left[\frac{f - f'}{f}\right] \approx \frac{2|R|}{\pi|\mu_r - \epsilon_r|(l/\lambda')} \tag{8.3-29}$$

and the bandwidth for $|R| \leqslant 0.1$ is

$$B = \frac{0.2}{\pi|\mu_r - \epsilon_r|(l/\lambda')} \tag{8.3-30}$$

Considering the cases where (l/λ') is small and only electric or magnetic losses are present, approximate expressions for the bandwidth in terms of the ratio (l/λ') or $(\epsilon' l/\lambda')$ can be obtained. For the case of electrical loss only, $\mu = 1$ and

$$B_e \approx \frac{0.2}{|\pi(l/\lambda') - (\pi/16)(l/\lambda')^{-1} - i|} \tag{8.3-31}$$

while for magnetic loss only, $\epsilon = \epsilon'$ and

$$B_m \approx \frac{0.2}{|-(2/3\pi)(\epsilon' l/\lambda') + (i/2\pi)|} \tag{8.3-32}$$

These are plotted in Figure 8-66, where it is apparent that B_e has a maximum of about 20% at $l/\lambda' \approx \frac{1}{4}$, and goes to zero as the layer becomes smaller. On the other hand, B_m increases as the layer becomes thinner, being always larger than B_e for the same layer thickness. The values of B_m, given in Figure 8-66, are not reliable for very small $(\epsilon' l/\lambda')$, since the original assumption that the bandwith is small is violated. However, the fact that the bandwidth should increase as the layer becomes thinner for a magnetically lossy layer is still valid, since in the limit an infinitesimally thin layer is equivalent to a magnetic Salisbury screen which has, in principle, an infinite bandwidth.

In the preceding analysis, ϵ and μ were assumed to be independent of frequency which, of course, is not true since the real and imaginary parts

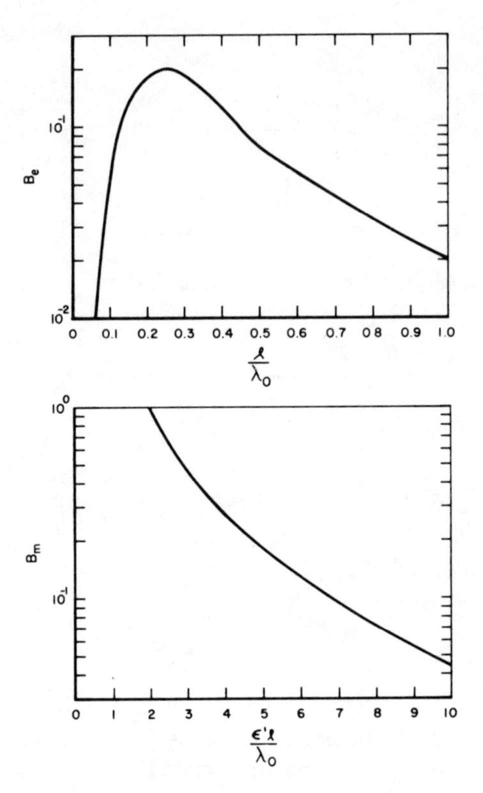

Fig. 8-66. Bandwidths of thin homogeneous layers versus the layer thickness for layers having only electric or only magnetic losses.

of both ϵ and μ are connected by the Kramers–Kronig relations.[40] Unfortunately, the dependence on wavelength of ϵ' and ϵ'' or μ' and μ'', as specified by the Kramers–Kronig relations, is not compatible with that required by Eq. (8.3-18) for zero reflection. Thus a very broadband homogeneous layer absorber is not, in general, possible to achieve, at least not with a layer having only electrical or magnetic loss.

Again as with the Salisbury screen, the bandwidth of a homogeneous layer-type absorber can be increased by using several layers of different dielectric parameters, adjusted to produce zeros of the reflection coefficient at different frequencies. The analytical evaluation of absorbers of this type is relatively difficult; however, using the general theory given in Section 7.1, machine programs can be easily developed to evaluate numerically various multilayer absorber configurations.

8.3.2.1.1.4. *Circuit-Analog Absorber.* Another resonant absorber, an example of which has already been encountered, is what might be designated

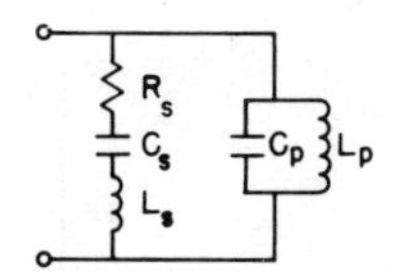

Fig. 8-67. Equivalent circuit for a dipole absorber.

a circuit-analog absorber. These absorbers are characterized by thin sheets of material having specific surface impedances, separated by relatively lossless dielectric layers. The analysis of this type absorber is most readily accomplished by using conventional network theory and the equivalent circuit for the absorber.

The Salisbury screen, discussed earlier, is perhaps the simplest example of a circuit-analog absorber. Another is a type called the dipole absorber, which consists of a sheet of lossy dipoles separated from a perfectly conducting surface by a dielectric spacer of $\lambda/4$ electrical length. The equivalent circuit for such a structure is (see Figure 8-67) a parallel combination of a series resonant circuit and a lossless parallel resonant circuit. Using this equivalent circuit, the reflection coefficient as a function of frequency can be computed for various values of R_S. Figure 8-68 gives computed curves of $|R|$ versus frequency for such an absorber. The greatest bandwidth occurs when $R_S < \eta_0$ and for $R_S = 0.8\,\eta_0$, $B \approx 70\,\%$ for $|R| < 0.1$.

If the lossy dipole sheet is constructed of similarly oriented linear dipoles, the reflection coefficient would obviously be highly polarization-dependent; however, crossed dipoles, discs, dipole–slot combinations, etc., can be used to reduce the polarization dependence.

Multiple-layer absorbers of this type can also be constructed with improved bandwidth over the single-layer type. Figure 8-69 compares the calculated frequency response of a single-layer Salisbury screen, a single-layer dipole absorber, and a two-layer dipole absorber.

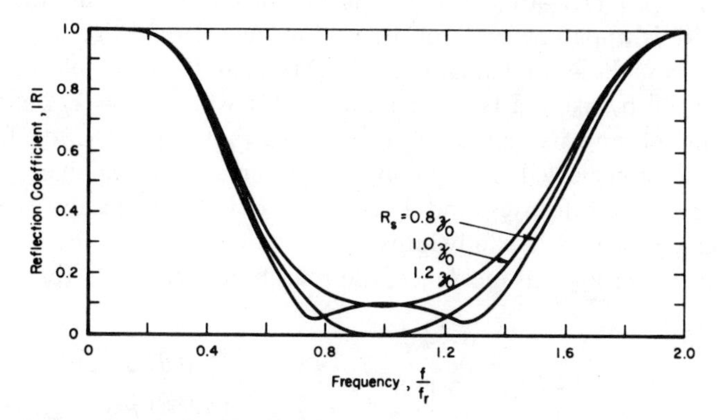

Fig. 8-68. Reflection coefficient versus frequency for a dipole absorber with several values of R_s.

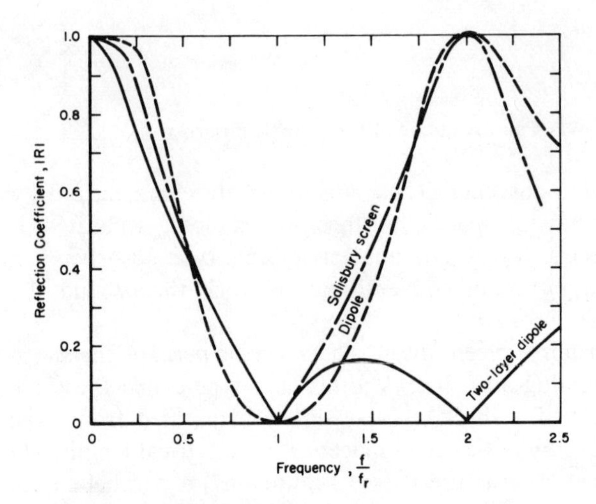

Fig. 8-69. Comparison of the frequency response of single-layer dipole, two-layer dipole, and single-layer Salisbury-screen absorbers.

By the use of modern filter or network theory, circuits can be designed which, to a large extent, provide specified impedance versus frequency characteristics. If the RF analog of these circuits can then be realized, absorbers can be designed having comparable reflection coefficient versus frequency characteristics. The primary difficulty, of course, lies in obtaining materials having the proper values and frequency dependence of μ and ϵ.

8.3.2.1.2. *Broadband RAM*

8.3.2.1.2.1. $\mu = \epsilon$ *Absorber.* Neglecting any variation of ϵ and μ with frequency, if Eq. (7.1-40) for the reflection coefficient of an interface is examined, it is apparent that the normal-incidence reflection coefficient is zero if $\mu_r = \epsilon_r$ in the second medium. A RAM which is, in principle, broadband can be devised by using a layer of material, for which $\mu_r = \epsilon_r$, and which is thick enough and has large enough loss so that any reflection from the backing can be neglected. In general, the only materials capable of accomplishing this are the ferrites, and a few $\mu_r \approx \epsilon_r$ type ferrite absorbers have been developed with fair bandwidths.

At angles off normal incidence, the reflection coefficients for a $\mu_r = \epsilon_r$ absorber are

$$|R_\perp| = |R_\parallel| = \left| \frac{\cos\theta_0 - \sqrt{1 - (\sin^2\theta)/\mu_r\epsilon_r}}{\cos\theta_0 + \sqrt{1 - (\sin^2\theta)/\mu_r\epsilon_r}} \right| \tag{8.3-33}$$

This is plotted as a function of θ for several values of $(\mu_r\epsilon_r)$ in Figure 8-70.

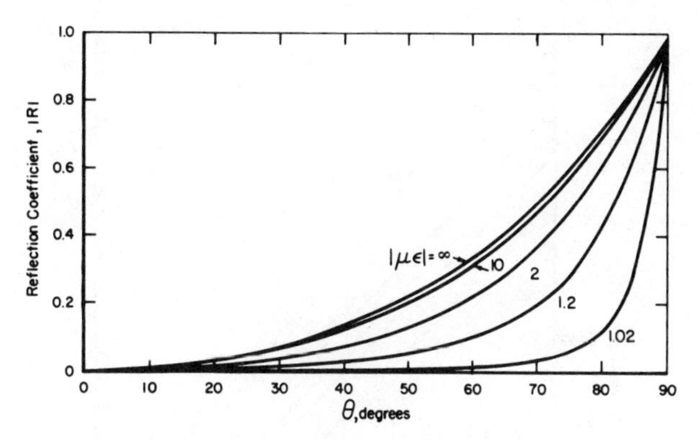

Fig. 8-70. Angular performance of a $\mu = \epsilon$ absorber.

In the event that $\mu_r' = \epsilon_r'$ and $\mu_r', \epsilon_r' \gg \mu'', \epsilon''$ with $\mu'' \neq \epsilon''$, then the normal-incidence reflection coefficient is approximately

$$| R | \approx \frac{1}{4} \left| \frac{\mu_r''}{\mu_r'} - \frac{\epsilon_r''}{\mu_r'} \right| \tag{8.3-34}$$

The frequency dependence of absorbers of this type is determined by the frequency dependence of μ and ϵ, and relatively thin broadband absorbers can, in principle, be constructed in this fashion.

For a $\mu_r = \epsilon_r$-type absorber of finite thickness, backed by a perfectly conducting surface, the reflection coefficient is given by Eq. (8.3-17) with $z = z_0$ or

$$R = \frac{-i \tan \alpha - 1}{-i \tan \alpha + 1} = -e^{2i\alpha} \tag{8.3-35}$$

At normal incidence, for the case where $\epsilon', \mu' \gg \epsilon'', \mu''$, the magnitude of the reflection coefficient becomes

$$| R | = \exp[-2k_0 l \epsilon_r''] \tag{8.3-36}$$

Figure 8-71 gives curves of minimum l/λ_0 versus $\tan \delta$ for various values of $| R |$.†

For an arbitrary material the wave number within the material, k, is complex with its real and imaginary parts given in terms of ϵ', ϵ'', μ', μ'' by

$$\operatorname{Re} k = \omega \left[\frac{\epsilon'\mu' - \epsilon''\mu''}{2} \left(1 + \sqrt{1 + \frac{(\mu'\epsilon'' + \epsilon'\mu'')^2}{(\epsilon'\mu' - \epsilon''\mu'')^2}} \right) \right]^{1/2} \tag{8.3-37}$$

and

$$\operatorname{Im} k = \omega \left[\frac{\epsilon'\mu' - \epsilon''\mu''}{2} \left(-1 + \sqrt{1 + \frac{(\mu'\epsilon'' + \epsilon'\mu'')^2}{(\epsilon'\mu' - \epsilon''\mu'')^2}} \right) \right]^{1/2} \tag{8.3-38}$$

† $\tan \delta$ is often called the "loss tangent" and is defined as $\tan \delta = \epsilon''/\epsilon'$.

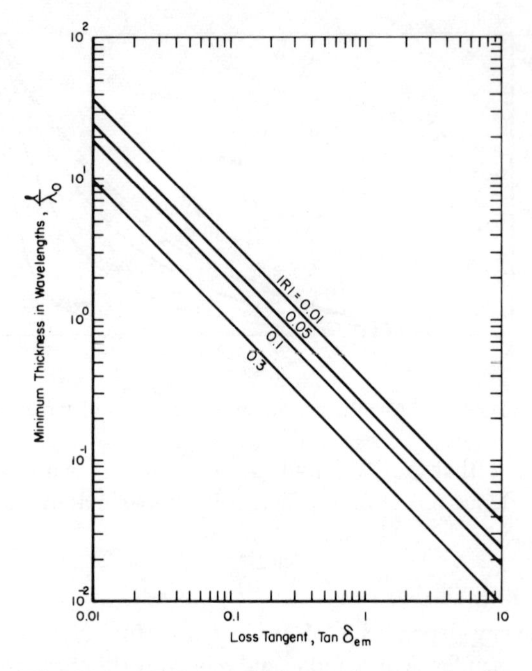

Fig. 8-71. Minimum thickness of $\mu = \epsilon$ absorber necessary to give a specific normal incidence reflection coefficient versus the loss tangent of the material.

In the event that $\mu_r = \epsilon_r$, but $\mu_r'\epsilon_r'$ is not much greater than μ_r'', ϵ_r'', then from the above equations the magnitude of the reflection coefficient at normal incidence is

$$|R| \approx e^{-2k_0 l}\left[\frac{\epsilon'^2 - \epsilon''^2}{2}\left(-1 + \sqrt{1 + \frac{4\epsilon'^2\epsilon''^2}{(\epsilon'^2 - \epsilon''^2)}}\right)\right]^{1/2} \quad (8.3\text{-}39)$$

8.3.2.1.2.2. *Inhomogeneous Layer.* Two other types of broadband RAM are based on the principle that the absorber should present a minimum discontinuity at the free space–absorber interface, and progressively increase the loss to absorb the transmitted field completely. This can be accomplished electrically by the use of an inhomogeneous layer, or mechanically by constructing the absorber of lossy cones, wedges, etc. Discussing initially the inhomogeneous layer, the goal is to achieve an impedance match between free space and a perfectly conducting surface. The optimum way to accomplish this would be to determine analytically the functional forms of $\mu(z)$ and $\epsilon(z)$ necessary to produce a specified maximum reflection coefficient over an indicated range of frequencies and angles of incidence with a specific thickness of material. This is a very difficult problem which has not been solved as yet. A considerably less difficult procedure is to assume some model for $\epsilon(z)$ and $\mu(z)$, and then solve for the resulting reflection coefficient. Even this procedure, as discussed in Section 7.2, in general requires that the equations for the reflection coefficient be evaluated numerically by computer. A number of authors have studied the reflection coefficients for a variety of forms of dielectric variation. Jacobs, using a criterion that the fractional change in $(1/\epsilon_r)\, d\epsilon_r/dz$ referred to the local wavelength, $\lambda = \lambda_0(\epsilon_r)^{-1/2}$, is a small constant at all points within the layer, or

$$\frac{\lambda_0}{2\pi}\frac{1}{(\epsilon_r)^{3/2}}\left(\frac{d\epsilon_r}{dz}\right) = a \ll 1 \tag{8.3-40}$$

obtained a dielectric variation of the form

$$\epsilon_r(z) = \left(1 - \frac{\pi a}{\lambda_0} z\right)^{-2} \tag{8.3-41}$$

as that which minimizes the reflection coefficient.[41] Throughout his analysis, Jacobs assumed $\mu_r(z)$ to be one. In the ideal case, $a = \lambda_0/\pi l$ and $|R| < 0.1$ for $l/\lambda_0 \gtrsim 0.3$. Any small amount of loss in the layer is sufficient to insure this performance. Unfortunately, with this value of a, Eq. (8.3-41) yields a dielectric constant which varies from one at $z = 0$ to infinity at $z = l$. For a more realistic type of dielectric variation, i.e., $\epsilon(0) = 1$, $\epsilon(l) = 400$, $\sigma/\omega_0\epsilon_0 = \frac{1}{2}$, then $|R| \leqslant 0.1$ for $l/\lambda_0 \gtrsim 0.42$.

The analysis by Jacobs assumes a matched interface at $z = l$, that is, interference effects due to reflections occurring at $z = l$ are neglected. When l/λ_0 is small, and $\epsilon(l)$ finite, it is to be expected that the presence of the metallic backing can be significant. In fact, in some cases, the presence of the metallic backing leads to better performance than that obtained from an ideal Jacobs layer.[42]

Among the other forms of dielectric variations which have been considered are linear and a variety of exponential variations.[43,44] Examples of some of the types of variation which have been considered and the layer thickness required for $|R| \leqslant 0.1$ are given in Table 8-6.

The angular dependence of the reflection coefficient for an arbitrary absorbing layer, subject to the condition $\mathrm{Re}\,(\mu_r\epsilon_r) \gg 1$ and l not too large, i.e., a thin dense absorber, is determined completely by the normal incidence reflection coefficient.[45] If $R(0) = |R(0)|\, e^{-i\psi}$, then

$$|R_\perp(\theta)|^2 \approx \frac{\tan^2(\theta/2) - 2\,|R(0)|\tan(\theta/2)\cos\psi + |R(0)|^2}{1 - 2\,|R(0)|\tan(\theta/2)\cos\psi + |R(0)|^2\tan^2(\theta/2)} \tag{8.3-42}$$

and

$$|R_\parallel(\theta)|^2 \approx \frac{\tan^2(\theta/2) + |R(0)|\tan(\theta/2)\cos\psi + |R(0)|^2}{1 + 2\,|R(0)|\tan(\theta/2)\cos\psi + |R(0)|^2\tan^2(\theta/2)} \tag{8.3-43}$$

For a thin dense absorber having a minimum reflection coefficient $|R(\theta_0)|_{\min}$

Table 8-6

Type of Variation	$\mu_r'(z)$	$\mu_r''(z)$	$\epsilon_r'(z)$	$\epsilon_r''(z)$	Minimum l/λ_0 for $R \leqslant 0.1$	Reference
1. Ideal Jacobs	1	0	$(1 - z/l)^{-2}$	Small constant $\ll 1$	0.3	46
2. Finite Jacobs	1	0	$(1 - 0.95\, z/l)^{-2}$	$\frac{1}{2}$	0.42	46
3. Linear	1	0	1	$3\, z/l$	0.55	48
4. Linear	1	0	1	$6\, z/l$	0.77	48
5. Exponential	1	0	1	$3.76\, (z/l)^{1.5}$	0.4	49
6. Exponential	1	0	1	$0.285\, e^{2.73z/l}$	0.35	49
6. Exponential	1	0	1	$\frac{0.25}{l/\lambda_0} [6^{z/l} - 1]$	0.35	49
8. Exponential	1	0	$2^{z/l}$	$5^{z/l} - 1$	0.56	49
9. Exponential	1	0	$10^{z/l}$	$7^{z/l} - 1$	0.68	49
10. Exponential	$3.3^{z/l}$	0	$50^{z/l}$	$0.3\, (3.3^{z/l} - 1)$	0.6	49
11. Exponential	$3.3^{z/l}$	$50^{z/l} - 1$	$50^{z/l}$	$0.3\, (3.3^{z/l} - 1)$	0.5	49
12. Three-layer discrete approximation to exponential	1	0	1	0.58 for $0.344l$ 1.16 for $0.359l$ 3.48 for $0.297l$	0.33	48

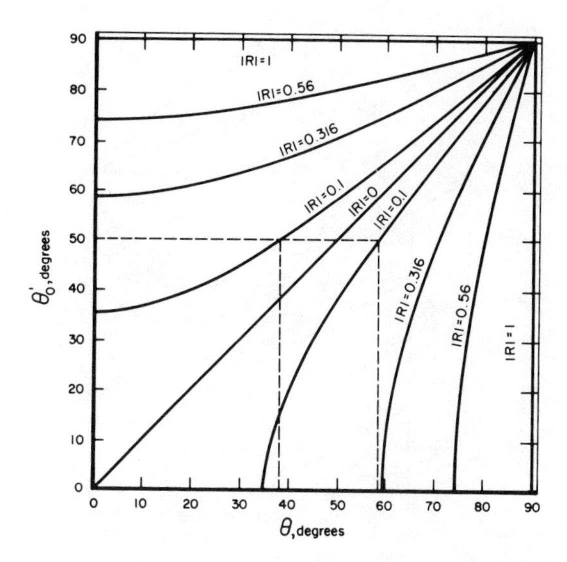

Fig. 8-72. Curves for determining the angular performance of a thin, dense absorber [after Redheffer[45]].

at some angle of incidence θ_0 and a particular polarization, if $|R(\theta_0)|_{\perp,\parallel} = \tan(q/2)$, then

$$|R(\theta)|^2_{\perp,\parallel} = \frac{\cos^2\theta_0 - 2\cos\theta_0\cos\theta\cos q + \cos^2\theta}{\cos^2\theta_0 + 2\cos\theta_0\cos\theta\cos q + \cos^2\theta} \tag{8.3-44}$$

and

$$|R(\theta)|^2_{\parallel,\perp} = \frac{\cos^2\theta\cos^2\theta_0 - 2\cos\theta_0\cos\theta\cos q + 1}{\cos^2\theta\cos^2\theta_0 + 2\cos\theta_0\cos\theta\cos q + 1} \tag{8.3-45}$$

For $R(\theta_0) = 0$, then Eqs. (8.3-44) and (8.3-45) reduce to (8.3-22) and (8.3-23). Curves of θ, θ_0 for various values of $|R|$ in decibels are given in Figure 8-72, which can be used to determine the range of incidence angles over which a specified minimum reflection coefficient can be obtained for a particular polarization; for example, if $\theta_0 = 50°$, then $|R| < 0.1$ for $38° < \theta < 58°$.

8.3.2.1.2.3. *Geometric Transition Absorbers.* Another class of broadband absorbers is that characterized by a geometrical transition from free space into a lossy media. This class of absorbers usually takes the form of pyramids or wedges of synthetic sponge rubber or plastic foam which is loaded with an electrically lossy material such as carbon particles. A sketch of such an absorber is shown in Figure 8-73. The geometrical transition may be also combined in some cases with an electrical transition by increasing the loss toward the base of the pyramids. Absorbers of this type have been built which exhibit reflection coefficients $|R| \leqslant 0.1$ for thicknesses $l/\lambda_0 \gtrsim 0.25$. In addition, better angular performance is generally available

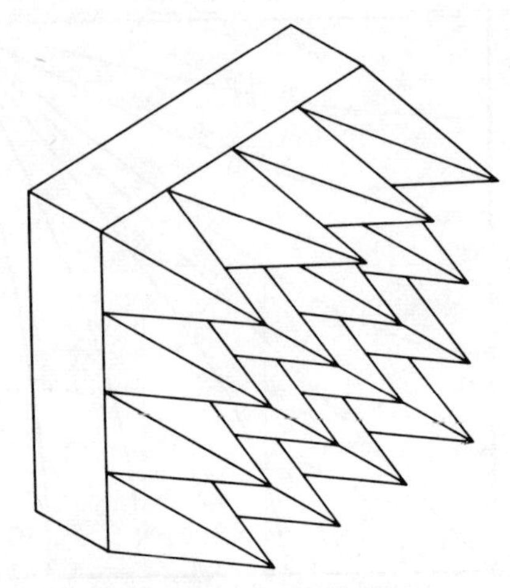

Fig. 8-73. Geometric transition absorber.

with geometric transition type absorbers than with those presenting the perfectly planar interface previously discussed. Reflection coefficients of less than 0.1 have been obtained for angles of incidence as large as 50–60° with pyramidal absorbers.

8.3.2.1.2.4. *Low-Density Absorbers.* The last type of broadband absorber to be discussed is one in which matching to free space is accomplished by using materials of very low density, i.e., materials for which $\epsilon \approx \epsilon_0$. Absorption is accomplished by incorporating a small amount of loss, and using relatively thick material in terms of wavelength. Typical examples of this type of absorber are the hair-mat types, consisting of a losely packed mat of lossy fibers, and the low-density plastic foams, such as styrofoam, in which small amounts of carbon particles have been incorporated. This type of absorber is most suitable for use in radar dark rooms and, due to its thickness and general mechanical properties, is not a likely candidate for application to most actual radar targets.

8.3.2.2. *Absorber-Covered Bodies*

In the previous section, various types of absorbing materials were discussed from the standpoint of their infinite-slab reflection coefficients. When absorbers are applied to actual radar targets their performance can, in some cases, be predicted if these "flat-plate" reflection coefficients are known. In particular, for targets large in terms of wavelength and for which the primary contributors to the radar cross section are one or more specular points, the absorber performance can be related to its flat-plate reflection coefficients. When the primary contributors are edges, surface discontinuities, traveling waves, or shadow-boundary phenomena, the effectiveness of a particular RAM is not necessarily closely related to its flat-plate performance.

It can be proven that for bodies of revolution, consisting of a material such that $\mu_r = \epsilon_r$, the backscatter cross section for a wave incident along the symmetry axis is zero.[4] In fact, the backscatter cross section for incidence along the axis of a body of revolution will be zero if at the surface of the body the tangential field components satisfy the following boundary condition:

$$\mathbf{E} - (\hat{\mathbf{n}} \cdot \mathbf{E})\,\hat{\mathbf{n}} = \sqrt{\frac{\mu_0}{\epsilon_0}}\,(\hat{\mathbf{n}} \times \mathbf{H}) \tag{8.3-46}$$

This is a form of a Leontovich boundary condition and could be satisfied, for example, by a perfectly conducting core of arbitrary shape surrounded by a lossy layer which was smoothly shaped, had $\mu_r = \epsilon_r$, and was sufficiently thick and lossy to prevent significant energy from penetrating to the core. A more exact statement of the conditions under which a curved surface is capable of satisfying a Leontovich boundary condition of the form,

$$\mathbf{E} - (\hat{\mathbf{n}} \cdot \mathbf{E})\,\hat{\mathbf{n}} = \eta_s(\hat{\mathbf{n}} \times \mathbf{H})^{\dagger} \tag{8.3-47}$$

has been given by Weston as follows[4]: (i) The index of refraction $\sqrt{\mu\epsilon}$ of the material is large and has a large imaginary part; (ii) the fields on the surface of the body vary slowly compared to the wavelength in the material; (iii) the radii of curvature of the surface are large compared to the wavelength in the material; and (iv) variations of μ and ϵ in a direction tangential to the surface are small.

8.3.2.2.1. *High-Frequency Performance.* In the high-frequency limit the specular-point contribution from a RAM-covered object will be $|R|^2\,\sigma_{\text{spec}}$ where R is the flat-plate reflection coefficient of the RAM. The effect of the curvature of the body can be taken into account, and a slight improvement over the above expression can be obtained by using a modification of physical-optics theory. Instead of assuming the magnetic field on the body surface to be twice the incident field, the following assumptions are made; for perpendicular polarization,

$$\hat{\mathbf{n}} \times \mathbf{E}^T = \hat{\mathbf{n}} \times (1 + R_{\perp})\,\mathbf{E}^i \tag{8.3-48}$$

and for parallel polarization,

$$\hat{\mathbf{n}} \times \mathbf{H}^T = \hat{\mathbf{n}} \times (1 + R_{\parallel})\,\mathbf{H}^i \tag{8.3-49}$$

where $\mathbf{E}^T$ and $\mathbf{H}^T$ are the total fields on the body surface. Using these assumptions for the total field values on the illuminated portions of the

† η_s is an equivalent surface impedance for the material. For a material such that no reflections from the interior of the body return to the surface, the surface impedance η_s is given by $\eta_s = \sqrt{\mu/\epsilon}$.

surface, and assuming the total fields are zero on the shadowed portion of the surface, a modified physical-optics result for the cross section can be obtained. In the event that the RAM coating satisfies the Leontovich boundary condition (8.3-47), the reflection coefficients $R_{\parallel}$ and $R_{\perp}$ simplify considerably thus allowing the integrals for the scattered fields to be evaluated more readily. As an example of this technique, in Figure 8-74 the radar cross section of a perfectly conducting sphere is compared with that of an absorbing sphere computed exactly by the Mie series, computed by $\sigma = |R|^2 \sigma_{\text{p.c.}}$, and computed by the modified physical-optics method discussed above. The absorbing sphere is assumed to satisfy the Leontovich boundary condition with $\eta_s = 1.5\,\eta_0$.

Since for a sphere in the resonance region the radar cross section is primarily due to the sum of a specular component and a creeping-wave component emanating from the shadow boundary, it is to be expected that using $\sigma = |R|^2 \sigma_{\text{p.c.}}$ would give better results than those predicted by modified physical optics. The physical-optics solution is incapable of accounting for the creeping-wave contribution, and introduces spurious effects at the shadow boundaries. Thus the expression $\sigma = |R|^2 \sigma_{\text{p.c.}}$

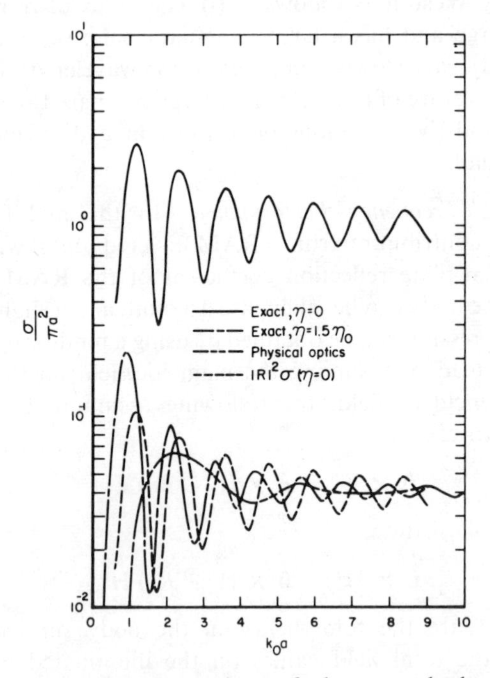

Fig. 8-74. Comparison of the exact backscatter cross sections for a perfectly conducting sphere and an absorbing sphere with various approximations.

retains the creeping-wave contributions, even though it implies that they are reduced by the absorber to the same extent as the specular contribution. A better approach to the estimation of the high-frequency effect of an absorber on a sphere would be to use the modified physical-optics technique to compute the effect of the absorber on the specular contribution, to discard the spurious shadow-boundary terms, and then to estimate the effect of the absorber on the creeping-wave contribution independently. Summing these with the proper phasing should then provide a good estimate of the effect of the absorber on the sphere cross section.

Thus to predict the performance of RAM on a specific body in the high frequency or upper resonance region, the effect of the absorber on the various analytic components of the cross section must be estimated. The influence on the specular contribution can be estimated on the basis of the flat-plate reflection coefficients, whereas the influence on the other components is not so easily determined. In the event that the Leontovich boundary condition is satisfied on the surface of the absorber-covered body, the influence of the absorber on the creeping-wave field may be partially estimated. This can be accomplished by using geometrical diffraction theory or Fock theory. Weston has computed the attenuation coefficients and the launch coefficients for a surface-diffracted ray, propagating on a surface, satisfying the Leontovich boundary condition with $\eta_s = \eta_0$, i.e., a perfect absorber.[4] Comparing these with the results for a perfectly conducting surface, the launch coefficient for the absorbing surface is of the order of one half that for the perfectly conducting surfacc, and the attenuation coefficient is approximately twice that for the perfect conductor. Thus a material which is a perfect flat-plate absorber will reduce a creeping wave contribution to the cross section of a sphere with $k_0a \approx 16$ in the order of 20 db, although it will not eliminate it completely. For absorbers which are not perfect yet provide good flat-plate performance, the above conclusions should also hold provided the Leontovich boundary condition is satisfied.

The question might arise at this time as to how a zero cross section results for axial incidence on a finite body of revolution for which $\mu_r = \epsilon_r$, when, although the specular contribution is zero, nonzero creeping waves can be supported by the body. Weston[4] has shown that this is due to the symmetry axis being a caustic of the diffracted field. For $\eta_s = \eta_0$, the launch and attenuation coefficients are the same for both TE and TM creeping-wave modes, and for axial-incidence backscatter the vector sum of all surface diffracted rays propagating toward the source is zero. Thus, as the direction of incidence moves off the symmetry axis, even though the specular contribution remains zero, a nonzero backscatter cross section would in general result due to the creeping-wave contribution alone.

For an absorber which satisfies the Leontovich boundary condition, there still remains the question of whether there exists a value of η_s which will eliminate entirely any creeping-wave contribution to the backscatter. This question cannot be completely answered at this time; however, Streifer[46] has studied the creeping-wave propagation coefficients for the case of arbitrary η_s and obtained expressions for ν_1. For an arbitrary relative surface impedance η, then†

$$\begin{aligned}\nu_1 = {} & x + \beta_1 e^{i\pi/3}\left(\frac{x}{2}\right)^{1/3} + \frac{\exp[i2\pi/3]}{10}\left[\frac{\beta_1^2}{6} + \frac{1}{\beta_1}\right]\left(\frac{2}{x}\right)^{1/3} \\ & + \frac{1}{50}\left[\frac{\beta_1^3}{28} + 1 + \frac{1}{4\beta_1^3}\right]\left(\frac{2}{x}\right) + \frac{i\eta}{\beta_1} e^{-i\pi/3}\left(\frac{x}{2}\right)^{1/3} - \frac{\eta^2}{2\beta_1^3}\left(\frac{x}{2}\right) \\ & - \frac{i\eta}{10}\left[1 + \frac{1}{\beta_1^3}\right] - i\eta^3 e^{-i2\pi/3}\left[\frac{1}{3\beta_1^2} - \frac{1}{2\beta_1^5}\right]\left(\frac{x}{2}\right)^{4/3} \\ & + \frac{\eta^2}{10} e^{-i2\pi/3}\left[\frac{1}{\beta_1^2} - \frac{3}{2\beta_1^5}\right]\left(\frac{x}{2}\right)^{1/3} + \frac{\eta^4}{4} e^{-i\pi/3}\left[\frac{7}{3\beta_1^4} - \frac{5}{2\beta_1^7}\right]\left(\frac{x}{2}\right)^{5/3} \\ & + \frac{i\eta}{20} e^{-i2\pi/3}\left[\frac{1}{5\beta_1^2} - \frac{41\beta_1}{126} - \frac{3}{10\beta_1^5}\right]\left(\frac{2}{x}\right)^{2/3} \\ & - \frac{i\eta^3}{2} e^{-i\pi/3}\left[\frac{1}{2\beta_1^7} - \frac{11}{45\beta_1} - \frac{7}{15\beta_1^4}\right]\left(\frac{x}{2}\right)^{2/3} \\ & + i\eta^5\left[\frac{21}{20\beta_1^6} - \frac{1}{5\beta_1^3} + \frac{7}{8\beta_1^9}\right]\left(\frac{x}{2}\right)^2 + O(x^{-5/3}) \end{aligned} \tag{8.3-50}$$

valid for $|\eta| \ll 1$; and

$$\begin{aligned}\nu_1 = {} & x + \alpha_1 e^{i\pi/3}\left(\frac{x}{2}\right)^{1/3} + \frac{\alpha_1^2 \exp[i2\pi/3]}{60}\left(\frac{2}{x}\right)^{1/3} - \frac{1}{140}\left(1 - \frac{\alpha_1^3}{10}\right)\left(\frac{2}{x}\right) \\ & - \frac{i}{\eta} - \left(\frac{1}{2} - \frac{1}{\eta^2}\right)\frac{i\alpha_1 \exp[i\pi/3]}{3\eta}\left(\frac{2}{x}\right)^{2/3} - \left(1 - \frac{1}{\eta^2}\right)\frac{1}{4\eta^2}\left(\frac{2}{x}\right) \\ & + \left(\frac{1}{72} + \frac{13}{18\eta^2} - \frac{1}{\eta^4}\right)\frac{i\alpha_1^2 \exp[i2\pi/3]}{5\eta}\left(\frac{2}{x}\right)^{4/3} + O(x^{-5/3}) \end{aligned} \tag{8.3-51}$$

valid for $1 \leqslant |\eta| \leqslant \infty$.‡ In Eqs. (8.3-50) and (8.3-51), $x = k_0a$; and β_1, α_1 are given in Table 2-3.

Curves of Im ν_1 and Re ν_1 versus η for η real and $k_0a = 10, 20$ are given in Figure 8-75 (from Reference 46). From these it appears that increasing η beyond one increases the attenuation coefficient a very small amount, while decreasing η from one always reduces the attenuation. These results are for η real. For η complex, Eqs. (8.3-50) and (8.3-51) are

† In the following equations and discussion η is the relative surface impedance; it has been normalized by dividing by $\sqrt{\mu_0/\epsilon_0}$.

‡ The root ν_1 corresponds to the root q_1, discussed in Section 8.1.1.1.1.4 as follows:

$$\nu_1 = k_0a + q_1 \frac{\exp[i\pi/3]}{3^{1/3}}\left[\frac{k_0a}{2}\right]^{1/3}$$

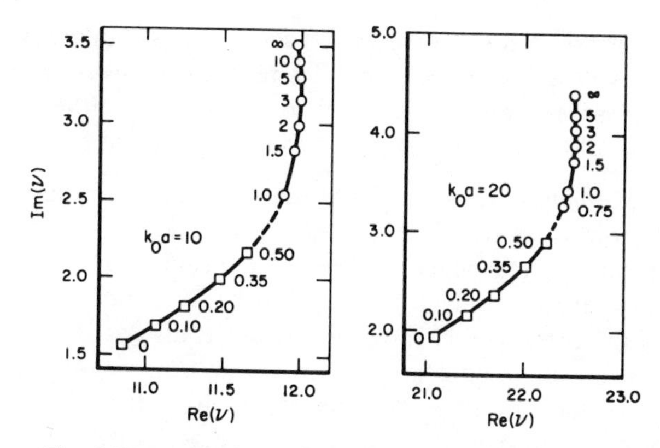

Fig. 8-75. The real and imaginary parts of ν_1 versus η for $k_0a = 10$ and $k_0a = 20$ [from Streifer[46]].

valid; however, the variation of the propagation coefficient with complex η has not been specifically examined. In addition, the launch coefficients have not been determined for values of η other than one, zero, and infinity.

In the event that the Leontovich boundary condition is not satisfied at the surface of a RAM-covered object, very little is known about the magnitude of the creeping-wave contribution to the backscattered field.

The effectiveness of RAM in reducing the contributions of edges to the radar cross section of a body could be partially estimated from results on the diffraction by infinite absorbing wedges and half-planes. Using the Sommerfeld–Macdonald technique of considering the local fields at an edge to be the same as for the corresponding infinite wedge and integrating along the edge to determine the total edge contribution to the scattered fields, as discussed previously in Section 2.3.1.1.3, should provide an estimate of the absorber effectiveness. Unfortunately, however, no results are as yet available for absorbing wedges or half-planes in forms which are suitable for the application of the Sommerfeld–Macdonald technique. Similarly, the diffraction coefficients for use with the geometrical theory of diffraction are not known for absorbing edges.

The influence of RAM on traveling-wave contributions to the radar cross section of an object is also difficult to assess. In general, the presence of a lossless dielectric sheath surrounding a body would tend to enhance traveling-wave contributions to the cross section. In addition, higher-order traveling-wave modes could be excited resulting in a nonzero end-fire echo. The effect of moderate amounts of loss in such a dielectric sheath is unknown in general. For highly conducting bodies, the effect of a reduction in conductivity is to decrease the traveling-wave echo. This is clearly illustrated by Peter's results in which identical long thin wires, one silver and one steel, produced maximum traveling-wave echoes of 28.2 square wavelengths and 9 square wavelengths, respectively.[47] Thus traveling-wave contributions to the radar cross section of a specific body may be either reduced or enhanced by the addition of RAM designed on a flat-plate basis.

8.3.2.2.2. *Resonance-Region Performance.* In the resonance region, it is very difficult to assess correctly the performance capabilities of a particular flat-plate RAM when applied to a specific body. At the upper (high-frequency) end, the effect of the absorber on the various analytic components of the cross section can be estimated as discussed in the previous section; however, in the middle- and low-frequency regions, little is available in the way of analytical results. Experimental results indicate that, in general the performance of a specific absorber begins to deteriorate as the resonance region is approached and continues to decline, until in the Rayleigh region it is no longer an effective absorber. An example of this is given in Figures 8-76 and 8-77, where the effect of an absorber, which gives a normal incidence reduction of 20 to 25 db on a 1.2 by 1.2 λ_0 square plate, is shown on smaller size plates.[48] It is apparent from this figure that as the plate size is reduced, the normal-incidence effectiveness of the absorber declines severely.

This effect holds for other than flat plates, as can be seen from

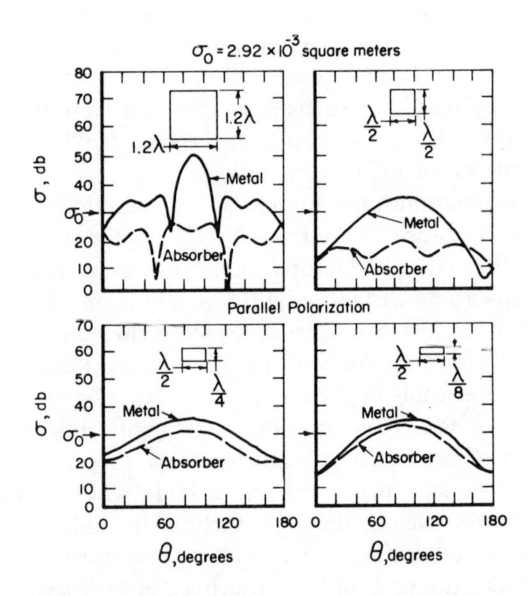

Fig. 8-76. The backscatter cross sections of perfectly conducting and absorbing small flat plates versus the angle of incidence, parallel polarization [from Sletten *et al.*[48]].

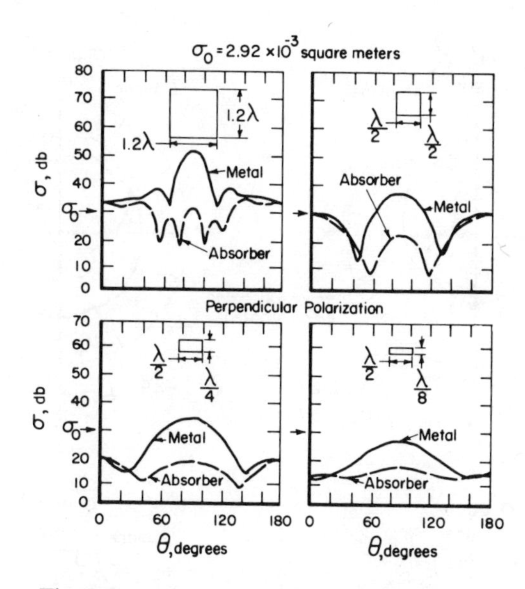

Fig. 8-77. The backscatter cross sections of perfectly conducting and absorbing small flat plates versus the angle of incidence, perpendicular polarization [from Sletten *et al.*[48]].

Figure 8-78, where the bistatic cross sections for six different size absorber-coated spheres are given.[49] Figure 8-79 summarizes the performance of these spheres. It appears that for backscatter, only minor differences in the cross section occur versus sphere size, with the absorber giving a reduction of 15–20 db, except for the $d/\lambda_0 = 0.212$ sphere where a reduction of less than 10 db resulted. It is also interesting to note from this figure that the absorbing coating produces an enhancement of the cross section at certain bistatic angles for the smaller spheres.

The above statements are not intended to imply that effective absorbers do not exist for the resonance region, but rather that present RAM design techniques do not lead to absorbers which are particularly effective in the resonance region, and any cross section reduction obtained in this region is generally the residual effect of a good high-frequency design.

8.3.2.2.3. *Rayleigh-Region Performance.* The major characteristic of Rayleigh-region scattering is the dependence of the cross section on the square of the body volume, and the relative independence of the cross section on the body shape and material properties. The cross section does, of course, depend upon the body material, however, the dependence is in general smoothly varying and exhibits no resonance effects. This is well

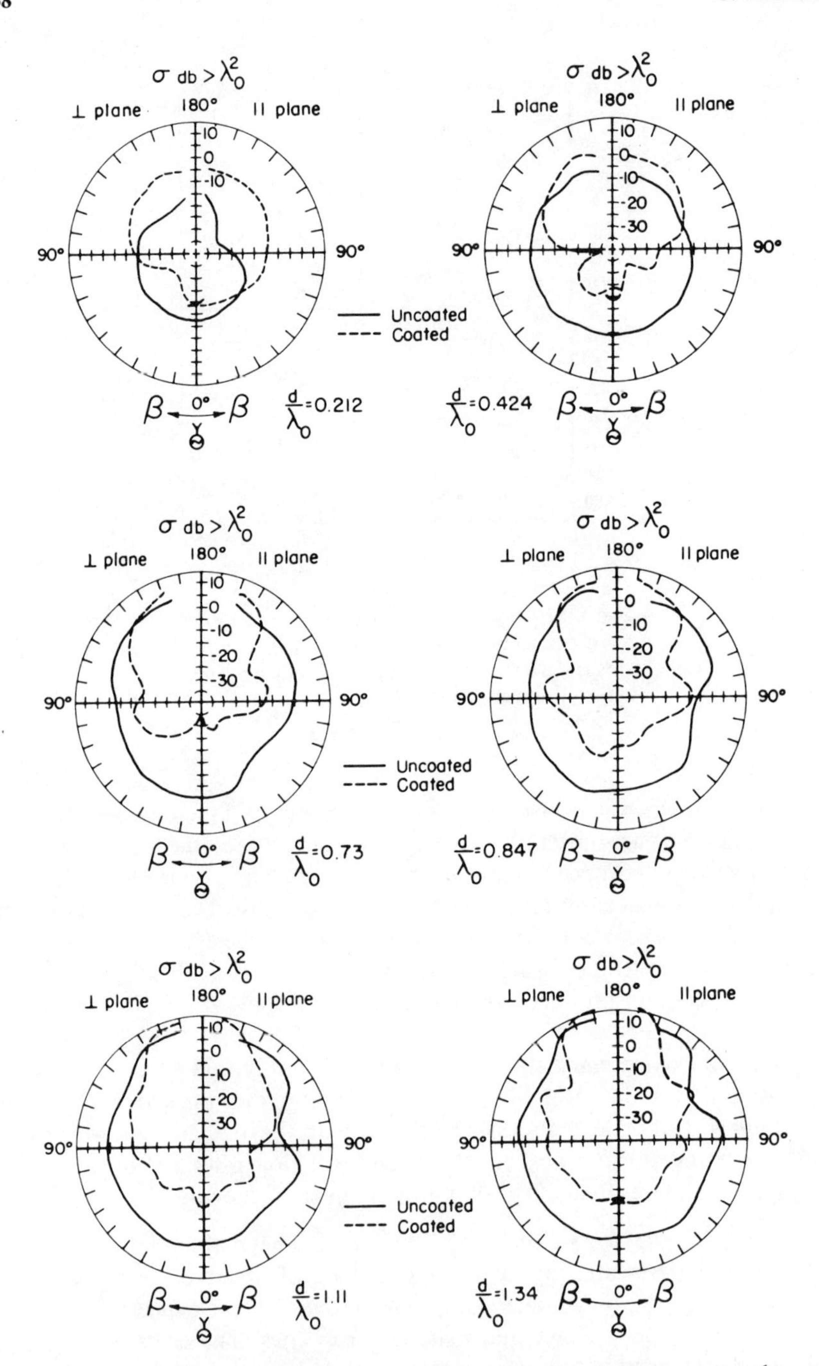

Fig. 8-78. Bistatic cross sections of small absorber-coated spheres versus the bistatic angle β [after Garbacz and Moffatt[49]].

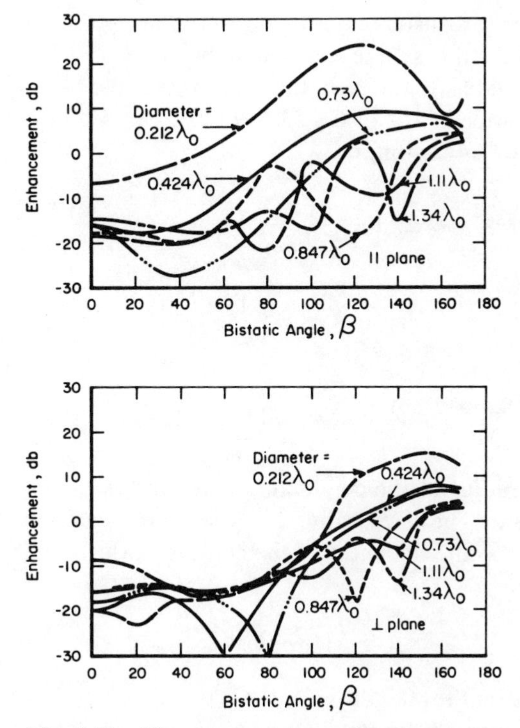

Fig. 8-79. The cross section enhancement produced by the absorbing coating on the spheres of Figure 8-78 versus the bistatic angle [after Garbacz and Moffatt[49]].

illustrated by the Rayleigh backscatter cross section for a homogeneous dielectric sphere. The Rayleigh-region cross section for such a sphere is

$$\sigma(0) = 4\pi k_0{}^4 a^6 \left[\frac{\epsilon_r - 1}{\epsilon_r + 2} - \frac{\mu_r - 1}{\mu_r + 2}\right]^2 \tag{8.3-52}$$

It is obvious from this equation that there is only one situation where $\sigma(0)$ can be zero, and this occurs when $\mu_r = \epsilon_r$. In this event $\sigma(0)$ is zero at all frequencies, not just in the Rayleigh region. However, $\sigma(0)$ can be small even if $\epsilon_r \neq \mu_r$; this occurs when $|\mu_r|, |\epsilon_r|$ are both much greater than one, and the cross section is approximately given by

$$\sigma(0) \approx 36\pi k_0{}^4 a^6 \left[\frac{1}{\mu_r} - \frac{1}{\epsilon_r}\right]^2 \tag{8.3-53}$$

with $\mu_r, \epsilon_r \gg 1$. It is apparent from Eq. (8.3-53) that $\sigma(0)$ is largely determined by the value of the smallest parameter, ϵ_r or μ_r.

In the more realistic case of an homogeneously coated perfectly conducting object, the sphere can again be used as a model since exact solutions are available. In the event that $k_0 a_0 \leqslant 0.2$ and $| k_0 \sqrt{\mu_r \epsilon_r} | a_0 \leqslant 0.2$, then the modal impedances of Eq. (3.4-18) can be used in (3.4-12) to obtain the backscatter cross section, or

$$\sigma(0) = \pi k_0{}^4 a_0{}^6 \left[\frac{1 - (2y_1/k_0 a_0)}{1 + (y_1/k_0 a_0)} - \frac{1 - (2z_1/k_0 a_0)}{1 + (z_1/k_0 a_0)} \right]^2 \tag{8.3-54}$$

with

$$z_1 = k_0 a_0 \mu_r \left[\frac{1 - \delta^3}{2 + \delta^3} \right] \tag{8.3-55}$$

and

$$y_1 = k_0 a_0 \epsilon_r / 2 \left[\frac{1 + 2\delta^3}{1 - \delta^3} \right] \tag{8.3-56}$$

where a_1 is the radius of the conducting core, and $a_0 = \delta a_1$ is the radius of the coated sphere. The backscatter cross section is zero in this case when $z_1 = y_1$. For a specific thickness of the coating δ, this requires

$$\frac{\epsilon_r}{\mu_r} = \frac{2(1 - \delta^3)}{(1 + 2\delta^3)(2 + \delta^3)} \tag{8.3-57}$$

and it is obvious that for $0 \leqslant \delta \leqslant 1$, the ratio $(\epsilon_r/\mu_r) \leqslant 1$. Thus, the coating material must be magnetic, and μ_r must be greater than ϵ_r. In principle then, it is possible by means of a suitable absorbing coating to reduce the Rayleigh-region cross section of a perfectly conducting sphere to zero. In practice, however, this can be accomplished only for small objects, since the requirement for $k_0 a_0 \leqslant 0.2$ and $| k_0 \sqrt{\mu_r \epsilon_r} | a_0 \leqslant 0.2$ imposes severe restrictions on the size of the object to be coated. For example, at 10 Mcps, using a coating of Ferramic *A* ($\epsilon_r = 8.87$, $\mu_r = 24.6$) in order for $| k_0 \sqrt{\mu_r \epsilon_r} | a_0 \leqslant 0.2$ to be satisfied, the total radius a_0 must be less than 0.65 meters, and the actual radius of the conducting core, a_1, is less than 0.4 meters.

8.3.3. Impedance Loading

Recently, the technique designated variously as "impedance" or "reactive" loading has received extensive investigation as a means for controlling the radar cross section of an object. In essence this technique consists of loading the body surface with distributed or lumped impedances. As a design problem, the question arises as to what values of impedance should be used, and where should the loading be placed in order to achieve the desired cross-section control. At the present time, no general solutions to this problem exist, and a detailed examination of the surface fields excited on each body shape by a particular incident field is necessary in order to estimate the influence of a particular load placed at a specific point.

Some general relationships can be obtained, however, which appear to be valid independent of the body shape and type of loading. These results are closely related to the theory of antenna scattering which will be discussed in Section 8.4.

8.3.3.1. *Theoretical Analysis*

A number of researchers have obtained expressions for the scattered field from an arbitrary perfectly conducting body loaded at a point by an arbitrary impedance as the sum of a component independent of the load value and a load-dependent component. Green[50] has obtained an expression for the scattered field as

$$\mathbf{E}^s(Z) = \mathbf{E}^s(Z_a^*) - \Gamma I(Z_a^*)\,\mathbf{E}^r \tag{8.3-58}$$

where $\mathbf{E}^s(Z_a^*)$ is the field scattered by the object with a conjugate matched load, $\mathbf{E}^r$ is the field radiated by the object if excited by a unit current source at the load point, Z_a^* is the conjugate of the equivalent impedance seen when looking into the load terminals, $I(Z_a^*)$ is the current flowing through a conjugate matched impedance at the load terminals when the object is in the presence of the incident field, and

$$\Gamma = \frac{Z_L - Z_a^*}{Z_a + Z_L} \tag{8.3-59}$$

Upon examining Eq. (8.3-58) it is apparent that the first term depends only on the body shape and the location of the load point, and is independent of the load value Z_L. Green calls this term the "structural" scattering component. The second term of Eq. (8.3-58) depends upon Z_L through the coefficient Γ, and is designated by Green as the "antenna mode" scattering component. For a given body and load location, the cross section can be controlled only to the extent that the value of Γ determines the value of the second term of Eq. (8.3-58). For many bodies, a location for the load point exists such that the antenna mode component can be made equal and opposite in phase to the structural component, thus resulting in zero scattered field for a particular frequency and aspect angle.

Harrington[51] has also examined loaded scatterers, considering the receiver, transmitter, and multiply loaded scatterer as an $n + 2$ port linear system, with n the number of load points. For a scatterer loaded at a single point, Harrington's formulation leads to a result equivalent to Eq. (8.3-58).

8.3.3.2. *Applications*

Numerous investigations of the application of impedance loading to specific body shapes have been undertaken recently. These have been both theoretical and experimental in nature, and in most cases consist of an examination of the optimum impedance required at a specific load point to minimize the backscattered field for various angles of incidence. Historically, one of the first objects investigated was the thin wire, and this work will be reviewed initially.

8.3.3.2.1. *Singly Loaded Cylinders.* Several authors[52,56] have studied thin perfectly conducting cylinders with and without central loads; however, Chen and Liepa[54] have obtained a general expression for the optimum load impedance and the scattered field subject only to the conditions that $k_0 a \ll 1$ and $\lambda_0/4 < 2h < 2\lambda_0$, with a the cylinder radius and $2h$ the cylinder length. For a plane wave incident at an arbitrary angle θ on a perfectly conducting, thin, centrally loaded cylinder, as illustrated in Figure 8-80, the backscattered field is

$$
\begin{aligned}
E_\theta(\theta) = -\frac{1}{k_0}\frac{\exp[ik_0 r]}{r}\Bigg\{&\left[\frac{\cos(k_0 h \sin\theta) - M_2 T_{\theta a} - N_2 T_{sa}}{\cos(k_0 h) - M_1 T_{ca} - N_1 T_{sa}}\right]\left[\frac{2M_1}{\cos^2\theta \sin\theta}\right] \\
&\times [\sin(k_0 h)\sin\theta\cos(k_0 h\sin\theta) - \cos(k_0 h)\sin(k_0 h\sin\theta)] \\
&+ \left[\frac{N_1\cos(k_0 h\sin\theta) - N_1 M_2 T_{\theta a} + M_1 N_2 T_{ca} - N_2\cos k_0 h}{\cos(k_0 h) - M_1 T_{ca} - N_1 T_{sa}}\right] \\
&\times \left[\frac{2[\cos(k_0 h\sin\theta) - \cos k_0 h]}{\cos^2\theta}\right] \\
&- \left[\frac{M_2}{2\sin\theta}\right][2k_0 h\sin\theta - \sin(2k_0 h\sin\theta)] \\
&+ \left[\frac{\tfrac{1}{2}\sin(k_0 h\sin\theta)\sec(k_0 h/2) - \sin[(k_0 h/2)\sin\theta)]}{T_a(h) - \tfrac{1}{2}\sec(k_0 h)\,T_b(h)}\right]\left[\frac{1}{2\cos^2\theta\sin\theta}\right] \\
&\times [\sin(k_0 h)[(1+\sin^2\theta)\sin(2k_0 h\sin\theta) - 2k_0 h\sin\theta\cos^2\theta] \\
&- 4\cos(k_0 h)\sin\theta\sin^2(k_0 h\sin\theta)]\Bigg\}
\end{aligned}
\qquad (8.3\text{-}60)
$$

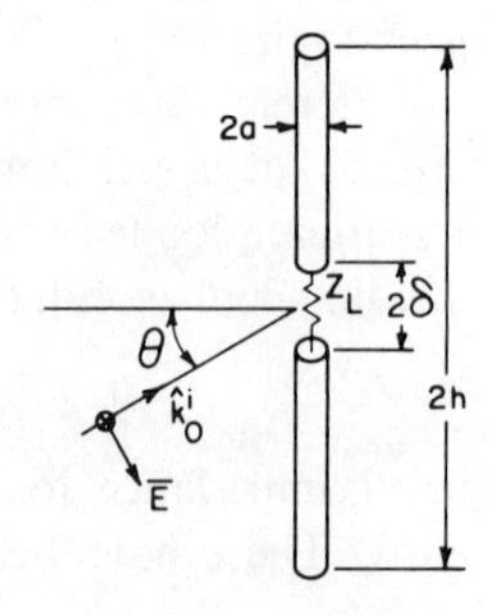

Fig. 8-80. A thin center-loaded cylinder.

where

$$M_1 = \frac{1}{T_{cd}}(1 - \cos k_0 h) \tag{8.3-61}$$

$$N_1 = \frac{-Z_L \sin k_0 h(1 - \cos k_0 h)^2}{T_{cd} Z_L \sin^2 k_0 h + i60 T_{cd} T_{sd} \cos k_0 h} \tag{8.3-62}$$

$$M_2 = \frac{1}{T_{\theta d}}[1 - \cos(k_0 h \sin \theta)] \tag{8.3-63}$$

$$N_2 = \frac{-Z_L \sin k_0 h[1 - \cos(k_0 h \sin \theta)]^2}{T_{\theta d} Z_L \sin^2 k_0 h + i60 T_{\theta d} T_{sd} \cos k_0 h} \tag{8.3-64}$$

and

$$\begin{aligned}
T_{cd} &= C'(0) - C'(h) - \cos(k_0 h)[E'(0) - E'(h)] \\
T_{sd} &= \sin(k_0 h)[C'(0) - C'(h)] - \cos(k_0 h)[S'(0) - S'(h)] \\
T_{\theta d} &= C_\theta'(0) - C_\theta'(h) - \cos(k_0 h \sin \theta)[E'(0) - E'(h)] \\
T_{ca} &= C'(h) - \cos(k_0 h)E'(h) \\
T_{sa} &= \sin(k_0 h)C'(h) - \cos(k_0 h)S'(h) \\
T_{\theta a} &= C_\theta'(h) - \cos(k_0 h \sin \theta)\, E'(h) \\
T_a(h) &= [\sin(k_0 h \sin \theta)\, S'(h/2) - \sin(k_0 h)S_\theta'(h/2)] \\
T_b(h) &= [\sin(k_0 h \sin \theta)\, S'(h) - \sin(k_0 h)S_\theta'(h)]
\end{aligned} \tag{8.3-65}$$

The functions C_θ, S_θ, C', S', and E' in the above equations are defined as

$$C_\theta'(x) + iS_\theta'(x) = \int_{-h}^{h} f(x, \tau, \pi/3) \exp[ik_0 \tau \sin \theta]\, d\tau \tag{8.3-66}$$

$$C'(x) + iS'(x) = \int_{-h}^{h} f(x, \tau, \pi/3) \exp[ik_0 \tau]\, d\tau \tag{8.3-67}$$

$$E'(x) = \int_{-h}^{h} f(x, \tau, \pi/3)\, d\tau \tag{8.3-68}$$

with

$$f(x, \tau, \alpha) = \frac{\exp[ik_0 \sqrt{(x - \tau)^2 + 4a^2 \sin^2(\alpha/2)}]}{\sqrt{(x - \tau)^2 + 4a^2 \sin^2(\alpha/2)}} \tag{8.3-69}$$

In Figures 8-81, 82 curves of the backscatter cross section computed from Eq. (8.3-60) are compared with experimental results for several cylinders and different values of the load impedance Z_L.[54] The agreement is seen to be quite good.

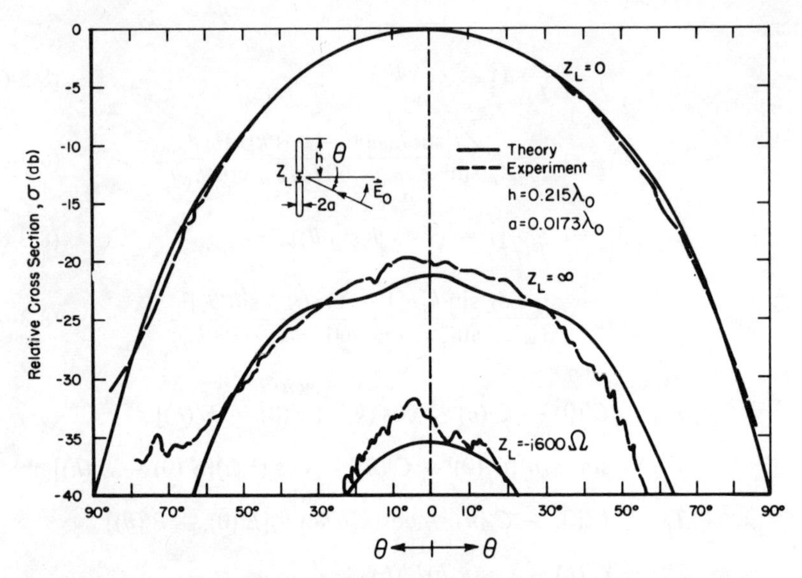

Fig. 8-81. Experimental and theoretical relative backscatter cross sections versus the angle of incidence for thin centrally loaded cylinders with several values of load, $h = 0.215\lambda_0$, $a = 0.0173\lambda_0$ [from Chen and Liepa[54]].

Chen and Liepa also determine the optimum value of impedance required to minimize the normal-incidence backscatter cross section to be

$$[Z_L]_{\min} = \frac{i60T_{sd}(1 - k_0h \cot k_0h)}{2 \cos k_0h - 2 + k_0h \sin k_0h} \tag{8.3-70}$$

In this particular case, the above value of Z_L will theoretically give zero backscatter at normal incidence. Figure 8-83 plots $[Z_L]_{\min}$ as a function of the cylinder length for a cylinder with a radius of $a = 0.0173\,\lambda_0$.

Chen has also investigated the optimum central loading for minimizing the backscatter cross section of a perfectly conducting cylinder which is not thin. In general, for this case, a different loading technique is necessary for each polarization, parallel or perpendicular. Considering the incident electric-field vector to lie in the plane formed by the axis of the cylinder and the incident field direction, the load must be of a type that the longitudinal currents on the cylinder are modified in order for the load to modify the backscattering properties effectively. Chen[55] considered a circumferential slot located at the center of the cylinder and loaded by means of a radial transmission line, as shown in Figure 8-84. The optimum load impedance for zero normal-incidence backscatter cross section from such a cylinder was determined by Chen to be

$$[Z_{Lm}]_0 = \frac{i60T_{sm}(1 - k_0h \cot k_0h)}{2 \cos k_0h - 2 + k_0h \sin k_0h} \tag{8.3-71}$$

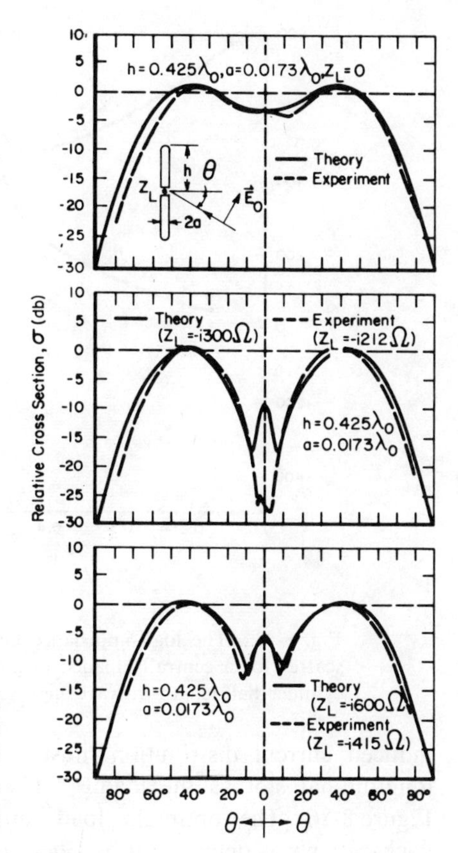

Fig. 8-82. Experimental and theoretical relative backscatter cross sections versus the angle of incidence for thin centrally loaded cylinders with several values of load, $h = 0.425\lambda_0$, $a = 0.0173\lambda_0$ [from Chen and Liepa[54]].

with

$$T_{sm} = \int_{-h}^{h} \sin k_0(h - \tau)\, d\tau \int_0^{2\pi} a \cos m\alpha[f(0, \tau, \alpha) - f(h, \tau, \alpha)]\, d\alpha \qquad (8.3\text{-}72)$$

where $f(x, \tau, \alpha)$ is defined by Eq. (8.3-69). The parameter m in the above equations corresponds to the induced surface current modes on the cylinder, where the total longitudinal surface current is the sum of an infinite number of such modes, or

$$\mathbf{K}(z, \phi) = \sum_{m=0}^{\infty} \mathbf{K}_m(z) \cos m\phi \qquad (8.3\text{-}73)$$

The optimum modal impedance for zero backscatter is plotted in Figure 8-85 versus the cylinder length for $m = 0, 1, 2$ and $k_0 a = 0.833$.[55]

For the incident electric-field vector perpendicular to the plane formed by the cylinder axis and the incident-field direction, the circumferential

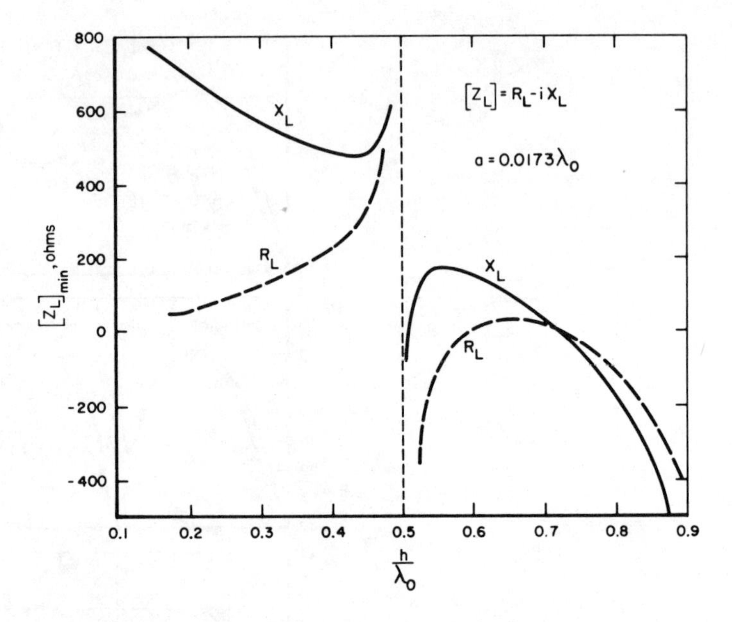

Fig. 8-83. The load impedance giving zero normal-incidence backscatter for a centrally loaded cylinder with $a = 0.0173\lambda_0$ versus the cylinder half-length in wavelengths.

induced current distribution must be modified; thus Chen considered a longitudinal slot as the load.[56] The configuration for this is shown in Figure 8-86. The optimum load impedance for zero normal-incidence backscatter was determined by Chen to be

$$[Z_L]_0 = \left[-i2\pi a\sqrt{\frac{\mu_0}{\epsilon_0}}\sum_{n=0}^{\infty}(-1)^n\,\epsilon_n\,\frac{J_n'(k_0a)}{H_n^{(1)\prime}(k_0a)}\right]$$

$$\times\left[-i\,\frac{2}{\pi k_0 a}\left(\sum_{n=0}^{\infty}(i)^n\,\frac{\epsilon_n\cos n\phi_0}{H_n^{(1)\prime}(k_0a)}\right)^2\right.$$

$$-\left(\sum_{n=0}^{\infty}\frac{2\epsilon_n\sin\left(\frac{n\delta}{2}\right)\cos^2 n\phi_0}{n\delta}\,\frac{H_n^{(1)}(k_0a)}{H_n^{(1)\prime}(k_0a)}\right)$$

$$\left.\times\left(\sum_{n=0}^{\infty}(-1)^n\,\epsilon_n\,\frac{J_n'(k_0a)}{H_n^{(1)\prime}(k_0a)}\right)\right]^{-1} \tag{8.3-74}$$

This is plotted in Figure 8-87 versus the cylinder radius k_0a for $\phi_0 = 180°$ and $\delta = 2°$. In Figure 8-88, $[Z_L]_0$ is plotted versus the slot location ϕ_0, for a cylinder of radius $k_0a = 0.5$ and $\delta = 2°$.

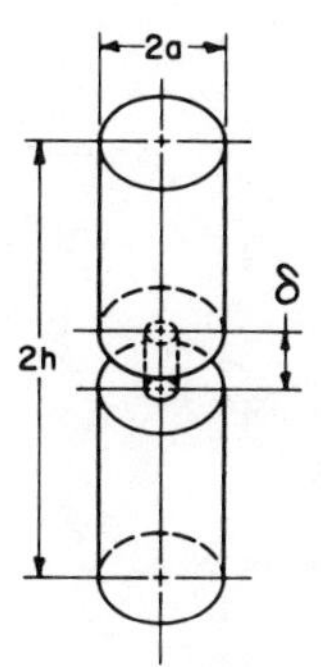

Fig. 8-84. A thick circumferentially loaded cylinder.

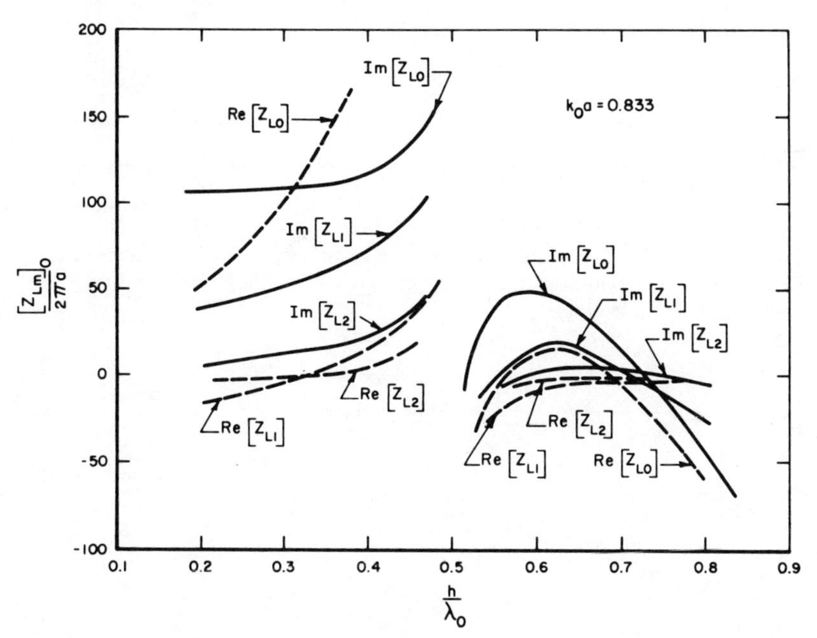

Fig. 8-85. The optimum load impedance for zero normal-incidence backscatter from a thick circumferentially loaded cylinder versus the cylinder half-length in wavelengths for $m = 0, 1, 2$ and $k_0 a = 0.833$.

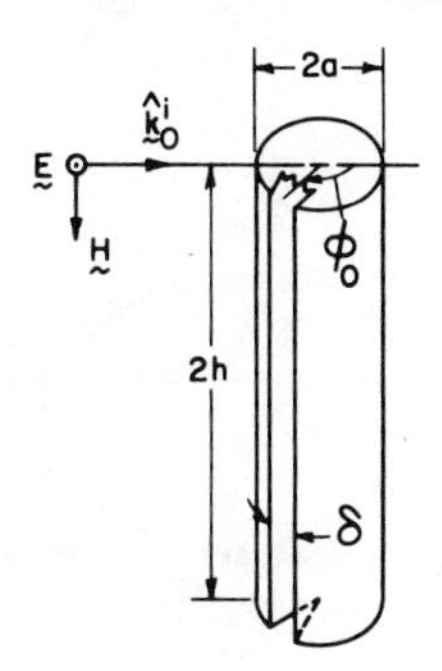

Fig. 8-86. Longitudinally loaded thick cylinder.

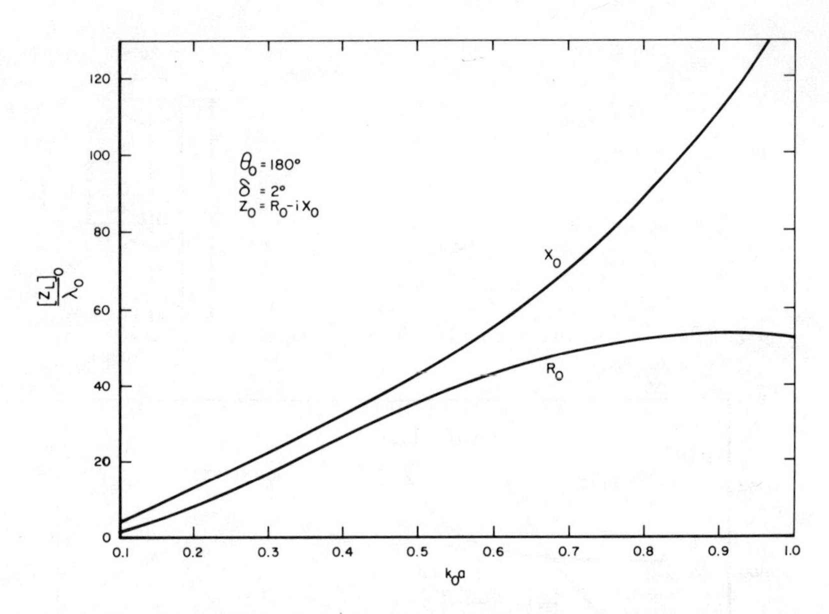

Fig. 8-87. The load impedance for zero normal-incidence backscatter cross section from an infinite longitudinal-slot-loaded cylinder versus the cylinder radius.

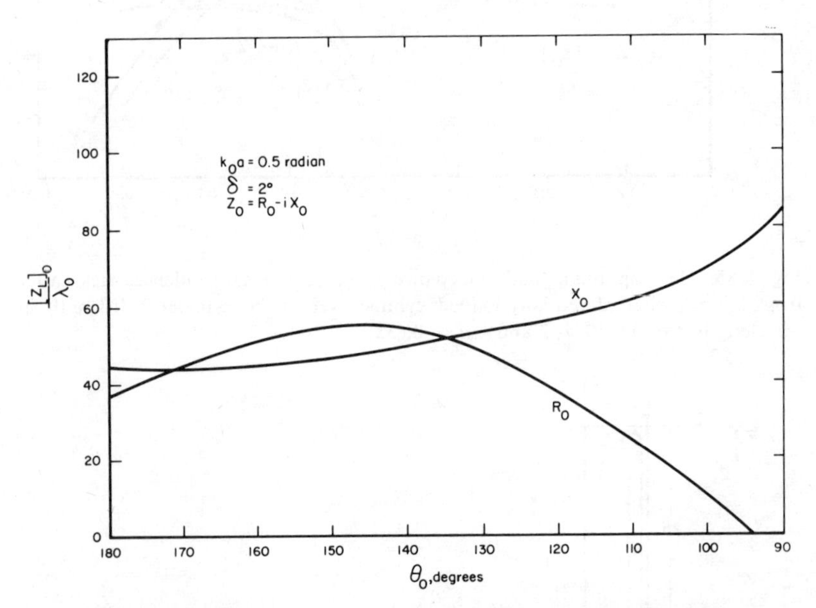

Fig. 8-88. The load impedance for zero normal-incidence backscatter cross section from an infinite longitudinal-slot-loaded cylinder versus the slot location.

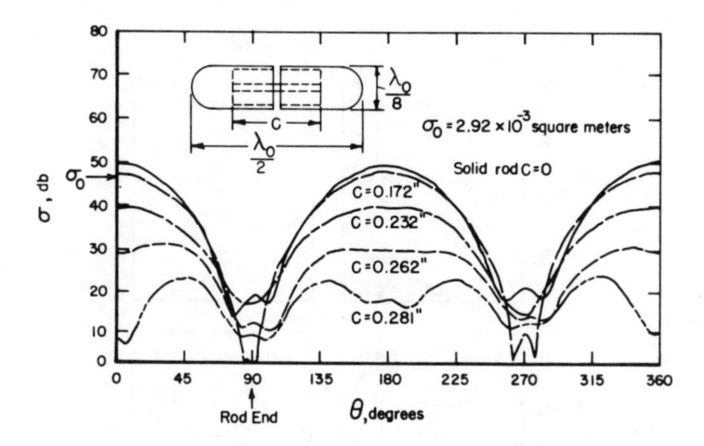

Fig. 8-89. Experimental backscatter cross sections of thick circumferentially loaded cylinders versus the angle of incidence for parallel polarization [after Sletten *et al.*[48]].

In Figure 8-89 are shown the results of experimental measurements of the backscatter cross section for parallel polarization of a thick cylinder reactively loaded by means of an internal, cavity-backed, circumferential slot. These measurements were made by Sletten *et al.*,[48] and reductions of the broadside backscatter cross section of greater than 30 db were obtained. The effects of the circumferential slot on a perpendicularly polarized incident field is negligiblc, as illustrated in Figure 8-90, where measurements of the backscatter cross section for a thick, reactively-loaded cylinder and double cone are shown for both polarizations.[48]

8.3.3.2.2. *Traveling-Wave Echo Reduction.* For long, thin, conducting bodies, an incident field with an electric-field component polarized parallel to the long dimension of the body excites current modes similar to those of a traveling-wave antenna, resulting in a large, near nose-on backscatter cross section. (This traveling-wave cross section is discussed in more detail in Section 2.3.3.) Hanson[57] demonstrated that impedance loading can be used to reduce the near-nose-on backscatter cross section of such bodies.

Some of Hanson's experimental results are shown in Figure 8-91. He loaded an ogive by placing a coaxially fed probe at the rear tip, and measured the backscatter cross section for the unloaded ogive and for the loaded ogive, with the load adjusted to give minimum backscatter at the near-nose-on peak. The value of load impedance required for minimizing the backscatter was determined to be $Z_L = 68 + i98$ ohms.

Chen[58] has theoretically examined the problem of minimization of the traveling-wave echo from a long, thin body by impedance loading, and

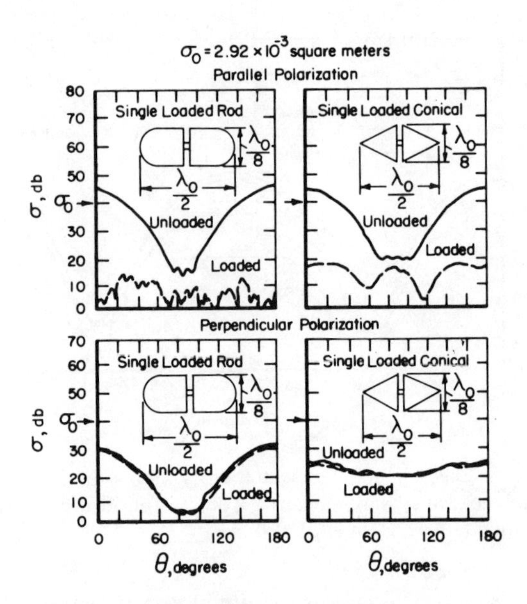

Fig. 8-90. Experimental backscatter cross sections of thick circumferentially loaded cylinders versus the angle of incidence for parallel and perpendicular polarization [after Sletten *et al.*[48]].

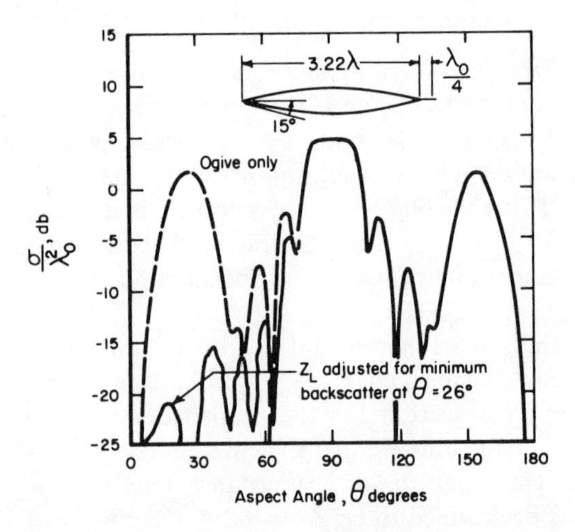

Fig. 8-91. Comparison of the backscatter cross sections of loaded and unloaded perfectly conducting ogives versus the angle of incidence [from Hanson[57]].

has determined the optimum value of the load impedance to be

$$[Z_L]_{\min} = \frac{-i30\Psi(\exp[ik_0h] - \exp[ik_0h\cos\theta])}{\sin k_0(h-d)(\exp[ik_0d] - \exp[ik_0d\cos\theta])} \tag{8.3-75}$$

with Ψ a complicated function of the body dimensions. A good approximation for Ψ is

$$\Psi \approx 2\left[\sinh^{-1}\left(\frac{h}{2a}\right) - C(2k_0a, k_0h) + iS(2k_0a, k_0h)\right] \tag{8.3-76}$$

where the functions S and C are generalized sine and cosine integrals defined by

$$C(\alpha, x) = \int_0^x \frac{1-\cos\sqrt{\alpha^2+\tau^2}}{\sqrt{\alpha^2+\tau^2}}\,d\tau \tag{8.3-77}$$

and

$$S(\alpha, x) = \int_0^x \frac{\sin\sqrt{\alpha^2+\tau^2}}{\sqrt{\alpha^2+\tau^2}}\,d\tau \tag{8.3-78}$$

The above equations were derived for a thin cylinder as shown in Figure 8-92; where a is the cylinder radius, h is the total cylinder length, d is the length from the forward end of the cylinder to the load point, and θ is the scattering angle measured from the cylinder axis. In Figure 8-93, a number of curves are given of the optimum load impedance as a function of the angle of incidence, cylinder length, cylinder radius, and frequency. These curves were computed from Eq. (8.3-75). Chen compares his theoretical results with the experimental results of Hanson, using several different values of cylinder radius a, since the equivalent radius to be used for comparison between the cylinder theory and the ogive measurements is not known precisely. The comparison is given in Table 8-7, and the agreement is seen to be good.

8.3.3.2.3. *Loops.* Chen[59,60] has examined the minimization of the radar cross section of perfectly conducting loops both rectangular and circular by impedance loading.

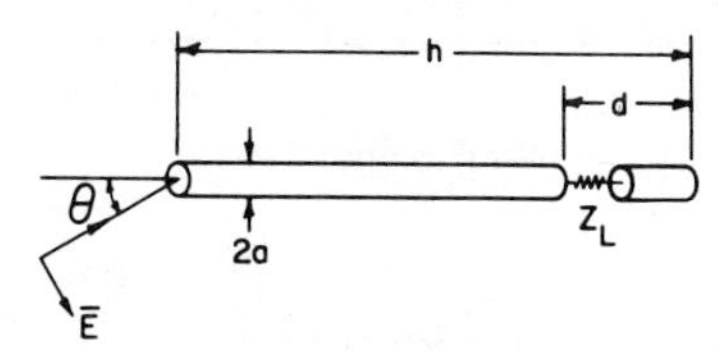

Fig. 8-92. A thin asymmetrically loaded cylinder.

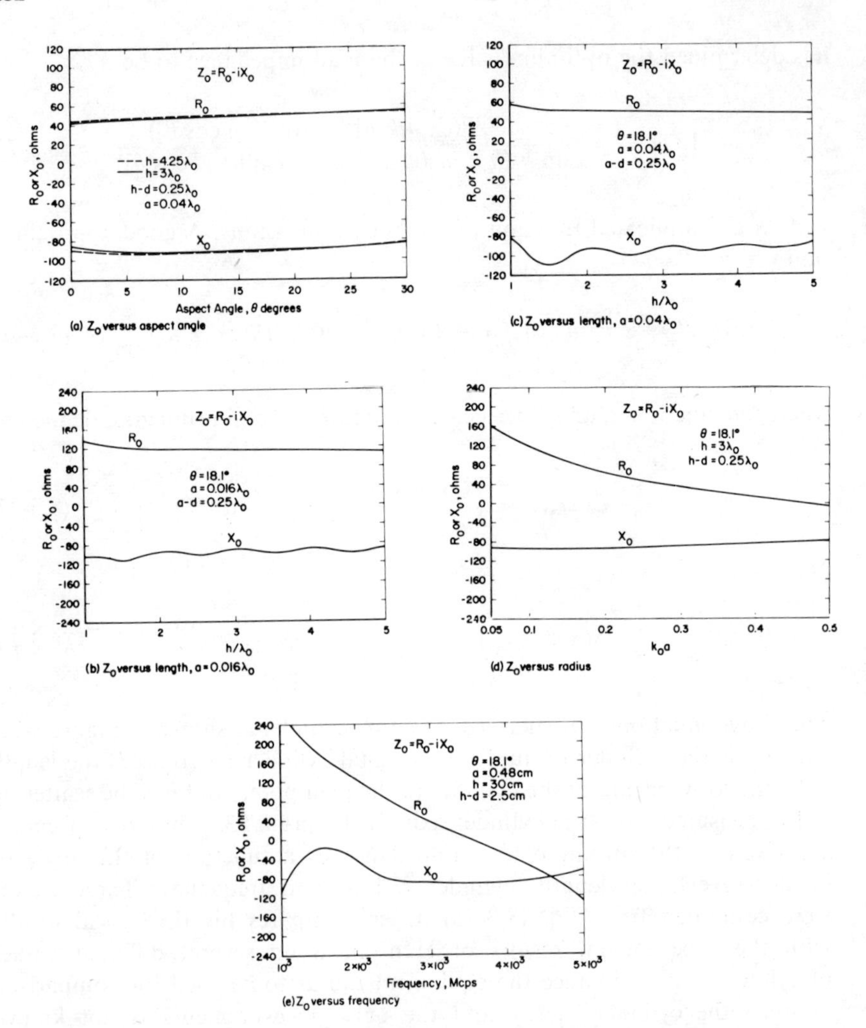

Fig. 8-93. Optimum load impedance for minimizing the traveling-wave cross section versus the angle of incidence, cylinder length, cylinder radius, and frequency.

Table 8-7

k_0a	Z_0 (theoretical) Chen[58]	Z_0 (experimental) Hanson[57]
0.15	53.2-i92.2 ohms	
0.20	67.9-i93.2 ohms	68-i98 ohms
0.25	84.9-i94.6 ohms	
$h = 3.47\lambda_0$, $h\text{-}d = 0.25\lambda_0$, $\theta = 23°$		

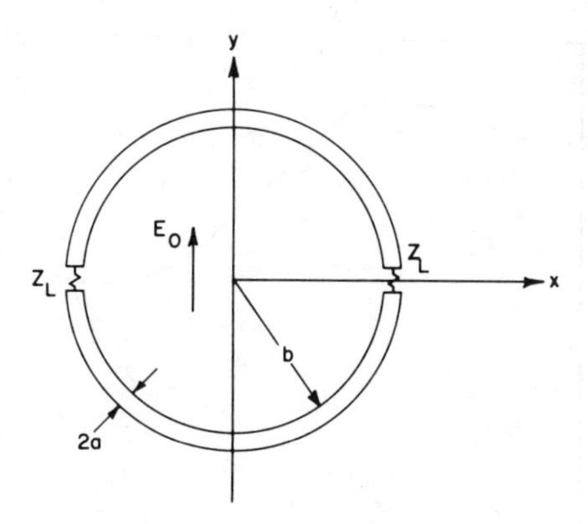

Fig. 8-94. Loaded circular loop.

For a circular loop, such as illustrated in Figure 8-94, the optimum impedance for minimizing the backscatter cross section for a normally incident plane wave is[60]

$$[Z_L]_0 = \frac{-i\sqrt{(\mu_0/\epsilon_0)}[(\Phi_1/\Phi_2)\,k_0{}^2b^2 - 1]\,\Phi_2}{2[4k_0b + \pi(k_0{}^2b^2\Phi_1/\Phi_2 - 1)\tan[(\pi k_0b/2)\ \sqrt{\Phi_1/\Phi_2})]\sqrt{\Phi_2/\Phi_1}]} \tag{8.3-79}$$

where the functions Φ_1, Φ_2 are given in Figures 8-95 and 8-96 for a loop

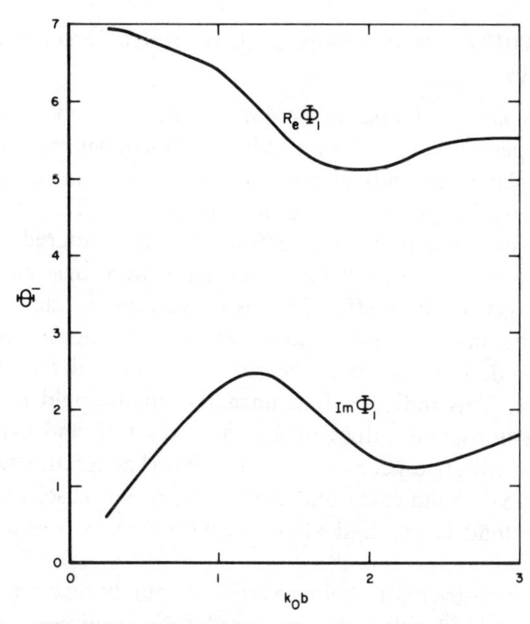

Fig. 8-95. The function Φ_1 versus the loop radius.

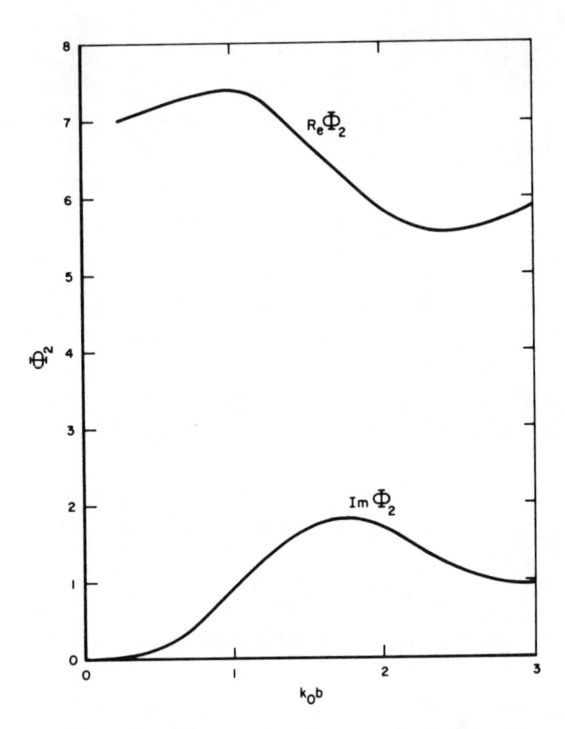

Fig. 8-96. The function Φ_2 versus the loop radius.

with $a^2/b^2 = 0.00179$. For this loop, $[Z_L]_0$ is plotted versus the loop radius k_0b in Figure 8-97.

8.3.3.2.4. *Spheres.* Liepa and Senior(61,62) have considered the scattering behavior of a perfectly conducting sphere loaded with a circumferential slot lying in a plane perpendicular to the direction of incidence at various angular positions, θ_0, as illustrated in Figure 8-98. The slot width was assumed to be small but finite. Expressions for the scattered far-field components and the total surface-field components were derived and used to determine the effect on the scattered fields by various slot admittances.

To reduce the backscattered field to zero, the slot admittance in general is complex and, depending upon the slot location and k_0a, the real part may be negative. This indicates that an active load would be required for certain slot positions and values of k_0a. Figures 8-99 and 8-100 show the values of slot admittance necessary to give zero backscatter when $\theta_0 = 45$ and 90°. Figure 8-100 indicates that for the $\theta_0 = 90°$ case, the k_0a regions where an active load is required alternate with those where a passive load is satisfactory.

If the load is restricted to being passive or purely reactive, the question arises as to what load values are required to minimize or maximize the

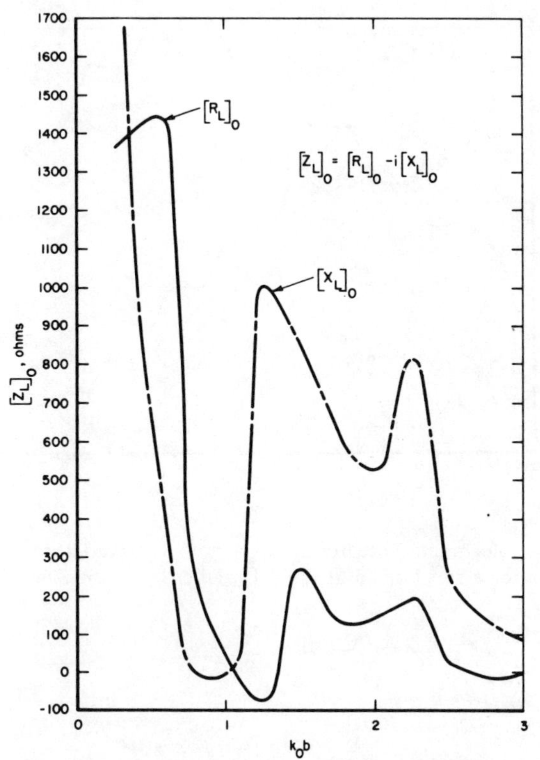

Fig. 8-97. Optimum load impedance for minimizing the normal-incidence backscatter cross section for a circular loop versus the loop radius.

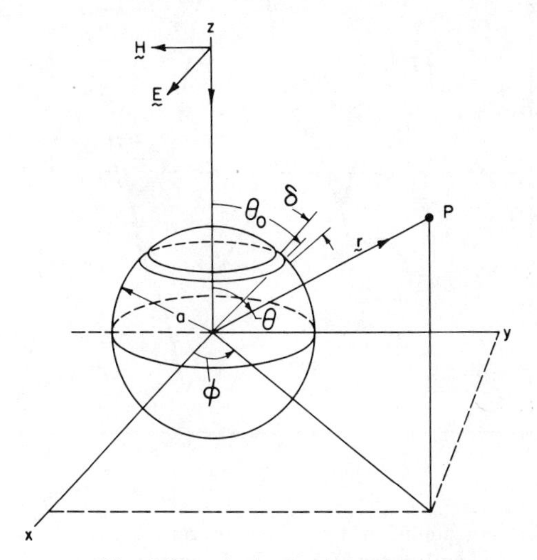

Fig. 8-98. A slot-loaded sphere.

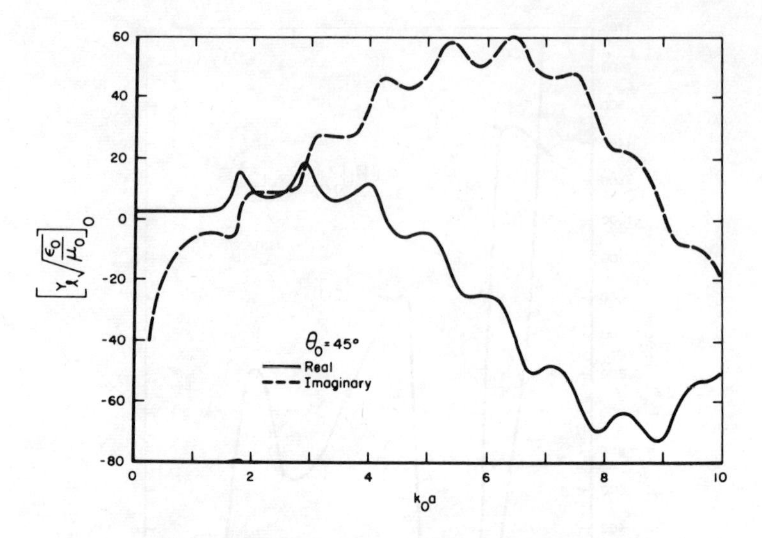

Fig. 8-99. Slot admittance required for zero backscatter cross section versus k_0a for a slot located at $\theta_0 = 45°$ [after Liepa and Senior[62]].

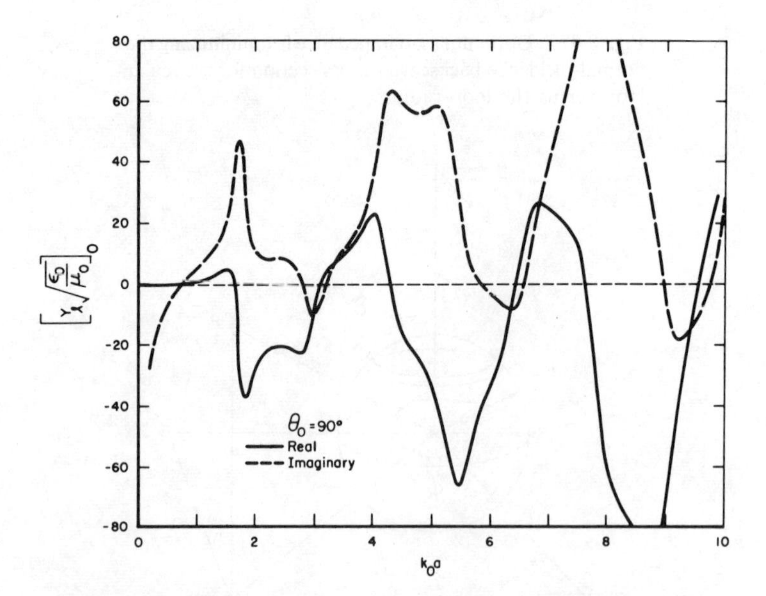

Fig. 8-100. Slot admittance required for zero backscatter cross section versus k_0a for a slot located at $\theta_0 = 90°$ [after Liepa and Senior[62]].

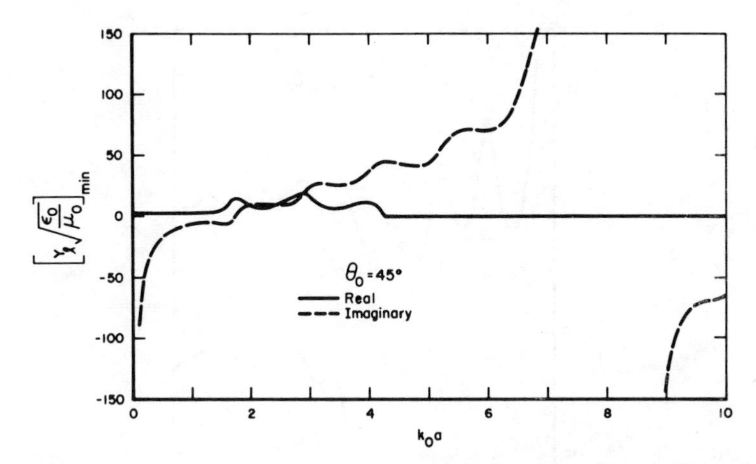

Fig. 8-101. Passive slot admittance required to minimize the backscatter cross section versus k_0a for a slot located at $\theta_0 = 45°$ [after Liepa and Senior[62]].

backscatter cross section for a particular load position θ_0 and various values of k_0a. In addition, the minimum or maximum values of the cross section attained with this load are of interest. This has also been investigated by Liepa and Senior,[61,62] and some of their results are given in Figures 8-101 to 8-103. In Figure 8-101, the theoretical admittance required to minimize the backscatter cross section for a slot located at $\theta_0 = 45°$ is given versus k_0a. Figure 8-102 gives the theoretical admittance required to maximize the cross section for the same slot. Figure 8-103 gives the theoretical minimum and maximum cross sections attainable by means of a passive reactively loaded slot at $\theta_0 = 45°$ versus the sphere size, k_0a.

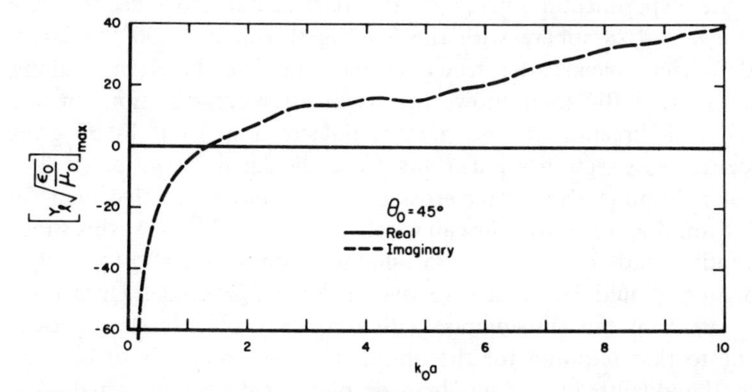

Fig. 8-102. Passive slot admittance required to maximize the backscatter cross section versus k_0a for a slot located at $\theta_0 = 45°$ [after Liepa and Senior[62]].

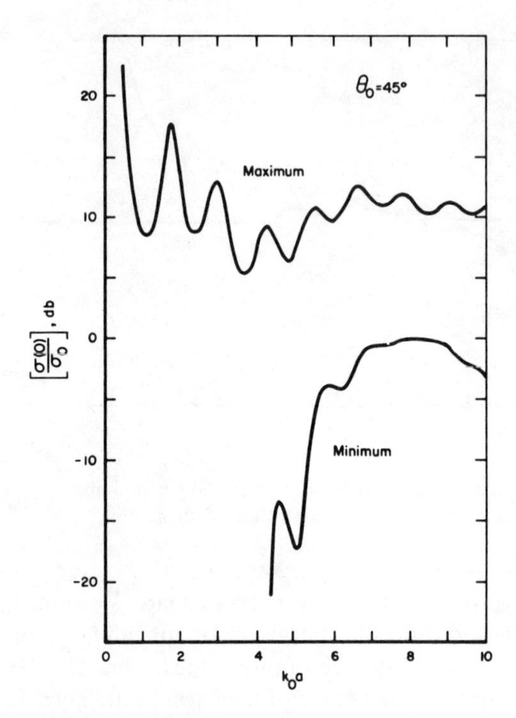

Fig. 8-103. Theoretical minimum and maximum obtainable backscatter cross sections versus k_0a for a passive reactively loaded slot located at $\theta_0 = 45°$ [after Liepa and Senior(62)].

In Figures 8-104 and 8-105, the effect of placing the loading slot in a plane other than one perpendicular to the incident field direction is shown. These are experimental curves of the backscatter cross section from a loaded $k_0a = 4.28$ sphere with the loading slot at $\theta_0 = 90°$, and the slot loaded for zero backscatter when the incident field direction is along the z-axis. Figure 8-104 then shows the backscatter cross section for various incident field directions θ and parallel polarization, while 8-105 gives the backscatter cross section versus θ for perpendicular polarization.

For minimum backscatter cross section a purely reactive load can be found. From Figure 8-101, one can see that for $1.7 \lesssim k_0a \lesssim 9$, this should be a susceptive load for the $\theta_0 = 45°$ slot position. Similarly, for a 90° slot the loading should be susceptive over a large k_0a range. Unfortunately, the variation of the susceptance with frequency of a lossless network is opposite to that required for the optimum load. Thus it is to be expected that the bandwidth of a singly loaded sphere will be quite small. Experimental results of Liepa and Senior(62) support this conclusion. In Figure 8-106 is shown the variation of the backscatter cross section with k_0a for a sphere

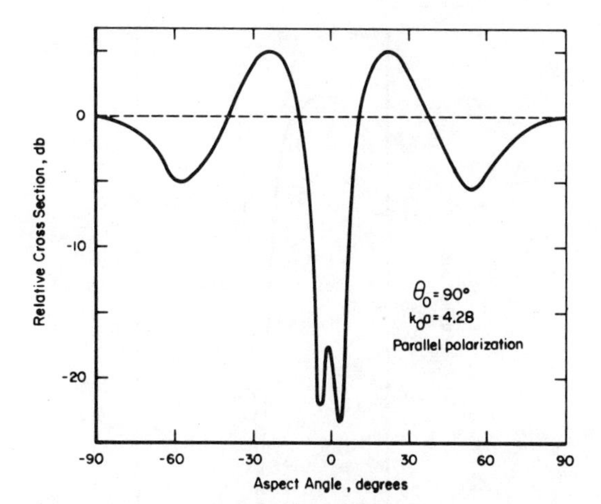

Fig. 8-104. Backscatter cross section of a loaded sphere versus the angle of incidence, parallel polarization [after Liepa and Senior[62]].

loaded at $\theta_0 = 90°$ for zero backscatter at $k_0a = 4.28$. From this figure one observes that for a 10 db reduction in the backscatter cross section, only a 6.5% bandwidth is achieved, and this decreases to 2% for a 20 db reduction. This difficulty can perhaps be alleviated by the use of multiple

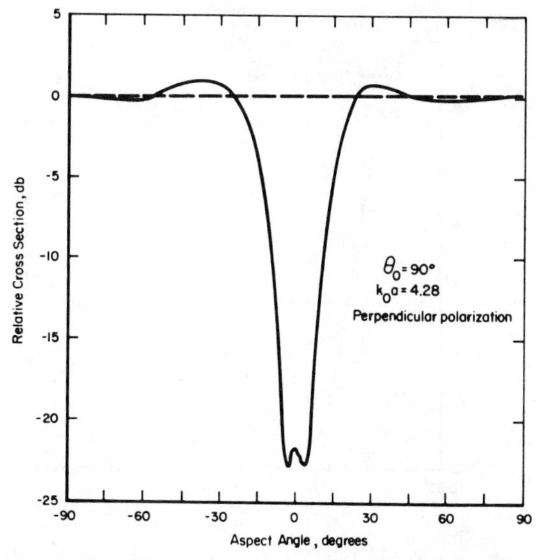

Fig. 8-105. Backscatter cross section of a loaded sphere versus the angle of incidence, perpendicular polarization [after Liepa and Senior[62]].

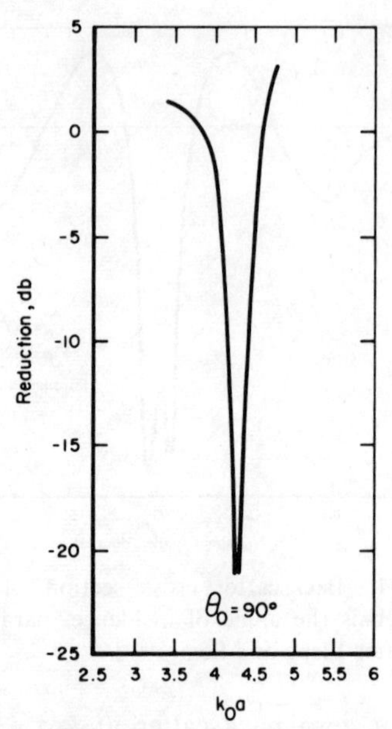

Fig. 8-106. Backscatter cross section of a sphere loaded for zero backscatter at $k_0a = 4.28$ versus k_0a [after Liepa and Senior[62]].

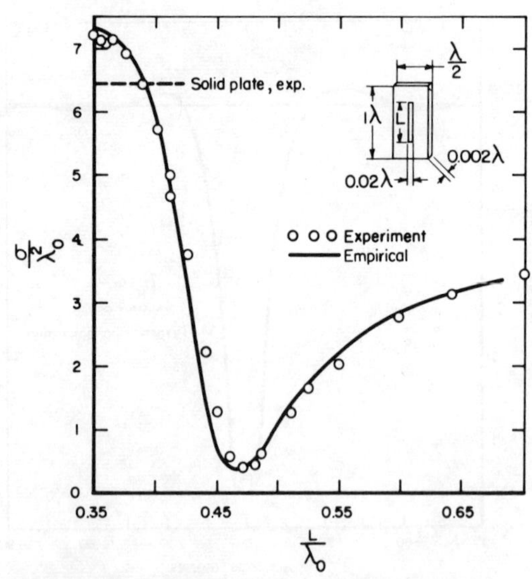

Fig. 8-107. Normal-incidence backscatter cross section of a slot-loaded rectangular plate versus the slot length in wavelengths [after Green[50]].

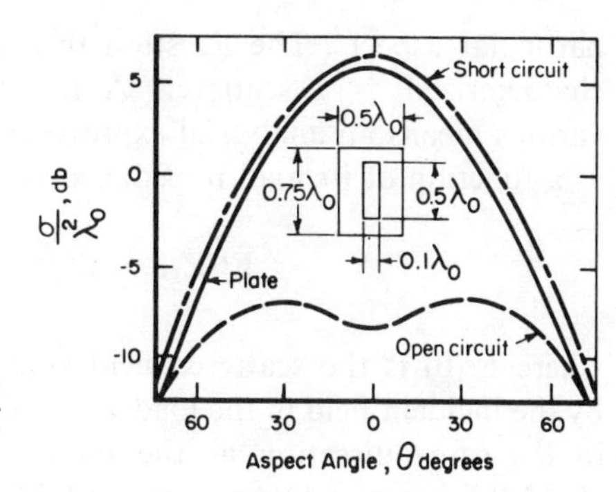

Fig. 8-108. Backscatter cross section of a slot-loaded plate versus the angle of incidence [after Green[50]].

loading or a lossy load; however, the analysis becomes much more difficult in these cases.

8.3.3.2.5. *Flat Plates.* Green[50] has examined, empirically, the effect of a single slot on the normal-incidence backscatter cross section of flat, rectangular plates. In Figure 8-107 is shown the normal-incidence backscatter cross section of a λ_0 by $\lambda_0/2$ plate with a narrow central slot versus the length of the slot. It is apparent that a slot of the proper length can result in considerable reduction of the cross section. Figure 8-108 gives the backscatter cross section versus the angle of incidence for a $0.75\lambda_0$ by $0.5\lambda_0$ plate with a centrally located, narrow, half-wave slot. Curves are given for the plate with no slot, and for the slotted plate with the slot short circuited and open circuited. The open circuit slot provides a cross-section reduction of greater than 10 db over an angular range of $\pm 30°$ from normal incidence.

8.3.3.2.6. *Other Shapes.* Flood and Field[63] have investigated reactive loading of several structures. Among the structures they have examined are flat plates, rods, thick cylinders, polyhedrons, a corner reflector, monopoles, dipoles, and a pyramid above a ground plane. The planar structures and the polyhedrons were constructed of small metal plates which were electrically connected to each other by means of wires connected to the center of each plate at the rear. The investigation was entirely experimental, and backscatter reductions were obtained over bandwidths exceeding 50%. The magnitudes of these reductions were generally less than 10 db over the above bandwidths, often averaging from 3 to 6 db.

8.4. ANTENNA SCATTERING

8.4.1. Theory

Many radar targets such as aircraft, satellites, etc., utilize various antennas located at one or more points on the target. Often these are large contributors to the bistatic or monostatic cross section of the body at particular aspects. The question then arises as to how an antenna can be characterized as a scatterer. A number of researchers have derived by various means an analytical expression for the field scattered by an antenna as a function of its load impedance, or

$$\mathbf{E}^s(Z_l) = \mathbf{E}^s(0) - \frac{Z_l I(0)}{Z_l + Z_a}\mathbf{E}^r \tag{8.4-1}$$

where $\mathbf{E}^s(0)$ is the scattered field with $Z_l = 0$, $I(0)$ is the current induced by the incident field in the load $Z_l = 0$, with the positive sense of the current in the same direction as the positive sense of the transmitting current, Z_a is the antenna impedance, and $\mathbf{E}^r$ is the electric field produced by the antenna when excited by a unit current source. A more convenient expression for the scattered field, developed by Green[50], is

$$\mathbf{E}^s(Z_a) = \mathbf{E}^s(Z_a^*) - \Gamma I(Z_a^*)\,\mathbf{E}^r \tag{8.4-2}$$

where, in this equation, $\mathbf{E}^s(Z_a^*)$ is the electric field scattered by the antenna with a conjugate matched load, $I(Z_a^*)$ is the current induced at the load terminals in a conjugate matched impedance due to the incident field, and

$$\Gamma = (Z_l - Z_a^*)/(Z_l + Z_a) \tag{8.4-3}$$

The parameter Γ in Eq. (8.4-2) always has a magnitude of one or less for passive loads, and a magnitude of one for purely reactive loads.

Equation (8.4-2) consists of two terms, the first of which has been designated by Green as the structural scattering term, and the second, the antenna scattering term. The structural term is independent of the antenna load impedance, and is due to the currents induced on the antenna surface by the incident field even though the antenna has been conjugate-matched. The second, or antenna mode term, is determined completely by the radiation properties of the antenna and this term vanishes when the antenna is conjugate-matched. This term is related to the energy absorbed in the load of a lossless antenna, as well as to the energy radiated by the antenna due to load mismatch. The scattering pattern for the antenna mode is precisely the square of the antenna radiation pattern.

Now designating the cross section due to the structural term as σ^s, and the cross section due to the antenna mode term as σ^a, then the scattering cross section for any antenna is bounded by

$$|\sqrt{\sigma^s} - \sqrt{\sigma^a}| \leqslant \sigma \leqslant |\sqrt{\sigma^s} + \sqrt{\sigma^a}| \tag{8.4-4}$$

If the antenna is terminated by a conjugate matched load, or $Z_l = Z_a^*$, then $\Gamma = 0$ and

$$\mathbf{E}^s = \mathbf{E}^s(Z_a^*) \tag{8.4-5}$$

Similarly, if the antenna is open-circuited, $\Gamma = 1$ and

$$\mathbf{E}^s = \mathbf{E}^s(Z_a^*) - I(Z_a^*)\,\mathbf{E}^r \tag{8.4-6}$$

This can be written as

$$\mathbf{E}^s = \mathbf{E}^s(Z_a^*) - \frac{i\eta_0}{4\lambda_0 R_a}\,\mathbf{h}_a{}^t(\mathbf{h}_a{}^r \cdot \mathbf{E}^i)\,\frac{\exp[ik_0 r]}{r} \tag{8.4-7}$$

where R_a is the antenna radiation resistance, $\mathbf{h}_a{}^t$ is the vector antenna height evaluated in the direction of the incident wave, and $\mathbf{h}_a{}^r$ is the vector antenna height evaluated in the direction of the scattered wave.

The scattering cross section can be obtained from Eq. (8.4-7), and has the form

$$\sigma = |\sqrt{\sigma^s} + \sqrt{\Delta\sigma_e}\,e^{i\psi}|^2 \tag{8.4-8}$$

where

$$\Delta\sigma_e = \frac{\lambda_0{}^2}{4\pi}\,G_r G_t P_r P_t \tag{8.4-9}$$

with G_r is the antenna gain in the transmitter direction, G_t is the antenna gain in the receiver direction, P_r is a factor accounting for the polarization mismatch between the scattering and receiving antennas, P_t is a factor accounting for the polarization mismatch between the scattering and transmitting antennas, and ψ is the phase difference between the structural and antenna mode scattered fields. Kennaugh[64] has called $\Delta\sigma_e$ the effective echo area of the antenna.

If the receiving, transmitting, and scattering antennas are polarization matched, then for backscattering $\Delta\sigma_e$ is

$$\Delta\sigma_e = \frac{\lambda_0{}^2}{4\pi}\,G^2 \tag{8.4-10}$$

In general, the analytical determination of the structural and antenna-mode components of the field scattered by an antenna is very difficult, and requires the solution of the electromagnetic boundary-value problem. Results have been obtained for only a very few antenna types, most of which have been discussed in Section 8.3.3. The structural and antenna mode contributions can be obtained experimentally, however, and Green[65] and Garbacz[66] have devised graphical representations of the scattering properties of antennas which utilize constructions on a Smith chart to relate various measurements to the antenna properties. If the quantity $I(Z_a^*)\,\mathbf{E}^r$ is factored out of Eq. (8.4-2) one component at a time, then

$$E_{ij}^s(Z_l) = I_j(Z_a^*)\,E_i{}^r\left[\frac{E_{ij}^s(Z_a^*)}{I_j(Z_a^*)\,E_i{}^r} - \Gamma\right] \tag{8.4-11}$$

and

$$\sigma_{ij}(Z_l) = \sigma_{ij}^a \mid A_{ij} - \Gamma \mid^2 \tag{8.4-12}$$

where

$$A_{ij} = \frac{E_{ij}^s(Z_a^*)}{I_j(Z_a^*)\, E_i^r} \tag{8.4-13}$$

The parameter A_{ij} is a complex number independent of the load impedance Z_l, thus the cross section for any receiver–transmitter polarization combination i, j is proportional to

$$\sigma_{ij} \propto \mid A_{ij} - \Gamma \mid^2 \tag{8.4-14}$$

and, in fact,

$$\sigma_{ij} = (\lambda_0^2/4\pi)\, G_r G_t P_i P_j \mid A_{ij} - \Gamma \mid^2 \tag{8.4-15}$$

By plotting A and Γ on a Smith chart for a particular polarization combination, the variation of the phasor sum $A - \Gamma$ as a function of Γ is easily seen. In addition, if σ is measured for several values of Γ, A can be determined by utilizing various Smith chart constructions. This is discussed in detail in References 65 and 66.

In general, any arbitrary antenna can be placed in one of three classes depending upon the location of the point A on a Smith chart.[65] If A lies outside the boundary of the Smith chart ($\mid A \mid > 1$), it is not possible to reduce the scattered field to zero through any choice of a passive load. A maximum or a minimum in the scattered field amplitude can, however, be obtained using a purely reactive load. This type of antenna is one which has a large structural scattering component due either to the antenna construction, or strongly excited antenna modes which are not coupled to the terminals of interest.

A second category of antennas are those for which A lies on the periphery of the Smith chart ($\mid A \mid = 1$). For this type of antenna a reactive load can be found which will reduce the antenna scattering to zero. Thin linear antennas with lengths of $\lambda_0/2$ or less are of this type.

The third class of antennas are those for which A is located within the Smith chart boundary ($\mid A \mid < 1$). In this case, a dissipative load can be found which will reduce the scattered field to zero, and maximum scattering can be obtained with a reactive load. A well-designed parabolic antenna is an example of an antenna of this type.

From Eq. (8.4-15), if all antennas are polarization matched, then

$$\sigma = \frac{\lambda_0^2}{4\pi} G^2 \mid A - \Gamma \mid^2 \tag{8.4-16}$$

For a thin dipole $A \approx 1$, thus

$$\sigma \approx \frac{\lambda_0^2}{4\pi} G^2 \left| 1 - \frac{Z_l - Z_a^*}{Z_l + Z_a} \right|^2 \tag{8.4-17}$$

or

$$\sigma \approx \frac{\lambda_0^2}{4\pi} G^2 \left| \frac{2R_a}{Z_l + Z_a} \right|^2 \tag{8.4-18}$$

A short-circuited $\lambda_0/2$ dipole has a gain factor at normal incidence of 1.64, and since $Z_l = 0$,

$$\sigma/\lambda_0^2 \approx \frac{G^2}{\pi} \approx 0.85 \tag{8.4-19}$$

This agrees very well with experimental results for the backscatter cross section of a half-wave dipole obtained by various workers.[67,68]

8.4.2. Experimental Results

Very few experimental results on the cross sections of antennas are available. Some were presented earlier in Section 8.3.3 where impedance loading as a means for cross section reduction was discussed. Figure 8-109 shows the measured normal incidence backscatter cross section of a standard gain horn antenna ($G = 15.1$ db) as a function of the position of a waveguide short which presents various reactive loads to the horn.[66] For this measurement, a constant signal was artificially introduced to make A effectively equal to one, or, in other words, the minimum backscatter cross section was made zero. Thus, from Eq. (8.4-16), the maximum cross section is given by

$$\sigma_{\max} = \lambda_0^2 G^2/\pi \tag{8.4-20}$$

or $\sigma_{\max} = 339\lambda_0^2$, which agrees well with the measured value. Figure 8-109 also shows the predicted variation of the cross section versus the local position, and again the agreement is quite good.

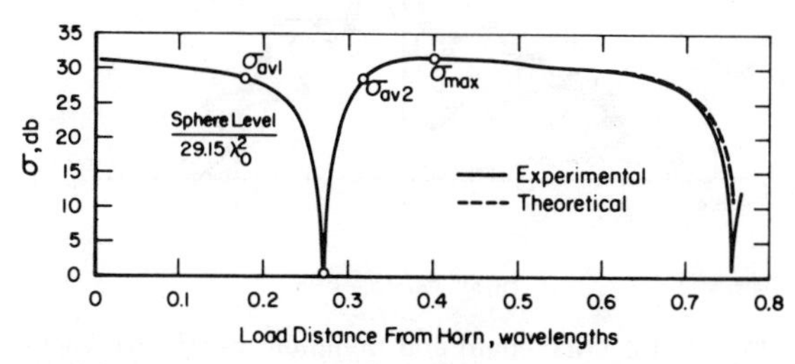

Fig. 8-109. Measured normal-incidence backscatter cross section of a standard gain horn antenna versus the position of a waveguide short circuit [after Garbacz[66]].

Figure 8-110a, b shows the maximum and minimum backscatter cross sections obtainable by means of a reactive load versus the angle of incidence for an X-band horn antenna and a 14-inch parabolic dish antenna, respectively.[69] It is interesting to note that enough information is contained in this figure to determine the structural scattering component for these antennas. From Eq. (8.4-15), one can determine that

$$|A| = -\frac{1 \pm \sqrt{\sigma_{\max}/\sigma_{\min}}}{1 \mp \sqrt{\sigma_{\max}/\sigma_{\min}}}, \qquad \begin{bmatrix} |A| > 1 \\ |A| < 1 \end{bmatrix} \tag{8.4-21}$$

when all antennas are polarization matched. Once $|A|$ has been obtained, then the structural scattering cross section is given by

$$\sigma^s = \frac{\lambda_0^2}{4\pi} G^2 |A|^2 \tag{8.4-22}$$

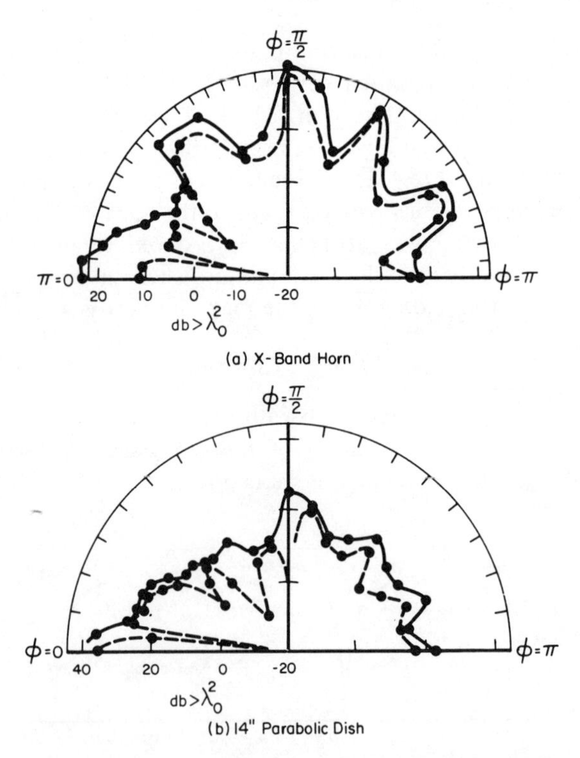

Fig. 8-110. Maximum and minimum obtainable backscatter cross sections versus the angle of incidence for (a) X-band horn; (b) 14-in. parabolic dish [after Moffatt[69]].

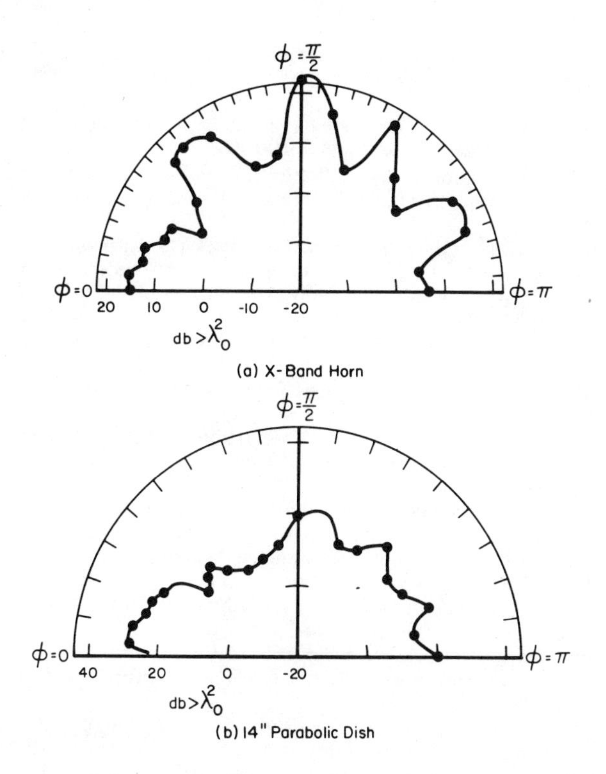

Fig. 8-111. Structural scattering cross sections versus the angle of incidence for (a) *X*-band horn; (b) 14-in. parabolic dish [after Moffatt[69]].

Figure 8-111 shows the structural scattering cross section obtained in this fashion for the antennas of Figure 8-110.

8.6. REFERENCES

1. Kawano, T., and Peters, L., An Extension of the Modified Geometrical Optics Methods for Radar Cross Sections of Dielectric Bodies, Antenna Laboratory, Ohio State University, Report No. 1116-38 (November 1963).
2. Kawano, T., and Peters, L., An Application of Modified Geometrical Optics Method for Bistatic Radar Cross Sections of Dielectric Bodies, Antenna Laboratory, Ohio State University, Report No. 1116-39 (January 1965).
3. Weston, V. H., The Effect of a Discontinuity in Curvature in High-Frequency Scattering, *IEEE Trans.* **AP-10**:776 (1962).
4. Weston, V. H.,Theory of Absorbers in Scattering, *IEEE Trans.* **AP-11**:578 (1963).
5. Logan, N. A., General Research in Diffraction Theory, Vol. I, Lockheed Aircraft Corp., Report No. LMSD-288087 (December 1959), AD 241 228.
6. Levy, B. R., and Keller, J. B., Diffraction by a Smooth Object, *Comm. Pure Appl. Math.* **XII**:159 (1959).

7. Adachi, S., The Nose-on Echo Area of Axially Symmetric Thin Bodies Having Sharp Apices, Antenna Laboratory, Ohio State University, Report No. 925-1 (March 1960), AD 240 651.
8. Weil, H. and Raymond, J. L., On Axis Far-Field Backscattering from Flared Axially Symmetric Bodies, *Radio Science* **1**:825 (1966).
9. Fock, V. A., Generalization of the Reflection Formulas to the Case of Reflection of an Arbitrary Wave from a Surface of Arbitrary Form, *Zh. Eksp. i Teor. Fiz.* **20**:961 (1950).
10. Sokolnikoff, I. S., *Tensor Analysis*, Wiley, New York (1951), p. 78.
11. Crispin, J. W., Goodrich, R. F., and Siegel, K. M., A Theoretical Method for the Calculations of the Radar Cross Sections of Aircraft and Missile, University of Michigan, Report No. 2591-1-M (July 1949).
12. Siegel, K. M., Far-Field Scattering from Bodies of Revolution, *Appl. Sci. Res.* **7B**:293 (1958).
13. Senior, T. B. A., A Survey of Techniques for Cross-Section Estimation, *Proc. IEEE* **53**:822 (1965).
14. Dawson, T. W. G., and Turner, W. R., Calculation of Radar Echoing Areas by the Cylindrical Current Method, Royal Aircraft Establishment, Technical Note RAD 788 (September 1960), AD 250 352.
15. Morse, B. J., Diffraction by Polygonal Cylinders, *J. Math. Phys.* **5**:199 (1964).
16. Abramowitz, M., and Stegun, I. A., *Handbook of Mathematical Functions*, National Bureau of Standards, Department of Commerce, Waschington D.C. (March 1965).
17. Blore, W. E., and Royer, G. M., The Radar Cross Section of Bodies of Revolution, Defense Research Telecommunications Establishment, DRTE 1105 (March 1963), AD 406 112.
18. Eberle, J. W., and St. Clair, R. W., Echo Area of Combinations of Cones, Spheroids, and Hemispheres as a Function of Bistatic Angle and Target Aspect, Antenna Labotory, Ohio State University, Report No. 1073-1 (June 1960), AF 19(604)-6157.
19. Blore, W. E., and Musal, H. M., Jr., The Radar Cross Section of Metal Hemispheres, Spherical Segments, and Partially Capped Spheres, *IEEE Trans.* **AP-13**:478 (1965).
20. Raybin, D. M., Radar Cross Section of Spherical Shell Segments, *IEEE Trans.* **AP-13**: 756 (1965).
21. Crispin, J. W., Jr., Hiatt, R. E., Sleator, F. B., and Siegel, K. M., The Measurement and Use of Scattering Matrices, University of Michigan, Radiation Laboratory, Report No. 2952-1-F, (May 1960).
22. Crispin, J. W., Jr., and Maffett, A. L., Radar Cross-Section Estimation for Complex Shapes, *Proc. IEEE* **53**:972 (1965).
23. Mack, R. B., and Gorr, B. B., Measured Radar Backscatter Cross Sections of the Project Mercury Capsule, Air Force Cambridge Research Laboratory, AFCRL 38 (February 1961).
24. Marlow, H. C., RAT SCAT Measurements of the BQM-34A Target Drone Q2c Series 2, Air Force Missile Development Center, MDC-TR-65-66 (February 1966).
25. Marlow, H. C., RAT SCAT Cross Section Measurements of the Agena Vehicle, Air Force Missile Development Center, MDC-TR-65-70 (December 1965).
26. Robertson, S. D., Targets for Microwave Radar Navigation, *Bell System Tech. J.* **26**:852 (1947).
27. Peters, L., Jr., Theory of Corner Reflectors as Bistatic Radar Enhancement Devices, Antenna Laboratory, Ohio State University, Report No. 768-5 (May 1959), AF 33(615)-5078.
28. Peters, L., Jr., Passive Bistatic Radar Enhancement Devices, *Proc. IEE (G.B.)* **109C**:1 (1962).
29. Bonkowski, R. R., Lubitz, C. R., and Schensted, C. E., Cross Sections of Corner Reflectors and Other Multiple Scatterers at Microwave Frequencies, University of Michigan, Willow Run Research Center, UMM-106 (October 1953).
30. Kay, A. F., Spherically Symmetric Lenses, *IRE Trans.* **AP-7**:32 (1959).
31. Armstrong Microwave Devices, Description and Technical Information, Sales Brochure of Armstrong Cork Co., Lancaster, Pa. (August 1965).

32. Chiron, B., and Holvoet-Vermant, F., Experimental study of Spherical Dielectric Lenses and Reflectors, *Onde Elect. (France)* **41**:481 (1961)
33. Holt, F. S., Comparison of Reflective Properties of Corner Reflector Clusters and Luneberg Lens Reflectors AFCRL-66-626 (August 1966).
34. Van Atta, L. C., Electromagnetic Reflector, U. S. Patent No. 2,908,002 (October 6, 1959).
35. Sharp, E. D., Properties of the Van Atta Reflector Array, Rome Air Development Center, RADC-TR-58-53 (April 1958), AD 148 684.
36. Pannell, J. H., Rheinstein, J., and Smith, A. F., Radar Scattering from a Conducting Cone-Sphere, MIT-Lincoln Laboratory, Report No. 349 (March 1964), AF 19(628)-500.
37. Smith, A. F., Radar Backscatter from Some Low Cross-Section Shapes, MIT-Lincoln Laboratory, Group 22, Report 1964-3 (January 1964).
38. Knott, E. F., and Senior, T. B. A., The Effect of Fins on the Nose-on Cross Section of Cone-Spheres, *IEEE Trans.* **AP-11**:504 (1963).
39. Salisbury, W. W., Absorbent Body for Electromagnetic Waves, U. S. Patent No. 2,599,944 (June 10, 1952).
40. Landau, L. D. and Lifshitz, E. M., *Electro-Dynamics of Continuous Media*, Pergamon Press, New York (1960), p. 256.
41. Jacobs, I., The Nonuniform Transmission Line as a Broadband Termination, *Bell System Tech. J.* **37**:913 (1958).
42. Franceschetti, G., Scattering from Plane Layered Media, *IEEE Trans.* **AP-12**:754 (1964).
43. Lenz, K. L., Leitungen mit ortsabhängiger Dämpfung zur reflexionsarmen Absorption elektromagnetischer Wellen, *Z. Angew. Phys.* **10**:17 (1958).
44. Walther, K., Reflection Factor of Gradual-Transition Absorbers for Electromagnetic and Acoustic Waves, *IRE Trans.* **AP-8**:608 (1960).
45. Redheffer, R., The Dependence of Reflection on Incidence Angle, *IRE Trans.* **MTT-7**: 423 (1959).
46. Streifer, W., Creeping Wave Propagation Constants for Impedance Boundary Conditions, *IEEE Trans.* **AP-12**:764 (1964).
47. Peters, L., Jr., End-Fire Echo Area of Long, Thin Bodies, *IRE Trans.* **AP-6**:133 (1958).
48. Sletten, C. J., Blacksmith, P., Holt, F. S., and Gorr, B. B., Scattering from Thick Reactively Loaded Rods. *In* Schindler, J. K., and Mack, R. B. (Eds.), *The Modification of Electromagnetic Scattering Cross Sections in the Resonant Region—A Symposium Record*, Vol. I, Air Force Cambridge Research Laboratory, AFCRL-64-727(1) (September 1964).
49. Garbacz, R. F., and Moffatt, D. L., An Experimental Study of Bistatic Scattering from Some Small Absorber Coated Metal Spheres, Antenna Laboratory, Ohio State University, Report No. 925-2 (May 1960).
50. Green, R. B., The General Theory of Antenna Scattering, Antenna Laboratory, Ohio State University, Report No. 223-17 (November 1963).
51. Harrington, R. F., Theory of Loaded Scatterers, *Proc. IEE (G.B.)* **111**:617 (1964).
52. As, B. O., and Schmitt, H. J., Backscattering Cross Section of Reactively Loaded Cilindrical Antennas, Cruft Laboratory, Harvard University, Scientific Report No. 18 (August 1958).
53. Hu, Y. Y., Backscattering Cross Section of a Center-Loaded Cylindrical Antenna, *IRE Trans.* **AP-6**:140 (1958).
54. Chen, K. M., and Liepa, V. V., The Minimization of the Radar Cross Section of a Cylinder by Central Loading, University of Michigan Radiation Laboratory, Report No. 5548-1-T (April 1964).
55. Chen, K. M., Minimization of Backscattering of a Cylinder with a Moderate Radius by Loading Method, Michigan State University, Technical Report No. 1 (December 1964), AF 33(615)-1656.
56. Chen, K. M., A Slotted Cylinder with Zero Backscattering, Michigan State University, Technical Report No. 3 (October 1965), AF 33(615)-1656.
57. Hanson, W. P., Jr., Backscatter Reduction of Long Thin Bodies by Impedance Loading.

In Schindler, J. K., and Mack, R. B. (Eds.), *The Modification of Electromagnetic Scattering Cross Sections in the Resonant Region—A Symposium Record*, Vol. I, Air Force Cambridge Research Laboratory, AFCRL-64-727(1) (September 1964).
58. Chen, K. M., Minimization of End-Fire Radar Echo of a Long Thin Body by Impedance Loading, Michigan State University, Technical Report No. 4 (October 1965), AF 33(615)-1656.
59. Chen, K. M., Backscattering of a Loop and Minimization of its Radar Cross Section, Michigan State University, Technical Report No. 2 (May 1965), AF 33(615)-1656.
60. Lin, J. L., and Chen, K. M., Minimization of Backscattering of a Metallic Loop by an Impedance Loading Method, Michigan State University, Scientific Report No. 2 May 20), 1966, AF 19(628)-5732.
61. Liepa, V. V., and Senior, T. B. A., Modification to the Scattering Behavior of a Sphere by Reactive Loading, University of Michigan Radiation Laboratory, Report No. 5548-2-T (October 1964), AF 19(628)-2374.
62. Liepa, V. V., and Senior, T. B. A., Theoretical and Experimental Study of the Scattering Behavior of a Circumferentially-Loaded Sphere, University of Michigan Radiation Laboratory, Report No. 5548-5-T (February 1966), AF 19(628)-2374.
63. Flood, D. P., and Field, J. C., Experimental Study of the Backscattering from Conducting Objects in the Range of One Quarter to One Wavelength in Size, Andrew Alford Consulting Eng., AFCRL-64-541 (April 1964).
64. Kennaugh, E. M., The Echoing Area of Antennas, Antenna Laboratory, Ohio State University, Report No. 601-14 (December 1957), AD 152 786.
65. Green, R. B., The Effect of Antenna Installations on the Echo Area of an Object, Antenna Laboratory, Ohio State University, Report No. 1109-3 (September 1961).
66. Garbacz, R. J., The Determination of Antenna Parameters by Scattering Cross-Section Measurements, III. Antenna Scattering Cross Section, Antenna Laboratory, Ohio State University, Report No. 1223-10 (November 1962).
67. Dike, S. H., and King, D. D., Absorption Gain and Backscattering Cross Section of the Cylindrical Antenna, *Proc. IRE* **40**:853 (1952).
68. Sevick, J., Experimental and Theoretical Results on the Backscattering Cross Section of Coupled Antennas, Cruft Laboratory, Harvard University, Technical Report No. 150 (May 1952).
69. Moffatt, D. L., Determination of Antenna Scattering Properties from Model Measurements, Antenna Laboratory, Ohio State University, Report No. 1223-12 (January 1964).

Chapter 9

ROUGH SURFACES

D. E. Barrick

9.1. ANALYTICAL MODELS

9.1.1. Introduction and Definitions

Most of the effort toward the analysis of scattering by rough surfaces has come in the past 20 years. It received its biggest impetus from the advent of radar and the need to know more about terrain and sea return. The analysis has been facilitated by recent developments in the area of applied statistics.

The problem of scattering from rough surfaces is inherently different in nature from that of scattering by other bodies. Normally in the rough-surface problem, an exact knowledge of the shape of the surface is neither available, nor is it of concern to the radar operator. Instead, only average properties of the surface shape enter into the problem. This last requirement rules out a boundary-value approach, since the exact boundary is not known. Rather, one is more interested in the relationship between the average scattered field or radar cross section and the average surface properties.

The average scattered field or power from a rough surface is not a difficult parameter to measure. When either the receiver–transmitter or the surface is in motion, one is essentially measuring the return from different surfaces (or different members of an ensemble of rough surfaces) at each point in time. Hence, the radar receiver in most cases, due to its finite bandwidth, averages the returning signal with time, both by averaging the signal within a given pulse (if pulses are used) and by averaging over several pulses in the display mechanism. One can assume that such a time average of the return signal is more or less equivalent to an ensemble average (i.e., in this case, an ensemble average means an average of the cw signals scattered from an ensemble of statistically similar surfaces).

9.1.1.1. *Definitions and General Assumptions*

Throughout this chapter, only the average scattered power and radar cross section will be discussed. Moreover, this cross section will be normalized with respect to the average area illuminated by the radar, A. This average

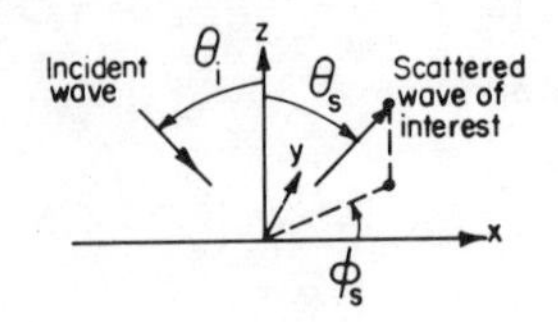

Fig. 9-1. Bistatic scattering geometry for a planar rough surface whose mean height coincides with the *xy*-plane.

radar cross section is, therefore, defined as†

$$\gamma(\theta_i , \theta_s , \phi_s) = \frac{\langle\sigma(\theta_i , \theta_s , \phi_s)\rangle}{A} \tag{9.1-1}$$

where θ_i, θ_s, ϕ_s are shown for the bistatic case in Figure 9-1. The rough surface is assumed to consist of height variations about a mean plane, which is taken as the *xy*-plane. The incident wave lies in the *xz*-plane at a polar angle θ_i. The brackets $\langle\ \rangle$ refer to an average. For the backscattering direction, $\theta_i = \theta_s$, $\phi_s = \pi$, and the normalized average backscattering cross section, are defined as‡

$$\gamma(\theta_i) = \gamma(\theta, \theta, \pi). \tag{9.1-2}$$

These average scattering cross sections are implied functions of the wavelength or frequency, since a given class of rough surfaces scatters differently at different frequencies.

The transmitting and receiving antennas are assumed to be sufficiently far from the rough surface, such that the relationship $R_T R_R \gg A$ is satisfied, where R_T, R_R are the distances from the transmitting and receiving antennas to the origin. The incident field striking the surface is then assumed to be locally plane over the surface, as is the scattered field in the aperture of the receiving antenna.§

The wavelength is assumed to be much shorter than the dimensions of the illuminated area in this chapter, i.e., $\lambda_0^2 \ll A$.

9.1.1.1.1. *Roughness Scale.* One can categorize the problems in rough surface scattering into one of two classes representing the extremes with regard to roughness scale or dimensions. If the rms height of the roughness,

† All curves of γ shown in this chapter (either theoretical or experimental) are plotted to the same scale. This facilitates comparison of the various curves.

‡ The notation γ differs from that of several other authors, such as Kerr,[1] who defines this quantity as σ^0. It should not be confused, however, with the γ used by some other authors, such as Cosgriff *et al.*,[2] who used γ to designate the average scattering cross section per unit projected area. This author's γ has an additional $\cos\theta_i$ factor beyond that of Reference 2.

§ The presence of roughness on the surface relaxes considerably the usual far-field requirement of $R_T > 2A/\lambda$; this inequality must be satisfied if the incident wave is to be essentially planar across the entire surface. This subject is discussed in detail in Appendix B of Reference 3.

h, is much less than the wavelength, the surface is called "slightly" rough.[†] At the opposite extreme, the surface is called "very" rough.

A slightly rough surface scatters as a perfectly smooth plate in the limit as the roughness height, h, approaches zero. This suggests the analysis of the scattering problem by a perturbation method, which is one of the approaches taken.

Treatment of very rough surfaces is generally done using an asymptotic method at some point in the analysis based upon k_0h as a largeness parameter.

9.1.1.1.2. *Goal of the Analysis.* Generally one has specific purposes or reasons for investigating scattering from a rough surface. Broadly these ultimate goals may be divided into three categories:

(1) *Direct Scattering Problem.* Here one wishes to know the average properties of the scattered signal or cross section when the surface properties of the rough surface are known, and the scattering information is expressed in terms of the surface properties. This approach is the most straightforward, and in this chapter the results are presented in such a form that this information is the most readily obtainable.

(2) *Inverse Scattering Problem.* In this case, one wishes to obtain statistical information about the rough surface from a knowledge of the average properties of the scattered field. This problem is more difficult in that there appear to be many classes of rough surfaces producing the same average scattering cross section as a function of the bistatic scattering angles and wavelength. If one has, however, some knowledge of the surface, it is possible to use the results of this chapter (both equations and measured curves from actual surfaces) to obtain more general information about the surface under consideration. Inverse scattering, for instance, is of interest in obtaining information about the lunar and planetary surfaces using radar measurements.

(3) *Clutter Problem.* In this case, return from terrain is not wanted. However, since in many cases it is unavoidably present along with the desired signal, one can detect and analyze the desired signal better if more is known about the properties of the clutter or noise produced by terrain scattering. In this case, surface information is generally known, and the properties of the scattered signal are sought. Hence, this problem is closely tied with the direct scattering problem discussed above, and some of the results of this chapter are applicable.

9.1.1.1.3. *Area Illuminated.* Four general situations with respect to the area illuminated are possible; these are illustrated in Figure 9-2 as follows:

[†] More precisely, $k_0h \cos \theta_i \ll 1$. This is discussed in References 4 and 5.

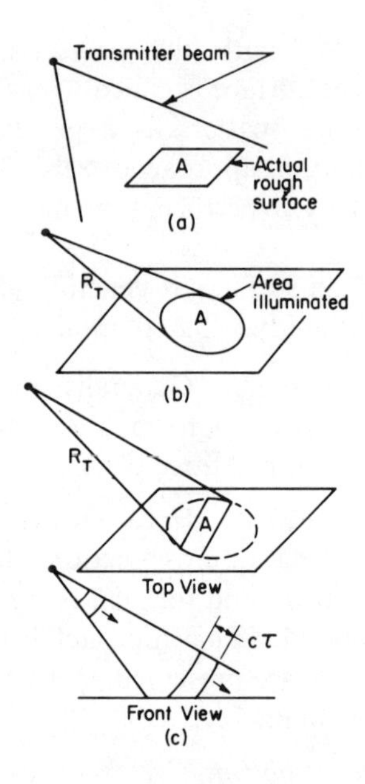

Fig. 9-2. Illuminated surface-area configurations for various radar situations.

(1) The actual area of the rough surface A is smaller than the beam of the transmitting antenna so that the entire area A is illuminated evenly by a plane wavefront [see Figure 9-2(a)].

(2) The extent of the rough surface is much larger than the area illuminated by the transmitting antenna beam (e.g., an airplane measuring the return from the ocean) [see Figure 9-2(b)]. In this case, the field strength over the illuminated area is not uniform but varies with the angle from the center of the beam according to the gain pattern of the antenna. As an approximation, one may assume that the beam has a uniform field within a given beamwidth and zero field outside that width. Then the illuminated area A may be treated in the same manner as in case (1).† Such a procedure is valid if the angular beamwidth is small, so that the angle of incidence within the area illuminated does not vary by more than a few degrees.

If the antenna beam pattern is circular and the half-power angular beamwidth is $2\nu_B$ radians, then one might model the field in the above manner by a beam having a uniform field over an angular extent $0 < \xi < \nu_B$,

† The result obtained in this manner is the scattering cross section per unit uniformly illuminated area. In order to then take into account the true, nonuniform antenna patterns of the transmitting and receiving antennas, either bistatically or monostatically located, one must average these patterns over the true illuminated area. This is done, and tables are derived for this pattern average in Reference 6.

and zero field for $\xi > \nu_B$. In this case, the illuminated region is an ellipse with area A given by

$$\frac{\pi R_T^2 \nu_B^2}{\cos \theta_i} \tag{9.1-3}$$

(3) The problem of the amount of surface area contributing at a given time for a pulsed radar is extremely complex in the bistatic case. Therefore only the backscatter case will be treated. For an airborne radar with a circular beam, the geometry of Figure 9-2(c) applies if the pulse width τ is sufficiently short, i.e., when $c\tau \gg R_T \nu_B \tan \theta_i$. The rough surface area which effectively contributes to the backscattered power (i.e., the area within the resolution cell) is not the illuminated area shown in the figure, but one-half of it, or

$$A = \tfrac{1}{2} A_I = \frac{c\tau R_T \nu_B}{\sin \theta_i} \sqrt{1 - \left[1 - \frac{c(t - t_0)}{2 R_T \nu_B \tan \theta_i}\right]^2} \tag{9.1-4}$$

In Eq. (9.1-4), t_0 refers to the time at the receiver when one first receives the signal from the edge of the ellipse closest to the radar. Also, if one wishes to generalize to a noncircular beam whose half-power azimuthal beamwidth is $2\nu_A$, while the elevation beamwidth remains $2\nu_B$, one merely replaces the ν_B in front of the radical by ν_A.

(4) The antennas for this situation point close to the horizontal direction.† This is common for a search radar. Again due to the complexity of the bistatic geometry, only the results for backscattering are given. The rough surface area contributing to the backscattered power, as in the preceding case, is one-half of the illuminated area, or

$$A = \tfrac{1}{2} A_I = c\tau R_T \nu_B \tag{9.1-5}$$

In any of the four cases above, the results for the normalized average scattering cross sections given as γ in the following sections, can be used to find the actual average scattering cross section of the appropriate scattering area, A, by multiplying γ by A. The resultant average cross section is then suitable for use in the standard radar equation (2.1-1), or may be used by itself as an absolute measure of the scattering ability of the illuminated rough surface.

9.1.1.1.4. *Use of Models in Analysis of Rough-Surface Return.* In order to obtain a mathematical result or equation relating the average scattering cross section of a rough surface to the surface properties and wavelength, one must invariably employ certain assumed mathematical or physical relationships or constraints at some point in the analysis; as such, one is said to be using a "model" for the surface. In this manner, one arrives at a result which is as valid as the model chosen for the analysis. The results

† This is the special case of Figure 9-2(c) (front view), where the source approaches grazing incidence.

for rough-surface scattering are categorized here in three different classes of models, and they are given in the next section. The three types or classes of models are:

(a) *Semi-empirical Models.* These models usually offer the simplest results. They involve very little analytical derivation, and often the entire result or "law" is chosen simply because it matches best a certain set of measured data, or because it provides a reasonable and physically explainable scattering behavior. All such models involve one or more arbitrary constants which are functions of the properties of the surfaces considered and are determined by matching the model to measured results. These semi-empirical models offer little insight into the scattering process and the surface properties affecting the scattered fields. They can aid in categorizing the surface according to the best fit of the radar signal with one of the model equations.

(b) *Geometrical Models.* These models employ a surface made up of deterministic or simple shapes arranged randomly over a planar area. Since the scattering behavior of the simple shape alone is known, the model is then usually analyzed by multiple scattering techniques or from another scattering theory approach. The model is chosen in order to ease the analytical solution of a difficult problem, or because the model surface may to some degree resemble a certain class of natural surfaces.

(c) *Statistical Models.* These models treat the surface height above the xy-plane as a random variable. As such, this class of models is the most general. In order to analyze the problem and obtain results, however, one must choose a form for the statistical properties of the surface-height random variable (e.g., assume that its statistics are gaussian and that its correlation coefficient has a certain mathematical form, etc.). Since they are the most general, they provide results which are written explicitly in terms of the average surface properties rather than arbitrary constants. In the analysis, one generally uses a scattering theory best suited to the scale of roughness considered.

The limitations, restrictions, assumptions, approximations, and applications of the various models will be discussed as the results for the various models are presented in the next section. Comparison with measured results will be made when convenient or informative in Section 9.2.

9.1.2. Models of Rough-Surface Scattering

9.1.2.1. *Semi-empirical Models*

The existence and usefulness of semi-empirical models to describe the average scattered power (or cross section) of a rough surface is based primarily upon their ability to compare favorably with measured results for certain types of surfaces. They may or may not have arisen from a physical analysis of the scattering process. They all contain arbitrary constants which are functions of the average properties of the surface, but since this functional relationship is not known, a different constant must be chosen for each class of surfaces so that one obtains agreement between the model and measured results.

The average scattering cross section predicted by various models are tabulated in Table 9-1, along with their origin and name (if any). The polarization of the bistatically scattered wave is unobtainable from these models, and it is assumed that the average scattering cross section represents all of the power in the scattered field in a given direction. Any conditions on the constants or equations are also given in the table. The scattering angles on the equations are as shown in Figure 9-1. Comments on the various models are made below.

9.1.2.1.1. *Model 9A1.* This model, proposed by Clapp,[7] has the form of the well-known Lambert's law. For most terrain and natural surfaces at radar wavelengths, it agrees with measured results rather poorly. There is no satisfactory scattering mechanism which can explain the behavior of this model. It may have applicability for certain surfaces at very short wavelengths compared with the roughness dimensions.

9.1.2.1.2. *Model 9A2.* This is a natural extension or generalization of the above model. The constant n_2 need not be an integer. For example, $n_2 = 0.75$ (such that $2n_2 = \frac{3}{2}$) was suggested in References 8 and 9 to explain the scattering from the moon at angles of incidence, θ_i, greater than 7° at $\lambda \approx 1.5$ meters. Although unexplainable analytically, it does compare favorably with backscatter somewhat away from normal incidence (referred to by many as diffuse scattering).

9.1.2.1.3. *Model 9A3.* This model, although able to explain scattering at high frequencies from a hypothetical surface made up of a layer of spheres, does not offer results which compare exceptionally well with return from natural surfaces. The complete lack of dependence upon incidence and scattering angles is the reason. However, certain types of natural surfaces, especially those covered by vegetation (such as wheat or oats in head), are relatively independent of the incidence and scattering angles over certain ranges of these angles.[4]

9.1.2.1.4. *Models 9A4, 9A5.* These models are generally in quite good agreement with much of the measured return from natural surfaces whose dimensions are considerably greater than wavelength. They are valid in many instances only for scattering angles considerably away from the specular direction (referred to as diffuse scattering). They concur with visual observations of the moon at optical frequencies, for example, which show the moon to be uniformly bright across the visible disc at full moon (corresponding to backscattering).

Table 9-1

Semi-empirical Models for Average Scattering Cross Section from Rough Surfaces. Here K_1, K_2, K_i, n_2, λ_2, are Arbitrary Constants

	Name of Model	Bistatic Cross Section, $\gamma(\theta_i, \theta_s, \phi_s)$	Backscattering Cross Section, $\gamma(\theta_i)$ (or σ^0)	Source of Model (References)
9A1	Lambert, Clapp's First Model	$K_1 \cos\theta_i \cos\theta_s$	$K_1 \cos^2\theta_i$	4 and 7
9A2	Generalized Lambert	$K_2 (\cos\theta_i \cos\theta_s)^{n_2}$	$K_2 (\cos\theta_i)^{2n_2}$	4
9A3	Clapp's Second Model	K_3	K_3	4 and 7
9A4	Lommel–Seeliger	$K_4 \dfrac{2\cos\theta_i \cos\theta_s}{\cos\theta_i + \cos\theta_s}$	Clapp's Third Model $K_{4,5} \cos\theta_i$	4 and 7
A95	Grass	$K_5 \dfrac{\cos\theta_i + \cos\theta_s}{2}$	$K_{4,5} \cos\theta_i$	4 and 7
9A6	—	—	$K_6 \dfrac{\sin 2\theta_i}{\sin\theta_i} (1 + \cot\theta_1)$	10
9A7	—	—	$(K_7/\lambda_0^2) e^{(-\tan^2\theta_i/2s_0^2)}$ for $\lambda_0 < \lambda_2$	11
9A7	—	—	$\dfrac{1}{\lambda_0^6} [K_8 e^{(-\tan^2\theta_i)\lambda_0/2s_0^2\lambda_2} + K_9]$ for $\lambda_0 > \lambda_2$	11
			s_0^2 = mean square slope of surface	
9A8	—	—	$K_{11} \exp[-K_{12} \sin\theta_i]$	8 and 9
9A9	—	—	$\dfrac{K_{13}}{1 + K_{14}\theta_i^2}$	9

9.1.2.1.5. *Model 9A6.* This model was proposed in Reference 10 because it matched quite well backscattered power from natural surfaces such as forests, meadows, and desert at *X*-band. There has been no attempt to explain analytically the scattering mechanism behind such a model.

9.1.2.1.6. *Model 9A7.* This is one model of several proposed in Reference 11, and the only one which is explicitly a function of the angle of incidence. It is also one of the few in which an explicit dependence upon wavelength, λ_0, appears. The constant λ_2 represents an arbitrary upper limit on the rms slope spectrum of the surface. The constant s_0^2 is the actual mean-square slope of the surface. The model is based upon a physical explanation of a possible scattering mechanism, and it does compare with return from certain types of natural surfaces; there is not, however, enough measured data at different frequencies on similar types of surface to confirm the proposed wavelength dependence of this model.

9.1.2.1.7. *Model 9A8.* This model has been proposed in References 8 and 9 to explain radar backscatter from the lunar surface (a surface whose roughness is large compared to wavelength). It is valid for incidence angles, θ_i, generally less than 60°. For the lunar surface, values for K_{12} between 7 (at $\lambda_0 = 3.6$ cm) and 15 (at $\lambda_0 = 68$ cm) were obtained by fitting the equation to the measured curve from the moon.[9]

9.1.2.1.8. *Model 9A9.* This model was chosen to match measured backscattered power from the moon.[9] It is not based upon any known scattering mechanism, but it seems to offer the best closed-form angular representation for different frequencies. Constants chosen for this model on a subjective best-fit basis are $K_{14} = 50$ at $\lambda = 3.6$ cm and $K_{14} = 460$ at $\lambda = 68$ cm for lunar radar return.[9]

9.1.2.2. *Geometrical Models*

9.1.2.2.1. *Model 9B1—Twersky's Two-Dimensional Model.* Twersky[12,13] has solved for the scattering cross sections for a model of a two-dimensional rough surface consisting of infinitely long, perfectly conducting, circular half-cylinders of equal radii, randomly spaced on an infinitely long, perfectly conducting plane (see Figure 9-3). Solutions have been obtained only for incidence and scattering in the *xz*-plane, which is perpendicular to the cylinder axes.

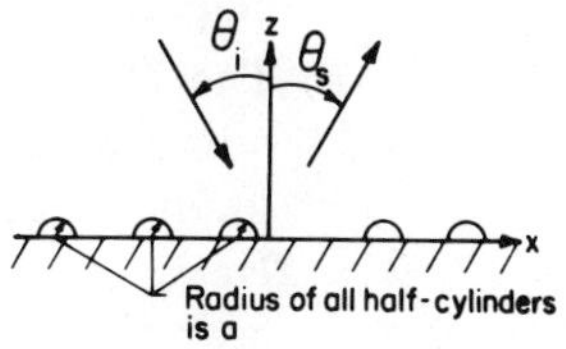

Fig. 9-3. Scattering geometry for a one-dimensionally rough surface consisting of half-circles (or half-cylinders) of radius *a* randomly spaced along the *x* direction.

The following are the assumptions or approximations, the advantages, and the limitations of the model.

(a) *Assumptions or restrictions:*

(i) The average spacing between cylinders is very large compared to a, the cylinder radius.

(ii) All surfaces are perfectly conducting.

(iii) The cylinders are randomly spaced with a uniform probability density function (i.e., it is as likely that the ith cylinder occurs within x_1 and $x_1 + \Delta x$, as it is that it occurs in any other range such as x_2 and $x_2 + \Delta x$). The average number of cylinders per unit length is then N_L.

(b) *Advantages of the model:*

(i) An exact scattering solution is employed using the method of imaging along with the exact solution for a single protuberance. As such, the method of solution is not restricted to certain asymptotic frequency ranges.

(ii) The model properly takes into account multiple scattering between various parts of the surface. Very few rough surface models take this effect into account.

(iii) The model takes into account shadowing of one portion of the surface by another.

(iv) As a consequence of two the preceding considerations, both of which become increasingly important near grazing incidence, this is one of the few models which offer results valid near grazing. It gives the correct dependence (according to data measured from the sea) of backscattering cross section upon grazing angles for horizontal and vertical polarizations.

(v) Results are available from the method for both large cylinders ($k_0a \gg 1$) and small cylinders ($k_0a \ll 1$).

(c) *Limitations and disadvantages of the model:*

(i) As with any geometrical model, one can seriously question how typical such a surface is of a rough natural surface. Hence, application of the results of this model to the interpretation of the scattering behavior of a natural surface may be helpful but always remains questionable.

(ii) The model predicts an extremely sharp increase in scattered power at the specular angle ($\theta_s = \theta_i$). This, of course, is due to the "flat-plate" effect of the perfectly conducting plane. Hence, the equations given here for this model are not useful near the specular direction. Thus the scattering cross sections predicted by this model describe better the incoherent scattered power rather than the total scattered power.†

(iii) This model is valid for power scattered only in the xz-plane, or across the roughness. Results in any other scattering or incidence plane are not available. Hence, the model might be useful in explaining scattering from the waves of the sea surface, but only when looking upwind or downwind.

For the bistatic scattering cross sections, the results of this model are:[12]

(a) $k_0a \ll 1$ (*small semicylinders*)

$$\gamma_{vv} = \frac{\pi N_L a (k_0 a)^3}{2} (1 \mp 2 \sin \theta_i \sin \theta_s)^2 \ddagger \tag{9.1-6}$$

$$\gamma_{hh} = \frac{\pi N_L a (k_0 a)^3}{2} (2 \cos \theta_i \cos \theta_s)^2 \tag{9.1-7}$$

In these equations the subscripts refer to the E-field polarization states (v—vertical, h—horizontal) of the scattered field (left subscript) and incident field (right subscript).

The above results do not give the correct power in the specular direction. The use of the upper or lower sign in (9.1-6) depends upon whether scattering is forward (into $+x$ quadrant) or backward (into $-x$ quadrant). θ_i and θ_s are always positive in these equations.

(b) $k_0a \gg 1$ (*large semicylinders*)

$$\gamma_{vv} = \gamma_{hh} = \frac{N_L a}{2(1 + N_L a \sec \theta_i)^2} \left[\left| \sin \frac{\pm \theta_s - \theta_i}{2} \right| + \left| \cos \frac{\pm \theta_s + \theta_i}{2} \right| \right] \tag{9.1-8}$$

and

$$\gamma_{hv} = \gamma_{vh} = 0$$

The results are not valid in the specular direction (i.e., where $\theta_i = \theta_s$, and where the scattering direction is in the $+x$ quadrant). The upper and lower signs have the same meaning as for (9.1-6), explained above.

† Incoherent scattered power is defined and discussed in Section 9.1.3.1.

‡ According to restriction (i) of the model, $N_L a \ll 1$; this is the ratio of cylinder radius to average distance between adjacent cylinders. The notation γ is equivalent to the σ^0 sometimes used in the rough-surface literature. See Eq. (9.1-1) for the definition of γ.

For the backscatter cross section, the results are:[12]

(a) $k_0 a \ll 1$ (*small semicylinders*)

$$\gamma_{vv} = \frac{\pi N_L a (k_0 a)^3}{2} (2 - \cos 2\theta_i)^2$$

$$\gamma_{hh} = \frac{\pi N_L a (k_0 a)^3}{2} (1 + \cos 2\theta_i)^2 \tag{9.1-9}$$

and

$$\gamma_{hv} = \gamma_{vh} = 0$$

This result is obtained by setting $\theta_i = \theta_s$ in Eqs. (9.1-6) and (9.1-7):

(b) $k_0 a \gg 1$ (*large semicylinders*)

$$\gamma_{vv} = \gamma_{hh} = \frac{N_L a}{2(1 + N_L a \sec \theta_i)^2} (1 + \sin \theta_i) \tag{9.1-10}$$

and

$$\gamma_{hv} = \gamma_{vh} = 0$$

The results (9.1-9) to (9.1-10) show that the backscatter cross section from this model may increase as a function of the incidence angle, θ_i, away from normal. In the latter result, this is due to the factor $(1 + \sin \theta_i)$. One must remember that these cross sections represent only the incoherent backscattered power, not the total backscattered power. Although the total power backscattered from natural rough surfaces is almost always observed to decrease with increasing angle from normal, the incoherent contribution may increase.

Since $\theta_i = \theta_s$ for backscattering, and this angle is close to $\pi/2$ near grazing incidence, define $\beta = \pi/2 - \theta_i$ as the grazing angle. Near grazing incidence, the backscatter cross sections must be derived separately; they do not come directly from Eqs. (9.1-9) to (9.1-10). The near grazing incidence backscatter cross sections are:[12]

(a) $k_0 a \ll 1$ (*small semicylinders*), $\beta^2 \ll 1$†

$$\gamma_{vv} \approx \frac{18 k_0}{\pi N_L} \beta^2 \tag{9.1-11}$$

$$\gamma_{hh} \approx 2\pi N_L a (k_0 a)^3 \beta^4 \tag{9.1-12}$$

$$\gamma_{hv} = \gamma_{vh} = 0$$

† β is to be measured in radians.

(b) $k_0 a \gg 1$ (*large semicylinders*), $\beta^2 \ll 1$

$$\gamma_{vv} \approx \frac{1}{2N_L a} \beta^2 \tag{9.1-13}$$

$$\gamma_{hh} \approx \frac{N_L a(k_0 a)^2}{2} \beta^4 \tag{9.1-14}$$

$$\gamma_{hv} = \gamma_{vh} = 0$$

One can see from the above equations that for a one-dimensionally rough surface and two-dimensional scattering, the backscattering cross section for horizontal polarization approaches zero more rapidly than does that for vertical, regardless of the scale of the roughness. This grazing angle polarization dependence is confirmed from measurements on the sea surface.

9.1.2.2.2. *Model 9B2—Twersky's Three-Dimensional Model.* Twersky[12] has obtained results for a three-dimensional rough-surface model by the same method discussed in the preceding section. His three-dimensional model consists of randomly spaced, perfectly conducting hemispheres, all of equal radii, on a perfectly conducting plane (see Figure 9-4). The perfectly-conducting plane is taken to be the xy-plane, in accordance with Figure 9-1, and commonly accepted definitions of scattering angles for rough surfaces.

All of the restrictions, advantages, and disadvantages listed for the two-dimensional model of the preceding section also apply here, except that the three-dimensional model offers a much more general solution to the scattering problem than does the two-dimensional model. One can consider scattering out of the plane of incidence, as well as within it. Almost all surfaces found in nature are actually rough in two dimensions, rather than just one. Hence, one can remove (iii) in the list of limitations mentioned previously.

Assumption (i) of the preceding section should be modified in the three-dimensional case to state that the average distance between hemisphere centers should be large compared to wavelength. Also, the hemispheres are to be randomly distributed over a surface area, rather than along a line.

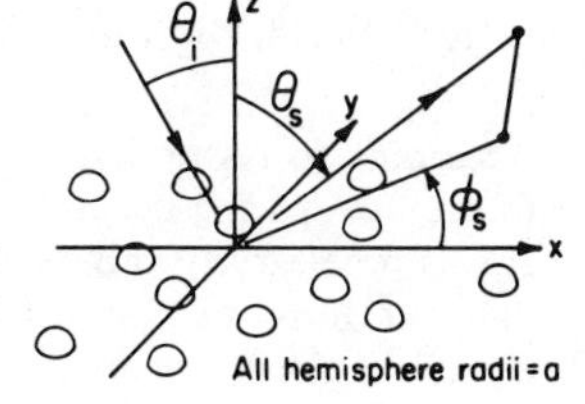

Fig 9-4 Scattering geometry for a two-dimensionally rough surface consisting of hemispheres of radius a randomly spaced on the xy-plane.

The appropriate parameter N_A gives the average number of hemispheres per unit surface area.†

For the bistatic scattering cross sections, the results of Twersky's three-dimensional model are:[(12)‡]

(a) $k_0 a \ll 1$ (*small hemispheres*)

$$\gamma_{vv} = N_A a^2 (k_0 a)^4 (\cos\phi_s - 2\sin\theta_i \sin\theta_s)^2 \tag{9.1-15}$$

$$\gamma_{hv} = N_A a^2 (k_0 a)^4 (\cos\theta_s \sin\phi_s)^2 \tag{9.1-16}$$

$$\gamma_{vh} = N_A a^2 (k_0 a)^4 (\cos\theta_i \sin\phi_s)^2 \tag{9.1-17}$$

$$\gamma_{hh} = N_A a^2 (k_0 a)^4 (\cos\theta_i \cos\theta_s \cos\phi_s)^2 \tag{9.1-18}$$

(b) $k_0 a \gg 1$ (*large hemispheres*)

$$\gamma_{vv} = \gamma_{hh} = \frac{\frac{1}{4} N_A a^2}{(1 + (\pi/2) N_A a^2 \sec\theta_i)^2} \times \left[\left(\frac{\cos\theta_i \cos\theta_s \cos\phi_s + \sin\theta_i \sin\theta_s - \cos\phi_s}{1 - \cos\theta_i \cos\theta_s - \sin\theta_i \sin\theta_s \cos\phi_s}\right)^2 + \left(\frac{\cos\theta_i \cos\theta_s \cos\phi_s - \sin\theta_i \sin\theta_s + \cos\phi_s}{1 + \cos\theta_i \cos\theta_s - \sin\theta_i \sin\theta_s \cos\phi_s}\right)^2\right] \tag{9.1-19}$$

$$\gamma_{hv} = \gamma_{vh} = \frac{\frac{1}{4} N_A a^2 \sin^2\phi_s}{(1 + (\pi/2) N_A a^2 \sec\theta_i)^2} \times \left[\left(\frac{\cos\theta_i - \cos\theta_s}{1 - \cos\theta_i \cos\theta_s - \sin\theta_i \sin\theta_s \cos\phi_s}\right)^2 + \left(\frac{\cos\theta_i + \cos\theta_s}{1 + \cos\theta_i \cos\theta_s - \sin\theta_i \sin\theta_s \cos\phi_s}\right)^2\right] \tag{9.1-20}$$

These results are not valid at the specular reflection direction or near grazing for either of the above ranges of hemisphere radius (or roughness scale). At the specular angle, there is also a very large coherent component of power due to direct reflection from the flat plane. One may, however, use or extrapolate the above results at the specular angle in the discussion of natural surfaces, understanding, of course, that the scattering cross sections given here represent only the incoherently scattered power. The coherent power dominates near the specular direction, especially when the hemisphere (or roughness) size is small compared to wavelength. See Section 9.1.3.1.1 and 9.1.3.1.2 for further discussion of coherent and

† According to restriction (i), $N_A a^2 \ll 1$. This is the ratio squared of hemisphere radius to average spacing between spheres.

‡ It was necessary to transform Twersky's results for his three-dimensional model into a different form to conform to the coordinate system used here. Consequently, the equations given here are not identical to those in Reference 12.

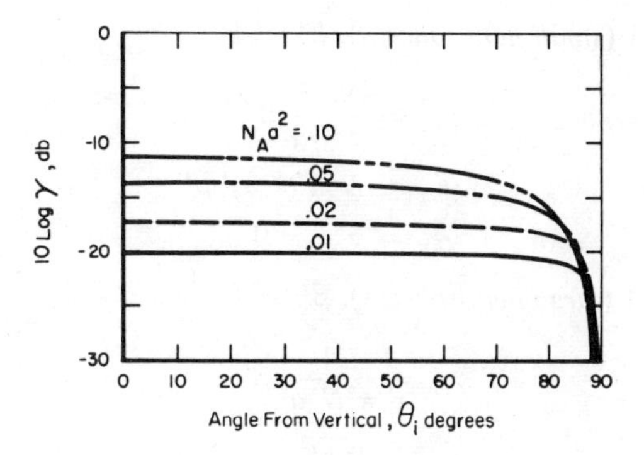

Fig. 9-5. Average backscattering cross section per unit area† for Twersky's rough-surface model consisting of hemispheres on a plane versus the incident angle for various hemisphere densities. Hemisphere size is greater than the wavelength ($k_0a > 1$). [Plotted from Eq. (9.1-23).]

incoherent scattered power. Results valid near grazing incidence are given below.

Equations (9.1-15) through (9.1-20) can be specialized to the case of backscattering by putting $\theta_i = \theta_s$, $\phi_s = -\pi$. The results are not valid at normal incidence, where $\theta_i = \pi/2$ (i.e., the specular direction in this case). For backscatter then, the results are:

(a) $k_0a \ll 1$ (*small hemispheres*)

$$\gamma_{vv} = N_A a^2 (k_0 a)^4 (2 - \cos 2\theta_i)^2 \tag{9.1-21}$$

$$\gamma_{hh} = N_A a^2 (k_0 a)^4 \cdot \tfrac{1}{4}(1 + \cos 2\theta_i)^2 \tag{9.1-22}$$

$$\gamma_{hv} = \gamma_{vh} = 0$$

(b) $k_0a \gg 1$ (*large hemispheres*)

$$\gamma_{vv} = \gamma_{hh} = \frac{N_A a^2}{[1 + (\pi/2)\, N_A a^2 \sec \theta_i]^2} \tag{9.1-23}$$

$$\gamma_{hv} = \gamma_{vh} = 0$$

Figure 9-5 shows curves of this average incoherent backscattering cross section per unit surface area† for large k_0a. These are plotted in decibels from Eq. (9.1-23) for various values of $N_A a^2$.

Again, let $\theta_i = \pi/2 - \beta$, where β is the grazing angle. The backscattering cross sections very near grazing do not come from Eqs. (9.1-21) through (9.1-23), but are obtained separately, and are given by:[12]

† The notation γ is equivalent to the σ^0 sometimes used in the rough-surface literature; see Eq. (9.1-1).

(a) $k_0 a \ll 1$ (*small hemispheres*), $\beta^2 \ll 1$

$$\gamma_{vv} \approx \frac{9k_0^2}{\pi^2 N_A} \beta^2 \tag{9.1-24}$$

$$\gamma_{hh} \approx N_A a^2 (k_0 a)^4 \beta^4 \tag{9.1-25}$$

$$\gamma_{hv} = \gamma_{vh} = 0$$

(b) $k_0 a \gg 1$ (*large hemispheres*), $\beta^2 \ll 1$

$$\gamma_{vv} \approx \frac{1}{\pi^2 N_A a^2} \beta^2 \tag{9.1-26}$$

$$\gamma_{hh} \approx \frac{N_A a^2}{4} (k_0 a)^2 \beta^4 \tag{9.1-27}$$

$$\gamma_{vh} = \gamma_{hv} = 0$$

In summary, this model may be recommended for its validity near grazing incidence. It is the only geometrical model which properly takes into account all of the multiple interaction phenomena which enter into the problem near grazing (e.g., shadowing of one portion of the surface by another, multiple reflections, etc.). It is intuitively reasonable to assume that near grazing, all rough surfaces tend to look the same, and scatter by the same mechanism. Thus, while a model consisting of hemispheres randomly spaced on a plane may be highly artificial at higher angles, near grazing it tends to look and scatter the same way as any rough surface. The scale of the roughness does not affect the angular dependence of the backscattering cross section; the cross section for vertical polarization always decreases slower than that for horizontal by factors β^2 and β^4, respectively. This behavior is evidenced experimentally for the sea surface.[13]

At higher incidence angles, there are other desirable features of this model, even though it is highly artificial. For one thing, it is strikingly simple in form and easy to visualize. The approximations used to obtain the results are held to a minimum. Solutions are available where the protuberances (hemispheres in this case) are both large and small compared to wavelength. Thus, for the same type of protuberance, one can get an idea of how the cross section behaves in the limits of large and small protuberance dimension, but where the portion of the smooth plane occupied by the spheres remains the same (i.e., $N_A a^2$ is constant).

A useful parameter for this model, when comparing its roughness to the roughness of other statistical models and natural surfaces, is the height variance, or mean square height of the composite surface about its average height, defined here as h^2. For this model of hemispheres on a flat plane, it is

$$h^2 = N_A a^2 \pi a^2 \left(\frac{1}{2} - \frac{4\pi}{9} N_A a^2\right) \tag{9.1-28}$$

It should be noted that even though the individual protuberance may be quite large ($k_0a \gg 1$), and even though there may be very many such protuberances in the illuminated area, the surface height variance h^2 may still be quite small compared to wavelength squared for this model, since $N_Aa^2 \ll 1$, according to restriction (i).

9.1.2.2.3. *Model 9B3—Macfarlane's Model.* Macfarlane[14,15] has proposed a model for a one-dimensionally rough surface consisting of separate, successive, and sinusoidal waves. Each separate complete wave is connected to the succeeding wave, as shown in Figure 9-6. The wavelength and amplitude of each wave are random variables which are Gaussian-distributed. Hence, the model may appear somewhat representative of the sea surface when looking upwind or downwind. Only backscattering is considered, and physical-optics methods are used to solve the problem.

The following are the approximations, advantages, and limitations of the model.

(a) *Approximations or restrictions*:

(i) The surface is rough in one dimension only, i.e., in the x direction. Only backscattering which takes place in the xz-plane is considered.

(ii) The wave amplitudes and wavelengths of successive waves are taken as uncorrelated random variables obeying a Gaussian law.

(iii) Physical optics methods are used to solve the problem. This requires that the surface radius of curvature at all points be large compared to wavelength. This condition, in turn, demands that

$$4\pi^2 A_i \lambda_0 / L_i^2 \ll 1$$

for the various waves. Physical optics also implies that shadowing of one portion of a surface by another, and multiple

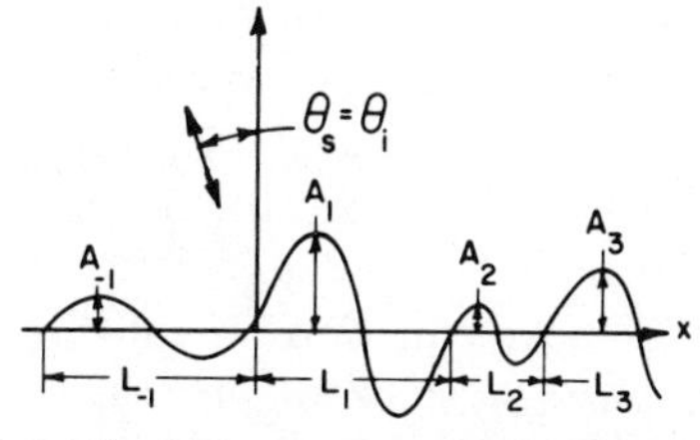

Fig. 9-6. Scattering geometry for a one-dimensionally rough surface consisting of separate full-cycle pieces of sine waves, each with random amplitude and period.

reflections of waves between various points on the surface, are neglected. These phenomena become more important near grazing incidence.

(b) *Advantages of the model*:

(i) The model approximates a one-dimensionally rough surface, such as a portion of the sea surface, much better than a purely deterministic constant amplitude and wavelength sinusoidal surface. Randomness is introduced in the two parameters of each surface wave, which brings the model closer to the realm of real life.

(ii) Relatively simple closed-form results are available in terms of the wavelength of the incident field, average amplitude and length of the successive surface wave elements, and angle of incidence, θ_i.

(c) *Limitations of the model*:

(i) The model is not valid near grazing incidence.

(ii) Solutions are obtained only for backscattering in a direction looking across the one-dimensional roughness, i.e., in the xz-plane.

(iii) Being a physical-optics model, the results are independent of polarization. Thus, the backscattering cross section given is the same for vertical incident–vertical scattered and horizontal incident–horizontal scattered polarization. Physical optics predicts no cross-polarized backscattered power.

The result for the average backscatter cross section per unit surface area† is

$$\gamma = \frac{L}{\lambda_0} \cos^2 \theta_i \exp \left[- \frac{(4\pi)^2 A^2}{\lambda_0{}^2} \cos^2 \theta_i \right] I_m \left(\frac{(4\pi)^2 A^2}{\lambda_0{}^2} \cos^2 \theta_i \right) \tag{9.1-29}$$

where L is the average length of one surface wave segment (i.e., $L = \langle L_i \rangle$), A^2 is the mean square amplitude of a surface wave ($A^2 = \langle A_i{}^2 \rangle$), and $m = (2L/\lambda_0) \sin \theta_i$. $I_\nu(z)$ is the modified cylindrical Bessel function with argument z and order ν (see Reference 16, Section 9.6 for definition and properties of this function).

One can simplify the above equation only if the argument of the Bessel function is either much larger or much smaller than the order. For instance, very close to normal incidence, $m = (2L/\lambda_0) \sin \theta_i \approx (2L/\lambda_0)\, \theta_i$ may be

† The notation γ is equivalent to the σ^0 sometimes used in the rough-surface literature; see Eq. (9.1-1).

much smaller than the argument, $[(4\pi)^2 A^2/\lambda_0^2]\cos^2\theta_i \approx (4\pi)^2 A^2/\lambda_0^2$. In this case, the cross section may be approximated as

$$\gamma = \frac{L\cos\theta_i}{2A(2\pi)^{3/2}} \tag{9.1-30}$$

In order to show the behavior of these cross sections throughout the entire range of θ, plots were made of the cross section of Eq. (9.1-29) for various parameters, A/λ_0 and L/λ_0. These are shown in Figure 9-7.

This model is useful for the analysis of radar sea return when looking upwind or downwind (i.e., across the waves).

Some further properties of this model may be useful in relating the statistical properties of this surface to those of other rough surfaces. These are the mean square slope along a given direction (the x direction in this case), s_x^2, and the mean square height, h^2, above the mean plane. These are

$$h^2 = \tfrac{1}{2}A^2$$

and

$$s_x^2 = \frac{2\pi^2 A^2}{L^2} \tag{9.1-32}$$

9.1.2.2.4. *Model 9B4—Peake's Model.* Peake[4,17,18] has proposed and investigated a model consisting of long, thin, dielectric cylinders, arranged randomly, but preferring a vertical orientation, or z direction, as shown in Figure 9-8. Thus this model resembles physically many natural surfaces consisting of vegetation, such as forests, grass, wheat fields, etc. Such surfaces would be extremely difficult to describe statistically by merely representing the height as a random variable. Besides, in the case of scattering from vegetation, the waves penetrate into the growth area and are multiply reflected between branches or stems; hence, it is extremely doubtful whether such a process can be compared to simple scattering from a single surface.

The assumptions and restrictions, advantages, and limitations of the model are given below.

(a) *Assumptions and restrictions*:

(i) All cylinders are taken to be semi-infinite in length. They are terminated at the upper end on the $z = 0$ plane, but extend downward infinitely far. All cylinders are circular and have the same radius and electrical properties.

(ii) The cylinders are much less than a wavelength in diameter. This permits use of a low-frequency approximation in the description of the fields inside the cylinder. Only dielectric cylinders are analyzed, and thus the component of electric field inside the cylinder parallel to the axis is taken to equal

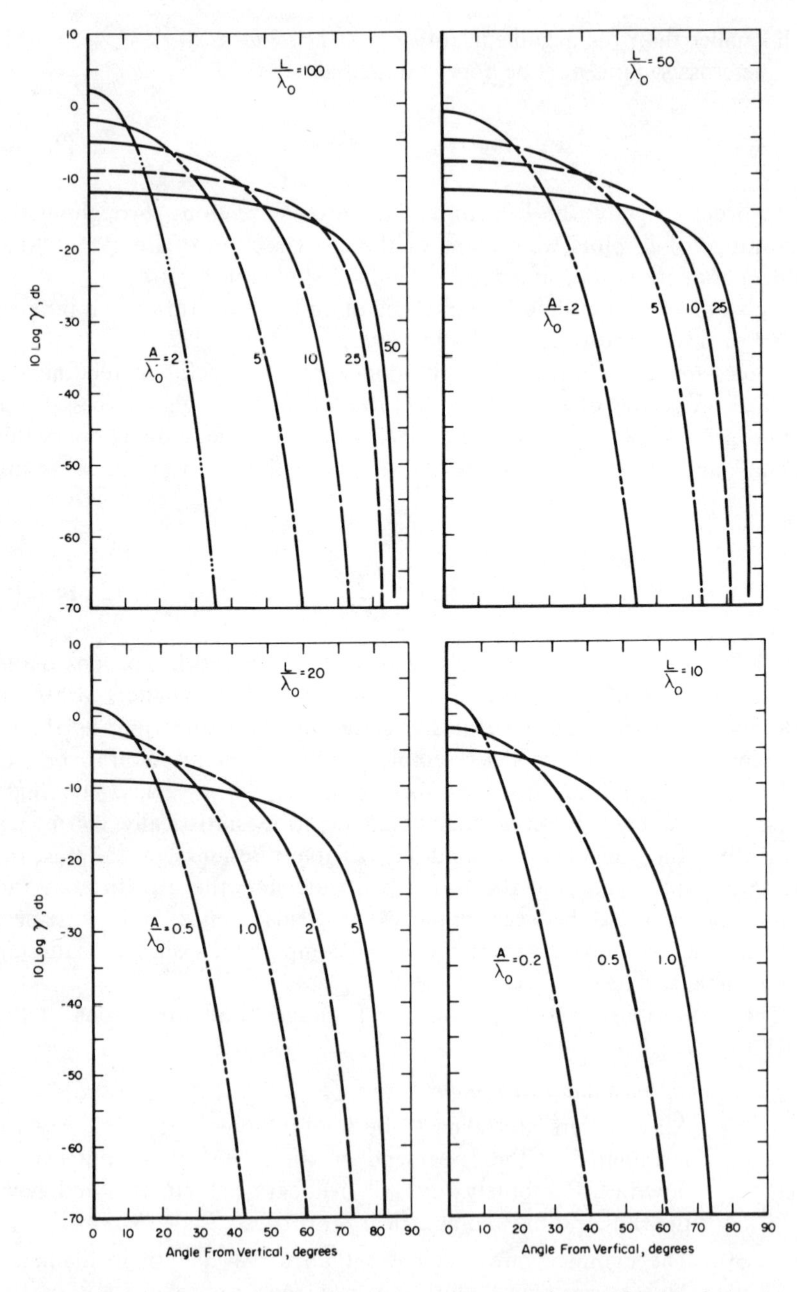

Fig. 9-7. Average backscattering cross section per unit area for Macfarlane's rough-surface model versus incidence angle. [Plotted from Eq. (9.1-29) for various values of L/λ_0 (average relative wave period) and A/λ_0 (average relative wave amplitude).]

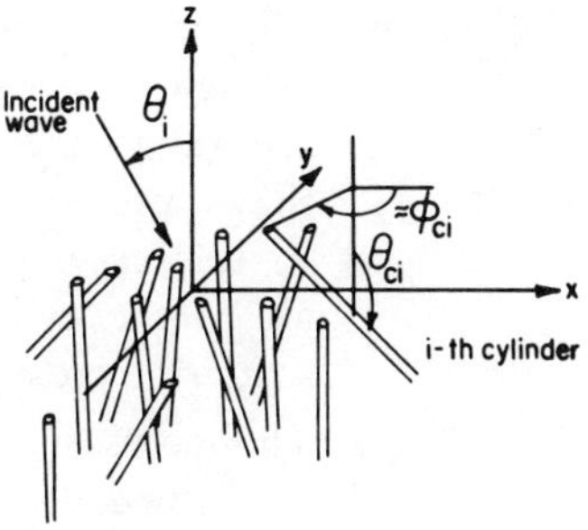

Fig. 9-8. Scattering geometry for a vegetation-like rough-surface model consisting of thin dielectric cylinders randomly arranged and oriented, but terminating at their upper end on the xy-plane.

that outside the cylinder, while the component inside perpendicular to the axis is taken to be $t = 2/(1 + \epsilon_r)$ times that outside the cylinder. The basic analysis is valid for more general cylinders (e.g., noncircular cross section, complex dielectric constant, etc.), but then t becomes more complicated and must generally remain an undetermined parameter in the result.

(iii) The cylinders are randomly spaced and oriented. The average number of cylinders per unit area of the $z = 0$ plane is taken to be N. The probability that a cylinder, e.g., the ith cylinder, is oriented at angle θ_{ci} from vertical, and ϕ_{ci} in azimuth is $p(\theta_{ci})\, d\Omega$, where the probability density assumed is $p(\theta_{ci}) = (3/2\pi) \cos^2 \theta_{ci}$, and $d\Omega$ is a unit solid angle defined by $d\Omega = \sin \theta_{ci}\, d\theta_{ci}\, d\phi_{ci}$ in a spherical coordinate system. Hence all azimuthal angles ϕ_{ci} are equally likely, while the cylinder is more likely to be closer to vertical. Such a preference for the vertical direction is exhibited by almost all types of vegetation. The range of ϕ_{ci} is $0 \leqslant \phi_{ci} \leqslant 2\pi$, and for θ_{ci} is $\pi/2 \leqslant \theta_{ci} \leqslant \pi$.

(iv) The incident wave entering the cylinder structure is assumed to be attenuated exponentially as it propagates deeper vertically into the medium. This attenuation is due both to absorption losses within the cylinders themselves, and to a continuous scattering and redirecting of the incident energy. The attenuation constant is defined as α, and may be considered as a part of the wave number of refractive index of the volume of space containing the cylinders. Thus the incident wave effectively never gets very far into the cylinder structure, and it is immaterial whether the cylinders extend down to infinity, or whether they terminate on a "ground" very far down. This fact is confirmed experimentally in Reference 17 for radar return from (two-inch-tall) grass at K_a band; it is observed that nearly all of the scattered power is due to the grass and not from the ground beneath it.

(v) The spacing between adjacent cylinders is considered large enough in terms of wavelength, and their relative orientations are random enough, so that the phase difference between the scattered fields from these cylinders is truly random. In this case, the total average scattering cross section from all of the cylinders is the sum of their individual average scattering cross sections.

(vi) The necessary averaging operation done in the analysis is performed after an approximation. This approximation involves averaging the numerator and denominator separately in a complicated rational expression. The error in this approximation is negligible near normal incidence, but increases near grazing.

(b) *Advantages*:

(i) It is the only model known which can describe qualitatively and quantitatively the scattering by vegetation-covered surfaces.

(ii) The scattering cross section predicted by this model is a function of incidence and scattering angles, wavelength, dielectric constant, and water content of the "vegetation," number of cylinder scatterers per unit area, and the cross sectional area per cylinder, A. All of these factors have been shown experimentally to affect the scattered power from vegetation.

(iii) The model sensibly accounts for the attenuation of the incident wave as it propagates into the vegetation layer. This attenuation is observed experimentally.

(iv) Results are available for both bistatic and backscattering. The predicted scattering cross sections are functions of polarization directions of incident and scattered fields. Such polarization dependence is observed in return from such vegetation-covered surfaces.

(c) *Disadvantages*:

(i) The model does not account rigorously for the multiple interaction between cylinders (i.e., multiple scattering and shadowing). As with other models, these effects may be negligible for angles near normal, but they become increasingly important near grazing. Hence results deviate from measurements near grazing.

(ii) The model involves many constants and assumptions. Many of these constants are difficult to obtain in practice (e.g., dielectric constant of a blade of grass at K_a band, its cross-sectional area, etc.).

The various average bistatic cross sections per unit surface area† predicted by this model are:[17,18]

$$\gamma_{hh}(\theta_i, \theta_s, \phi_s) = B_p[(3 - 2\sin^2\phi_s) + t(8 - 10\sin^2\phi_s) + t^2(24 - 23\sin^2\phi_s)] \tag{9.1-33}$$

$$\gamma_{vh}(\theta_i, \theta_s, \phi_s) = B_p[(1 + 2\sin^2\theta_s) + 2\cos^2\theta_s\sin^2\phi_s) + t(-2 - 4\sin^2\theta_s + 10\cos^2\theta_s\sin^2\phi_s) + t^2(1 + 2\sin^2\theta_s + 23\cos^2\theta_s\sin^2\phi_s)] \tag{9.1-34}$$

$$\gamma_{hv}(\theta_i, \theta_s, \phi_s) = B_p[(1 + 2\sin^2\theta_i + 2\cos^2\theta_i\sin^2\phi_s) + t(-2 - 4\sin^2\theta_i + 10\cos^2\theta_i\sin^2\phi_s) + t^2(1 + 2\sin^2\theta_i + 23\cos^2\theta_i\sin^2\phi_s)] \tag{9.1-35}$$

$$\begin{aligned}\gamma_{vv}(\theta_i, \theta_s, \phi_s) = B_p[&(1 - t)^2(3 - 2\cos^2\theta_i\cos^2\theta_s\sin^2\phi_s) \\ &+ (\sin\theta_i\sin\theta_s - \cos\theta_i\cos\theta_s\cos\phi_s) \\ &\times [12(1 - t^2)\sin\theta_i\sin\theta_s \\ &+ 14t(1 - t)(3\sin\theta_i\sin\theta_s - \cos\theta_i\cos\theta_s\cos\phi_s) \\ &+ 35t^2(\sin\theta_i\sin\theta_s - \cos\theta_i\cos\theta_s\cos\phi_s)]]\end{aligned} \tag{9.1-36}$$

The symbols appearing in the above equations are defined as follows:

$$B_p = \frac{(AN)(Ak_0^2)[(\epsilon_r' - 1)^2 + \epsilon_r''^2]}{\left[28\pi[\frac{3}{5}(\alpha/k_0)^2 + 3(\cos\theta_i + \cos\theta_s)^2 + (\sin^2\theta_i + \sin^2\theta_s - 2\sin\theta_i\sin\theta_s\cos\phi_s)]\right]} \tag{9.1-37}$$

and N is the average number of cylinders per unit area of the $z = 0$ surface, A is the average cross-sectional area of a cylinder, and $\epsilon_r = \epsilon_r' + i\epsilon_r''$ is the average relative permittivity of the cylinder material, including loss effects; while

$$t = 2/(1 + \epsilon_r') \tag{9.1-38}$$

for thin (compared to wavelength) dielectric cylinders. The parameter α is the attenuation factor of the incident field as it propagates through the cylinders. Generally, α as a parameter is not easily related to the other scattering parameters, but an estimate of its value when the incident wave is horizontally or vertically polarized has been made by Peake.[17] These values are

$$\alpha_h = k_0[\tfrac{3}{16}AN\,\epsilon_r''\sec\theta_i](1 + 3t^2) \tag{9.1-39}$$

and

$$\alpha_v = k_0[\tfrac{3}{16}AN\,\epsilon_r''\sec\theta_i][(1 + 3t^2) + \sin^2\theta_i(1 - t^2)] \tag{9.1-40}$$

† The notation γ is equivalent to the σ^0 sometimes used in the rough-surface literature; see Eq. (9.1-1).

where α_h and α_v are to be used for α when the incident wave is polarized horizontally or vertically, respectively.

For typical vegetation containing water, the estimate for the dielectric constant is given in Reference 17 as

$$\epsilon_r = \epsilon_r + i\epsilon_r'' = 2.5(1 - f) + f\epsilon_{rw} \tag{9.1-41}$$

where f is the fraction by weight of water present in the vegetation, and ϵ_{rw} is the dielectric constant of water which can be approximated in the microwave region as

$$\epsilon_{rw} = 5 + \frac{75}{1 - i(1.85/\lambda_0)} \tag{9.1-42}$$

with λ_0, the free space wavelength in centimeters.

As noted in Reference 17, the radar return from vegetation-covered areas in practice depends significantly upon the season of the year and the time of the day. The factors which vary with the season and time of day are moisture content and the number of stems or blades (of grass, etc.) per unit area; both of these factors are taken into account in the above model. Observations indicate, as mentioned previously, that the depth of the vegetation cover does not influence significantly the radar return so long as the depth is greater than the wavelength.

Typical values of some of these constants have been measured for ordinary grass, such as a lawn. They are $A \approx 1.3 \times 10^{-2}\ \mathrm{cm}^2$ for the blade cross-sectional area. The product AN for grass lies in the range $0.01 < AN < 0.1$, depending upon the time of the year. The moisture content f varies between 30% for dry grass to about 90% for very moist green grass.

When backscattering is considered such that $\theta_i = \theta_s$ and $\phi_s = \pi$, Eqs. (9.1-33) through (9.1-36) simplify to become

$$\gamma_{hh} = \frac{(AN)(Ak_0^2)[(\epsilon_r' - 1)^2 + \epsilon_r''^2][3 + 8t + 16t^2]}{28\pi[3(\alpha_h/k_0)^2 + 4(1 + 2\cos^2\theta_i)]} \tag{9.1-43}$$

$$\gamma_{vv} = \frac{(AN)(Ak_0^2)[(\epsilon_r' - 1)^2 + \epsilon_r''^2][3 + 8t + 16t^2 + \sin^2\theta_i(12 + 4t - 16t^2)]}{28\pi[3(\alpha_v/k_0)^2 + 4(1 + 2\cos^2\theta_i)]} \tag{9.1-44}$$

$$\gamma_{hv} = \gamma_{vh} = \frac{(AN)(Ak_0^2)[(\epsilon_r' - 1)^2 + \epsilon_r''^2](1 - t)^2(1 + 2\sin^2\theta_i)}{28\pi[3(\alpha/k_0)^2 + 4(1 + 2\cos^2\theta_i)]} \tag{9.1-45}$$

where the choice for α of α_h or α_v in the last equation depends upon whether the incident field is horizontally or vertically polarized. The other constants are as defined previously.

Plots of predicted backscatter cross sections for this model using Eqs. (9.1-43) to (9.1-45) are shown in Figure 9-9 to 9-11 for various values of the parameters A, N, λ_0, and f.

9.1.3. Statistically-Rough Planar Surface Models

9.1.3.1. *Introduction and Physical Description of Effect of Roughness on Scattering*

This type of model is the most general. It treats the surface height above a mean planar surface as a random variable. The mean planar surface is the xy-plane of Figure 9-1. The model then consists of assigning to this random variable a given general behavior called statistics. Certain statistical

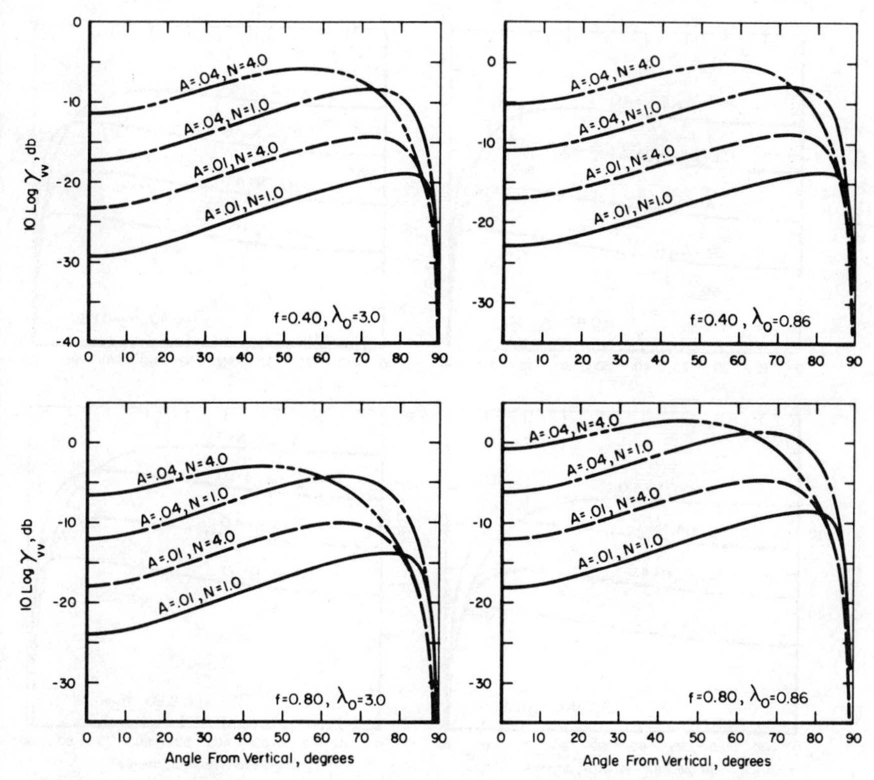

Fig. 9-9. Average backscattering cross section per unit area for Peake's vegetation-like rough-surface model (consisting of thin dielectric cylinders) versus incidence angle for vertical incident and vertical received polarization. [Plotted from Eq. (9.1-44) for various values of the parameters A, N, λ_0, and f. (A is average cross-sectional area of cylinder in square centimeters, N is average number of cylinders per square centimeter, λ_0 is wavelength in centimeters, and f is the fractional water content.)]

properties must then be chosen to obtain closed-form solutions for average scattering cross sections. Even when this is done, however, solutions can be obtained generally only when rms surface height is either much less or much greater than the wavelength. The former case is referred to here as a slightly rough surface. Its scattering properties are best derived by a perturbation approach. The latter case is referred to as a very rough surface. Its scattered fields are nearly always obtained using an optics or high-frequency approach.

In order to understand better the scattering process from a rough surface and the terminologies employed, consider the following description. In the absence of all roughness, a smooth planar surface of infinite extent reflects an incident plane wave into the specular direction ($\theta_s = \theta_i$); the

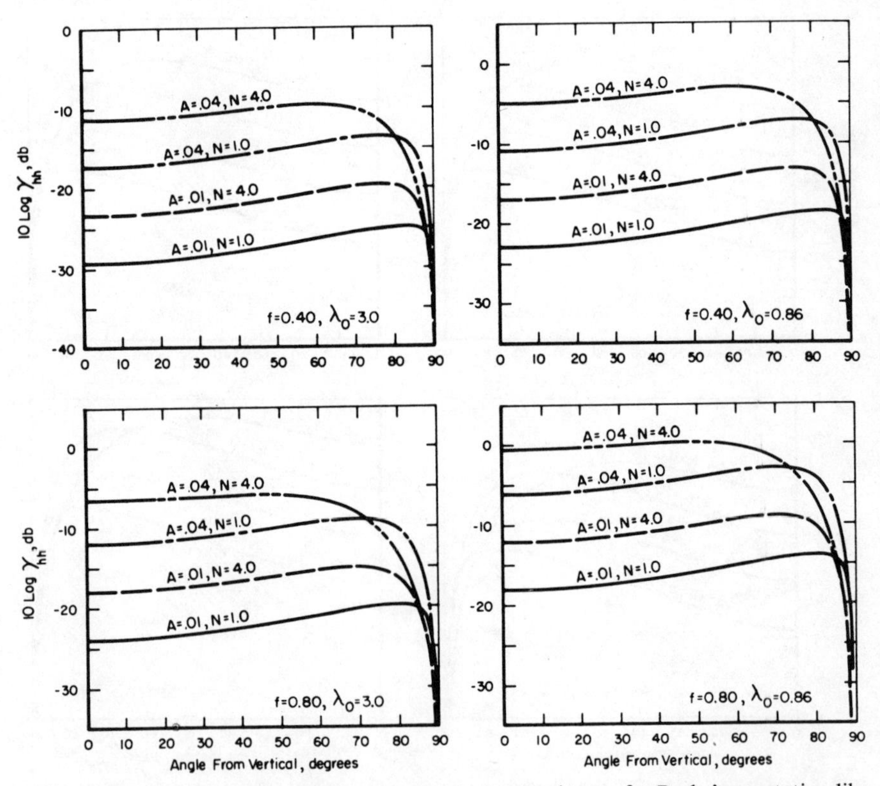

Fig. 9-10. Average backscattering cross section per unit area for Peake's vegetation-like rough-surface model (consisting of thin dielectric cylinders) versus incidence angle for horizontal incident and horizontal received polarization. [Plotted from Eq. (9.1-43) for various values of the parameters A, N, λ_0, and f. (A is average cross-sectional area of cylinder in square centimeters, N is average number of cylinders per square centimeter, λ_0 is wavelength in centimeters, and f is the fractional water content.)]

amplitudes of the reflected fields are equal to the incident field times the proper Fresnel reflection coefficient, $R_{\parallel}(\theta_i)$ or $R_{\perp}(\theta_i)$, depending upon whether the incident E-field lies in the plane of incidence (i.e., the xz-plane) or perpendicular to it.

Now, if instead of an infinite surface, one has either a finite-sized surface or finite uniform plane-wave illumination of an infinite surface, then the situation changes. In the far zone of the surface, the scattered power decreases as $1/R^2$, where R is the distance to the surface from the scattered field observation point. If, for example, the illuminated area A is rectangular, then the maximum scattered field will still exist in the specular direction, but a lobe structure will exist around the specular direction. At angles far from the specular direction, there will be very little scattered

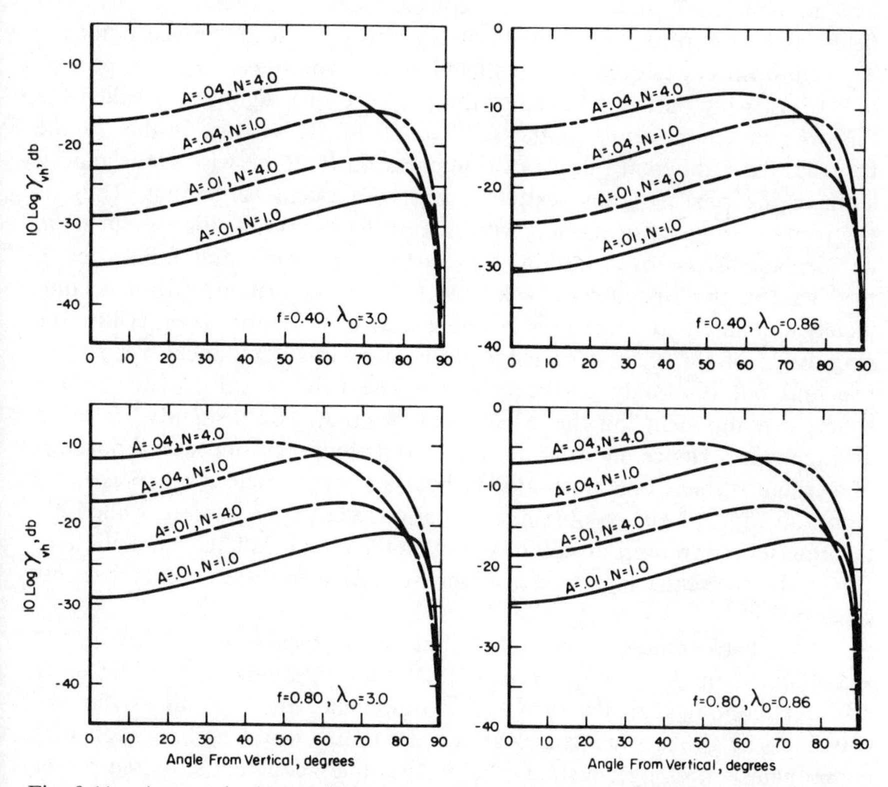

Fig. 9-11. Average backscattering cross section per unit area for Peake's vegetation-like rough-surface model (consisting of thin dielectric cylinders) versus incidence angle for vertical incident and horizontal received polarization (and conversely). [Plotted from Eq. (9.1-45) for various values of the parameters A, N, λ_0, and f. (A is average cross-sectional area of cylinder in square centimeters, N is average number of cylinders per square centimeters, λ_0 is wavelength in centimeters, and f is the fractional water content.)]

power. This scattering pattern is predicted adequately by physical optics (Section 7.5).

If now a very fine roughness appears on the otherwise smooth planar surface (whose mean-square height is still much less than wavelength), the first visible effect is the scattering of a small amount of power (small compared to that in the specular main lobe) into other directions; this power "fills in" the nulls between the original lobes due to the finite size of the smooth planar surface. At angles significantly different from the specular angle, however, the lobe structure is extremely small in the first place, and the latter power, due to the presence of the slight roughness, can dominate it entirely.

The former power component, due to the surface without roughness, is often referred to as the "coherent component" of the scattered field. This nomenclature arises because a constant, readily predictable, scattered field amplitude, phase angle, and pattern always exist at a given point for a given surface position. The scattered power component in nonspecular directions, which arises because of the presence of roughness, is called the "incoherent component" in the scattered field. This name is due to the fact that for a different sample of roughness on the otherwise same smooth surface, the phase of this scattered field component is random. Thus, if the rough surface is in motion with respect to the observation point (e.g., an airplane flying over terrain, the sea surface rising and falling, etc.), the phase of the "incoherent" scattered field at one instant of time is completely unrelated to the phase at an instant considerably later. Thus, the magnitude of the electric or magnetic field of this component appears to rise and fall randomly or incoherently. The field magnitude due to the coherent component, on the other hand, fluctuates only slightly about an average value. Hence, in order to obtain contributions from several portions of a rough surface, one must add the field quantities themselves (preserving amplitude and phase) to obtain the total coherent contribution, while only the intensities or powers are added to obtain the total incoherent contribution. Generally, the scattering theories of this section keep these two components separate.

As the roughness or surface turbulence increases in size, less power is contained in the lobe pattern of the coherent component, while more power is contained in the incoherent component. In the limit as the rms roughness height becomes somewhat greater than a wavelength, the coherent component is no longer distinguishable, and the incoherent scattered power dominates in all directions. In this limit, all of the scattered power is properly predicted by optics; i.e., the power scattered in a given direction is related to the number and properties of the surface facets oriented so that they reflect specularly in that direction. Such is the phenomenon when one sees the "shimmering of the moon on a choppy sea."

As examples, a flat field of short grass appears smooth at VHF frequencies (wavelength of three meters); its scattered field is predictable and nearly all coherent. At K_a band (wavelength of 1 cm), the scattered field is unpredictable (unless one knows the position of each blade of grass) and, consequently, is incoherent.

One other type of rough surface often appears and is discussed here; this is the composite surface.[5,19] In fact, this is probably the most common type of rough surface at microwave frequencies. Considering mountains and valleys as roughness, they represent a large-scale roughness or deviation from the smooth spherical earth. Grass and foliage represent a smaller-scale roughness. If one were to look at such a portion of the earth's surface from a satellite at L-band ($\frac{1}{3}$-meter wavelength), the mountains and valleys would scatter specularly in a facet-like manner. The return from them alone would be almost completely incoherent at the above wavelength. The slight (with respect to wavelength), grassy roughness on top of this further redistributes the scattered energy; this latter energy might appear noticeable in directions where specular scattered power from the mountains is a minimum. It is sometimes referred to as the "diffuse" component by radar experimentalists.

In practice, nearly all natural rough surfaces (e.g., sea surface, terrain, planetary surfaces, polished metals at optical frequencies, etc.) have composite roughnesses. There is generally a "very rough" surface profile along with a "slightly rough" profile. The latter is smaller in scale, while the former is larger in scale than wavelength. Both types of roughness scatter only incoherently since the coherent component is suppressed by the large-scale random roughness. Both types of roughness are generally produced by different natural phenomena and are for the most part statistically independent in nature. Hence the composite rough surface model, constructed in a subsequent section, consists of the average scattered power (or scattering cross section) from a very rough surface alone, added to the average scattered power from a slightly rough surface alone.

For backscattering, the large-scale roughness of such a composite produces a strong effect near the mean normal to the planar surface, while the slight roughness produces the dominant effect close to grazing. The reason for this is that the largest scales of roughness in natural surfaces are usually the most gently sloping, and their local normal deviates little from the mean surface normal (e.g., 10–20°). Optics theories predict that such surfaces backscatter predominantly near the mean normal. The smaller scale roughnesses, on the other hand, are generally more sharply sloping (such as stones, rubble, grass, etc.). Optics theories no longer apply to those scales whose rms heights and radii of curvature are smaller than the wavelength. Low-frequency techniques, such as the perturbation approach used below, must be employed. Such theories show that the scattered power is not as much due to the specularly oriented facets or the shape of the protuberances, as it is to their spacing and volume. These more sharply curving, small-scale protuberances scatter incident power into directions considerably away from the specular. This behavior will be observed in subsequent sections.

9.1.3.2. *Coherent Reflection Coefficients for Rough Planar Surface of Infinite Extent*

The fields reflected from a perfectly smooth planar surface of infinite extent are related to the incident plane waves by the well-known Fresnel reflection coefficients.† The energy redirected by the surface obeys Snell's law so that the direction of propagation makes an angle with the normal, θ_s, which is equal to the angle of incidence, θ_i (see Section 7.1 for further discussion of reflection from smooth planar surfaces).

The incident and reflected components of the *E*-field in the plane of incidence (i.e., the *xz*-plane of Figure 9-1) are related by

$$\hat{\mathbf{k}}_0{}^r \times \mathbf{E}^r = R_\parallel(\theta_i)\, \hat{\mathbf{k}}_0{}^i \times \mathbf{E}^i \qquad \text{or} \qquad \mathbf{H}_y{}^r = R_\parallel(\theta_i)\, \mathbf{H}_y{}^i \tag{9.1-46}$$

with

$$R_\parallel(\theta_i) = \frac{\epsilon_r \cos\theta_i - \sqrt{\epsilon_r\mu_r - \sin^2\theta_i}}{\epsilon_r \cos\theta_i + \sqrt{\epsilon_r\mu_r - \sin^2\theta_i}}$$

and

$$\hat{\mathbf{k}}_0{}^i = \sin\theta_i \hat{\mathbf{x}} - \cos\theta_i \hat{\mathbf{z}}, \qquad \hat{\mathbf{k}}_0{}^r = \sin\theta_i \hat{\mathbf{x}} + \cos\theta_i \hat{\mathbf{z}} \tag{9.1-47}$$

The incident and reflected components of the *E*-field normal to the plane of incidence (i.e., E_y or E_ϕ) are related by

$$\mathbf{E}_\perp{}^s = R_\perp(\theta_i)\, \mathbf{E}_\perp{}^i \tag{9.1-48}$$

with

$$R_\perp(\theta_i) = \frac{\mu_r \cos\theta_i - \sqrt{\epsilon_r\mu_r - \sin^2\theta_i}}{\mu \cos\theta_i + \sqrt{\epsilon_r\mu_r - \sin^2\theta_i}} \tag{9.1-49}$$

Here μ_r and ϵ_r are taken to be the relative permeability and permittivity of the surface medium.

Now if the smooth planar surface is made rough, the coherent reflected field given above is modified. It decreases in magnitude with the increase in roughness. The coherent nature of the specularly reflected field is not destroyed so long as the roughness is not too large. The average Fresnel reflection coefficients given above now become

$$\bar{R}_\parallel{}^R = R_\parallel(\theta_i) \exp[-2(k_0 h \cos\theta_i)^2] \tag{9.1-50}$$

† The term reflection, as used in this chapter, denotes the specular redirection of rays by a surface; thus Snell's law always holds for the reflected field. Scattering, on the other hand, refers to redirection of energy into directions other than the specular.

and

$$\bar{R}_{\perp}^{R} = R_{\perp}(\theta_i) \exp[-2(k_0 h \cos\theta_i)^2] \tag{9.1-51}$$

where $\bar{R}_{\parallel}^{R}$ and $\bar{R}_{\perp}^{R}$ are the modified average reflection coefficients to be used in Eqs. (9.1-46) and (9.1-48) above, and $R_{\parallel}(\theta_i)$ and $R_{\perp}(\theta_i)$ are given in (9.1-47) and (9.1-49). Again h^2 is the mean-square roughness height, i.e., $h^2 = \langle\zeta^2(x, y)\rangle$. The above results are taken from References 20 and 21. The restrictions used in their derivation are (a) surface slopes everywhere must be somewhat less than unity and (b) radii of curvature everywhere must be larger than the wavelength.

From the above equations, one can see that the average specularly-reflected coherent fields decrease exponentially with roughness height until they are negligible compared to the incoherent fields to be given subsequently.

9.1.3.3. *Average Coherent Scattering Cross Section of a Finite Surface*

A finitely-illuminated, flat, slightly rough surface reflects most of the incident energy into the specular direction in a pattern which depends in the far field upon a combination of the diffraction pattern and the surface-roughness statistics. At, and near the specular direction, there is a nonzero average scattered field (as discussed in Section 9.1.3.2), which is called the coherent component. The power in this average coherent field is proportional to the square of the average of the scattered field, $|\langle \mathbf{E}^s(P)\rangle|^2$.† As mentioned previously, the pattern produced by the coherent component depends mainly upon the diffraction pattern of the surface in the absence of roughness. The effect of a slight roughness is to reduce the magnitude of this pattern.

The following model[18,21] is used to determine the coherent contribution to the scattering cross section per unit area from a rough planar rectangular surface, as shown in Figure 9-1. The linear average dimensions of the illuminated area are L_x and L_y. The assumptions under which the model is applicable are the following:

(a) The tangent-plane approximation is employed, demanding that the radius of curvature at all points be greater than the wavelength. This also neglects shadowing and multiple scattering between two points of the surface.
(b) Surface slopes at all points should be considerably less than unity.
(c) The surface height is Gaussian-distributed.
(d) The illuminated area is rectangular, and the incident field is plane and uniform over this area.

No restriction is made concerning the mean-square roughness height, h^2, so that the coherent component derived for this model is valid for any h^2.

† The word average here is used interchangeably to refer to an average at a given instant of time over an ensemble of rough surfaces, or to an average in time for the radar return from a surface whose roughness is in motion with respect to the radar.

The analysis proceeds from the Chu–Stratton vector integral equation for the scattered E-field, Eq. (2.2-19), and employs the tangent-plane approximation to write the total H-field at the surface in terms of the incident field times the Fresnel reflection coefficients. The resulting average coherent radar cross section per unit area is

$$\gamma_{pq}^{c} = \frac{4\pi R_R^2}{L_x L_y} \frac{|\hat{\mathbf{p}} \cdot \langle \mathbf{H}^s(P') \rangle|^2}{|H_q^i|^2}$$
$$= \frac{k_0^2}{\pi} L_x L_y \left(\frac{\sin k_0 \xi_x L_x/2}{k_0 \xi_x L_x/2} \right)^2 \left(\frac{\sin k_0 \xi_y L_y/2}{k_0 \xi_y L_y/2} \right)^2 |\alpha_{pq}^c|^2 \exp[-k_0^2 h^2 \xi_z^2] \tag{9.1-52}$$

where

$$\xi_x = \sin\theta_i - \sin\theta_s \cos\phi_s \qquad (2 \sin\theta_i \text{ for backscatter}) \tag{9.1-53}$$

$$\xi_y = \sin\theta_s \sin\phi_s \qquad (\text{zero for backscatter}) \tag{9.1-54}$$

and

$$\xi_z = -\cos\theta_i - \cos\phi_s \qquad (-2\cos\theta_i \text{ for backscatter}) \tag{9.1-55}$$

Davies[21] derives and uses a similar model, but starts with a scalar integral equation and treats only the perfectly conducting surface. Hence, his results are independent of polarization, but their functional dependence is similar to γ_{hh}^c and γ_{vv}^c, obtainable from Eq. (9.1-52).

The p and q subscripts of Eq. (9.1-52) refer to the polarization states of the scattered and incident fields, respectively; p is a unit vector representing the desired polarization direction of the scattered field. The α_{pq}^c are directly proportional to the scattering matrix elements, and are defined for vertical and horizontal polarization states (i.e., p and q are equal to v and h) as

$$\alpha_{vv}^c = -\cos\theta_s \cos\phi_s R_\parallel(\theta_i) \tag{9.1-56}$$

$$\alpha_{hv}^c = \sin\phi_s R_\parallel(\theta_i) \tag{9.1-57}$$

$$\alpha_{vh}^c = \cos\theta_i \cos\theta_s \sin\phi_s R_\perp(\theta_i) \tag{9.1-58}$$

$$\alpha_{hh}^c = \cos\theta_i \cos\phi_s R_\perp(\theta_i) \tag{9.1-59}$$

where $R_\parallel(\theta)$ and $R_\perp(\theta)$ are the Fresnel reflection coefficients for the surface, as defined in Eqs. (9.1-47) and (9.1-49). When the surface is perfectly conducting, these reflection coefficients become

$$R_\parallel(\theta_i) = 1 = -R_\perp(\theta_i) \tag{9.1-60}$$

From the above matrix elements for the vertical and horizontal polarization states, the matrix elements for the circular polarization states can be determined from Eqs. (9.1-93) to (9.1-96). In this case, the p and q represent L and R (standing for left and right circular polarization). The average coherent cross sections can then be determined from Eq. (9.1-52) for the various combinations of circular polarization states.

The matrix element and average coherent scattering cross section for arbitrarily oriented linear transmitting and receiving polarization directions can be determined using Eq. (9.1-111). In this case, the p and q represent η_i and η_s; these are the angles from verticals, $\boldsymbol{\theta}_i$ and $\boldsymbol{\theta}_s$, at which the desired linear polarization states lie. See Section 9.1.3.4.4 for a more thorough discussion for these polarization orientations.

The form of the above bistatic scattering cross sections is identical to those for a large, smooth, rectangular plate, except for the factor $\exp[-k_0^2h^2\xi_z^2]$. Hence, in the limit, as the roughness vanishes ($h^2 \to 0$), this factor goes to unity as it should. This exponential factor is identical to that appearing in Eqs. (9.1-50) and (9.1-51) for the rough-surface reflection coefficients; here these coefficients are squared because reflected field intensity or power is considered. As the roughness scale becomes relatively large ($k_0^2h^2 \gg 1$), this factor causes the coherent component to disappear. Notice the usual diffraction terms [i.e., $(\sin u)/u$]; these depend only upon the shape of the illuminated area.

A more realistic illuminated area is an ellipse, since a radar with a circular beam will illuminate an elliptical area when tilted away from normal. If the illumination is considered uniform inside the cone, whose total angle ν_B is determined by the 3 db points of the main lobe and zero outside the cone, the major and minor semi-axes of the illuminated ellipse will be $L_x = R_t(\nu_B/2) \sec \theta_i$ and $L_y = R_T(\nu_B/2)$, where ν_B, the beamwidth, is small and expressed in radians. In this case, the coherent cross section will exhibit roughly the same lobe pattern as for the rectangular area. There will appear diffraction factors not too different from the $(\sin u)/u$ factors of Eq. (9.1-52).† Their exact form is not really too critical, since the assumption of uniform illumination over the ellipse and zero illumination outside the ellipse is not too realistic in the first place. For a reasonable estimate, however, Eqs. (9.1-52) to (9.1-59) can be applied to such an elliptical area when L_x and L_y have the above values in terms of incidence angle, θ_i, beamwidth, ν_B, and transmitter distance, R.

9.1.3.4. *Average Incoherent Scattering Cross Sections for a Slightly Rough Surface*

9.1.3.4.1. *Introduction.* Perhaps the best and most general solution available for the incoherent scattering cross sections from a slightly rough surface has been formulated by Rice[22] and derived by Peake.[17,18] This model is valid both for bistatic and backscattering and applies for both perfectly conducting and imperfectly conducting surfaces. The model preserves the polarization sensitivity of the scattering cross sections, and yields different results for different polarization combinations; this polariza-

† See Section 7.5.3 for the high-frequency diffraction coefficient for an elliptical area [Eq. (7.5-55)].

tion behavior, unexplained analytically for years, has been confirmed by measurements.

Others[21,23] have treated slightly rough surfaces, but their results are less general and more restrictive (i.e., they apply to perfectly conducting surfaces, and polarization dependence is lacking due to solution by scalar methods). The most serious restriction of these latter works, however, is the limitation of the tangent-plane approximation. This approximation is not realistic when applied to most slightly rough surfaces found in nature, and it is not valid in the low-frequency limit. The reason will be discussed more fully subsequently.

The Rice model employs a perturbation technique, in contrast to the tangent-plane (or physical-optics) approach. Assumptions employed in the analysis of the model are the following:

(a) $k_0 h < 1.0$, i.e., roughness height is small.

(b) $|\partial\zeta/\partial x|$, $|\partial\zeta/\partial y| < 1.0$, i.e., surface slopes are relatively small.

(c) $\langle(\partial\zeta/\partial x)^2\rangle = \langle(\partial\zeta/\partial z)^2\rangle$, i.e., roughness is isotropic.

The tangent-plane approximation requires that the radius of curvature everywhere on the surface be large compared to wavelength. This approximation is not required for the model considered here. A natural surface, smoothly curving and satisfying (b) above, will have all radii of curvature large in the high-frequency limit. However, the same surface will eventually fail this requirement, as frequency is decreased. Stated another way, the large radius of curvature requirement means that the average distance between hills must be large in terms of wavelength. At lower frequencies this requirement cannot be met for most natural surfaces. Hence, an intrinsically low-frequency technique is more meaningful than a high-frequency approach like the tangent-plane approximation.

The average incoherent scattering cross section per unit surface area† (taken from Reference 5) is

$$\gamma_{pq}^{I} = \frac{4}{\pi} k_0^4 h^2 \cos^2\theta_i \cos^2\theta_s \,|\alpha_{pq}|^2 I \tag{9.1-61}$$

where p and q represent the polarization states of the scattered and incident E-fields, respectively. The quantity α_{pq} is directly proportional to the scattering matrix element, and will be given below for particular polarization cases. The quantity I is defined as

$$I = 2\pi \int_0^\infty r\rho(r) J_0(k_0 \sqrt{\xi_x^2 + \xi_y^2}\, r)\, dr = \frac{\pi^2}{h^2} w(k_0 \sqrt{\xi_x^2 + \xi_y^2}) \tag{9.1-62}$$

† The notation γ is equivalent to the σ^0 sometimes used in the rough-surface literature; see Eq. (9.1-1).

The parameters ξ_x and ξ_y are defined in Eqs. (9.1-53) and (9.1-54), and $J_0(x)$ is the cylindrical Bessel function of zero order and argument x. The quantity $\rho(r)$ is the surface-height correlation coefficient for an isotropically rough surface, and it is a function only of the separation between two points on the surface, r, thus

$$\rho(r) = \frac{\langle \zeta(x, y)\, \zeta(x', y') \rangle}{h^2}$$

where

$$r = \sqrt{(x - x')^2 + (y - y')^2} \tag{9.1-63}$$

The quantity $w(u)$ is physically the roughness spectral density of the slightly rough surface. Hence the scattering cross sections are directly proportional to the surface roughness spectral density. Additional physical interpretation of the scattering process is given in Reference 5.

In order to obtain explicit results for I, one must know the surface-height correlation coefficient, $\rho(r)$, for the rough surface in question. In the absence of an exact knowledge of this quantity, one must assume a form or model for this function. Two of the most popular models are the Gaussian and the exponential, and the values of I for these two models are as follows:

(a) *Gaussian correlation coefficient model,* $\rho(r) = e^{-r^2/l^2}$. With the Gaussian correlation coefficient model, the integration in Eq. (9.1-62) can be carried out giving

$$I = \pi l^2 \exp\left[\frac{-k_0{}^2 l^2(\xi_x{}^2 + \xi_y{}^2)}{4}\right] \tag{9.1-64}$$

Surfaces having a Gaussian correlation coefficient are smoothly curving with derivatives at all points. The total mean-square slope of the surface at any point with this correlation coefficient is

$$s^2 = \left\langle \left(\frac{\partial \zeta}{\partial x}\right)^2 + \left(\frac{\partial \zeta}{\partial y}\right)^2 \right\rangle = \frac{4h^2}{l^2} \tag{9.1-65}$$

The quantity l is called the correlation length.

(b) *Exponential correlation coefficient model,* $\rho(r) = e^{-|r|/l}$. The integral of Eq. (9.1-62) for this cases becomes

$$I = 2\pi l^2 \frac{1}{[1 + k_0{}^2 l^2(\xi_x{}^2 + \xi_y{}^2)]^{3/2}} \tag{9.1-66}$$

Surfaces with exponential correlation coefficients are jagged and have many vertical facets. The surface slopes and all higher surface derivatives are infinite, undefined, and discontinuous at many surface points. It can be shown that such surfaces have infinite mean-square slopes and higher derivatives (i.e., $s^2 = \infty$). As an example, an urban area, having buildings and houses as the scattering surface, has a surface-height correlation coefficient which behaves like the exponential, especially near the origin (i.e., $\rho(r) \to 1 - \frac{|r|}{l}$ for $r \to 0$).

It should be noted from the above that for $2k_0 l < 1$, the behavior of I for the two correlation coefficients is not significantly different except for a factor of 2.

It is significant that only the surface-height correlation coefficient enters into Eq. (9.1-61) for the incoherent scattering cross section, and not the probability density function. It will be seen that for very rough surfaces, the joint probability density function for the surface height must be known or assumed.

9.1.3.4.2. *Results for Vertical and Horizontal Polarization States.* For this section, a vertically polarized incident wave has its E-field vector along the $\hat{\boldsymbol{\theta}}_i$ direction (Figure 9-1), while a horizontally polarized wave has its E-field along the $\hat{\boldsymbol{\phi}}_i$ direction (or the $-\hat{\mathbf{y}}$ direction in this case, since $\phi_i = 180°$). The vertically polarized component of the scattered field in any arbitrary direction has its E-field in the $\hat{\boldsymbol{\theta}}_s$ direction, while the horizontal scattered component is in the $\hat{\boldsymbol{\phi}}_s$ direction.

9.1.3.4.2.1. *Bistatic Scattering from a Surface of Homogeneous Material.* From Reference 5, the elements α_{pq}, to be used in Eq. (9.1-61) for the average scattering cross section, are

$$\alpha_{hh} = -\frac{\left[\begin{array}{r}(\mu_r - 1)(\mu_r \sin\theta_i \sin\theta_s - \cos\phi_s \sqrt{\epsilon_r\mu_r - \sin^2\theta_i}\sqrt{\epsilon_r\mu_r - \sin^2\theta_s}) \\ + \mu_r^2(\epsilon_r - 1)\cos\phi_s\end{array}\right]}{[\mu_r \cos\theta_i + \sqrt{\epsilon_r\mu_r - \sin^2\theta_i}][\mu_r\cos\theta_s + \sqrt{\epsilon_r\mu_r - \sin^2\theta_s}]} \tag{9.1-67}$$

$$\alpha_{vh} = \sin\phi_s \cdot \frac{\epsilon_r(\mu_r - 1)\sqrt{\epsilon_r\mu_r - \sin^2\theta_i} - \mu_r(\epsilon_r - 1)\sqrt{\epsilon_r\mu_r - \sin^2\theta_s}}{[\mu_r\cos\theta_i + \sqrt{\epsilon_r\mu_r - \sin^2\theta_i}][\epsilon_r\cos\theta_s + \sqrt{\epsilon_r\mu_r - \sin^2\theta_s}]} \tag{9.1-68}$$

$$\alpha_{hv} = \sin\phi_s \cdot \frac{\mu_r(\epsilon_r - 1)\sqrt{\epsilon_r\mu_r - \sin^2\theta_i} - \epsilon_r(\mu_r - 1)\sqrt{\epsilon_r\mu_r - \sin^2\theta_s}}{[\epsilon_r\cos\theta_i + \sqrt{\epsilon_r\mu_r - \sin^2\theta_i}][\mu_r\cos\theta_s + \sqrt{\epsilon_r\mu_r - \sin^2\theta_s}]} \tag{9.1-69}$$

$$\alpha_{vv} = \frac{\left[\begin{array}{r}(\epsilon_r - 1)(\epsilon_r \sin\theta_i \sin\theta_s - \cos\phi_s \sqrt{\epsilon_r\mu_r - \sin^2\theta_i}\sqrt{\epsilon_r\mu_r - \sin^2\theta_s}) \\ + \epsilon_r^2(\mu_r - 1)\cos\phi_s\end{array}\right]}{[\epsilon_r \cos\theta_i + \sqrt{\epsilon_r\mu_r - \sin^2\theta_i}][\epsilon_r\cos\theta_s + \sqrt{\epsilon_r\mu_r - \sin^2\theta_s}]} \tag{9.1-70}$$

In the above equations, ϵ_r is the relative permittivity of the surface and μ_r is the relative permeability. They may be either real or complex (for a lossy surface).

These quantities, along with the form chosen for I (Gaussian or exponential coefficient model), give the average bistatic-scattering cross section per unit surface area according to Eq. (9.1-61).

9.1.3.4.2.2. *Backscattering from a Surface of Homogeneous Material.* In the case of backscattering, $\theta_s = \theta_i$ and $\phi_s = 180°$, the above matrix elements become

$$\alpha_{hh} = -\frac{(\mu_r - 1)[(\mu_r - 1)\sin^2\theta_i + \epsilon_r\mu_r] - \mu_r^2(\epsilon_r - 1)}{[\mu_r\cos\theta_i + \sqrt{\epsilon_r\mu_r - \sin^2\theta_i}]^2} \tag{9.1-71}$$

$$\alpha_{vh} = \alpha_{hv} = 0 \tag{9.1-72}$$

$$\alpha_{vv} = \frac{(\epsilon_r - 1)[(\epsilon_r - 1)\sin^2\theta_i + \epsilon_r\mu_r] - \epsilon_r^2(\mu_r - 1)}{[\epsilon_r\cos\theta_i + \sqrt{\epsilon_r\mu_r - \sin^2\theta_i}]^2} \tag{9.1-73}$$

The average backscattering cross sections per unit area from Eq. (9.1-61) for each correlation coefficient model are

(a) *Gaussian surface-height correlation coefficient*

$$\gamma_{vv}^I = 4k_0^4h^2l^2\cos^4\theta_i \left| \frac{(\epsilon_r - 1)[(\epsilon_r - 1)\sin^2\theta_i + \epsilon_r\mu_r] - \epsilon_r^2(\mu_r - 1)}{[\epsilon_r\cos\theta_i + \sqrt{\epsilon_r\mu_r - \sin^2\theta_i}]^2} \right|^2 \times \exp[-k_0^2l^2\sin^2\theta_i] \tag{9.1-74}$$

$$\gamma_{hh}^I = 4k_0^4h^2l^2\cos^4\theta_i \left| \frac{(\mu_r - 1)[(\mu_r - 1)\sin^2\theta_i + \epsilon_r\mu_r] - \mu_r^2(\epsilon_r - 1)}{[\mu_r\cos\theta_i + \sqrt{\epsilon_r\mu_r - \sin^2\theta_i}]^2} \right|^2 \times \exp[-k_0^2l^2\sin^2\theta_i] \tag{9.1-75}$$

$$\gamma_{hv}^I = \alpha_{vh}^I = 0 \tag{9.1-76}$$

(b) *Exponential surface-height correlation coefficient*

$$\gamma_{vv}^I = 8k_0^4h^2l^2\cos^4\theta_i \left| \frac{(\epsilon_r - 1)[(\epsilon_r - 1)\sin^2\theta_i + \epsilon_r\mu_r] - \epsilon_r^2(\mu_r - 1)}{[\epsilon_r\cos\theta_i + \sqrt{\epsilon_r\mu_r - \sin^2\theta_i}]^2} \right|^2 \times [1 + 4k_0^2l^2\sin^2\theta_i]^{-3/2} \tag{9.1-77}$$

$$\gamma_{hh}^I = 8k_0^4h^2l^2\cos^4\theta_i \left| \frac{(\mu_r - 1)[(\mu_r - 1)\sin^2\theta_i + \epsilon_r\mu_r] - \mu_r^2(\epsilon_r - 1)}{[\mu_r\cos\theta_i + \sqrt{\epsilon_r\mu_r - \sin^2\theta_i}]^2} \right|^2 \times [1 + 4k_0^2l^2\sin^2\theta_i]^{-3/2} \tag{9.1-78}$$

$$\gamma_{hv}^I = \gamma_{vh}^I = 0 \tag{9.1-79}$$

According to the above results, the cross-polairzed terms of the backscattering cross sections are zero, at least to the lowest order of the perturbation theory. The actual cross-polarized terms are of the orders $k_0^2h^2$, $\langle(\partial\zeta/\partial x)^2\rangle$, and $\langle(\partial\zeta/\partial y)^2\rangle$, with respect to the vertical and horizontal terms shown here; since these parameters are assumed small, only the zero-order terms are shown where they exist.

It should be noted that the backscattering cross sections for vertically and horizontally polarized fields differ. Those for vertical polarization are greater than for horizontal polarization at angles away from normal incidence and for $\epsilon_r > \mu_r$. This has been observed experimentally. Theories based upon the tangent-plane approximation do not show any polarization dependence.

Figures 9-12 to 9-13 show curves of $\gamma_{vv}^I/k_0^2h^2$ and $\gamma_{hh}^I/k_0^2h^2$ for various values of dielectric constant ϵ_r and parameter $k_0^2l^2$. These are based upon the Gaussian surface-height correlation coefficient model, and were computed from Eqs. (9.1-74) and (9.1-75) for $\mu_r = 1$.

9.1.3.4.2.3. *Bistatic Scattering from a Perfectly Conducting Surface.* When the surface becomes perfectly conducting, Eqs. (9.1-69) to (9.1-70) can be used by permitting ϵ_r to approach infinity, giving

$$\alpha_{vv} = \frac{\sin\theta_i \sin\theta_s - \cos\phi_s}{\cos\theta_i \cos\theta_s} \tag{9.1-80}$$

$$\alpha_{hv} = \frac{\sin\phi_s}{\cos\theta_i} \tag{9.1-81}$$

$$\alpha_{hh} = -\cos\phi_s \tag{9.1-82}$$

$$\alpha_{vh} = -\frac{\sin\phi_s}{\cos\theta_s} \tag{9.1-83}$$

These quantities, along with the form chosen for I (Gaussian or exponential coefficient model), give the average bistatic scattering cross section per unit surface area according to Eq. (9.1-61).

9.1.3.4.2.4. *Backscattering from a Perfectly Conducting Surface.* For backscattering, $\theta_i = \theta_s$ and $\phi_s = 180°$, the above matrix elements become

$$\alpha_{vv} = \frac{1 + \sin^2\theta_i}{\cos^2\theta_i} \tag{9.1-84}$$

$$\alpha_{hh} = 1 \tag{9.1-85}$$

$$\alpha_{vh} = \alpha_{hv} = 0 \tag{9.1-86}$$

The average backscattering cross sections per unit area for each of the two surface correlation coefficient models are then:

(a) *Gaussian surface-height correlation coefficient*

$$\gamma_{vv}^{I} = 4k_0^4h^2l^2(1 + \sin^2\theta_i)^2 \exp[-k_0^2l^2 \sin^2\theta_i] \tag{9.1-87}$$

$$\gamma_{hh}^{I} = 4k_0^4h^2l^2 \cos^4\theta_i \exp[-k_0^2l^2 \sin^2\theta_i] \tag{9.1-88}$$

$$\gamma_{vh}^{I} = \gamma_{hv}^{I} = 0 \tag{9.1-89}$$

(b) *Exponential surface-height correlation coefficient*

$$\gamma_{vv}^{I} = 8k_0^4h^2l^2(1 + \sin^2\theta_i)^2(1 + 4k_0^2l^2 \sin^2\theta_i)^{-3/2} \tag{9.1-90}$$

$$\gamma_{hh}^{I} = 8k_0^4h^2l^2 \cos^4\theta_i(1 + 4k_0^2l^2 \sin^2\theta_i)^{-3/2} \tag{9.1-91}$$

$$\gamma_{vh}^{I} = \gamma_{hv}^{I} = 0 \tag{9.1-92}$$

The above results show that while for horizontal polarization the backscattered power approaches zero at grazing incidence, for vertical polarization the backscattered power does not approach zero. Natural surfaces, however, are not perfectly conducting. It can be seen from Eqs. (9.1-74) to (9.1-79) for dielectric surfaces, that γ_{vv}^{I} does approach zero at grazing for any ϵ_r so long as it is finite. For example, Figures 9-12 and 9-13 show the backscattering cross sections for $\epsilon_r = 55\,(1 + i.\,55)$. Although $|\,\epsilon_r\,| \approx 63$, which may be large enough to consider the surface as perfectly conducting in most situations, the behavior of γ_{vv}^{I} differs greatly near grazing from that of the perfectly conducting surface.

Figures 9-12 to 9-13 also show curves of $\gamma_{vv}^{I}/k_0^2h^2$ and $\gamma_{hh}^{I}/k_0^2h^2$ for the perfectly conducting surface along with curves for dielectric surfaces. These are based upon the Gaussian surface-height correlation coefficient model, and are constructed from Eqs. (9.1-87) and (9.1-88).

9.1.3.4.3. *Results for Circular Polarization States.* Since the α_{pq} of the preceding section are proportional to the scattering matrix elements, the α_{pq} for circular polarization states can be derived in terms of the α_{pq} for vertical and horizontal polarization (see Section 2.1.4). For circular polarization, let p and q represent R and L for right and left circularly polarized waves, respectively.

It should be noted that for backscattering, a right-circularly polarized wave is normally reflected from a large smooth object as a left-circular wave. The presence of any depolarization means that a portion of the incident wave returns right-circularly polarized. Hence, the term cross-polarized backscattered component, when referring to circular-polarization states, will refer to the same type of polarization as the incident wave.

Whereas for the vertical and horizontal polarization states it was observed that the cross-polarized terms for backscattering were zero (γ_{hv}^{I} and γ_{vh}^{I}), this will not be true for the cross-polarized circularly polarized terms. Thus, γ_{RR}^{I} and γ_{LL}^{I} are not generally zero.

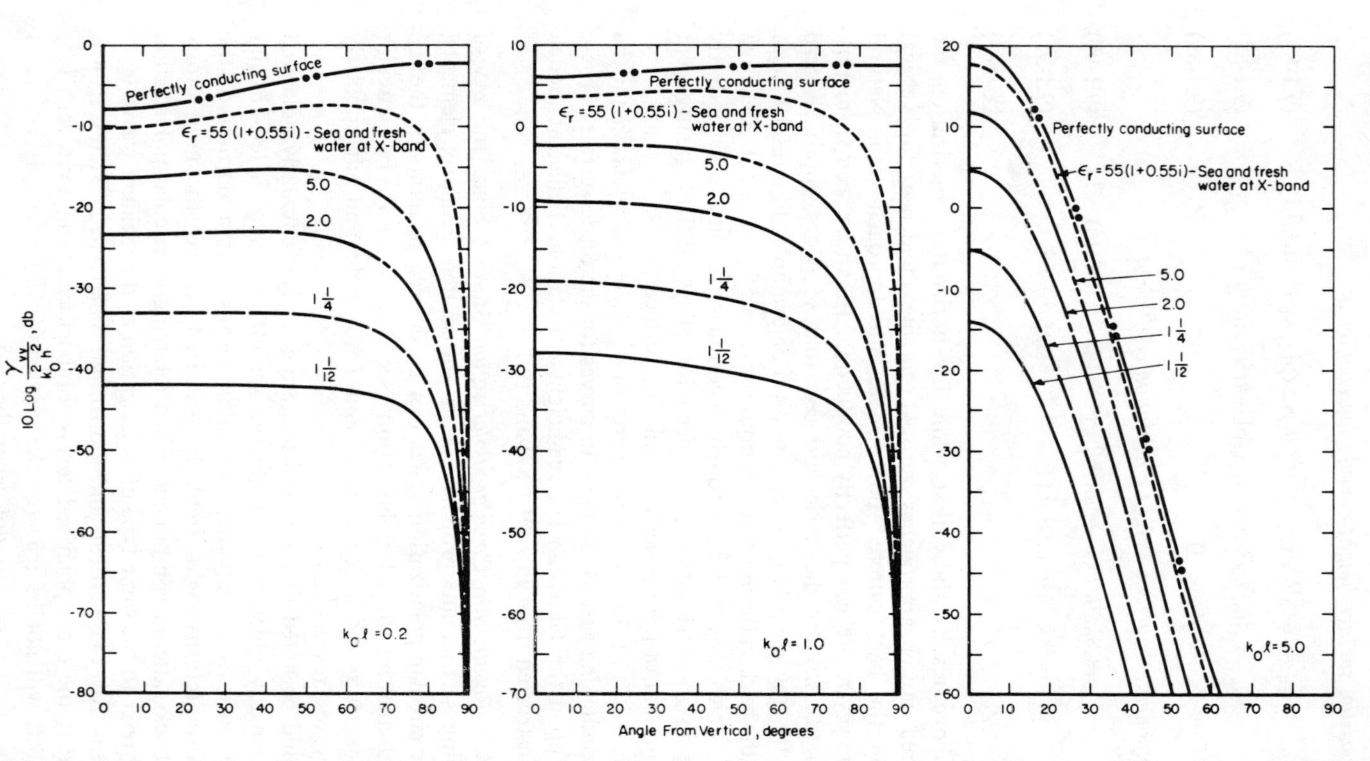

Fig. 9-12. Average incoherent backscattering cross section per unit area for slightly-rough-surface statistical model versus incidence angle for vertical incident and vertical received polarization. [Plotted from Eq. (9.1-74) and (9.1-87) for Gaussian surface-height correlation coefficient and for various values of $k_0 l$, where l is the surface-roughness correlation length. The cross section is divided by $k_0^2 h^2$, where h^2 is the mean-square surface height ($k_0^2 h^2 < 1$).]

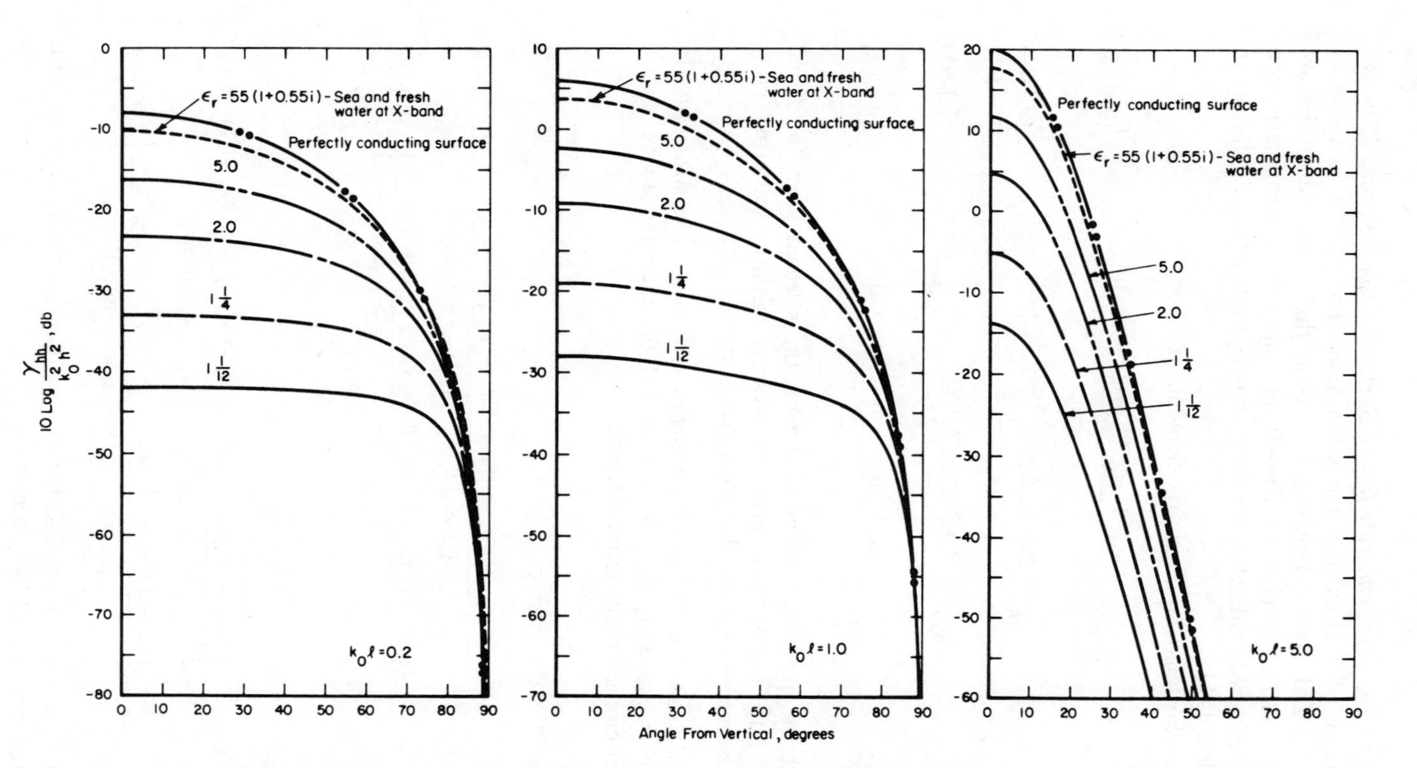

Fig. 9-13. Average incoherent backscattering cross section per unit area for slightly-rough-surface statistical model versus incidence angle for horizontal incident and horizontal received polarization. [Plotted from Eq. (9.1-75) and (9.1-88) for Gaussian surface-height correlation coefficient and for various values of $k_0 l$, where l is the surface-roughness correlation length. The cross section is divided by $k_0^2 h^2$, where h^2 is the mean-square surface-roughness height ($k_0^2 h^2 < 1$).]

9.1.3.4.3.1. *Bistatic Scattering from a Surface of Homogeneous Material.* The element α_{pq} for circular polarization states are written in terms of α_{vv}, α_{hv}, α_{vh}, and α_{hh} for linear polarization; these latter quantities are defined in Eqs. (9.1-67) to (9.1-70). The left subscript refers to the polarization state of the scattered field, while the right subscript refers to the polarization of the incident field.

The circular polarization matrix elements are then

$$\alpha_{LR} = \frac{\alpha_{hh} + \alpha_{vv} + i(\alpha_{hv} - \alpha_{vh})}{2} \tag{9.1-93}$$

$$\alpha_{RR} = \frac{\alpha_{hh} - \alpha_{vv} + i(\alpha_{hv} + \alpha_{vh})}{2} \tag{9.1-94}$$

$$\alpha_{RL} = \frac{\alpha_{hh} + \alpha_{vv} - i(\alpha_{hv} - \alpha_{vh})}{2} \tag{9.1-95}$$

$$\alpha_{LL} = \frac{\alpha_{hh} - \alpha_{vv} - i(\alpha_{hv} + \alpha_{vh})}{2} \tag{9.1-96}$$

The above quantities, when used with the selected form for I and substituted into Eq. (9.1-61), give the average bistatic scattering cross sections per unit area for circular polarization states.

9.1.3.4.3.2. *Backscattering from a Surface of Homogeneous Material.* In the case of backscatter, $\theta_i = \theta_s$ and $\phi = 180°$, then the above matrix elements for circular polarization become

$$\begin{aligned} \alpha_{LR} = \alpha_{RL} &= \frac{\alpha_{hh} + \alpha_{vv}}{2} \\ &= \frac{1}{2}\Big[-\frac{(\mu_r - 1)[(\mu_r - 1)\sin^2\theta_i + \epsilon_r\mu_r] - \mu_r^2(\epsilon_r - 1)}{[\mu_r\cos\theta_i + \sqrt{\epsilon_r\mu_r - \sin^2\theta_i}]^2} \\ &\quad + \frac{(\epsilon_r - 1)[(\epsilon_r - 1)\sin^2\theta_i + \epsilon_r\mu_r] - \epsilon_r^2(\mu_r - 1)}{[\epsilon_r\cos\theta_i + \sqrt{\epsilon_r\mu_r - \sin^2\theta_i}]^2}\Big] \end{aligned} \tag{9.1-97}$$

and

$$\begin{aligned} \alpha_{LL} = \alpha_{RR} &= \frac{\alpha_{hh} - \alpha_{vv}}{2} \\ &= -\frac{1}{2}\Big[\frac{(\mu_r - 1)[(\mu_r - 1)\sin^2\theta_i + \epsilon_r\mu_r] - \mu_r^2(\epsilon_r - 1)}{[\mu_r\cos\theta_i + \sqrt{\epsilon_r\mu_r - \sin^2\theta_i}]^2} \\ &\quad + \frac{(\epsilon_r - 1)[(\epsilon_r - 1)\sin^2\theta_i + \epsilon_r\mu_r] - \epsilon_r^2(\mu_r - 1)}{[\epsilon_r\cos\theta_i + \sqrt{\epsilon_r\mu_r - \sin^2\theta_i}]^2}\Big] \end{aligned} \tag{9.1-98}$$

Based upon these quantities, the average backscattering cross sections per unit area from Eq. (9.1-61) become

(a) *Gaussian surface-height correlation coefficient*

$$\gamma_{LR}^I = \gamma_{RL}^I = k_0^4 h^2 l^2 \cos^4\theta_i \times \left| -\frac{(\mu_r - 1)[(\mu_r - 1)\sin^2\theta_i + \epsilon_r\mu_r] - \mu_r^2(\epsilon_r - 1)}{[\mu_r\cos\theta_i + \sqrt{\epsilon_r\mu_r - \sin^2\theta_i}]^2} + \frac{(\epsilon_r - 1)[(\epsilon_r - 1)\sin^2\theta_i + \epsilon_r\mu_r] - \epsilon_r^2(\mu_r - 1)}{[\epsilon_r\cos\theta_i + \sqrt{\epsilon_r\mu_r - \sin^2\theta_i}]^2} \right|^2 \times \exp[-k_0^2 l^2 \sin^2\theta_i] \tag{9.1-99}$$

and

$$\gamma_{LL}^I = \gamma_{RR}^I = k_0^4 h^2 l^2 \cos^4\theta_i \left| \frac{(\mu_r - 1)[(\mu_r - 1)\sin^2\theta_i + \epsilon_r\mu_r] - \mu_r^2(\epsilon_r - 1)}{[\mu_r\cos\theta_i + \sqrt{\epsilon_r\mu_r - \sin^2\theta_i}]^2} + \frac{(\epsilon_r - 1)[(\epsilon_r - 1)\sin^2\theta_i + \epsilon_r\mu_r] - \epsilon_r^2(\mu_r - 1)}{[\epsilon_r\cos\theta_i + \sqrt{\epsilon_r\mu_r - \sin^2\theta_i}]^2} \right|^2 \times \exp[-k_0^2 l^2 \sin^2\theta_i] \tag{9.1-100}$$

(b) *Exponential surface-height correlation coefficient*

$$\gamma_{LR}^I = \gamma_{RL}^I = 2k_0^4 h^2 l^2 \cos^4\theta_i \times \left| -\frac{(\mu_r - 1)[(\mu_r - 1)\sin^2\theta_i + \epsilon_r\mu_r] - \mu_r^2(\epsilon_r - 1)}{[\mu_r\cos\theta_i + \sqrt{\epsilon_r\mu_r - \sin^2\theta_i}]^2} + \frac{(\epsilon_r - 1)[(\epsilon_r - 1)\sin^2\theta_i + \epsilon_r\mu_r] - \epsilon_r^2(\mu_r - 1)}{[\epsilon_r\cos\theta_i + \sqrt{\epsilon_r\mu_r - \sin^2\theta_i}]^2} \right|^2 \times [1 + 4k_0^2 l^2 \sin^2\theta_i]^{-3/2} \tag{9.1-101}$$

$$\gamma_{LL}^I = \gamma_{RR}^I = 2k_0^4 h^2 l^2 \cos^4\theta_i \times \left| \frac{(\mu_r - 1)[(\mu_r - 1)\sin^2\theta_i + \epsilon_r\mu_r] - \mu_r^2(\epsilon_r - 1)}{[\mu_r\cos\theta_i + \sqrt{\epsilon_r\mu_r - \sin^2\theta_i}]^2} + \frac{(\epsilon_r - 1)[(\epsilon_r - 1)\sin^2\theta_i + \epsilon_r\mu_r] - \epsilon_r^2(\mu_r - 1)}{[\epsilon_r\cos\theta_i + \sqrt{\epsilon_r\mu_r - \sin^2\theta_i}]^2} \right|^2 \times [1 + 4k_0^2 l^2 \sin^2\theta_i]^{-3/2} \tag{9.1-102}$$

The above results show that at normal incidence, the cross-polarized components, γ_{RR}^I and γ_{LL}^I, become zero. At other angles, however, these components may be significant.

Curves of $\gamma_{LR}/k_0^2h^2$ and $\gamma_{RR}/k_0^2h^2$ for various values of ϵ_r and $k_0^2l^2$ are shown in Figures 9-14 and 9-15. These were obtained from Eqs. (9.1-99) and (9.1-100) for the Gaussian correlation coefficient model.

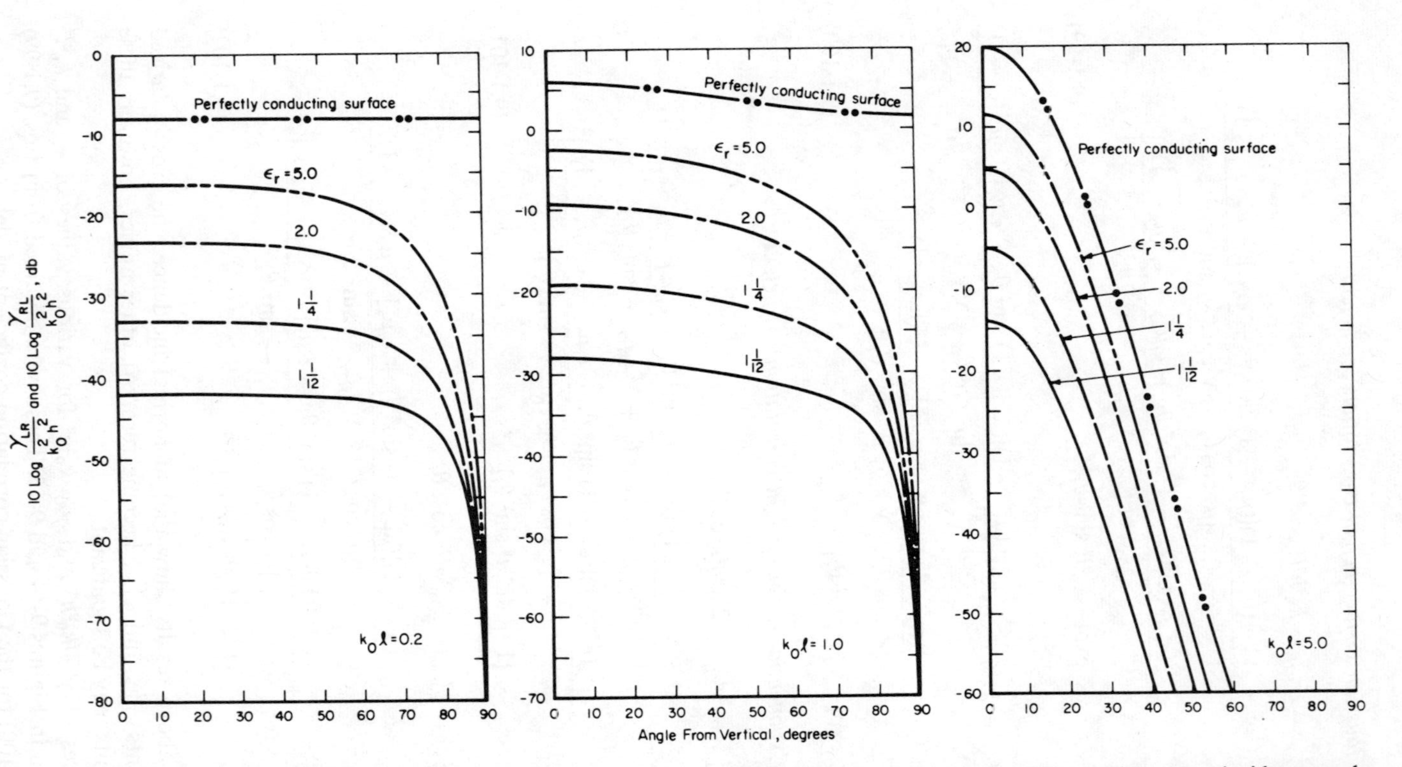

Fig. 9-14. Average incoherent backscattering cross section per unit area for slightly-rough-surface statistical model versus incidence angle for right (or left) circular incident and opposite-sense circular received polarization. [Plotted from Eq. (9.1-99) and (9.1-107) for Gaussian surface-height correlation coefficient and for various values of $k_0 l$ where l is the surface-roughness correlation length. The cross section is divided by $k_0^2 h^2$, where h^2 is the mean-square surface-roughness height ($k_0^2 h^2 < 1$).]

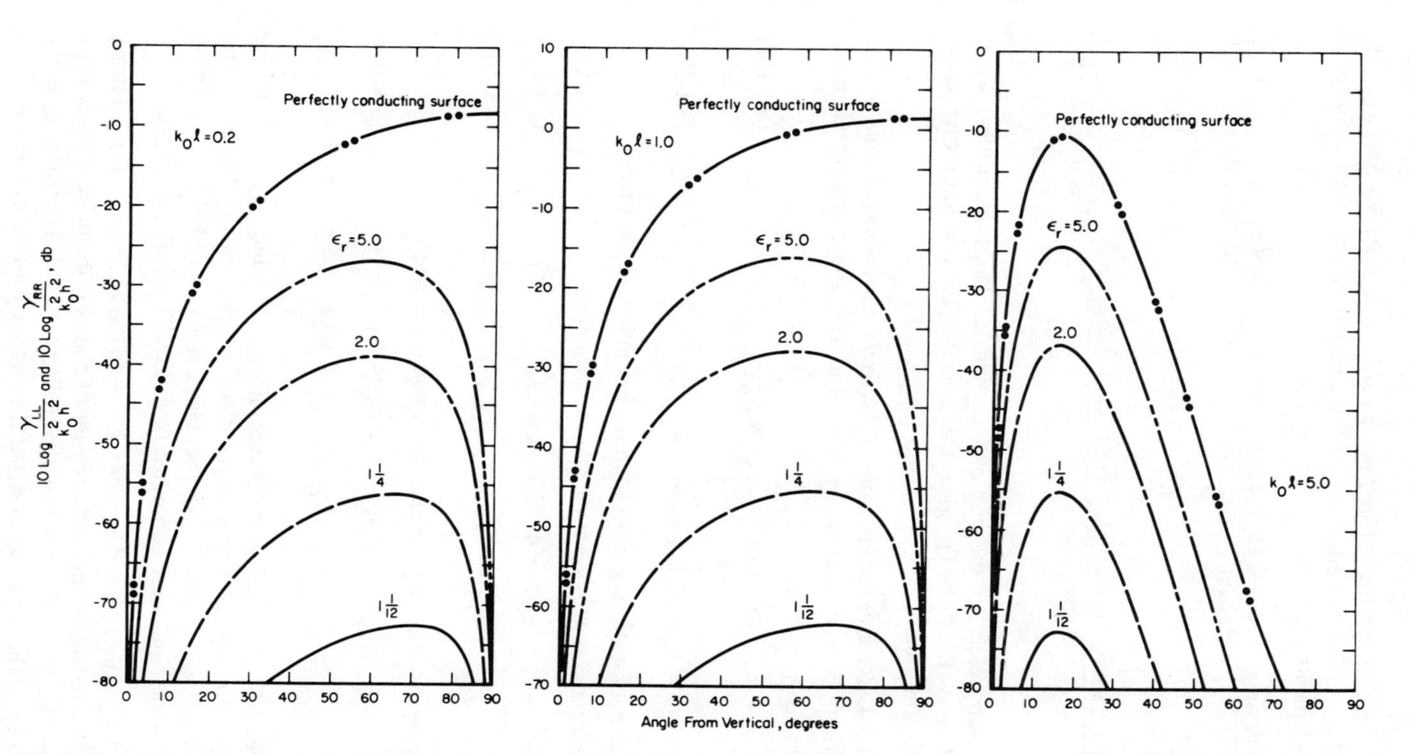

Fig. 9-15. Average incoherent backscattering cross section per unit area for slightly-rough-surface statistical model versus incidence angle for right (or left) circular incident and same-sense circular received polarization. [Plotted from Eq. (9.1-100) and (9.1-108) for Gaussian surface-height correlation coefficient and for various values of $k_0 l$, where l is the surface-roughness correlation length. The cross section is divided by $k_0^2 h^2$, where h^2 is the mean-square surface-roughness height ($k_0^2 h^2 < 1$).]

9.1.3.4.3.3. *Bistatic Scattering from a Perfectly Conducting Surface.* In the case of a perfectly conducting surface, letting ϵ_r approach infinity in the matrix elements gives

$$\alpha^{LR} = \frac{\sin\theta_i \sin\theta_s - \cos\phi_s(1+\cos\theta_i\cos\theta_s)}{2\cos\theta_i\cos\theta_s} + i\frac{\sin\phi_s}{2}\left(\frac{1}{\cos\theta_i} + \frac{1}{\cos\theta_s}\right)$$
$$= \alpha^*_{RL} \tag{9.1-103}$$

and

$$\alpha^{RR} = \frac{-\sin\theta_i \sin\theta_s + \cos\phi_s(1-\cos\theta_i\cos\theta_s)}{2\cos\theta_i\cos\theta_s} + i\frac{\sin\phi_s}{2}\left(\frac{1}{\cos\theta_i} - \frac{1}{\cos\theta_s}\right)$$
$$= \alpha^*_{LL} \tag{9.1-104}$$

where the star refers to the complex conjugate. These quantities, when substituted into Eq. (9.1-61), give the average bistatic scattering cross sections per unit area.

9.1.3.4.3.4. *Backscattering from a Perfectly Conducting Surface.* For backscattering, $\theta_s = \theta_i$ and $\phi_s = 180°$, and the above matrix elements become

$$\alpha_{LR} = \alpha_{RL} = \frac{1}{\cos^2\theta_i} \tag{9.1-105}$$

and

$$\alpha_{RR} = \alpha_{LL} = -\tan^2\theta_i \tag{9.1-106}$$

The average backscattering cross sections per unit area are then:

(a) *Gaussian surface-height correlation coefficient*

$$\gamma^I_{LR} = \gamma^I_{RL} = 4k_0{}^4h^2l^2 \exp[-k_0{}^2l^2 \sin^2\theta_i] \tag{9.1-107}$$

$$\gamma^I_{LL} = \gamma^I_{RR} = 4k_0{}^4h^2l^2 \sin^4\theta_i \exp[-k_0{}^2l^2 \sin^2\theta_i] \tag{9.1-108}$$

(b) *Exponential surface-height correlation coefficient*

$$\gamma^I_{LR} = \gamma^I_{RL} = 8k_0{}^4h^2l^2[1 + 4k_0{}^2l^2 \sin^2\theta_i]^{-3/2} \tag{9.1-109}$$

$$\gamma^I_{LL} = \gamma^I_{RR} = 8k_0{}^4h^2l^2 \sin^4\theta_i[1 + 4k_0{}^2l^2 \sin^2\theta_i]^{-3/2} \tag{9.1-110}$$

Figures 9-14 and 9-15 show curves of $\gamma_{LR}/k_0{}^2h^2$ and $\gamma_{RR}/k_0{}^2h^2$ for the perfectly conducting surface compared to similar curves for dielectric surfaces. These are based upon the Gaussian surface-height correlation coefficient model, and were computed from Eqs. (9.1-107) and (9.1-108).

9.1.3.4.4. *Results for Arbitrary Linear Polarization States.* For arbitrary linearly polarized receiving and transmitting states, define η_i as the angle from $\hat{\boldsymbol{\theta}}_i$ (in the direction of $\hat{\boldsymbol{\phi}}_i$) of the incident *E*-field. Likewise, η_s is the angle from $\hat{\boldsymbol{\theta}}_s$ (in the direction of $\hat{\boldsymbol{\phi}}_s$) of the scattered *E*-field of interest. Then the matrix element $\alpha_{\eta_i\eta_s}$ can be written in terms of the previously defined symbols as

$$\alpha_{\eta_i\eta_s} = \alpha_{vv} \cos \eta_i \cos \eta_s + \alpha_{vh} \sin \eta_i \cos \eta_s + \alpha_{vh} \cos \eta_i \sin \eta_s + \alpha_{hh} \sin \eta_i \sin \eta_s \tag{9.1-111}$$

From the above definition and a knowledge of α_{vv}, α_{vh}, α_{hv}, α_{hh} as defined in Eqs. (9.1-67) to (9.1-70), one can readily determine $\gamma^I_{\eta_i\eta_s}$ from Eq. (9.1-61). The latter quantity is the average bistatic or backscattering cross section per unit surface area for a pair of linearly polarized, but arbitrarily oriented antennas.

9.1.3.4.4.1. *Backscattering with Aligned Linearly Polarized Antennas.* Probably the most common situation is backscattering in which the transmitting and receiving antennas are linearly polarized and aligned in the same direction, but not necessarily vertical or horizontal. Here, $\eta_s = \eta_i$, and the matrix element for a dielectric surface becomes

$$\begin{aligned}\alpha_{\eta_i\eta_i} &= \alpha_{vv} \cos^2 \eta_i + \alpha_{hh} \sin^2 \eta_i \\ &= \left[\frac{(\epsilon_r - 1)[(\epsilon_r - 1) \sin^2 \theta_i + \epsilon_r\mu_r] - \epsilon_r^2(\mu_r - 1)}{[\epsilon_r \cos \theta_i + \sqrt{\epsilon_r\mu_r - \sin^2 \theta_i}]^2} \cos^2 \eta_i\right] \\ &\quad - \left[\frac{(\mu_r - 1)[(\mu_r - 1) \sin^2 \theta_i + \epsilon_r\mu_r] - \mu_r^2(\epsilon_r - 1)}{[\mu_r \cos \theta_i + \sqrt{\epsilon_r\mu_r - \sin^2 \theta_i}]^2} \sin^2 \eta_i\right]\end{aligned} \tag{9.1-112}$$

For a perfectly conducting surface, this reduces to

$$\alpha_{\eta_i\eta_i} = \left(\frac{1 + \sin^2 \theta_i}{\cos^2 \theta_i}\right) \cos^2 \eta_i + \sin^2 \eta_i \tag{9.1-113}$$

These quantities can be substituted into Eqs. (9.1-61) to obtain $\gamma^I_{\eta_i\eta_i}$, the average backscattering cross section.

9.1.3.4.4.2. *Backscattering with Crossed Linearly Polarized Antennas.* In the event that the receiving antenna is linearly polarized but rotated 90° with respect to the transmitting antenna, then $\eta_s = \eta_i + \pi/2$. The quantity $\alpha_{\eta_i\eta_i+\pi/2}$, for a dielectric surface, is then

$$\begin{aligned}\alpha_{\eta_i\eta_i+\pi/2} &= -(\alpha_{vv} - \alpha_{hh}) \sin \eta_i \cos \eta_i \\ &= -\left[\frac{(\epsilon_r - 1)[(\epsilon_r - 1) \sin^2 \theta_i + \epsilon_r\mu_r] - \epsilon_r^2(\mu_r - 1)}{[\epsilon_r \cos \theta_i + \sqrt{\epsilon_r\mu_r - \sin^2 \theta_i}]^2}\right. \\ &\quad \left. + \frac{(\mu_r - 1)[(\mu_r - 1) \sin^2 \theta_i + \epsilon_r\mu_r] - \mu_r^2(\epsilon_r - 1)}{[\mu_r \cos \theta_i + \sqrt{\epsilon_r\mu_r - \sin^2 \theta_i}]^2}\right] \\ &\quad \times \sin \eta_i \cos \eta_i\end{aligned} \tag{9.1-114}$$

For a perfectly conducting surface, this becomes

$$\alpha_{\eta_i\eta_i+\pi/2} = -2 \tan^2 \theta_i \sin \eta_i \cos \eta_i \tag{9.1-115}$$

The above quantities, when substituted into Eq. (9.1-61), give the average backscattering cross section, $\gamma^I_{\eta_i\eta_i+\pi/2}$.

9.1.3.4.4.3. *Backscattering with Aligned Linearly Polarized Antennas Rotated over All Angles.* If the transmitting–receiving antenna is rotating about its direction of propagation, then one would be interested in the backscattering cross section averaged over all equally likely angular orientations of η_i. This might be the case, for examplc, for a satellite antenna.

Another application utilizing this average is the backscattering by planetary surfaces when linear polarization states are employed. Here, the radar return at a given time arises from a ring on the sphere surface; the polarization at different locations on the ring is linear and makes an angle η_i with the vertical at that point. This angle varies around the ring from point to point because the direction of vertical polarization varies. Since the return from one portion of the ring is incoherent with respect to that from any other portion, one adds (and averages over η_i) the backscattered power from all portions of the ring.

Averaging $|\alpha_{\eta_i\eta_i}|^2$ by integrating over η_i from 0 to 2π, and dividing by 2π gives

$$\overline{|\alpha_{\eta_i\eta_i}|^2} = \frac{1}{2\pi}\int_0^{2\pi} |\alpha_{\eta_i\eta_i}|^2 \, d\eta_i = \frac{1}{8}\,[3|\alpha_{vv}|^2 + 2\,\mathrm{Re}[\alpha_{vv}\alpha_{hh}] + 3|\alpha_{hh}|^2] \tag{9.1-116}$$

The above result, when substituted into Eq. (9.1-61), gives $\bar{\gamma}^I_{\eta_i\eta_i}$, the backscattering cross section per unit area averaged over all angles, η_i, when the antennas are aligned. This applies both for dielectric and perfectly conducting surfaces.

9.1.3.4.4.4. *Backscattering with Crossed Linearly Polarized Antennas Rotated over All Angles.* The case in which the transmitting and receiving antennas are at right angles to each other and averaged over all angles, η_i, is considered here. It is similar in nature to that of the preceding section except that this represents the cross-polarized term. The result is

$$\overline{|\alpha_{\eta_i\eta_i+\pi/2}|^2} = \frac{1}{2\pi}\int_0^{2\pi} |\alpha_{\eta_i\eta_i+\pi/2}|^2 \, d\eta_i = \frac{|\alpha_{vv} - \alpha_{hh}|^2}{8} \tag{9.1-117}$$

This quantity can be substituted into Eq. (9.1-61) to give $\bar{\gamma}^I_{\eta_i\eta_i+\pi/2}$, the backscattering cross section per unit area averaged over all angles, η_i, when the antennas are crossed. This applies to both dielectric and perfectly conducting surfaces.

9.1.3.5. *Average Incoherent Scattering Cross Sections for a Very Rough Surface*

9.1.3.5.1. *Introduction.* As discussed in the introduction to this section, as the surface roughness increases in size with respect to wavelength, the scattered power becomes less coherent and more incoherent. For $k_0 h > 5$, the coherent component of the scattered field has disappeared for all practical purposes, and all of the scattered power is incoherent.† Thus, only the incoherent average scattering cross sections need be considered.

Solution of the scattering problem in this region is based upon an optics approach. The most rigorous approach proceeds from an integral equation for the scattered field [Eq. (2.2-19)]. Upon application of the tangent-plane approximation, the total field at every point on the surface is related to the incident field by the Fresnel reflection coefficients. Thus, the integral equation is converted into a definite integral, i.e., a physical-optics integral. The height of the surface is an unknown random variable, and consequently, the integral for the scattered power or intensity must be averaged; this is done by multiplying by the joint probability density function of the surface height and integrating.

However, before this is done, based upon a stationary phase analysis, the complicated factor in the integrand involving surface slopes and Fresnel reflection coefficients is removed from the integrand as a constant. This is justified because of the exponential $\exp[ik_0 \boldsymbol{\xi} \cdot \mathbf{r}]$ in the integrand; the vector, $\mathbf{r}$, contains the surface height, $\zeta(x, y)$, as the random z component. If this is expanded in a series as a function of x and y, then one can perform a stationary-phase integration, if the radii of curvature of the surface are everywhere large compared to wavelength. (These radii of curvature terms are contained in the second derivatives of $\zeta(x, y)$, upon which the quadratic stationary-phase integration is performed.)

This analysis is based upon the contention that all of the scattering from a very rough surface under these assumptions comes from areas which specularly reflect, such as one sees, for instance, when looking at moonlight reflected from specular points on a rough sea surface. Hagfors'[(24,25)] and Barrick's[(26)] papers support this point of view. Other optics methods based upon the same principle of specular reflection give nearly identical results. For instance, ray optics[(27)] and geometrical optics[(28)] have been applied recently, and the same results and interpretations are obtained.

The backscattered power predicted by any of these specular-point models is perfectly reflected; hence, there is no predicted cross-polarized component. Since for backscattering the specularly reflecting facets must be normal to the propagation direction, the backscatter cross section from these facets is polarization insensitive.

† The meanings of the terms coherent and incoherent, as used in these sections, are defined and discussed in Section 9.1.3.1.

The assumptions inherent in this model are the following:

(a) The tangent-plane approximation is employed, demanding that the surface radius of curvature at all points be considerably larger than the wavelength.

(b) Roughness is isotropic in both surface dimensions.

(c) $l \ll \sqrt{A}$, i.e., the correlation length of the surface roughness is much less than the length or width of the illuminated area. This precludes the possibility of looking at only a portion of a hill or wave, but rather it demands that several such hills or waves be included within the illuminated area.

(d) Multiple scattering and shadowing are neglected.

Under these assumptions, the average bistatic scattering cross section per unit surface area†, as derived in the above references and also Reference 29, is

$$\gamma_{pq}^{I} = |\beta_{pq}|^2 J \tag{9.1-118}$$

In the above equation, p and q refer to the polarization states of the scattered and incident E-fields, respectively. The quantity β_{pq} is directly proportional to the scattering matrix element, and will be given below for particular polarization cases. It is analogous to α_{pq} for the slightly rough surface. The quantity J, analogous to I of the preceding section, is defined as

$$\begin{aligned} J &= 2k_0^2 \int_0^\infty r J_0(k_0 \sqrt{\xi_x^2 + \xi_y^2}\, r)\, M_{\zeta\zeta'}(ik_0\xi_z, -ik_0\xi_z; r)\, dr \\ &= \frac{4\pi}{\xi_z^2}\, p\Big(-\frac{\xi_x}{\xi_z}, -\frac{\xi_y}{\xi_z}\Big) \end{aligned} \tag{9.1-119}$$

Here, ξ_x and ξ_y are as defined in Eqs. (9.1-53) and (9.1-54). $J_0(x)$ is the cylindrical Bessel function of zero order and argument x. The quantity $M_{\zeta\zeta'}(iu, iv; r)$ is the joint characteristic function of arguments u and v for the surface-height random variables ζ and ζ' which are separated by a distance, r. This joint characteristic function is merely the two-dimensional Fourier transform of the joint probability density function, $P_J(\zeta, \zeta')$.

An alternate form for J involves $p(\zeta_x, \zeta_y)$, the joint probability density function for the surface slopes ζ_x and ζ_y. The significance of this expression, as derived in Reference 26, is to show that the surface slopes which contribute to the scattered field are those oriented specularly between the incidence and scattering directions, i.e., where $\zeta_x = -\xi_x/\xi_z$ and $\zeta_y = -\xi_y/\xi_z$.

In order to obtain explicit results for J, one must know the joint probability density function for the surface heights, $\zeta(x, y)$ and $\zeta(x', y')$, separated by a distance, r (i.e., $r = \sqrt{(x - x')^2 + (y - y')^2}$), or the surface

† The notation γ is equivalent to the σ^0 sometimes used in the rough-surface literature; see Eq. (9.1-1).

slope probability density. In the absence of an exact knowledge of these quantities, one must assume a form or model for these functions. The most popular model used is the Gaussian. For the sake of comparison, an exponential function will also be used here.

Included implicitly in the joint-probability-density-function model is the surface-height correlation coefficient, $\rho(r)$. However, it turns out in the evaluation that only the first two terms of the power-series expansion of the correlation coefficient are of importance in this high-frequency model. The first term must always be unity, since statistical considerations demand that for $r = 0$, the correlation coefficient equal one. Since the correlation coefficient must be an even function, its form can be either $1 - |r/l|$ or $1 - r^2/l^2$ for the first two terms. Unfortunately, the former situation cannot be handled as a Maclaurin series because it is nonanalytic at $r = 0$. Consequently, only the second situation is considered here. Incidentally, the latter form is also the first two series terms of the Gaussian correlation coefficient model, $\rho(r) = \exp[-r^2/l^2]$.

The most general continuously curving surface is the quadric surface (upon which stationary phase models are based), and it can be shown that a quadric surface always has a parabolic correlation coefficient near the origin, i.e., $\rho(r) \approx 1 - r^2/l^2$ for r small. Hence, the latter is the form used here for the correlation coefficient as $r \to 0$.

Using the Gaussian surface-height probability density function,

$$P_J(\zeta, \zeta') = \frac{1}{2\pi h^2[1 - \rho^2(r)]^{1/2}} \exp\left[-\frac{[\zeta^2 - 2\zeta\zeta'\rho(r) + \zeta'^2]}{2h^2[1 - \rho^2(r)]}\right] \tag{9.1-120}$$

along with the small argument expansion mentioned in the preceding paragraph for $\rho(r)$ in Eq. (9.1-119); after integration, one arrives at

$$J = \frac{4}{s^2\xi_z^{\,2}} \exp\left[-\frac{1}{s^2}\left(\frac{\xi_x^{\,2} + \xi_y^{\,2}}{\xi_z^{\,2}}\right)\right] \tag{9.1-121}$$

The quantity s^2 in this equation is defined as $s^2 = 4h^2/l^2$, where h^2 is the mean-square roughness height and l is the surface correlation length. For the parabolic surface-height correlation coefficient (i.e., $\rho(r) \approx 1 - r^2/l^2$), s^2 is identically the mean-square total slope of the rough surface.

Using an exponential surface-height probability density function,

$$P_E(\zeta, \zeta') = \frac{3}{2\pi h^2[1 - \rho^2(r)]^{1/2}} \exp\left\{-\left[\frac{\zeta^2 - 2\zeta\zeta'\rho(r) + \zeta'^2}{\frac{1}{3}h^2[1 - \rho^2(r)]}\right]^{1/2}\right\} \tag{9.1-122}$$

then the quantity J of Eq. (9.1-119), becomes

$$J = \frac{12}{s^2\xi_z^{\,2}} \exp\left[-\frac{\sqrt{6}}{s}\sqrt{\frac{\xi_x^{\,2} + \xi_y^{\,2}}{\xi_z^{\,2}}}\right] \tag{9.1-123}$$

where the parameter, s, has the same meaning as discussed in the previous paragraph. This joint-probability-density-function model was chosen for the sake of comparison with the Gaussian; it satisfies all the statistical requirements for a joint probability density function, and also the physically obvious requirement that it be symmetric in ζ and ζ'.

9.1.3.5.2. *Bistatic Scattering for Vertical and Horizontal, Circular, and Arbitrarily Oriented Linear Polarization States.* The matrix elements β_{pq} will be defined here for the vertical and horizontal polarization states. A vertically polarized incident or scattered wave is taken here to mean the E-fields in the $\hat{\boldsymbol{\theta}}_i$ and $\hat{\boldsymbol{\theta}}_s$ directions, respectively; a horizontally polarized incident or scattered wave is taken to mean E-fields in the $\hat{\boldsymbol{\phi}}_i$ (or $-\hat{\mathbf{y}}$) and $\hat{\boldsymbol{\phi}}_s$ directions, respectively.

The matrix elements for linear polarization states are

$$\beta_{vv} = \frac{a_2 a_3 R_\parallel(\iota) + \sin\theta_i \sin\theta_s \sin^2\phi_s R_\perp(\iota)}{a_1 a_4} \tag{9.1-124}$$

$$\beta_{hv} = \sin\phi_s \frac{-\sin\theta_i a_3 R_\parallel(\iota) + \sin\theta_s a_2 R_\perp(\iota)}{a_1 a_4} \tag{9.1-125}$$

$$\beta_{hv} = \sin\phi_s \frac{\sin\theta_s a_2 R_\parallel(\iota) - \sin\theta_i a_3 R_\perp(\iota)}{a_1 a_4} \tag{9.1-126}$$

$$\beta_{hh} = \frac{-\sin\theta_i \sin\theta_s \sin^2\phi_s R_\parallel(\iota) - a_2 a_3 R_\perp(\iota)}{a_1 a_4} \tag{9.1-127}$$

where $R_\parallel(\iota)$ and $R_\perp(\iota)$ are the Fresnel reflection coefficients, as defined in Eqs. (9.1-47) and (9.1-49) for argument ι (angle of incidence). This angle and the remaining quantities are defined as

$$\cos\iota = \frac{1}{\sqrt{2}} \sqrt{1 - \sin\theta_i \sin\theta_s \cos\phi_s + \cos\theta_i \cos\theta_s} \tag{9.1-128}$$

$$a_1 = 1 + \sin\theta_i \sin\theta_s \cos\phi_s - \cos\theta_i \cos\theta_s \tag{9.1-129}$$

$$a_2 = \cos\theta_i \sin\theta_s + \sin\theta_i \cos\theta_s \cos\phi_s \tag{9.1-130}$$

$$a_3 = \sin\theta_i \cos\theta_s + \cos\theta_i \sin\theta_s \cos\phi_s \tag{9.1-131}$$

$$a_4 = \cos\theta_i + \cos\theta_s \tag{9.1-132}$$

Employing the β's defined above, along with one of the forms of J derived previously, one can write the average bistatic scattering cross sections per unit area γ_{vv}^I, γ_{hv}^I, γ_{vh}^I, γ_{hh}^I for the horizontal and vertical-polarization states from Eq. (9.1-118).

If the surface is perfectly conducting, then the above Fresnel reflection coefficients simplify to

$$R_\parallel(\iota) = 1 \quad \text{and} \quad R_\perp(\iota) = -1 \tag{9.1-133}$$

The bistatic cross sections for circular polarization states are written in terms of matrix elements β_{LR}, β_{RL}, β_{RR}, and β_{LL}. These can be expressed in terms of β_{vv}, β_{vh}, β_{hv}, and β_{hh} above, according to Eqs. (9.1-93) to (9.1-96) (β's substituted for α's). The average bistatic scattering cross sections per unit area for the circular polarization states, γ^I_{LR}, γ^I_{RL}, γ^I_{RR}, and γ^I_{LL}, are then obtained using Eq. (9.1-118).

The bistatic cross sections for arbitrarily oriented linear polarization directions (making angles η_i and η_s from the verticals) are expressed in terms of $\beta_{\eta_i\eta_s}$. This, in turn, is obtained from β_{vv}, β_{vh}, β_{hv}, and β_{hh}, using Eq. (9.1-111) (β's substituted for $\alpha'a$). The average bistatic scattering cross section per unit area for the arbitrarily oriented linear polarization state, $\gamma^I_{\eta_i\eta_s}$, can then be written using Eq. (9.1-118).

9.1.3.5.3. *Backscattering for Vertical, Horizontal, Circular, and Arbitrarily Oriented Linear Polarization States.* The matrix elements, β_{vv}, β_{hh}, β_{vh}, and β_{hv}, given previously, can be specialized to the case of backscattering. One must take the limits of the numerators and denominators of these quantities as $\theta_s \to \theta_i$ and $\phi_s \to 180°$. They become

$$\beta_{vv} = \beta_{hh} = \sec\theta_i R_{\parallel}(0) = -\sec\theta_i R_{\perp}(0) \tag{9.1-134}$$

and

$$\beta_{hv} = \beta_{vh} = 0 \tag{9.1-135}$$

From these, the β_{pq} for the circular and arbitrarily oriented linear polarization states may be written as

$$\beta_{LR} = \beta_{RL} = \sec\theta_i R_{\parallel}(0), \qquad \beta_{LL} = \beta_{RR} = 0 \tag{9.1-136}$$

and

$$\beta_{\eta_i\eta_s} = \sec\theta_i \cos(\eta_i - \eta_s)\, R_{\parallel}(0) \tag{9.1-137}$$

The above relations indicate that all of the power backscattered according to the specular-point model is polarized, and the cross-polarized component is identically zero for any polarization states. Hence, only γ^I will be written with the understanding that it refers to γ^I_{LR}, γ^I_{RL}, γ^I_{vv}, γ^I_{hh}, or $\gamma^I_{\eta_i\eta_i}$.

The quantities given previously for J are also specialized to backscattering, and the backscattering cross section for each probability-function model is given below:

(a) *Gaussian surface-height probability function.* For backscattering, the quantity J, given in Eq. (9.1-121), simplifies to

$$J = \frac{1}{s^2}\sec^2\theta_i \exp\left[-\frac{1}{s^2}\tan^2\theta_i\right] \tag{9.1-138}$$

where the quantity s^2 is defined and discussed under Eq. (9.1-121).

Using this expression in Eq. (9.1-118), the average backscattering cross section per unit area becomes

$$\gamma^I = \frac{\sec^4 \theta_i}{s^2} \mid R_\parallel(0) \mid^2 \exp\left[-\frac{1}{s^2} \tan^2 \theta_i\right] \tag{9.1-139}$$

Thus, the backscattered power of this model is related to the mean-square surface slope, s^2. The above equation has also been derived from ray optics models, instead of the physical-optics method used here.

Plots of $\gamma^I/\mid R_\parallel(0)\mid^2$ for various values of mean-square surface slope, s^2, are shown in Figure 9-16.

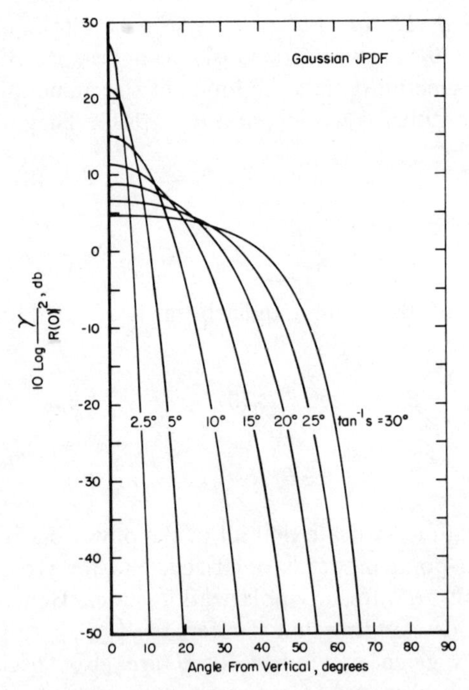

Fig. 9-16. Average incoherent backscattering cross section per unit area for very-rough-surface statistical model versus incidence angle for any polarization state. [Plotted from Eq. (9.1-139) for Gaussian surface-height joint probability density function and for various values of rms total surface slope, s. The cross section is divided by $\mid R(0) \mid^2$, the Fresnel reflection coefficient squared for the surface material for normal incidence.]

(b) *Exponential surface-height probability function.* For backscattering, the quantity J, given in Eq. (9.1-123), simplifies to

$$J = \frac{3}{s^2} \sec^2 \theta_i \exp \left[-\frac{\sqrt{6}}{s} \tan \theta_i \right] \tag{9.1-140}$$

The average backscattering cross section per unit area of Eq. (9.1-118) then becomes

$$\gamma^I = \frac{3}{s^2} \sec^4 \theta_i \mid R_{\parallel}(0) \mid^2 \exp \left[-\frac{\sqrt{6}}{s} \tan \theta_i \right] \tag{9.1-141}$$

Plots of $\gamma^I / | R_{\parallel}(0)|^2$ for this model are shown in Figure 9-17.

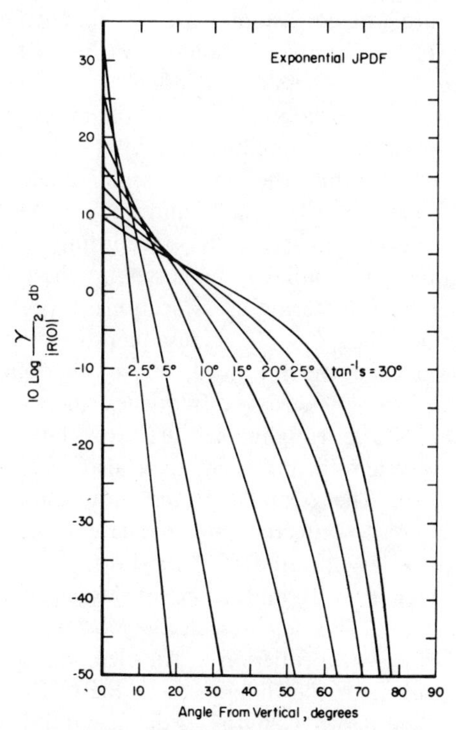

Fig. 9-17. Average incoherent backscattering cross section per unit area for very-rough-surface statistical model versus incidence angle for any polarization states. [Plotted from Eq. (9.1-141) for exponential surface-height joint probability density function and for various values of rms total surface slope, s. The cross section is divided by $| R(0) |^2$, the Fresnel reflection coefficient squared for the surface material for normal incidence.]

Both of the above models for backscattering show that more power is scattered near the mean surface normal (i.e., $\theta_i = 0$); the Gaussian model is parabolic at $\theta_i = 0$, while the exponential model is linear. At higher incidence angles, the power for the Gaussian model falls off more rapidly. As the mean square slope, s^2, of the surface increases, less power is scattered near normal incidence and more power is scattered out near grazing.

Another distinctive feature of all of the above specular-point models is the lack of dependence of the scattered power upon the wavelength. This must always be the case in the high-frequency limit for a smoothly curving surface with finite second derivatives (or radii of curvature) at the specular points; the power scattered from such a specular point is proportional only to its Gaussian curvature. It is often pointed out, however, that such models are not consistent with experimental observations from natural surfaces. There are two reasons why this is the case, neither of which reflect upon the soundness of the model.

Most natural surfaces are composed of many different scales of roughness. The largest scale roughnesses are usually the smoothest (e.g., mountains and valleys), while the smaller scale roughnesses are the most precipitous (e.g., small gravel, stones, and rocks). At lower frequencies, only the largest-scale roughness scatters according to the specular-point theory of this section; the smaller roughnesses, as shown in Section 9.1.3.4, have much smaller backscattering cross sections, especially near the mean normal to the surface (i.e., $\theta_i = 0$). Consequently, most of the incoherent backscattered power exists near $\theta_i = 0$, due to the gently sloping nature of the largest-scale roughnesses. As frequency increases, much more of the previously small-scale roughnesses are now large in dimensions as compared to the wavelength. As such, they also scatter according to the specular-point theory. The overall surface now scattering, according to this theory, has a greater effective mean-square slope, s^2, at the shorter wavelength. Hence, a larger value of s^2 at shorter wavelengths means less scattered power near $\theta_i = 0$ and more at larger incidence angles (see Figures 9-16 and 9-17). This is, in fact, the observed scattering behavior of most natural surfaces as a function of wavelength.

The second reason for the differences is due to the effect of the slight roughnesses at different wavelengths. For a natural composite surface made up of roughnesses larger and smaller than wavelength in scale, the smaller (or slight) roughnesses can affect significantly the scattering behavior at angles considerably different from the normal. As was shown in Section 9.1.3.4, these slight roughnesses are also wavelength-dependent in contrast to the larger roughnesses treated by the specular-point model of this section. Looking at Figures 9-12 to 9-15, one can see that as the wavelength is decreased, $k_0 l$ increases and the curves become steeper and larger. Hence, it appears that as k_0 becomes larger at higher frequencies, the curves for the slightly rough surfaces are beginning to look like those of this section, as one would expect. Composite surfaces will be discussed in more detail in the following section.

9.1.3.6. *Composite Rough Surfaces*

The term "composite rough surfaces" is used here to mean surfaces which have at least two classes of roughness; one class of roughness is significantly larger than the wavelength in its dimensions, while the second class is significantly smaller than the wavelength. Taken alone, the first class of roughness could be treated by the specular-point theory of the preceding section (Section 9.1.3.5). The second class, or slight roughness, can be treated by the model of Section 9.1.3.4. When both types of roughness are present on the same surface, however, they combine to produce a scattered power having some of the characteristics of each. In fact, to a first-order approximation, one can merely add the incoherent average scattering cross sections of the two preceding sections to obtain the average incoherent scattering cross section of the composite surface. A brief heuristic explanation will be given below as proof of this statement.

Consider first a flat surface consisting of only a slight roughness whose scale is less than a wavelength. Such a surface has a large coherent component in the scattered power which appears near the specular direction for the flat surface in the absence of roughness. In addition, there is an incoherent component of scattered power due to the slight roughness; this incoherent power may be negligible near the specular direction where the coherent component is so strong, but in other directions it can be the dominant contribution.

Now, let a larger scale of roughness, whose radii of curvature are much larger than wavelength, evolve under the slight roughness. One can picture this as a slight buckling and distending of the slightly rough, flat surface. Now, the coherent component of the scattered field which was present before disappears due to the growth of the large-scale roughness. In its place appears another incoherent component which is related to the slopes and specular areas of the large roughness. This component is, of itself, incoherent because the average distance between neighboring specular points is random and much larger than wavelength. Hence, the phase difference between these neighboring contributions is random and uniformly distributed (between 0 and 2π), and the scattered powers from neighboring hills add incoherently when averaged over all possible hill positions.

Two averaging processes must take place. The slight-roughness height variable and the large-roughness height variable are each averaged separately. This is due to the fact that, in general, the two different scales of roughness are produced by separate processes and are considered statistically independent. Hence, assume first that the large roughness is kept the same in shape, but the slight roughness on top of it is allowed to vary (in time or over an ensemble of surfaces, all with the same large-roughness features). In this manner, one can consider the contribution from the large roughness as "coherent" throughout the averaging process because it remains fixed. The incoherent power contribution from the slight roughness varies, however, and is averaged. Now consider an ensemble of such surfaces, already averaged with respect to their slight roughness, but whose large-roughness features vary. When this is done, the "coherent" component, mentioned above, for one surface of the ensemble adds incoherently to the same component for the other surfaces of the ensemble, because the large-scale hill and valley shifts from one surface to another produce random phase differences.

Thus, after both averaging processes, the average scattered-power contributions from each roughness scale, being each incoherent and independent of the other, may be added to give the total average scattered power. This explanation is meant to be intuitive rather than rigorous; if the reader desires a rigorous exposition of this analysis, he should consult References 30 and 31. The conclusions arrived at rigorously there are essentially the same as discussed here.

The composite surface model considered here involves two scales of roughness: a slight roughness whose height parameter, k_0h, is somewhat less than unity; and a large scale roughness whose parameter is considerably greater than unity. Neglected is any roughness scale whose height parameter, k_0h, is of the order of unity. Since most natural surfaces have a roughness component whose scale is of this order, the model here must be expected to give results which deviate somewhat from measured values of scattering cross section. This deviation is expected at angles of incidence and scattering where the scattering cross section for the slight roughness is approximately equal to that for the large-scale roughness. Hence, for backscattering, the results predicted by this composite surface model are expected to be valid over a range of values of θ_i near zero and near 90° (i.e., normal and grazing incidence).

Rather than reproduce the sums of the equations of the preceding two sections, curves will be shown here which represent these sums for back scattering. This is done for both horizontal and vertical polarization states, as well as for the circular polarization states.

Figures 9-18 to 9-21 show these curves for horizontal and vertical polarization states. They are made simply by laying the curves of Figures 9-12 or 9-13 over those of 9-16 or 9-17, and adding. One can see the effects of both roughnesses. The large scale roughness produces a strong component near normal ($\theta_i = 0$); this component is identical for both vertical and horizontal polarization. It is sometimes referred to by experimentalists as the "specular" of "quasi-specular" component. The small-scale roughness adds a "tail" to the curve; it is the dominant contribution further away from normal. This latter component is greater for vertical than for horizontal polarization. It is often called the "diffuse" component.

Figures 9-22 to 9-24 show similar curves for the circular polarization states. Near normal for this case, there is no cross-polarization component predicted by the specular-point model for the large-scale roughness. Hence, these curves consist only of the small-scale cross-polarized component.

It should be noted before leaving this section that any failure of measured scattering cross sections to agree with these predicted by these models is a measure of the inability of the models to truly represent natural types of roughnesses. The results presented in this section are accurate for the models for which they were derived, under the restrictions mentioned for the models.

9.1.4. Statistically-Rough Spherical Surface Models

9.1.4.1. *Introduction*

In this section, statistical models are given for the total average backscattering cross section of a rough sphere. In practice, one is often confronted with a spherical reflector or scatterer whose surface is not perfectly spherical. Examples are planets and the moon, satellites such as the ECHO's, balloons, etc. *The assumption is implicit that the entire sphere is illuminated by the incident plane wave simultaneously.* Thus if pulses illuminate the sphere, the pulse length is taken to be considerably greater than the sphere radius, A_r.

Throughout this section it is assumed that the mean sphere radius, A_r, is much greater than wavelength.[†] Hence, scattering from the sphere follows the optics approximations in the absence of roughness. Next, it is assumed that if the sphere is not perfectly conducting, it is large enough, compared to wavelength, and lossy enough so that rays penetrating the surface are attenuated before they reach the other side; hence, multiple internal reflection does not occur. With these restrictions, the backscattering cross section in the absence of roughness becomes

$$\sigma = \pi A_R^2 \mid R(0)|^2 \tag{9.1-142}$$

where $R(0)$ is the Fresnel reflection coefficient for normal incidence. It is given by

$$R(0) = \frac{\sqrt{\epsilon_r} - \sqrt{\mu_r}}{\sqrt{\epsilon_r} + \sqrt{\mu_r}} \tag{9.1-143}$$

where ϵ_r and μ_r are the complex relative permeability and permittivity of the sphere material. Stationary-phase methods show that, under these assumptions, all of the backscattering comes from that portion of the

[†] A rough sphere, whose average radius is small compared to wavelength, lies in the Rayleigh region as an overall scatterer. The effect of roughness on such a sphere is not visible so long as the mean spherical shape is still discernible, and the result for the smooth sphere may be used without significant error. Therefore, only the large rough sphere is of importance.

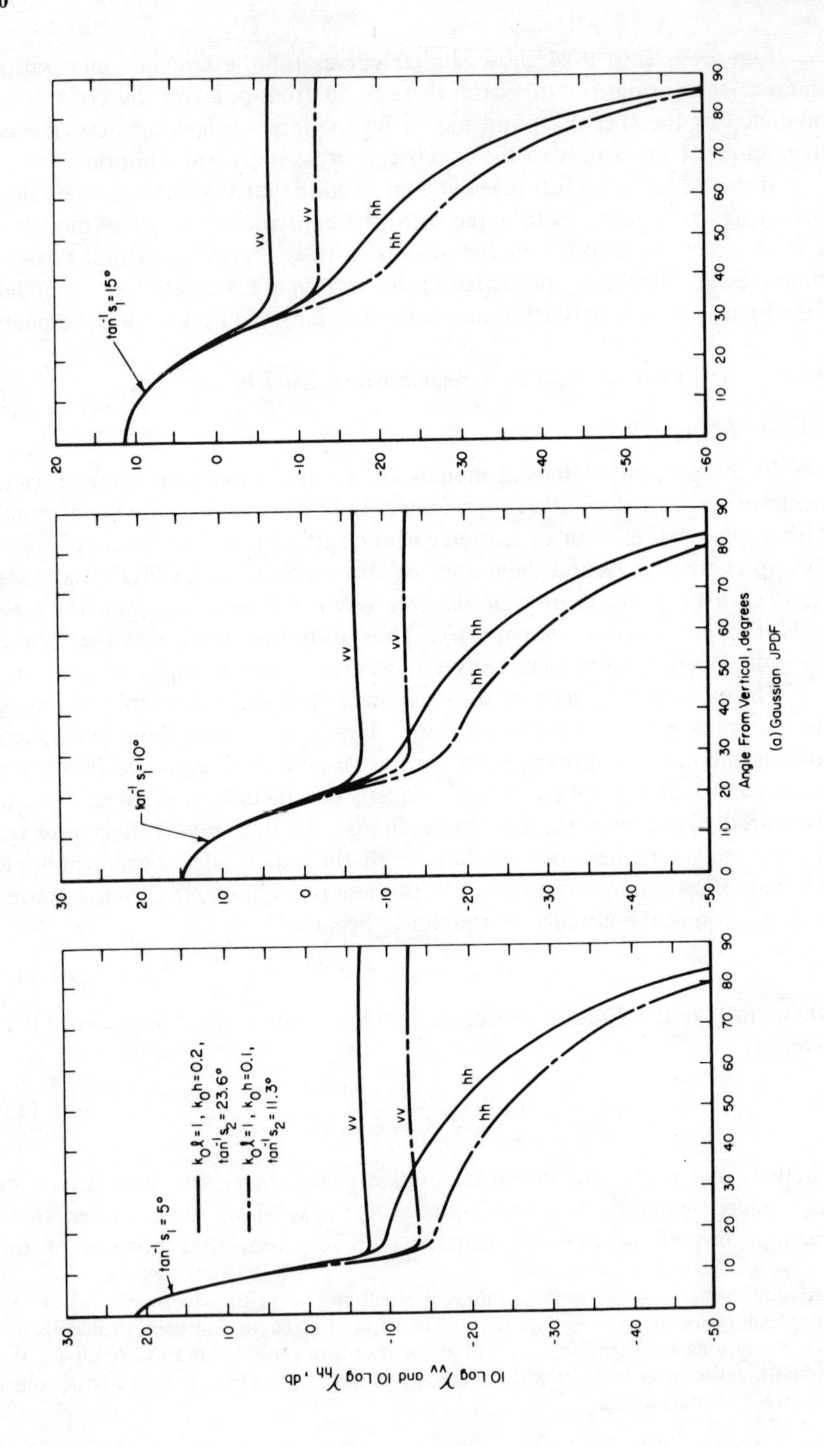

(a) Gaussian JPDF

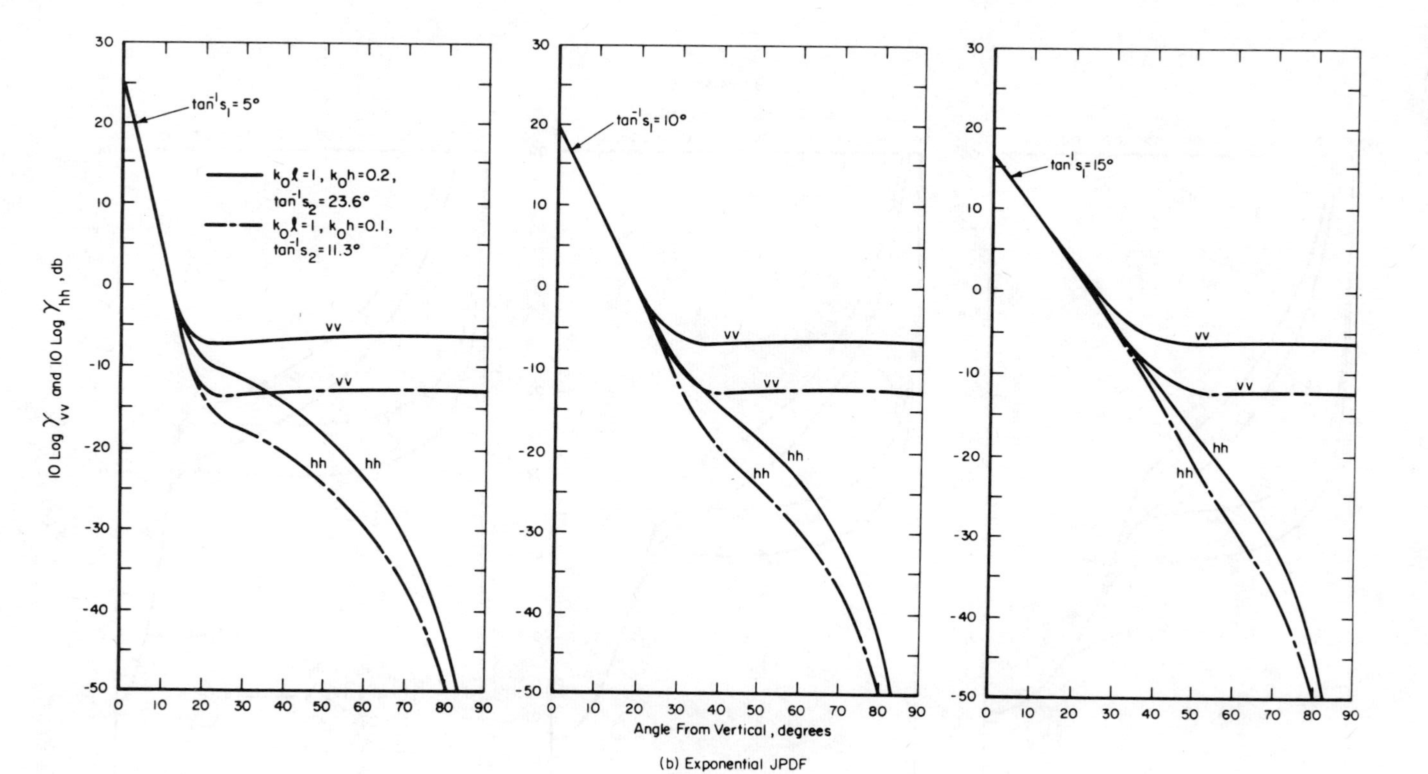

Fig. 9-18. Average backscattering cross section per unit area for composite rough-surface model (i.e., both large- and small-scale roughness present) versus incidence angle for various large-scale roughness parameters, s_1, and small-scale roughness parameters, k_0l and k_0h (or s_2). Linear polarization states. Perfectly conducting surface. (a) Gaussian JPDF; (b) Exponential JPDF.

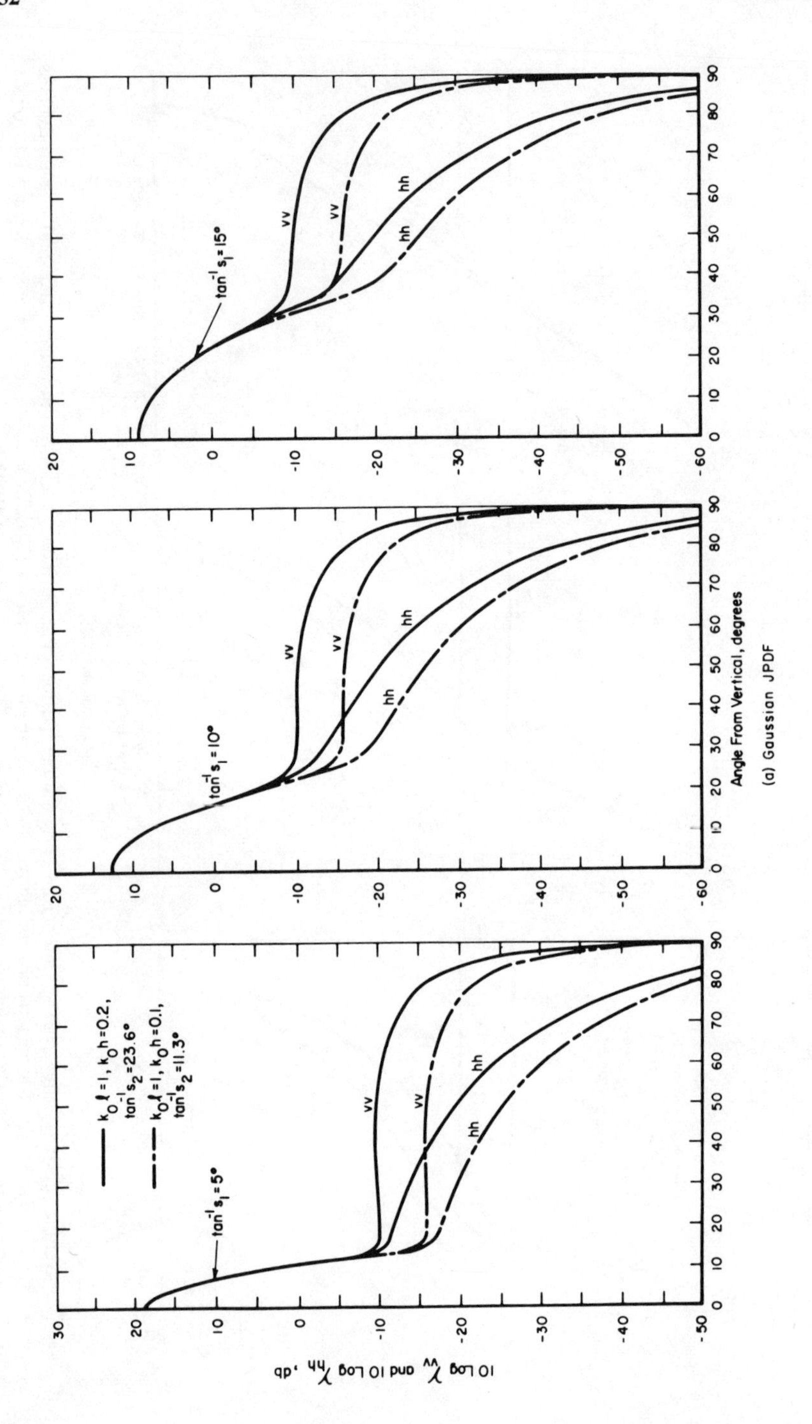

(a) Gaussian JPDF

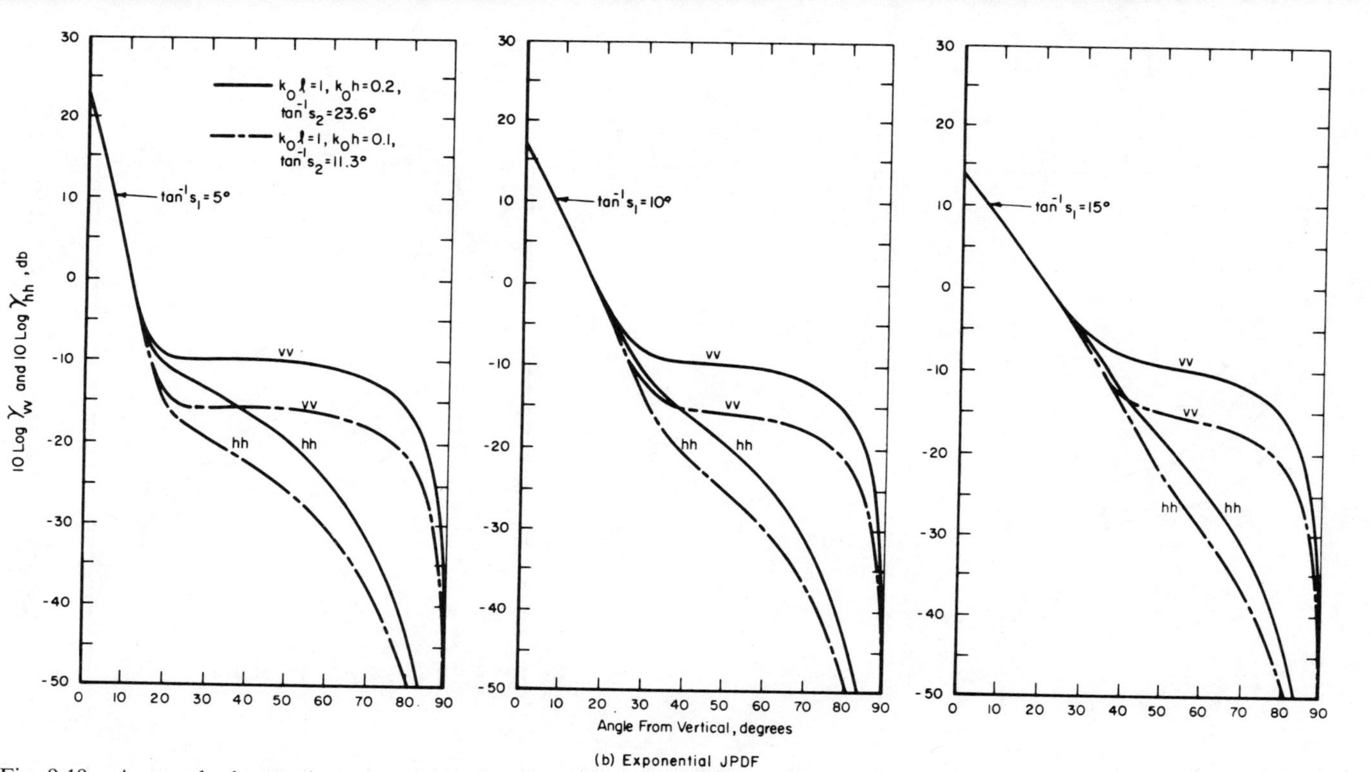

Fig. 9-19. Average backscattering cross section per unit area for composite rough-surface model (i.e., both large- and small-scale roughness present) versus incidence angle for various large-scale roughness parameters, s_1, and small-scale roughness parameters, $k_0 l$ and $k_0 h$ (or s_2). Linear polarization states. Surface material dielectric constant $\epsilon_r = 55(1 + i0.55)$. (Sea or fresh water at X-band). (a) Gaussian JPDF; (b) Exponential JPDF.

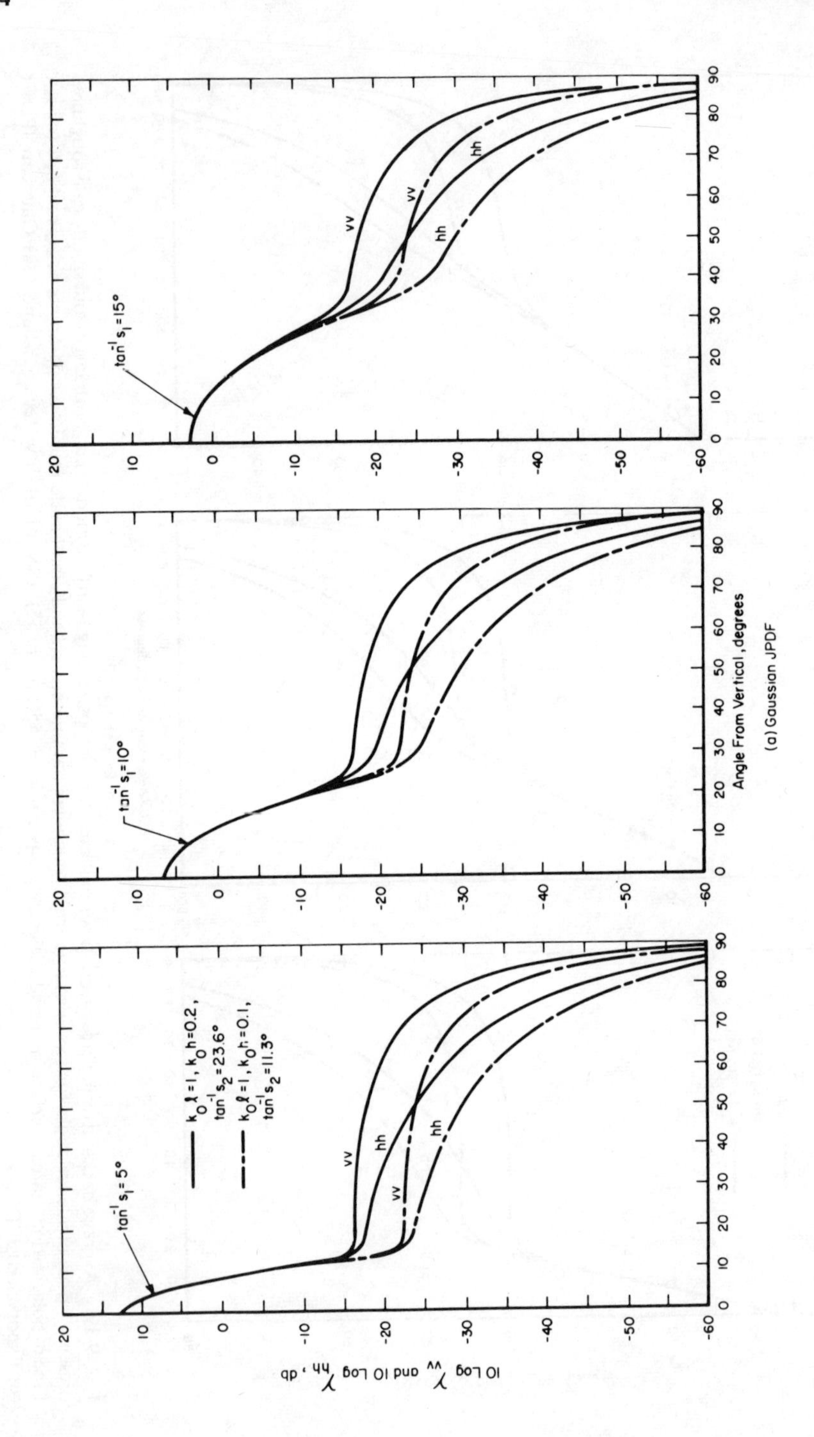

(a) Gaussian JPDF

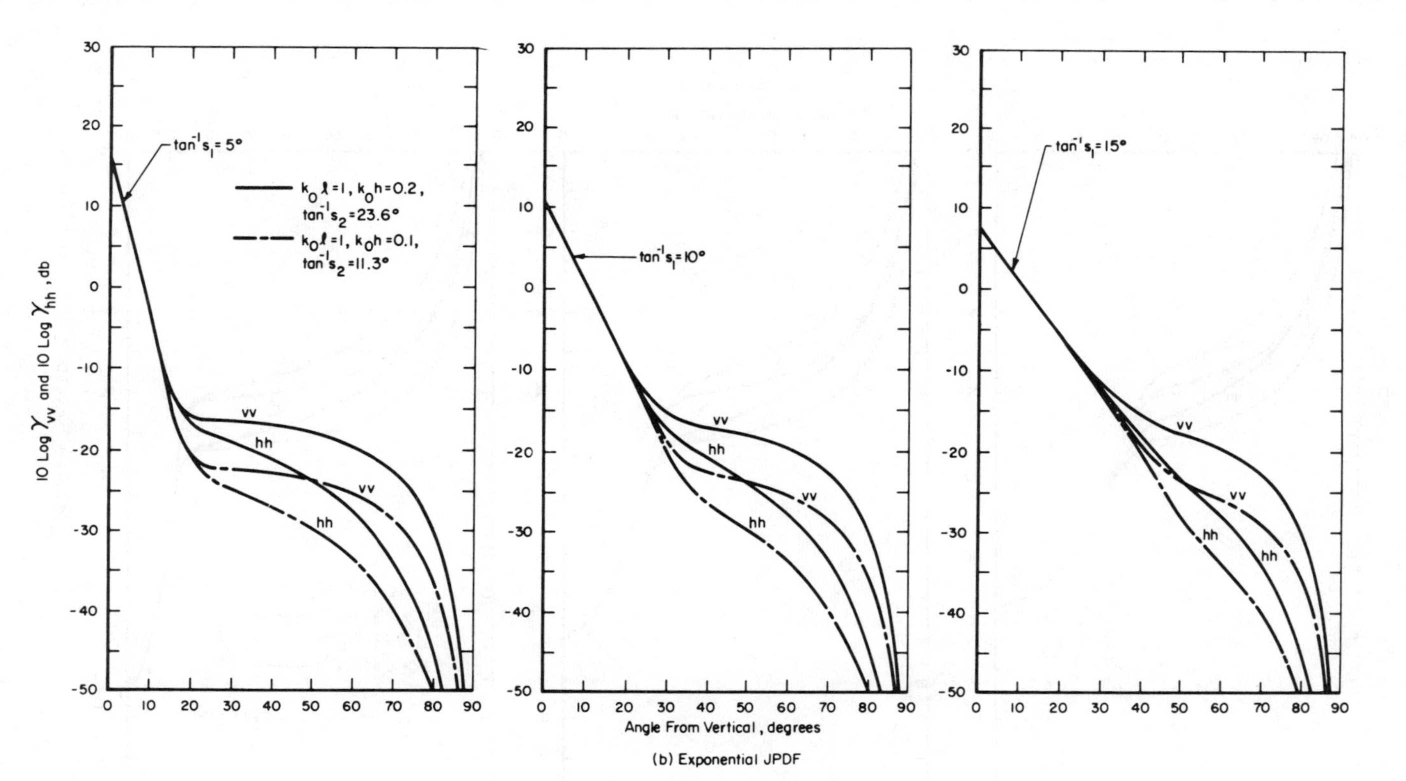

Fig. 9-20. Average backscattering cross section per unit area for composite rough-surface model (i.e., both large- and small-scale roughness present) versus incidence angle for various large-scale roughness parameters, s_1, and small-scale roughness parameters, $k_0 l$ and $k_0 h$ (or s_2). Linear polarization states. Surface material dielectric constant $\epsilon_r = 5.0$. (a) Gaussian JPDF; (b) Exponential JPDF.

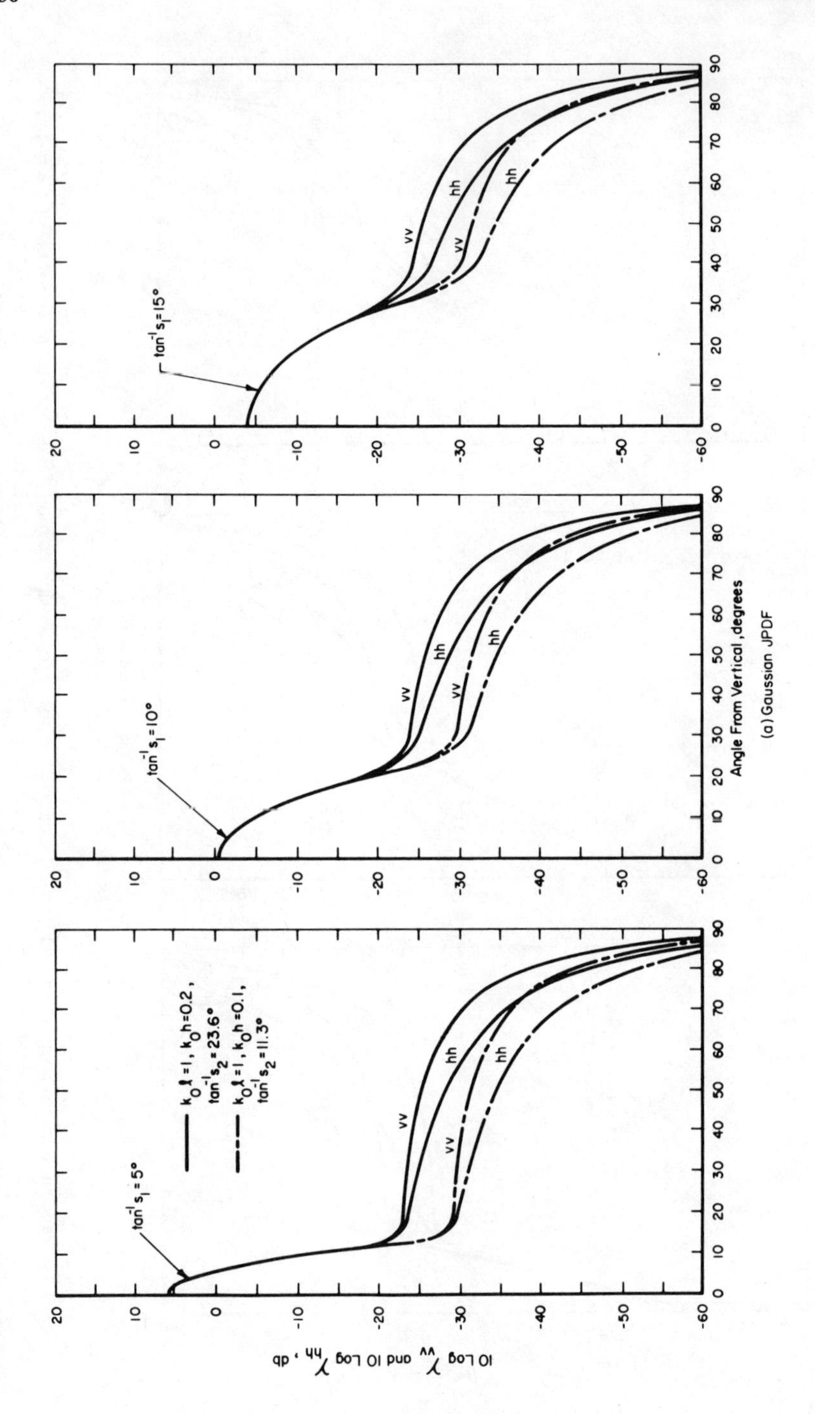

(a) Gaussian JPDF

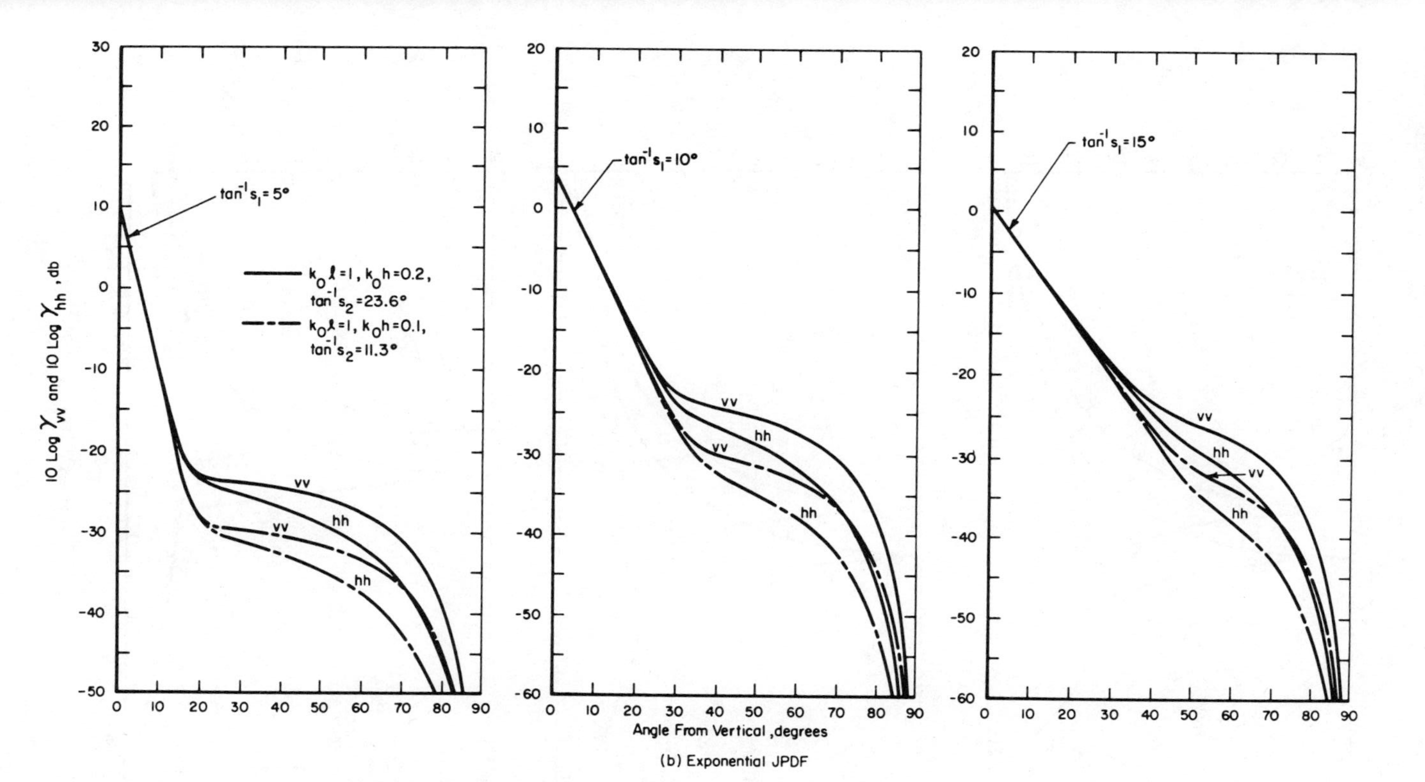

Fig. 9-21. Average backscattering cross section per unit area for composite rough-surface model (i.e., both large- and small-scale roughness present) versus incidence angle for various large-scale roughness parameters, s_1, and small-scale roughness parameters, $k_0 l$ and $k_0 h$ (or s_2). Linear polarization states. Surface material dielectric constant, $\epsilon_r = 2.0$. (a) Gaussian JPDF; (b) Exponential JPDF.

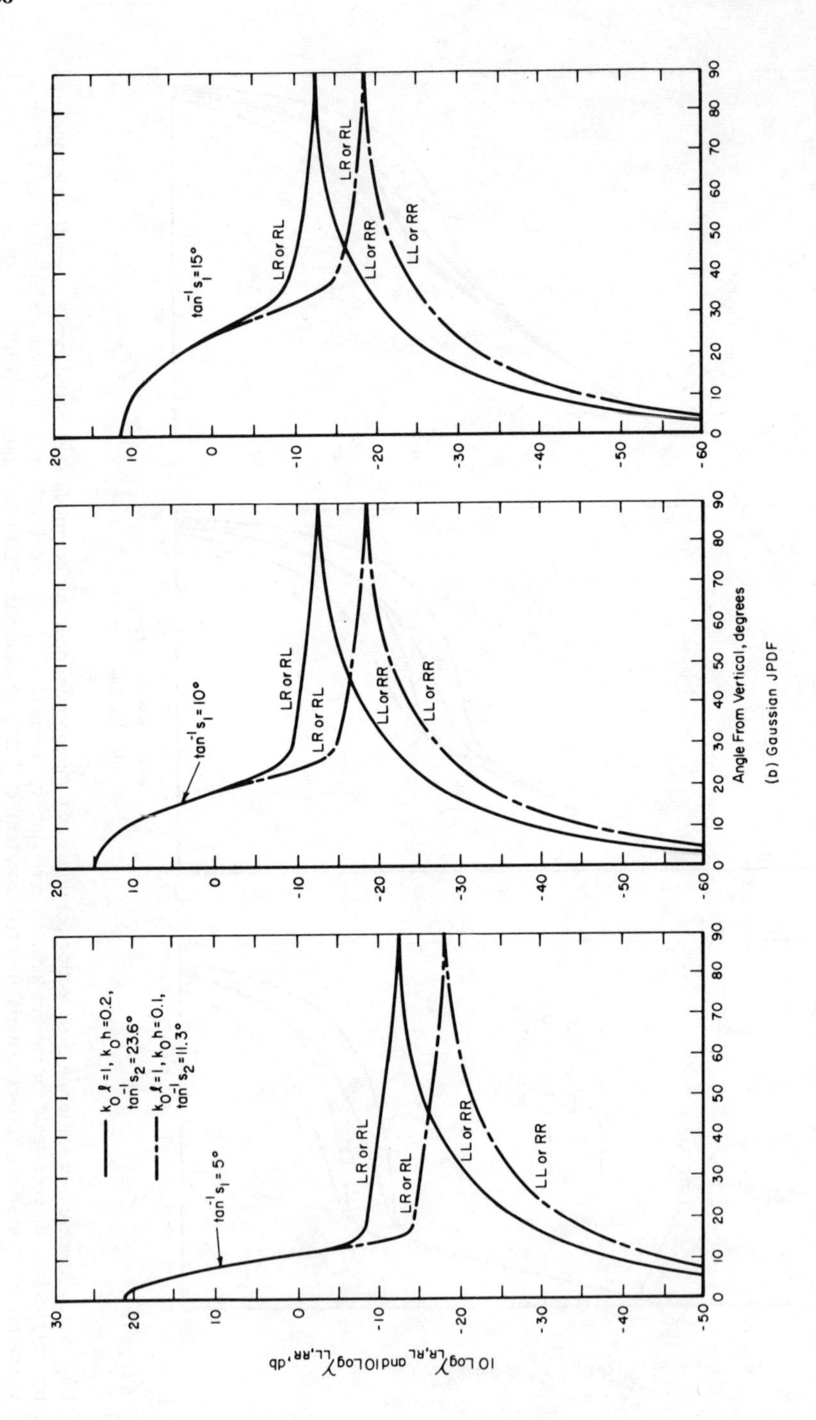

(b) Gaussian JPDF

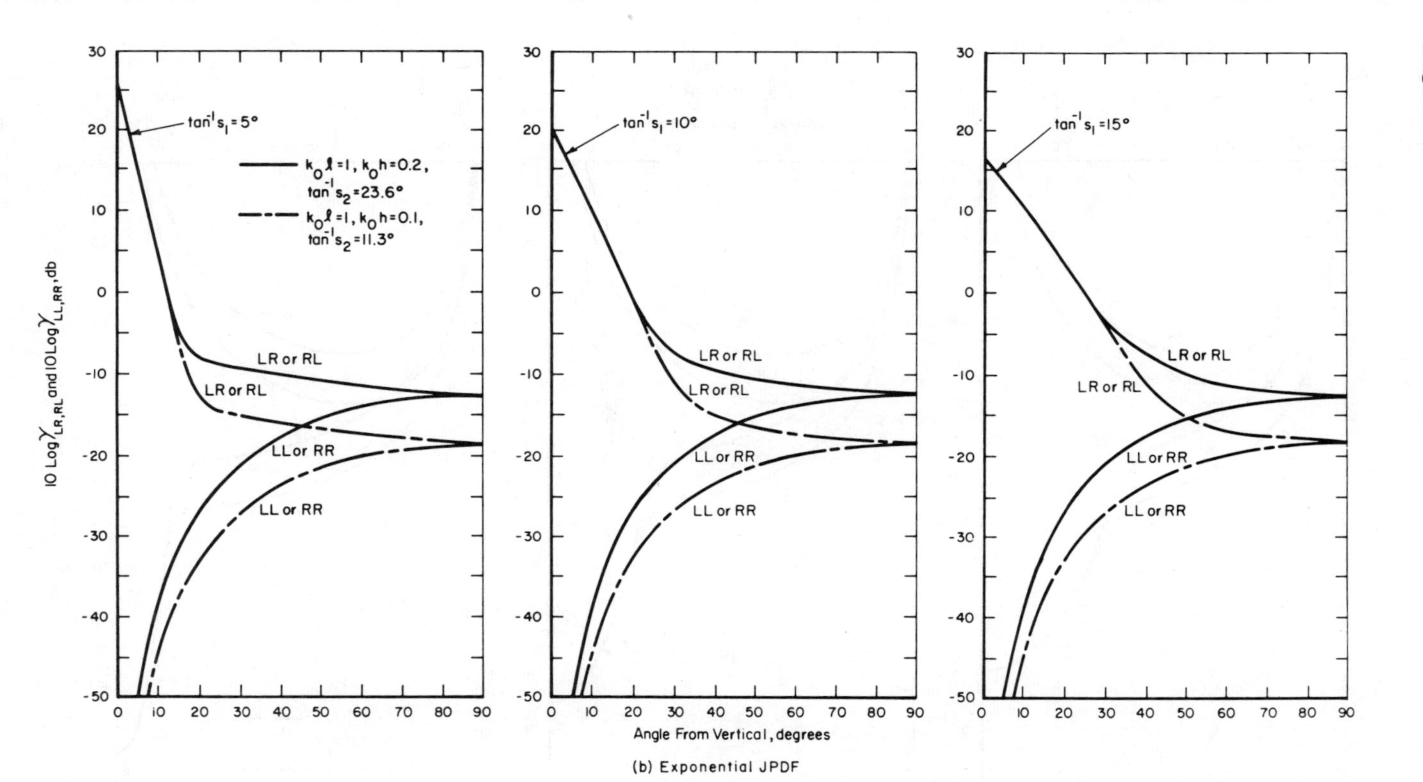

Fig. 9-22. Average backscattering cross section per unit area for composite rough-surface model (i.e., both large- and small-scale roughnesses present) versus incidence angle for various large-scale roughness parameters, s_1, and small-scale roughness parameters, $k_0 l$ and $k_0 h$ (or s_2). Circular polarization states. Perfectly conducting surface. (a) Gaussian JDPF; (b) Exponential JPDF.

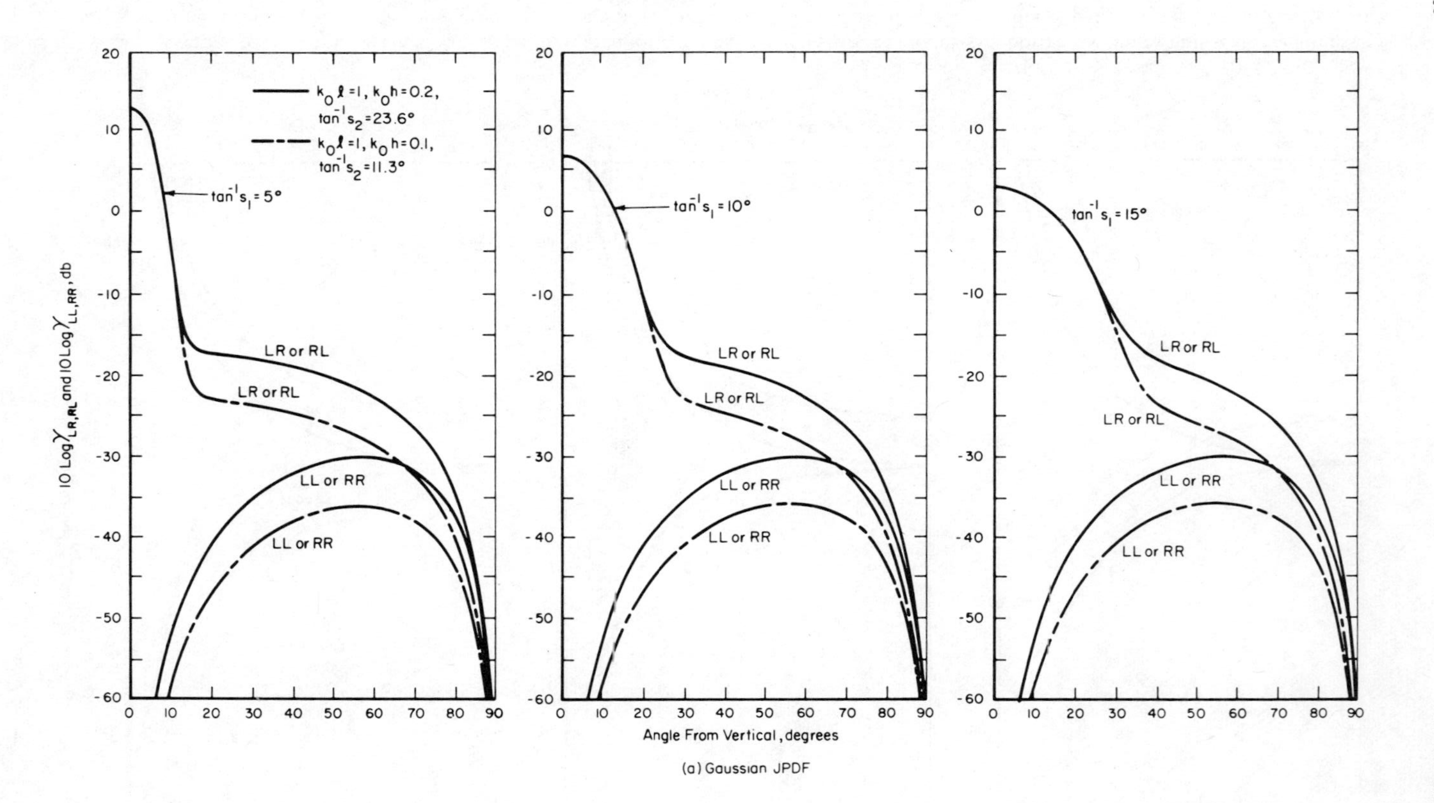

(a) Gaussian JPDF

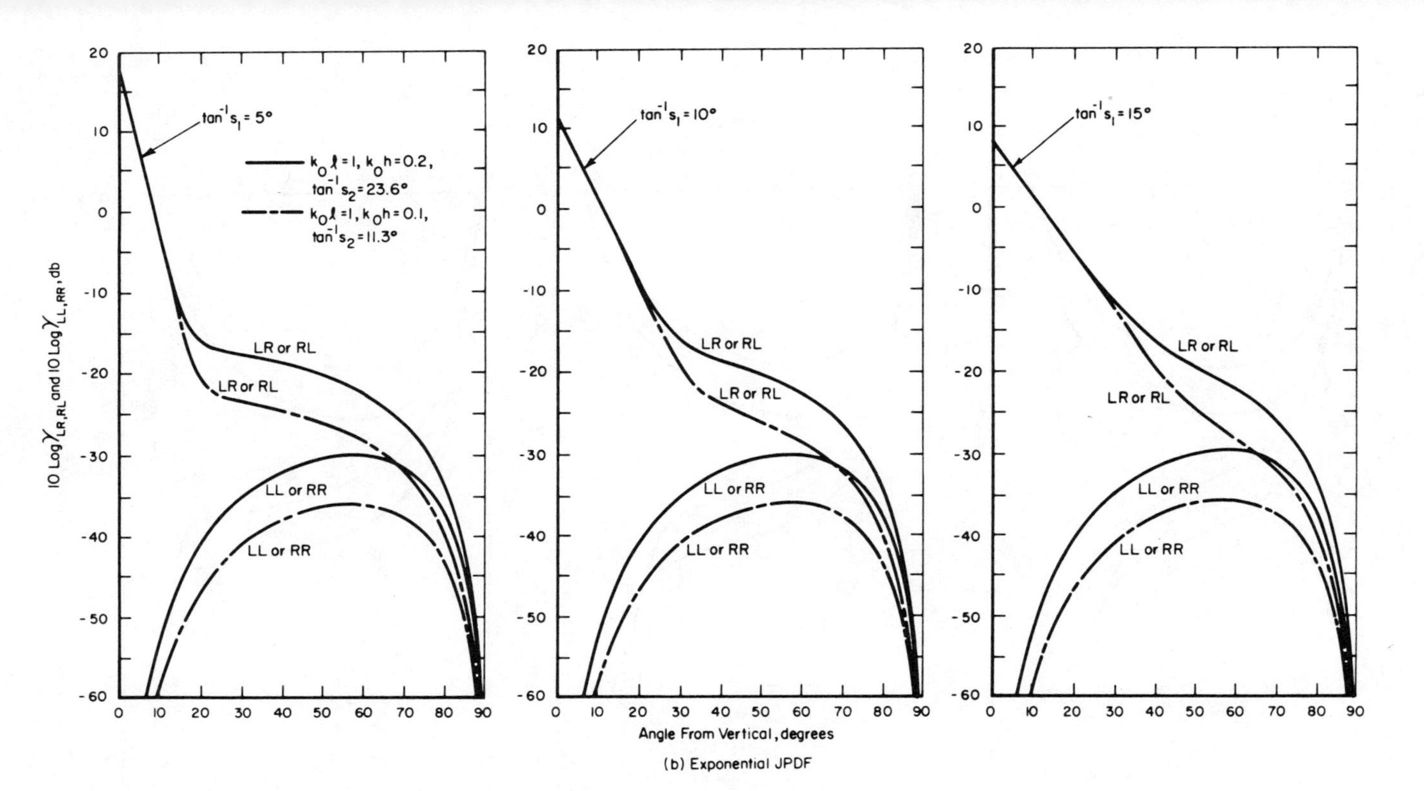

Fig. 9-23. Average backscattering cross section per unit area for composite rough-surface model (i.e., both large- and small-scale roughnesses present) versus incidence angle for various large-scale roughness parameters, s_1, and small-scale roughness parameters, $k_0 l$ and $k_0 h$ (or s_2). Circular polarization states. Surface material dielectric constant $\epsilon_r = 5.0$. (a) Gaussian JPDF; (b) Exponential JPDF.

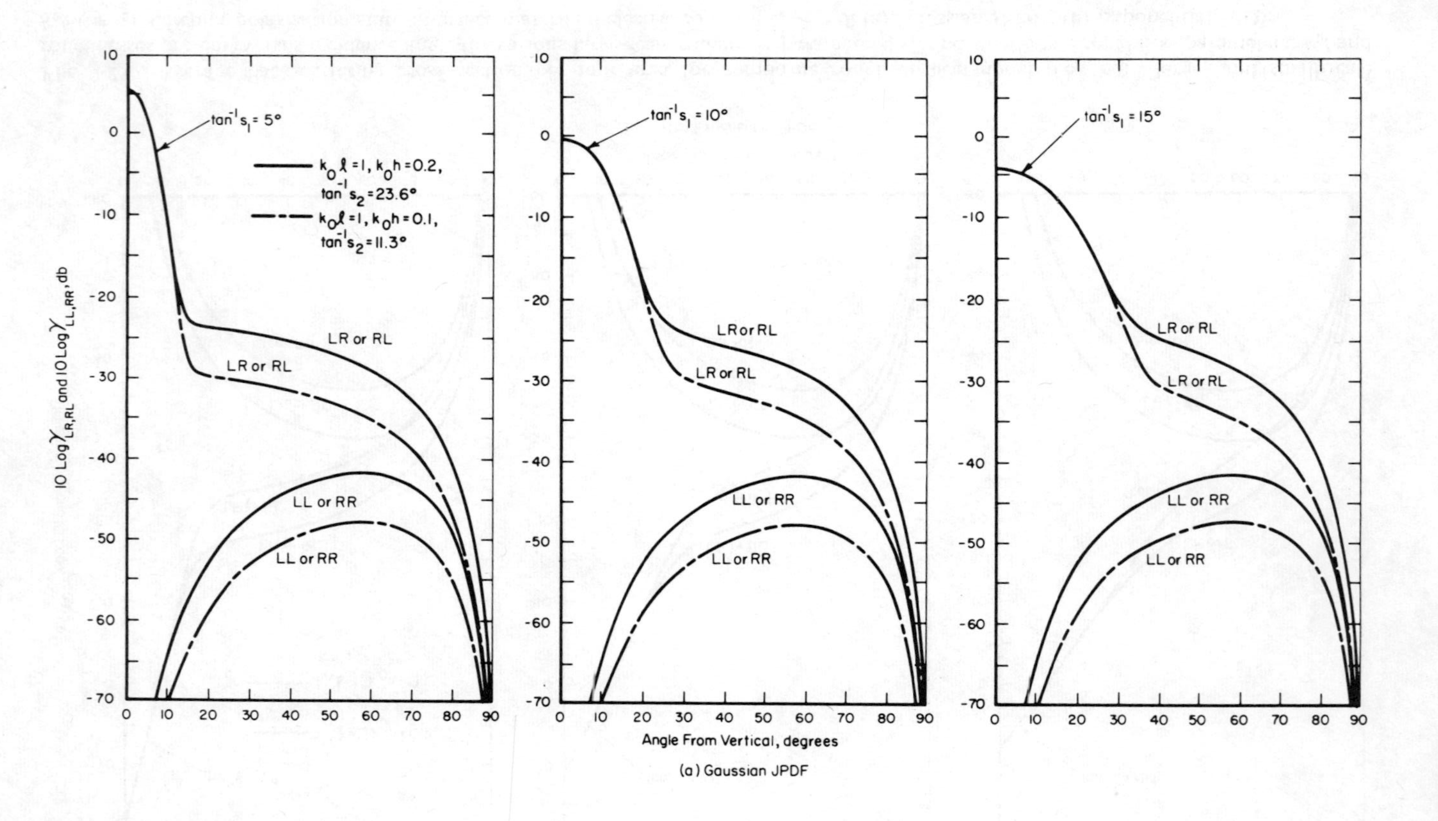

(a) Gaussian JPDF

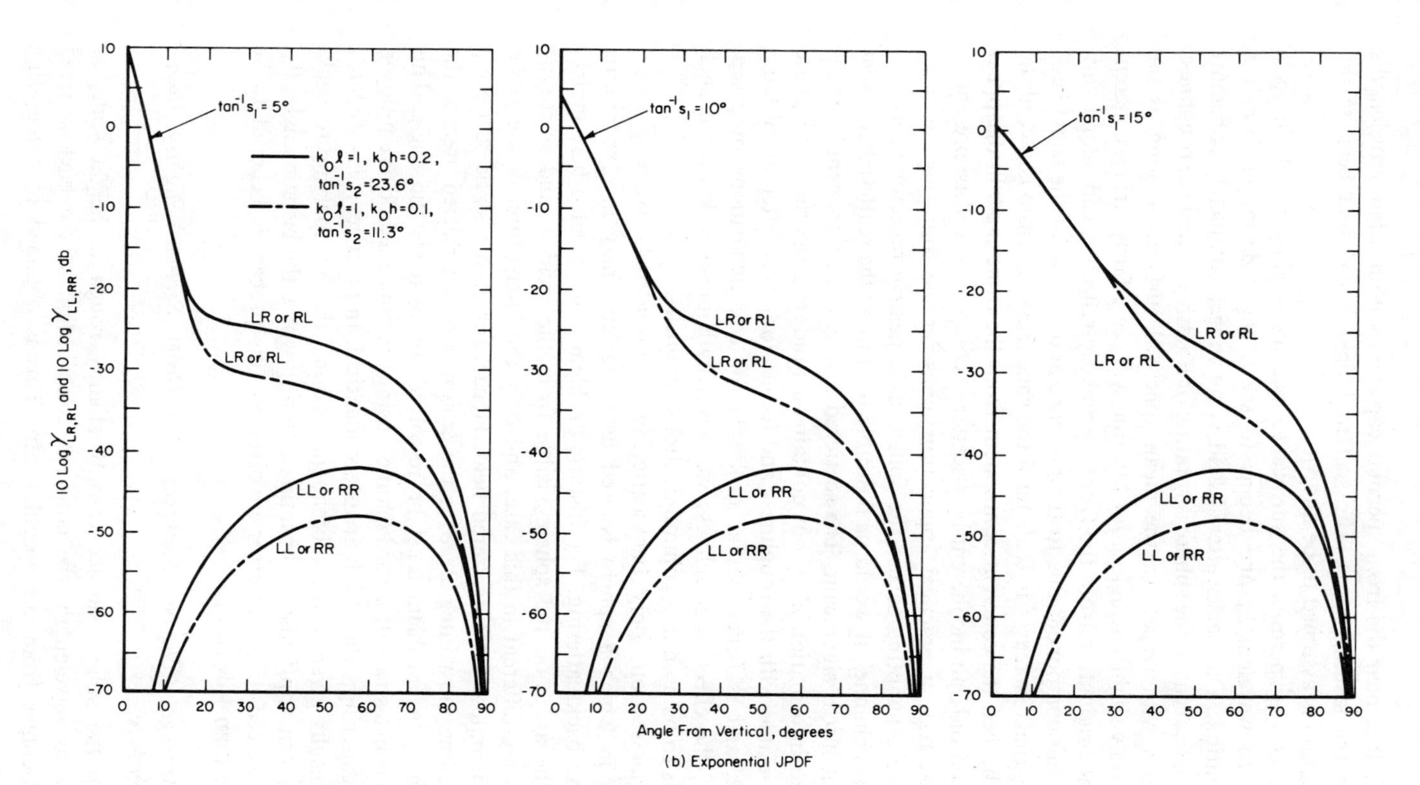

Fig. 9-24. Average backscattering cross section per unit area for composite rough-surface model (i.e., both large- and small-scale roughnesses present) versus incidence angle for various large-scale roughness parameters, s_1, and small-scale roughness parameters, $k_0 l$ and $k_0 h$ (or s_2). Circular polarization states. Surface material dielectric constant $\epsilon_r = 2.0$. (a) Gaussian JPDF; (b) Exponential JPDF.

sphere's surface near the front specular cap (i.e., within a few wavelengths deep from the leading edge). The scattered field is coherent because the sphere location is assumed to be known.

Now, as one permits the surface to become slightly rough in scale compared to wavelength, one begins to see a slight deviation from the above equation. The backscattered field is no longer completely coherent; in most practical cases, the sphere is rotating or librating so that the roughness is in motion with respect to the mean spherical surface. So long as the surface is only slightly rough, however, one will see a fairly strong coherent component originating from the front, specularly-reflecting cap, along with an incoherent component due to the presence and motion of the roughness. If the roughness increases in scale until the roughness dimensions exceed the wavelength, then the coherent component from the front cap will disappear entirely, and only an incoherent backscattered component will be present.

To illustrate a practical application, consider the following example. If one were attempting to use the sphere as a passive reflector in a communications channel, it would be desirable to employ the scattered coherent component if it is significant. The scattered incoherent component contains the desired information also, but contains a random amplitude and phase angle associated with the roughness and its motion. This latter signal fades with time, and is hence not as desirable for communications purposes. Therefore, the effect of roughness vs. wavelength must be known if such a sphere is to be used in a communications channel.

In this section, only backscattering is considered. Many practical situations in which a sphere is used as a scatterer may involve bistatic rather than backscattering. The bistatic problem is not treated here mainly because the areas on the sphere visible from the incident and scattering directions are different in this case, and only the intersection of these two areas contributes to the scattered field. Hence, for each bistatic scattering angle, θ_i, one must integrate over the common or intersected area of the sphere; thus, the problem must be re-solved for each bistatic angle. This can be done in a straightforward manner, using the same method as employed for backscattering, when both areas are identical. In practice, it is estimated that the results given in the subsequent sections, though applying to backscattering, can apply also to bistatic scattering where the bistatic angle, θ_i, does not exceed 45°. The error involved in this process is believed to be no greater than 1 db at $\theta_i = 45°$.

9.1.4.2. *Average Coherent Backscattering Cross Section—Slightly Rough Surface*

When the sphere surface is only slightly rough in height scale, as compared to wavelength, the coherent contribution of the backscattered field still comes from the specular cap. This is obtained by integrating

Eq. (9.1-52) over the illuminated hemisphere. (Actually, the square root of this quantity, i.e., $\langle \hat{\mathbf{p}} \cdot \mathbf{H}^s(P') \rangle$, is integrated, because the phase dependence must be preserved for the coherent component.)

The following assumptions apply to this model:

(a) The tangent-plane approximation is employed, demanding that the radius of curvature at all points be greater than the wavelength. Also neglected in this approximation are shadowing and multiple scattering between different surface points.

(b) The surface height above the mean sphere is Gaussian-distributed.

(c) The mean sphere radius, A_r, is considerably greater than the wavelength.

Under these assumptions, the coherent backscattering cross section of a rough sphere is

$$\sigma = \pi A_R^2 \mid R(0)|^2 \exp[-4k_0^2h^2] \tag{9.1-144}$$

where $R(0)$ is the Fresnel reflection coefficient for normal incidence of the sphere material, and is defined in Eq. (9.1-143). The quantity h, in this case, is the rms deviation of the roughness height about its mean spherical shape.

The above equation is valid whether $k_0^2h^2$ is large or small. It can be seen that as this parameter becomes large (i.e., as roughness increases), the coherent component decreases to zero.

The polarization state of the coherent component as given in the above equation is the same as that backscattered from a smooth sphere; i.e., the coherent component is not depolarized.

9.1.4.3. *Average Incoherent Backscattering Cross Section—Slightly Rough Surface*

In Section 9.1.3.4, the dimensionless average incoherent scattering cross sections per unit illuminated area for a rough planar surface were given. Since those results represented the incoherent scattered power or intensity, one merely multiplies these times an increment of area, and integrates over the illuminated area to obtain the total average scattering cross section. This is possible because scattered fields from various portions of the surface add incoherently, and consequently, it is merely necessary to add (or integrate) the scattered power or intensity.

The illuminated surface in this case is a hemisphere, and the increment of area is $A_R^2 \sin \theta_i \, d\theta_i \, d\phi$, where for backscattering θ_i is the angle of incidence. The expressions for γ^I, given in Section 9.1.3.4, which must be integrated over θ_i are complicated; for this reason the integration is performed numerically, and the results are given below for two polarization situations.

The assumptions implicit in this model are the same as those listed in Section 9.1.3.4.1 with three additions:

(a) The roughness-correlation length is assumed to be considerably less than the sphere radius, A_r.

(b) Sphere radius, A_r, is much larger than wavelength.

(c) The sphere is large and/or lossy enough so that rays entering do not exit again.

9.1.4.3.1. *Linear Polarization States.* When a linearly polarized plane wave is incident upon a rough spherical surface, the local polarization direction for each point around the surface varies between vertical and horizontal. The integration must take this into account.

The first case considered is the one in which the desired received polarization is in the same linear direction as the incident polarization direction. In this case, Eqs. (9.1-112) and (9.1-113) give the necessary relationships. The angle η_i there corresponds to ϕ here, i.e., the azimuth angle. Figures 9-25 and 9-26 show the quantities, $\sigma/(\pi A_R^2 k_0^2 h^2)$ vs. $k_0 l$, for different values of surface dielectric constant, and for both the Gaussian and exponential surface-height correlation coefficients. The other assumptions implicit in this model are given in Section 9.1.3.4.1.

The second case considered is the one in which the desired received polarization direction is perpendicular to the incident polarization direction; this represents the cross-polarized component. The solution is obtained in the same manner, except that now Eqs. (9.1-114) and (9.1-115) are used. The results of the numerical integration are also shown in Figures 9-25 and 9-26 vs. $k_0 l$ for the two surface-height correlation coefficient models. As can be seen, as $k_0 l$ becomes larger, the cross-polarized component becomes smaller because the surface, in effect, is becoming smoother.

9.1.4.3.2. *Circular Polarization States.* When the incident and desired scattered polarization states are circular, the backscattering cross sections per unit illuminated area to be used in the integrands are given in Sections 9.1.3.4.3.2 and 9.1.3.4.3.4. The integrations were performed numerically for the direct and cross-polarized backscattering cross section, and the results for the two surface-height correlation coefficients are shown in Figures 9-27 and 9-28. It is of interest to note that the cross-polarized component for this case is exactly twice that for the crossed linear polarization states.

One can see from these figures the effect of varying the rms surface height, h, or the roughness height correlation length, l, on the incoherent scattering cross sections. Also shown is the dependence upon wavelength (or k_0) and upon the surface dielectric constant, ϵ_r.

9.1.4.4. *Average Total Backscattering Cross Section—Very Rough Surface*

Assume that the surface of the sphere now becomes very rough so that the roughness height dimension, h, is large in terms of wavelength. Then backscattering from the surface appears to come entirely from specular

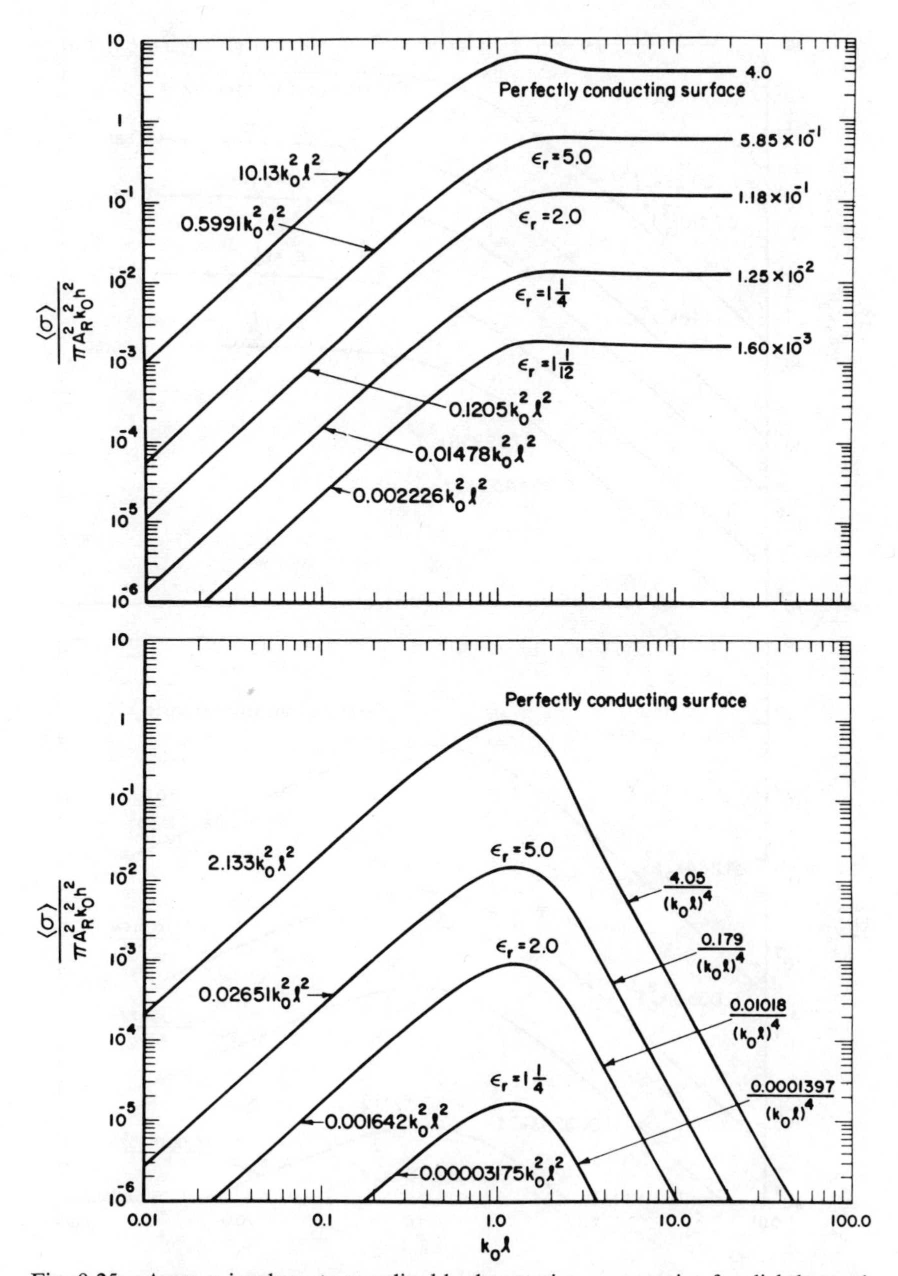

Fig. 9-25. Average incoherent normalized backscattering cross section for slightly rough spherical surface of mean radius A_r as a function of $k_0 l$, where l is surface-roughness correlation length. Cross section is divided by $k_0{}^2h^2$, where h^2 is the mean-square roughness height ($k_0{}^2h^2 < 1$). Linear polarization states. Gaussian JPDF. Same-sense polarizations (top); orthogonal polarizations (bottom).

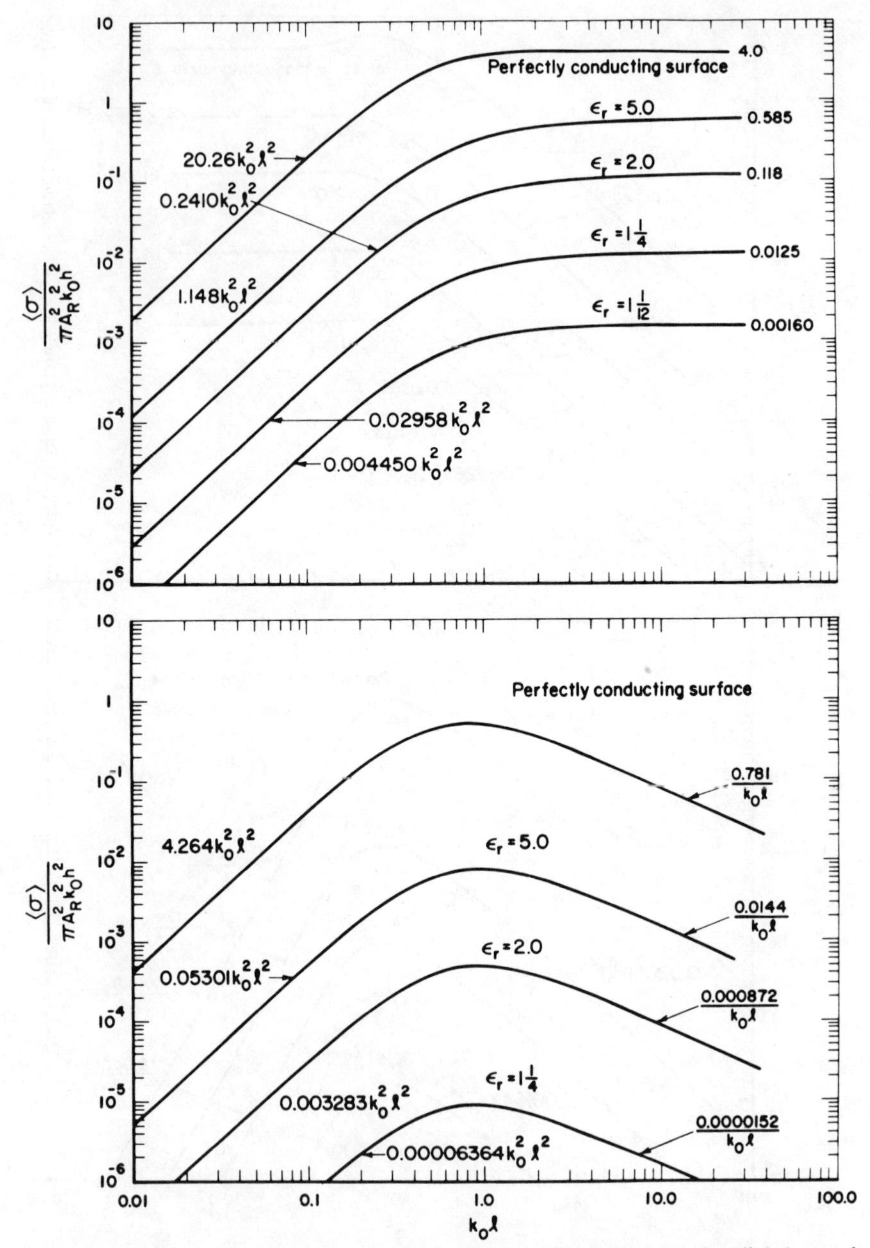

Fig. 9-26. Average incoherent normalized backscattering cross section for slightly rough spherical surface of mean radius A_r as a function of $k_0 l$, where l is surface-roughness correlation length. Cross section is divided by $k_0^2h^2$, where h^2 is the mean-square roughness height ($k_0^2h^2 < 1$). Linear polarization states. Exponential JPDF. Same-sense polarizations (top); orthogonal polarizations (bottom).

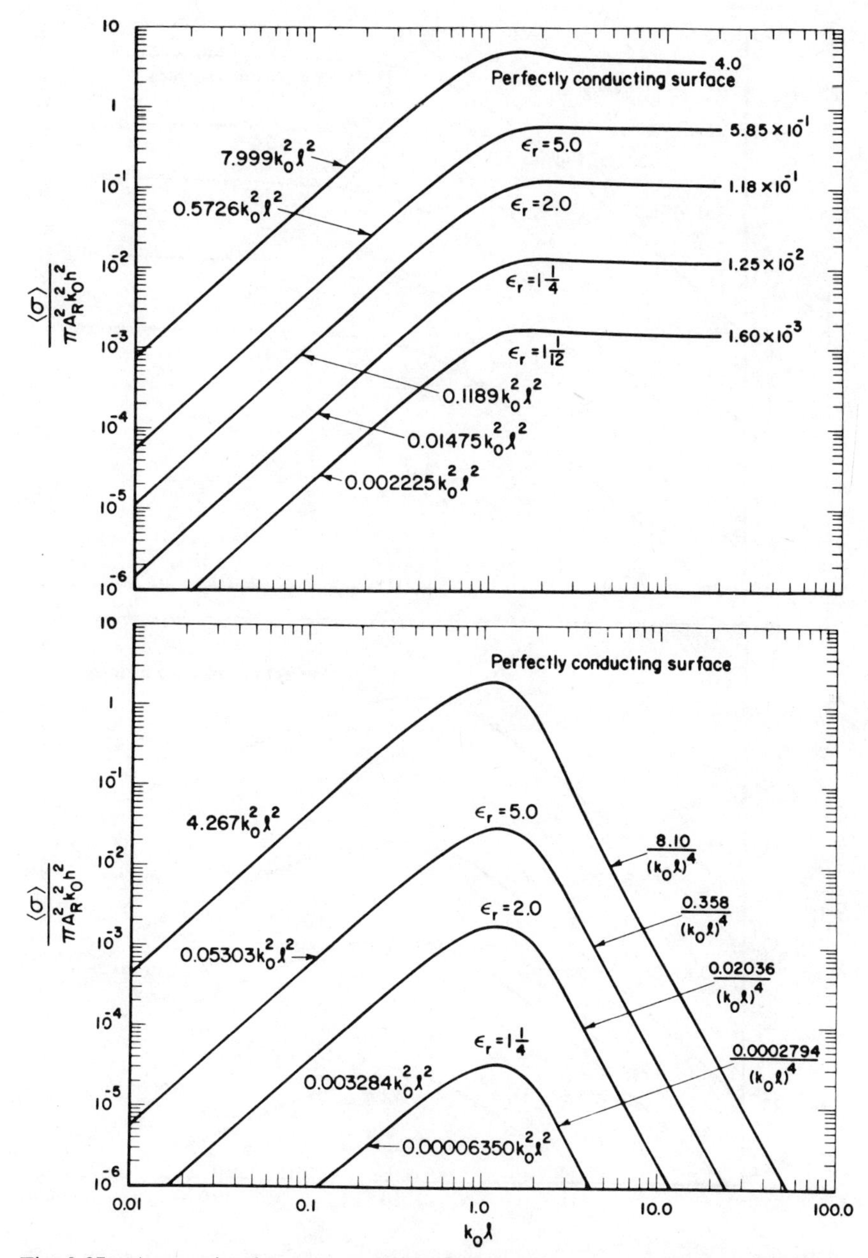

Fig. 9-27. Average incoherent normalized backscattering cross section for slightly rough spherical surface of mean radius A_r as a function of $k_0 l$, where l is surface roughness-correlation length. Cross section is divided by $k_0^2 h^2$, where h^2 is the mean-square roughness height ($k_0^2 h^2 < 1$). Circular polarization states. Gaussian JPDF. Orthogonal polarizations (top); same-sense polarizations (bottom).

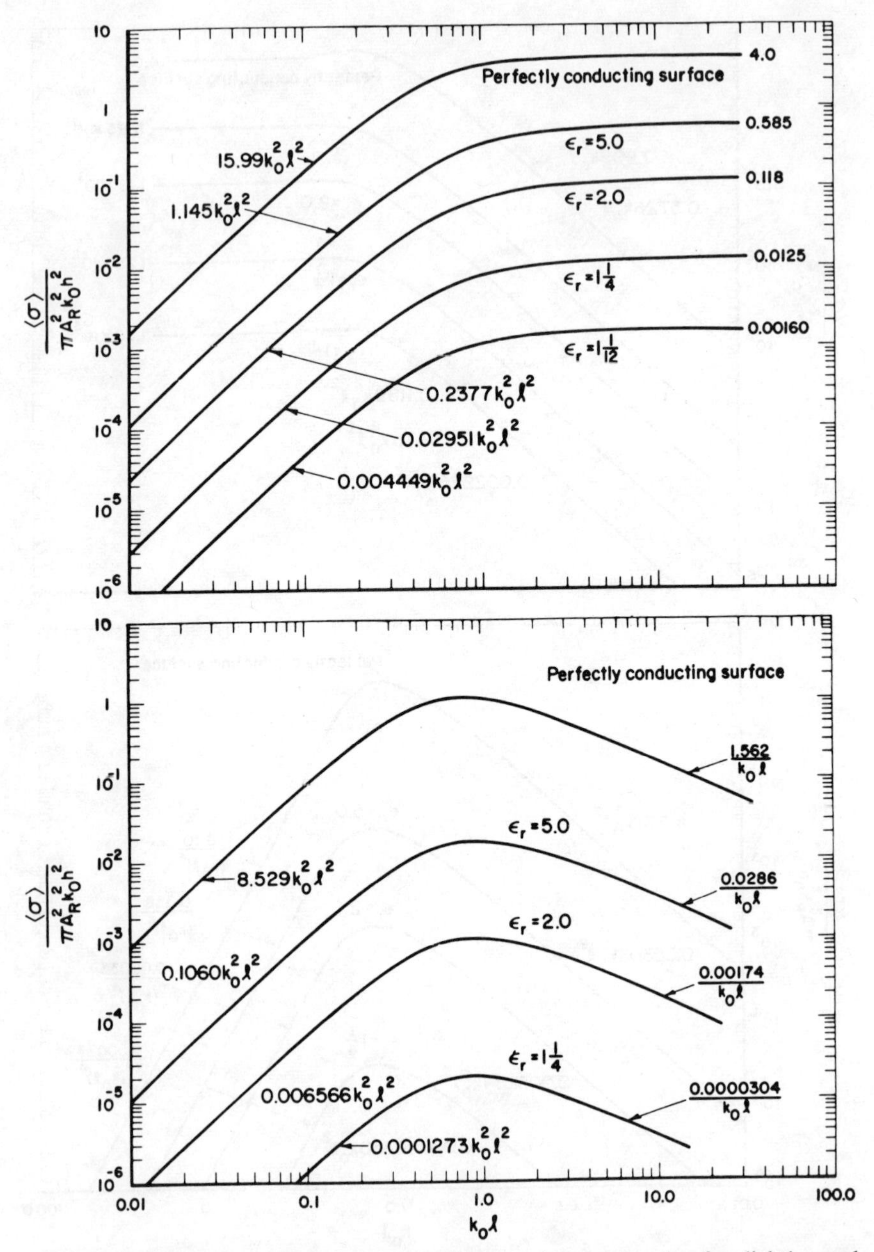

Fig. 9-28. Average incoherent normalized backscattering cross section for slightly rough spherical surface of mean radius A_r as a function of $k_0 l$, where l is surface roughness-correlation length. Cross section is divided by $k_0{}^2h^2$, where h^2 is the mean-square roughness height ($k_0{}^2h^2 < 1$). Circular polarization states. Exponential JPDF. Orthogonal polarizations (top); same-sense polarizations (bottom).

points, i.e., from surface areas oriented with their normals in the direction of incidence. Such specular areas do not depolarize, and the predicted cross-polarized component is always zero when such a specular-point model is employed. The direct polarized scattered component is the same for linear, circular, or any other polarization states.

The average backscattering cross section for such a specular-point model is obtained by integrating the average incoherent backscattering cross section per unit illuminated area over the illuminated hemisphere. This incoherent backscattering cross section per unit area to be used in the integrand is discussed in Section 9.1.3.5 and given in Eqs. (9.1-139) and (9.1-141). Two surface-height probability-density-function models are considered, the Gaussian and the exponential. The equations are numerically integrated. The assumptions implicit in this model are as listed in Section 9.1.3.5.1 with three additions: (a) roughness correlation length, l, is assumed small compared to the sphere radius, A_r ; (b) sphere radius, A_r , is much larger than wavelength; and (c) the sphere is large and/or lossy enough so that rays entering do not exit again.

The resulting backscattering cross sections, as functions of the arctangent of rms roughness slope, s ($s = 2h/l$), are presented in Figure 9-29 for the two probability-density-function models. They are normalized and plotted as $\sigma/[\pi A_R^2 \mid R(0)|^2]$.

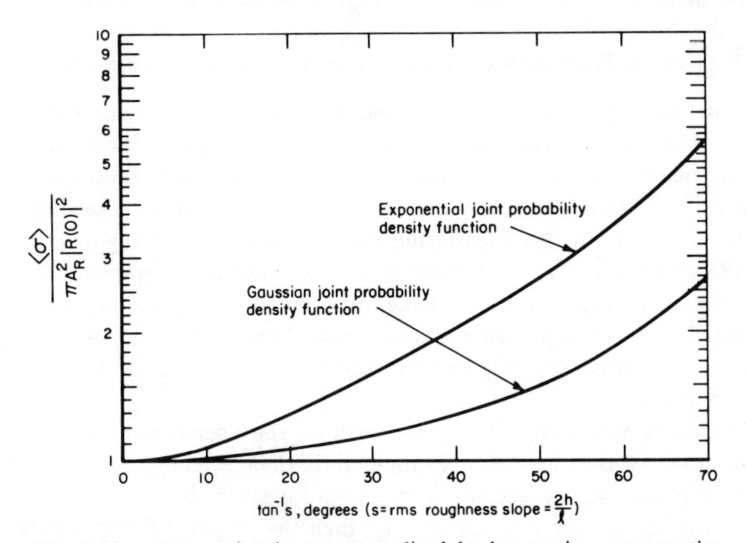

Fig. 9-29. Average incoherent normalized backscattering cross section for a very rough spherical surface of mean radius A_r as a function of $\tan^{-1} s$, where s is the rms roughness slope. Cross section is divided by $\mid R(0) \mid^2$, the Fresnel reflection coefficient for the surface material at normal incidence. Arbitrary polarization states.

It can be seen from this figure that for s very small (i.e., the surface essentially smooth), the backscattering cross sections approach the geometrical limit for a smooth sphere, or

$$\sigma = \pi A_R^2 \mid R(0)|^2 \tag{9.1-145}$$

As roughness increases, the average scattering cross section (becoming incoherent as roughness increases) actually becomes somewhat larger. However, it should be cautioned that the tangent-plane approximation used here is valid only for rms surface slope, s, small compared to unity. Consequently, the results shown in these figures are of questionable validity for $\tan^{-1}s > 45°$. In the region where $\tan^{-1}s < 45°$, one should note that the average scattering cross section does not vary appreciably as a function of roughness slope from its limit for the smooth sphere.

This important conclusion is of interest for planetary scattering. It implies that for planets, surface roughness for all practical purposes can be neglected, and the total average backscattering cross section can be predicted by Eq. (9.1-145). Consequently, knowing the approximate radius of the body, its surface dielectric properties can be approximated from measurement of its backscattering cross section by using the Fresnel coefficient. For example, measurements of $\sigma/\pi A_R^2$ for the moon at $5 \text{ cm} < \lambda_0 < 100 \text{ cm}$ yield approximately 0.07. Such a value implies an effective dielectric constant over these frequencies of approximately $\epsilon_r \approx 2.9$.

9.1.4.5. *Average Total Backscattering Cross Section—Composite Surface*

Most natural rough surfaces consist of roughnesses both larger and smaller than wavelength. By the methods discussed in Section 9.1.3.6, one can merely add the incoherent backscattering cross sections of the preceding two sections to obtain the effective total backscattering cross section. However, the backscattering cross section for the slight-roughness scale (Section 9.1.4.3) is proportional to $k_0^2h^2$, and this parameter is small according to the restrictions of that model. Therefore, only the very large-scale roughness of the preceding section contributes to the directly polarized backscattering cross section, and the effect of the smaller-scale roughness is not apparent.

The cross-polarized terms predicted by the specular-point model of the preceding section are zero. Consequently depolarization is due entirely to the slight-roughness scale. The cross-polarized backscattering cross section for a composite surface is then as given in the figures of Section 9.1.3.4.

Any deviation of the measured values of scattering cross sections from a rough surface from those predicted by the above models is due to the inability of those models to truly represent natural rough surfaces.

9.2. MEASURED RADAR RETURN FROM NATURAL SURFACES

In this section, measured curves and tables of the average backscattering cross section per unit area for several natural surfaces will be shown. These surfaces will include terrain, vegetation-covered surfaces, developed land areas (cities), sea and water surfaces, the lunar surface, and also a rough aluminum surface from which laser-generated, coherent light is scattered. Where possible and meaningful, curves from one of the models discussed previously will be compared with the measured results.

This section will in no way attempt to include all of the radar measurements made from natural surfaces. Only typical curves for the various classes of surfaces will be shown in an attempt to familiarize the reader with the general level and behavior of the power backscattered from these natural surfaces. Finally, a set of tables has been compiled in Section 9.2.6 of representative radar backscatter measurements from the earth and sea surfaces at various frequencies and incidence angles.

9.2.1. Backscattering from Terrain

Probably the most complete, readily available set of measured data on microwave-radar terrain return is found in Reference 2. Several curves from this reference are reproduced below. The normalization employed in this reference for the curves of average backscattering cross section is different from that used here, and the curves given below have been renormalized.

Figure 9-30 shows γ_{vv} and γ_{hh}, as defined in this chapter in Eq. (9.1-1) (i.e., the average backscattering cross section per unit surface area), for an asphalt surface at two frequencies (X- and K_a-band). The mean-square height of the roughness is $h^2 = 1.6 \times 10^{-3}\,\text{cm}^2$, giving values of $k_0 h = 0.085$ and 0.30 at X- and K_a-band frequencies. The measured surface-height correlation coefficient of the surface was approximated by the function $\rho(r) = 1/(1 + 20r^2)^{3/2}$, where r is the separation distance between two surface points measured in centimeters. From this the parameter $k_0 l$, becomes 0.48 and 1.7 at X- and K_a-bands, respectively. The complex dielectric constant for this surface is ϵ_r (X-band) $= 4.3 + i0.1$ and ϵ_r (K_a-band) $= 2.5 + i0.65$.

It is obvious from the above parameters that this surface falls into the category "slightly rough." As such, there is a coherent and an incoherent component in the backscattering cross section. Both are included in the measured curves of Figure 9-30. The coherent component dominates near normal incidence, while the incoherent component is dominant at angles greater than about 25° from normal. Shown in these figures are curves from the slightly-rough-surface model of Section 9.1.3.4 for the incoherent component. The above empirical form for the surface-height correlation coefficient is employed in this model. As seen, the agreement is reasonable.

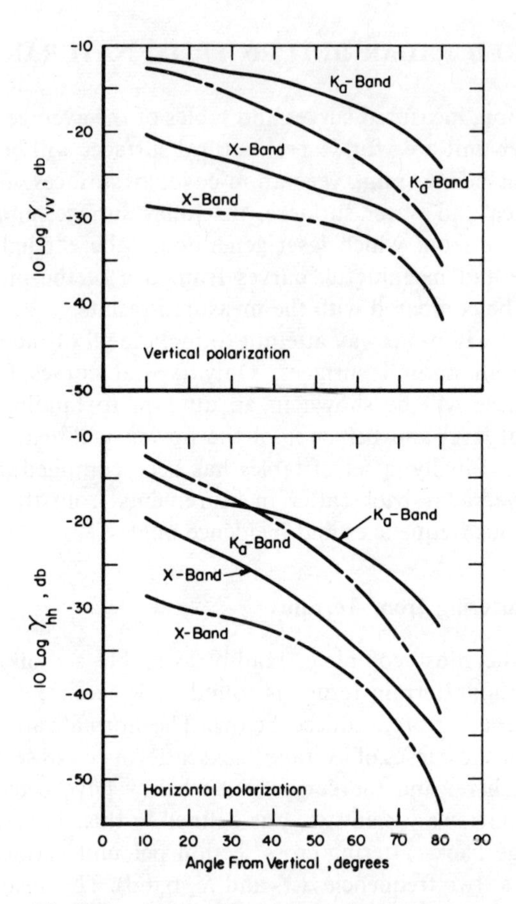

Fig. 9-30. Measured average backscattering cross section per unit area from slightly rough asphalt surface at *X*- and K_a-bands versus incidence angle for linear polarization states. Dashed curve is slightly-rough-surface model, shown for comparison. (a) Vertical polarization; (b) Horizontal polarization [after Cosgriff *et al.*[2]].

Figures 9-31 and 9-32 show the measured backscattering cross sections, γ_{vv} and γ_{hh}, for disked and plowed fields. Here the roughness is considerably greater than that of the asphalt road. Since the exact roughness and dielectric properties of these fields were not measured, no attempt is made to compare theoretical results with these curves. Nonetheless, an intuitive understanding of the scattering behavior can be based upon reasonable estimates of the roughness parameters for these fields.

Figure 9-33 shows the average backscattering cross section per unit surface area for gently sloping, smooth sand, and desert sand, occasionally

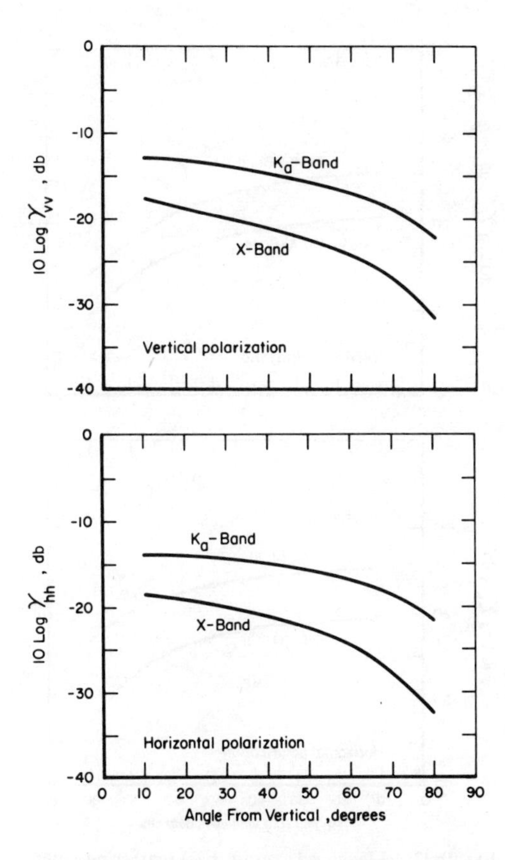

Fig. 9-31. Measured average backscattering cross section per unit area from disked field at X- and K_a-bands versus incidence angle for linear polarization states. (a) Vertical polarization; (b) Horizontal polarization [after Cosgriff *et al.*[2)].

broken by patches of vegetation and rocks.[32] The frequency for the measurements was 10 kMcps (X-band). It was found for the measured data of this figure that no apparent difference existed between returns for vertical or horizontal polarization. The absolute level of the curves is accurate only to within about ± 6 db. No data were available as to the surface roughness details. One can see a strong resemblance between these curves and certain of those of Section 9.1.3.6. From this comparison, one could estimate some of the surface roughness parameters.

Figure 9-34 shows the average backscattering cross section of a built-up or developed area containing homes and other buildings at X-band.[32] Again, the horizontally polarized return was observed to be the same as vertical.

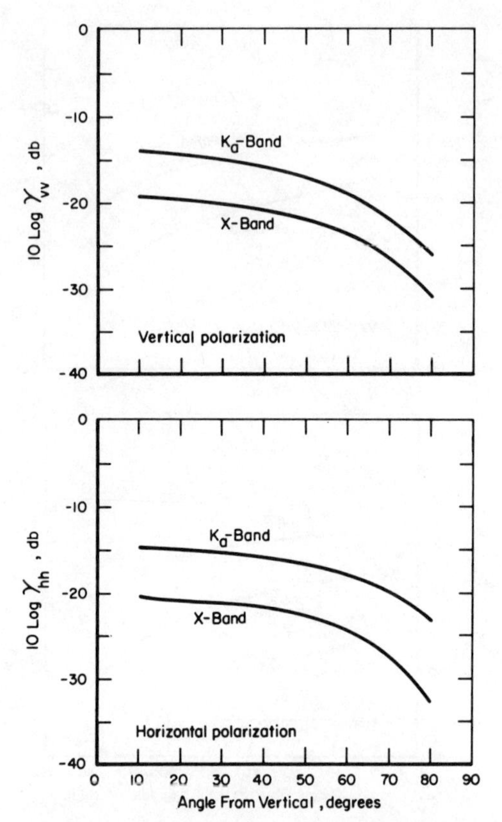

Fig. 9-32. Measured average backscattering cross section per unit area from plowed field at X- and K_a-bands versus incidence angle for linear polarization states. (a) Vertical polarization; (b) Horizontal polarization [after Cosgriff *et al.*[2]].

9.2.2. Backscattering from the Sea and Fresh Water

Figures 9-35 and 9-36 show the measured average backscattering cross section of the sea surface for two different sea states at X-band. Radar return for both vertical and horizontal polarization is shown.[32]

In Figure 9-35, the sea state is relatively calm; the wind is 3 knots, and the wind scale is Beaufort 1. The sea surface still appears smooth with waves about 5 inches high and ripples about $\frac{8}{10}$ inch high.

Figure 9-36 was made when the sea was considerably rougher. Wind was about 30 knots and the wind scale was Beaufort 7. The average waves height was about 4–5 feet and ripples averaged about 3.5 inches high. In both of these curves, the absolute level is not known exactly and may be as much as ± 6 db from that shown.

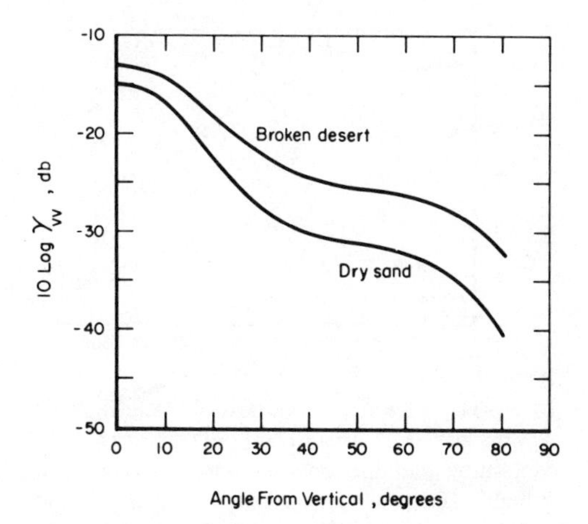

Fig. 9-33. Measured average backscattering cross section per unit area from smooth, gently sloping, dry desert sand, and also for desert broken by patches of vegetation and rocks versus incidence angle at X-band. Returns for vertical and horizontal polarization states were approximately the same.

Figure 9-35, for the smoother sea condition, should be compared to the composite-rough-surface-model curves of Section 9.1.3.6. The waves might be thought of as comprising the large-scale roughness, while the ripples are the slight roughness. One cannot tell exactly where to break the roughness scales in order to use the proper parameters for the model. For example, some of the larger ripples averaging $\frac{8}{10}$ inch give $k_0h = 5$, this parameter value is large enough to place the ripple-type roughness as well as the larger wave-type roughness in the "very rough" category. The latter type of surfaces are normally analyzed by optics approximations. However, there is undoubtedly a smaller ripple structure present also, having $k_0h < 1$. This latter type of slight roughness on water is often referred to as "capillary" waves. Such a slight roughness produces the "tail" of the curve, which is often referred to as the diffuse component. Nonetheless, the resemblance between these measured curves and the sea-surface model curves of Section 9.1.3.6 is apparent.

Figure 9-37 shows the measured average backscattering cross section of the sea surface for different wind conditions (producing different surface roughnesses) and for three different frequencies (X-, K_u-, and K_a-band). Only vertical polarization is considered.[33] From these figures one can obtain an estimate of expected sea return for different wind conditions and wavelengths. The behaviors of the curves as a function of wind speed (hence surface roughness) are as expected intuitively; as the roughness increases,

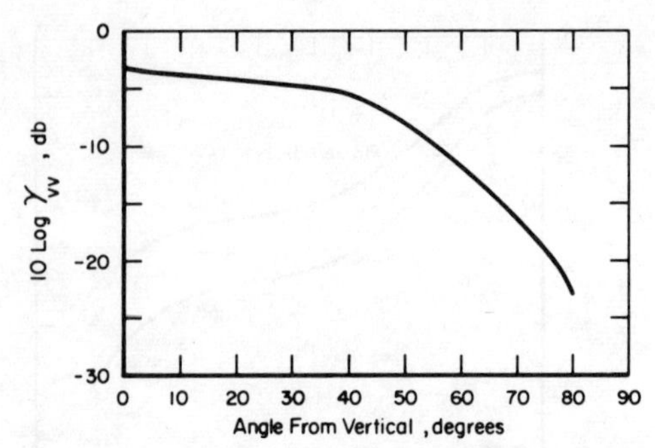

Fig. 9-34. Measured average backscattering cross section per unit area from built-up area (city) containing buildings versus incidence angle at *X*-band. Returns for vertical and horizontal polarization states were approximately the same.

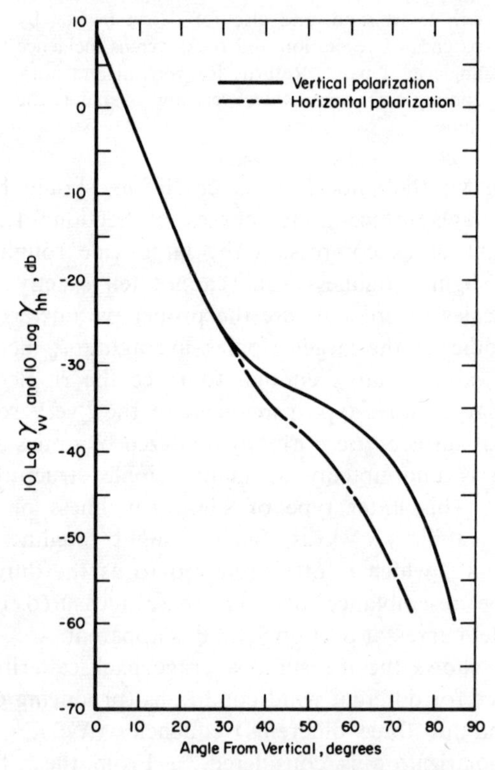

Fig. 9-35. Measured average backscattering cross section per unit area from relatively calm sea surface at *X*-band versus incidence angle for vertical and horizontal polarization states.

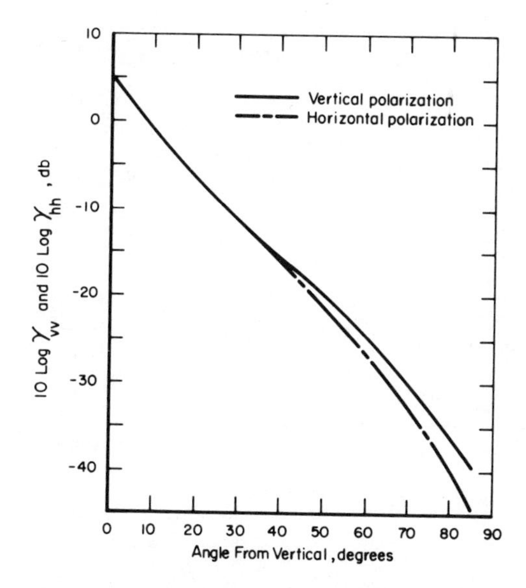

Fig. 9-36. Measured average backscattering cross section per unit area from relatively rough sea surface at X-band versus incidence angle for vertical and horizontal polarization states.

the power backscattered at directions different from normal increases, while that near normal incidence ($\theta_i = 0$) decreases. This behavior is in accordance with the statistical models and explanations of the preceding sections.

Figure 9-38 shows the measured normalized average backscattering cross section for fresh water at X-band.(34)† The return for both the vertical and horizontal polarization states are shown. For this surface, large-sized waves were not present; the measurements were made in a wave tank under controlled conditions. Hence the surface roughness was kept small. The surface under these conditions can be considered slightly rough. The author of Reference 34 estimates $k_0 h = 0.05$ and $k_0 l = 2$ for the water surface being measured. Curves plotted from Eqs. (9.1-74) and (9.1-75) are shown for comparison in Figure 9-36 for $\epsilon_r = 55\,(1 + i0.55)$, which is the dielectric constant of water at X-band. As seen in that figure, the agreement is quite good. Hence the validity of the slightly-rough-surface theory and model of Section 9.1.3.4 appears well established from comparison with measured results.

† The measured results in this case were not plotted in terms of average backscattering cross section per unit area, but in terms of average backscattering cross section in square meters. Since the area of the illuminated surface was not given, the absolute value for γ_{vv} and γ_{hh} cannot be ascertained. Hence, the magnitudes shown are only relative.

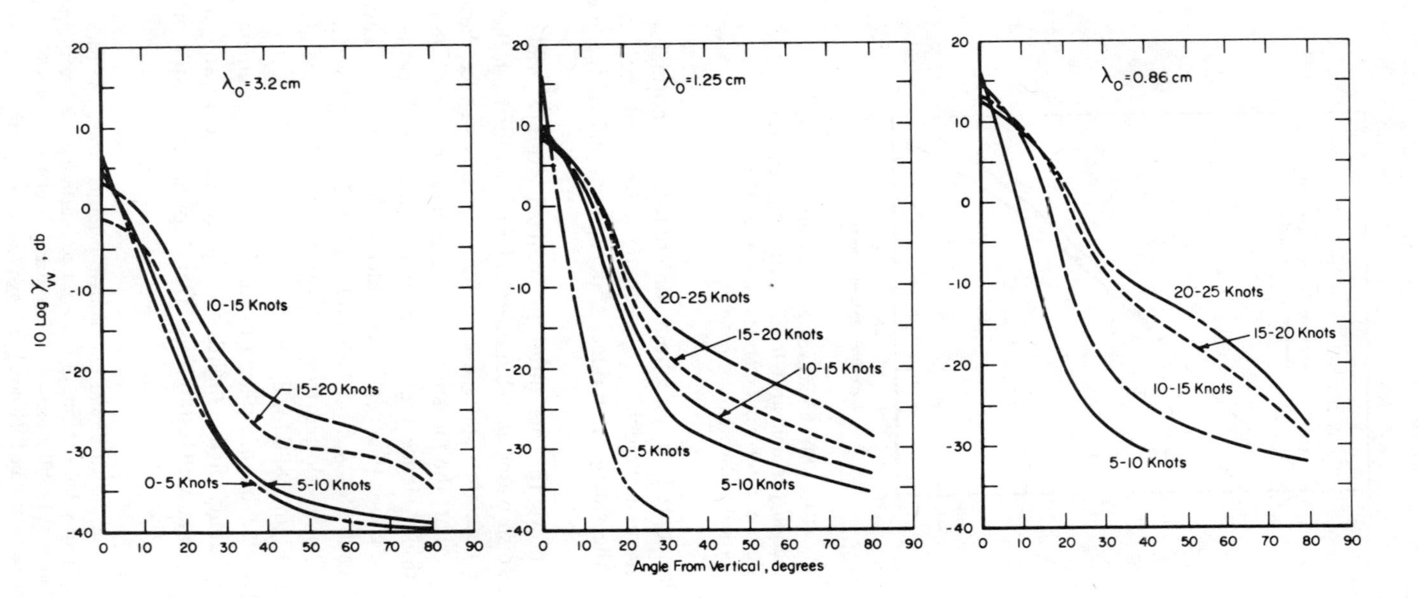

Fig. 9-37. Measured average backscattering cross section per unit area from sea surface for different sea states at three frequencies versus incidence angle; only vertical polarization states were measured [after Grant and Yaplee[33]].

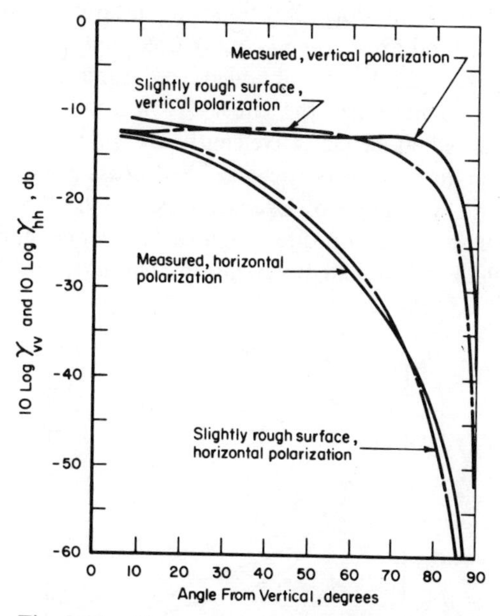

Fig. 9-38. Measured average backscattering cross section per unit area from very slightly rough fresh-water surface at X-band versus incidence angle. Dashed curves represent slightly rough surface model. [Plotted from Eq. (9.1-74) and (9.1-75); measured data adapted from Wright.(34)]

9.2.3. Backscattering from the Lunar Surface

In recent years, much attention has been given to obtaining radar measurements from the lunar and other planetary surfaces. Since the lunar surface is known to be rough, these measurements provide rough-surface radar return which can be compared to similar returns from known terrain samples on earth. Many of the predicted returns, using the previously mentioned statistical models, have been compared with the lunar returns in the hope of determining more information about the moon's surface.

Actually, one can very easily convert measured lunar radar data to the form given here, i.e., average backscattering cross section per unit illuminated surface area as a function of the angle of incidence, θ_i. The measurement employs a radar pulse much shorter in length than the radius of the moon. As this pulse moves past the moon, it illuminates a ring of constant surface area at an angle of incidence which varies in a known manner as the pulse moves along the surface. The measured returned power as a function of time is thus easily converted to backscattering cross section per unit illuminated area as a function of incidence angle. Since the surface of the moon undergoes motion (libration and nutation), this return varies from pulse to pulse, and an average is easily obtained for several pulses.

Most of the measured data obtained so far have employed circular polarization states to minimize the effect of the earth's atmosphere. Two such curves measured at 68 cm wavelength are shown in Figures 9-39(a) and (b). The latter figure shows the measured cross-polarized component. These measurements were made by Evans and Pettingill.[35] Shown along with these measured curves are composite rough surface model curves taken from Figures 9-22 to 9-24 for circular polarization. The latter model curves were

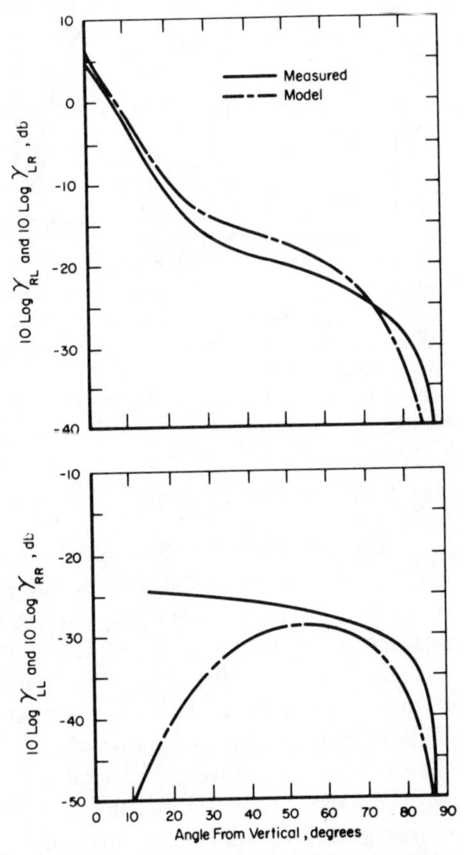

Fig. 9-39. Measured average backscattering cross section per unit area from lunar surface at UHF ($\lambda_0 = 68$ cm) versus incidence angle. Dashed curves represent composite rough-surface model, presented for comparison. Circular polarization states [after Evans and Pettengill[35]].

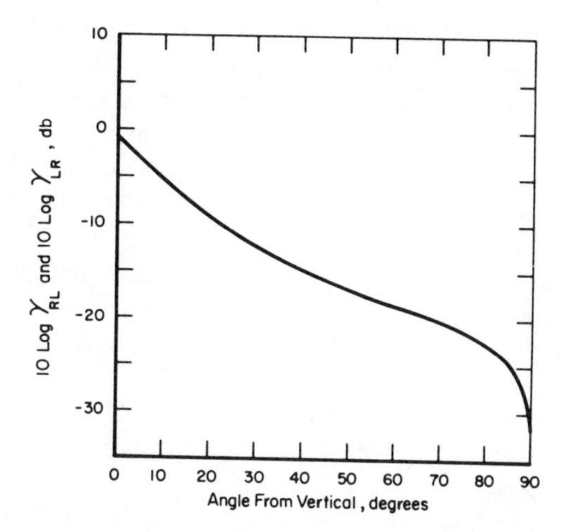

Fig. 9-40. Measured average backscattering cross section per unit area from lunar surface at X-band ($\lambda_0 = 3.6$ cm) versus incidence angle. Circular polarization states with orthogonal states transmitted and received.

selected simply on the basis of best fit, and their parameters are shown in the figures. It should be pointed out that the measured cross-polarized return is probably not accurate near normal incidence ($\theta_i \to 0$), because the polarized component is more than 20 db stronger. The lack of isolation between polarization states of the antennas was more likely responsible for the measured cross-polarized signal, where they differ by large amounts, than the true cross-polarized component from the lunar surface.

To give an indication of the frequency dependence of a lunar-type surface, Figure 9-40 shows the average backscattering cross section per unit area measured at a wavelength of 3.6 cm (X-band). The measurements were made by Evans, and the normalization done in Reference 36. As can be seen by comparison with the composite rough-surface model curve, the effective parameters of the surface are different at higher frequencies because more of the roughness moves into the "very rough" category.

9.2.4. Backscattering at Optical Frequencies

Recently, measurements of backscattered power per unit illuminated area as a function of incidence angle have been made using a laser as the source.[37,38] The surface was aluminum and was optically very rough; most likely the roughness was composite, consisting of both very rough and slightly rough scales. The measured results using vertical and horizontal polarization states for three such surfaces are plotted in the form of average backscattering cross section per unit area and displayed in Figure 9-41. The directly-polarized components are taken from Reference 37, while the cross-polarized terms were obtained from Reference 38. In each case, they had to be renormalized with a $\cos \theta_i$ factor inserted. In addition, the cross-polarized term, referred to in Reference 38 as the depolarized component is renormalized by dividing by two in order to conform to the scattering cross-section definitions used here.

The advantages of obtaining such measurements at optical frequencies rather than at radar frequencies are: (a) the convenience of smaller, more compact equipment size; (b) the ability to manufacture and control the surface roughness to a large extent; and (c) the ability to measure the roughness statistics easily by simple scaling of all dimensions (including wavelength). The results obtained are also meaningful and applicable at microwave frequencies.

It was shown in Reference 37 that the surface-height probability density functions of the three surfaces measured resemble the Gaussian. The laser wavelength of the figures shown here is $\lambda_0 = 0.63 \times 10^{-6}$ m. The three surfaces were machined to have the following large-scale roughness parameters: (a) Surface 1: $k_0 h = 70$, $k_0 l = 1250$, $s = 2h/l = 0.224 =$ rms slope, $\tan^{-1} s = 12.6°$; (b) Surface 2: $k_0 h = 30$, $k_0 l = 550$, $s = 2h/l = 0.218 =$ rms slope, $\tan^{-1} s = 12.3°$; (c) Surface 3: $k_0 h = 10$, $k_0 l = 200$, $s = 2h/l = 0.204 =$ rms slope, $\tan^{-1} s = 11.40°$.

The refractive index of these aluminum surfaces at the above optical frequency is $m = \sqrt{\epsilon_r} = 1.44 + i5.32$; it is evident that aluminum is not a perfect or even a very good conductor at optical frequencies as it is at microwave frequencies.

The surface-roughness statistics measured for these surfaces, and the resulting parameters given above, apply only to the observed largest-scale roughness. Smaller-scale (or slight) roughness, undoubtedly also present, would account for the scattering cross section at angles significantly away from normal incidence.

9.2.5. Backscattering from Vegetation-Covered Surfaces

The last class of natural rough surfaces from which radar return has been obtained is vegetation-covered surfaces. Radar returns from two types of vegetation are shown in Figures 9-42 and 9-43 for vertically and horizontally polarized waves at X- and K_a-band.[2] The first type of vegetation is two-inch green grass (Figure 9-42). This vegetation cover, although seeming to be very thin, produces nearly isotropic scatter, as can be observed from Figure 9-42. It should be noted that even for so thin a cover, very little of the scattered power at K_a-band came from the ground below the grass. This fact was verified with probes. Consequently, the grass blades themselves

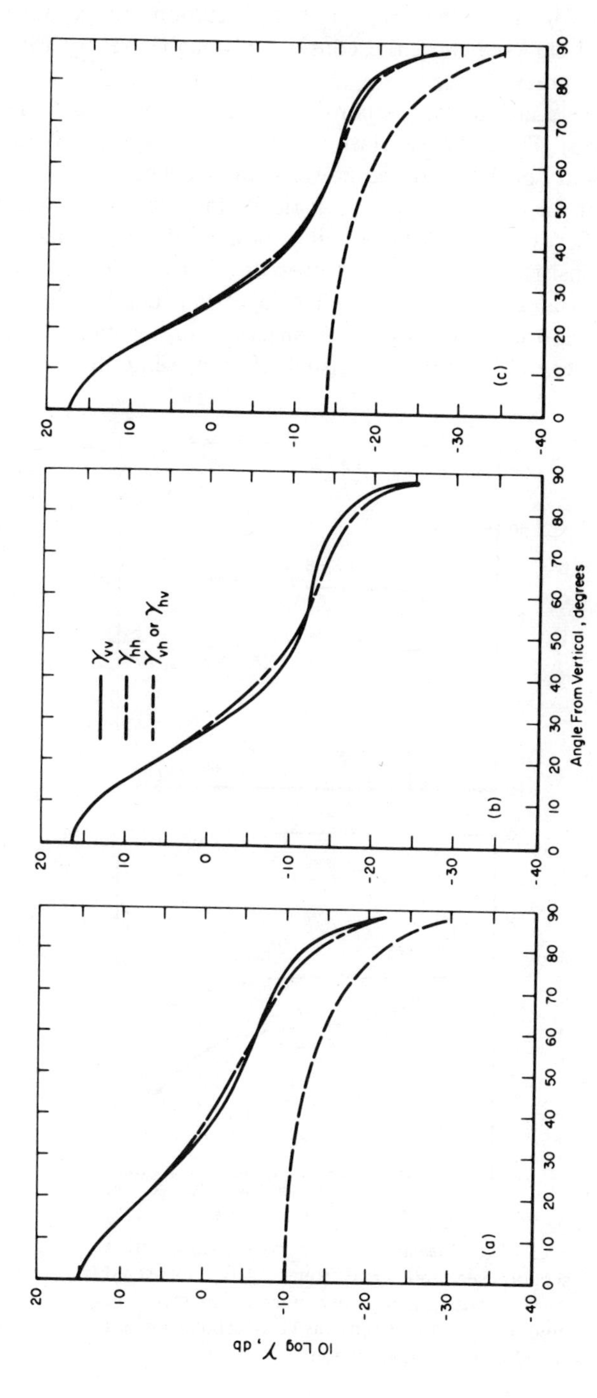

Fig. 9-41. Measured average backscattering cross section per unit area from aluminum surface at optical frequencies ($\lambda_0 = 0.63 \times 10^{-6}$ m). Linear polarization states and different roughness scales (see text for roughness parameters).

were responsible for most of the return. The vegetation in this case consists of nearly vertical stalks; hence, the considerably higher return for vertical polarization near grazing.

The second vegetation cover consists of green soy beans, three feet deep (Figure 9-43). Here the vegetation consists mainly of leaves. Again, the isotropic quality of the scattered power is in evidence.

This class of surfaces cannot be realistically modeled by the statistically rough surface (Section 9.1.3) whose height is a random variable. Instead, the roughness consists of many stalks, leaves, or other scatterers. The incident plane wave penetrates into the medium to a certain depth. The model of Section 9.1.2.2.4 is the only one based on scattering elements resembling a vegetation cover. That model consisted of thin, semi-infinite, dielectric

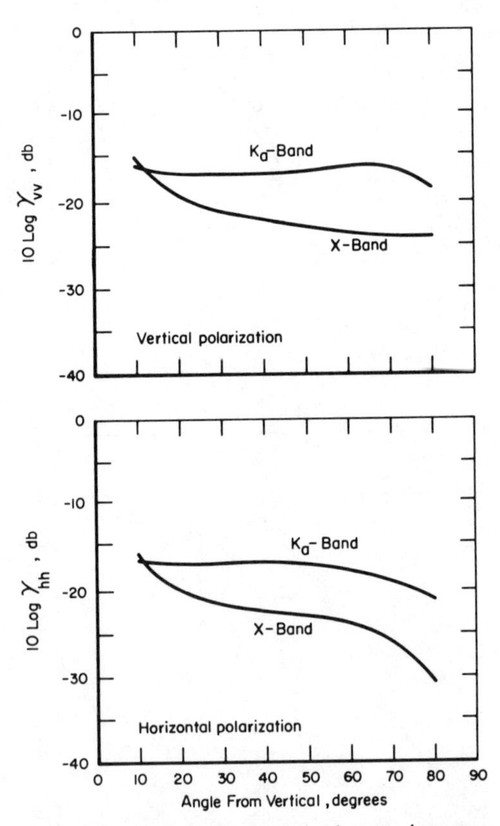

Fig. 9-42. Measured average backscattering cross section per unit area from vegetation-covered surface consisting of 2-inch grass versus incidence angle at X- and K_a-bands. (a) Vertical polarization; (b) Horizontal polarization.

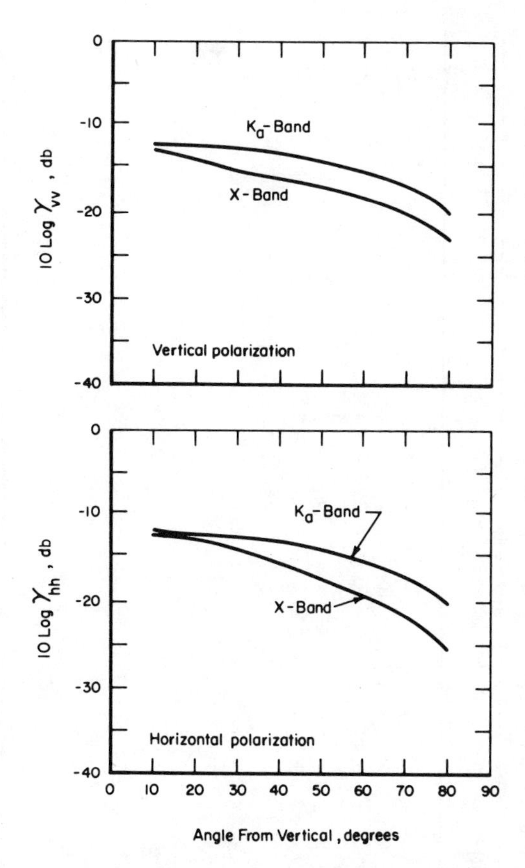

Fig. 9-43. Measured average backscattering cross section per unit area from vegetation-covered surface consisting of 3-feet-deep soybeans at X- and K_a-bands. (a) Vertical polarization; (b) Horizontal polarization.

cylinders (resembling stalks), randomly arranged, but generally preferring vertical orientation (see that section for more details of the model).

9.2.6. Compilation of Representative Radar Backscatter Measurements

It is often necessary for the radar systems engineer to obtain a rough estimate of the radar backscatter (commonly called clutter) from the earth's surface. For this purpose, an assortment of representative measurements is presented in Tables 9-2 and 9-3. The measurements are for various types of terrain, including cities, desert areas, cultivated land, open woods, wooded hills, and small houses. Measurements for the ocean surface at several

Table 9-2

Measurement of Average Radar Cross Section per Unit Surface Area for Various Terrain Types

Frequency	γ (or σ^0) in db for Various Terrain Types and Grazing Angles											
	Desert		Cultivated Land		Open Woods		Wooded Hills		Small Houses		Cities*	
	0-1°	10°	0-1°	10°	0-1°	10°	0-1°	10°	0-1°	10°	0-1°	10°
HF 32.8 MHz		−65 to −55		−75 to −65	−60	−65 to −55	−50 to −30	−60 to −45				−8
UHF 0.5 GHz		−40	−38 to −30	−36		−23	−34 to −24	−22		−23	−30 to −22	−9
S-band 3.0 GHz		−25	−25 to −35	−21	−33	−23	−47 to −32		−35	−23		−15
X-band 9.3 GHz	−38	−26	−36	−20	−30	−23	−36 to −30	−25	−36 to −30	−23	−24	−12
K_a-band 35 GHz		−22	−20	−20	−25	−19	−17	−19				

* The radar cross sections of cities have been observed to vary over several orders of magnitude as a function of the street directions and plane of incidence. At the lower frequencies especially, the building walls are the predominant scatterers, and the right-angled corner between the walls and ground produces a corner reflector effect. Thus, highly regular cities with nearly all streets either parallel or perpendicular to each other show several orders increase in cross section when the plane of incidence coincides with one of the two street directions. The values in the table represent, in general, the *maximum* reliable observed cross section per unit area.

Table 9-3

Measurements of Average Radar Cross Section per Unit Area for Sea Surface

Frequency	Polarization States	γ (or σ^0) in db for Various Grazing (or Depression) Angles					
		0.3°	1.0°	3.0°	10°	30°	60°
Sea State 1 (5 knot wind)							
HF—See Middle Section							
UHF	VV		−70	−60		−38	−23
0.5 GHZ	HH		−84	−70			−22
S-band	VV	−62	−56	−52		−40	−24
3.0 GHz	HH	−74	−65	−59			−25
X-band	VV	−58	−50	−45	−42	−39	−28
9.36 GHz	HH	−66	−51	−48	−51		−26
K_a-band	VV			−41	−38	−37	−26
35 GHz	HH		−40	−43			
Sea State 3 (14 knot wind)							
HF*	VV		−40	−41	−42	−40	−40
32.8 MHz	HH				−47		
UHF	VV		−58	−43	−34	−28	−18
0.5 GHz	HH		−76	−61	−50	−40	−21
S-band	VV	−55	−48	−43	−34	−29	−19
3.0 GHz	HH	−58	−48	−46	−46	−38	
X-band	VV	−45	−39	−38	−32	−28	−17
9.36 GHz	HH	−46	−49	−39	−37	−34	−21
K_a-band	VV		−34	−34	−31	−23	−14
35 GHz	HH		−36	−37	−31		

* HF sea return has not been distinguished as to the sea state. The numbers given here are believed to represent an average state, somewhere between 1 and 3.

Sea State 5 (22 knot wind)

Frequency	Polarization States	0.3°	1.0°	3.0°	10°	30°	60°
HF—See Middle Section							
UHF	VV	−75			−25		
0.5 GHz	HH		−65	−53	−46		
S-band	VV	−50	−38	−35	−28		
3.0 GHz	HH	−44	−42	−37	−38		
X-band	VV	−39	−33	−31	−26	−20	−10
9.36 GHz	HH	−39	−33	−32	−31	−24	−12
K_a-band	VV		−31	−30	−26	−20	−4
35 GHz	HH				−27	−20	

sea states are also included. The data are presented at a variety of frequencies from *HF* to K_a band and at several grazing (or depression) angles.

The data compiled in these tables were gathered from many sources, some of which have been referred to in previous paragraphs. The principal sources are an excellent collection by Nathanson[39] appearing in a report on radar clutter by Barrick,[40] a data compilation by the University of Michigan,[41] and a review by Hagn of *HF* backscatter.[42] Several techniques were employed in making the measurements, corresponding to the frequency region and the equipment available to the individual investigator. Consequently, single entries in the table have been found to very by as much as 5 db from those measured elsewhere. When measurements were available from more than one source, an average was used for the table. Where no measurements could be found for a particular frequency, gaps were left in the tables.

In many cases, investigators presented their measurements normalized differently from that employed here. Thus, it was necessary to renormalize those data. The parameter presented is the average backscattering cross section per unit surface area, γ, in decibels. In the rough-surface literature, this is sometimes referred to as σ^0 or σ_0 .

Table 9-2 covers terrain return for several types of surfaces. Few data were found except for near-grazing incidence. Many investigators who examined the polarization dependence of the return observed that there was no significant difference between the results for vertical and horizontal polarizations, and hence, only one number was reported. In some cases, the investigators averaged their returns for both polarizations. Hence, there are not enough data available to include the polarization dependence in the table.

Table 9-3 covers sea return for three surface conditions (or sea states). Since many investigators were able to obtain data at higher incidence angles, this broader parameter range is included in the table. Data are given in the table for both horizontal and vertical polarization states when available.

9.3. REFERENCES

1. Goldstein, H., Sea Echo. *In* Kerr, D. E. (Ed.), *Propagation of Short Radio Waves*, Vol. 13, MIT Radiation Laboratory Series, McGraw-Hill, New York (1957); also Dover, New York (1965).
2. Cosgriff, R. L., Peake, W. H., and Taylor, R. C., *Terrain Scattering Properties for Sensor System Design* (Terrain Handbook II), Ohio State University, Engineering Experiment Station Bulletin No. 181 (May 1960).
3. Barrick, D. E., A More Exact Theory of Backscattering from Statistically Rough Surfaces, Antenna Laboratory, Ohio State University, Report No. 1388-18, (1965).
4. Peake, W. H., Scattering from Rough Surfaces Such as Terrain. *In: Antenna and Scattering Theory: Recent Advances*, Short Course Notes, The Ohio State University (August 1965).

5. Barrick, D. E., and Peake, W. H., Scattering from Surfaces with Different Roughness Scales: Analysis and Interpretation, Battelle Memorial Institute, Report No. BAT-197-A-10-3 (November 1967), AD 662 751.
6. Barrick, D. E., Normalization of Bistatic Radar Return, Antenna Laboratory, Ohio State University, Report No. 1388-13 (January 1964).
7. Clapp, R. E., A Theoretical and Experimental Study of Radar Ground Return, Radiation Laboratory, Massachusetts Institute of Technology, Report No. 1024 (1946).
8. Hughes, V. A., Radio Wave Scattering from the Lunar Surface, *Proc. Phys. Soc.* **73**:988 (1961).
9. Evans, J. V., and Pettengill, G. H., Scattering Behavior of the Moon at Wavelengths of 3.6, 68, and 784 Centimeters, *J. Geophys. Res.* **68**:423 (1963).
10. Rashid, A., A New Aspect Function to Study Radar Signal Return from the Ground, *IEEE Trans.* **AP-13**:969 (1965).
11. Spetner, L. M., and Katz, I., Two Statistical Models for Radar Terrain Return, *IRE Trans. Ant. Prop.* **AP-8**:242 (1960).
12. Twersky, V., On Multiple Scattering and Reflection of Waves by Rough Surfaces, *IRE Trans. Ant. Prop.* **5**:81 (1957).
13. Twersky, V., On Multiple Scattering and Reflection of Waves by Rough Surfaces, Sylvania Electronic Defense Laboratory, Report No. EDL-E9 (1955).
14. Goldstein, H., Sea Echo. *In* Kerr, D. E., (Ed.), *Propagation of Short Radio Waves*, Vol. 13, MIT Radiation Laboratory Series, McGraw-Hill, New York (1957); also Dover, New York (1965).
15. Macfarlane, G. G., Sea Returns and the Detection of Schnorkel, Telecommunications Research Establishment, Report No. T1787 (February 1945).
16. *Handbook of Mathematical Functions with Formulas, Graphs, and Mathematical Tables*, Abramowitz, M., and Stegun, I. A. (Eds.), U.S. Department of Commerce (March 1965).
17. Peake, W. H., The Interaction of Electromagnetic Waves with Some Natural Surfaces, Antenna Laboratory, Ohio State University, Report No. 898-2 (1959).
18. Peake, W. H., Theory of Radar Return from Terrain, *IRE Internat. Conv. Record* **7**, Part 1:27 (1959).
19. Barrick, D. E., and Peake, W. H., A Review of Scattering from Surfaces with Different Roughness Scales, *Radio Science* **3** (New Series):865 (1968).
20. Ament, W. S., Toward a Theory of Reflection by a Rough Surface, *Proc. IRE* **41**:142 (1953).
21. Davies, H., The Reflection of Electromagnetic Waves from a Rough Surface, *Proc. IEE (GB)* **101 (Part IV)**:209 (1954).
22. Rice, S. O., Reflection of Electromagnetic Waves by Slightly Rough Surfaces. *In* Kline, M. *The Theory of Electromagnetic Waves*, John Wiley (Interscience), New York (1951); also Dover, New York (1963).
23. Beckmann, P., and Spizzichino, A., *The Scattering of Electromagnetic Waves from Rough Surfaces*, Macmillan, New York (1963).
24. Hagfors, T., Backscattering from an Undulating Surface with Applications to Radar Returns from the Moon, *J. Geophys. Res.* **69**:3779 (1964).
25. Hagfors, T., Scattering and Transmission of Electromagnetic Waves at a Statistically Rough Boundary Between Two Dielectric Media. *In* J. Brown (Ed.), *Electromagnetic Wave Theory*, Part 2:997, Pergamon Press, New York (1967).
26. Barrick, D. E., Relationship Between Slope Probability Density Function and the Physical Optics Integral in Rough Surface Scattering, *Proc. IEEE* **56**:1728 (1968).
27. Muhleman, D. O., Radar Scattering from Venus and the Moon, *Astron. J.* **69**:34 (1964).
28. Barrick, D. E., Rough Surface Scattering Based on the Specular Point Theory, *IEEE Trans.* **AP-16**:449 (1968).
29. Semenov, B. I., Scattering of Electromagnetic Waves from Restricted Portions of Rough Surfaces with Finite Conductivity, *Radiotekhnika i Elektronika* **10**:1952 (1965).
30. Semenov, B. I., An Approximate Calculation of Scattering of Electromagnetic Waves from a Slightly Rough Surface, *Radiotekhnika i Elektronika* **11**:1351 (1966).

31. Fuks, I. M., Contribution to the Theory of Radio Wave Scattering on the Perturbed Sea Surface, *Izvestia Vysshikh Uchebnykh Zavedenii Radiofizika* **5**:876 (1966).
32. Wilmot, T. A. W. (unpublished work).
33. Grant, C. R., and Yaplee, B. S., Backscattering from Water and Land at Centimeter and Millimeter Wavelengths, *Proc. IRE* **45**:976 (1957).
34. Wright, J. W., Backscattering from Capillary Waves with Application to Sea Clutter, *IEEE Trans. Ant. Prop.* **AP-14**:749 (1966).
35. Evans, J. V., and Pettengill, G. H., The Scattering Behavior of the Moon at Wavelengths of 3.6, 68, and 784 Centimeters, *J. Geophys. Res.* **68**:423 (1963).
36. Ott, R. H., A Theoretical Model for Scattering from Rough Surfaces with Applications to the Moon and Sea, Antenna Laboratory, Ohio State University, Report No. 1388-1 (November 1961).
37. Renau, J., and Collinson, J. A., Measurements of Electromagnetic Backscattering from Known, Rough Surfaces, *Bell Syst. Tech. J.* **44**:2203 (1965).
38. Renau, J., Cheo, P., and Cooper, H., Depolarization of Linearly Polarized Electromagnetic Waves Backscattered from Rough Metals and Inhomogeneous Dielectrics, *J. Opt. Soc. Am.* **57**:459 (1967).
39. Nathanson, F. E., *Radar and Clutter*, Short Course Notes, University of Alabama in Huntsville (1967); also, *Radar Design Principles—Signal Processing and the Environment*, McGraw-Hill, New York (to be published).
40. Barrick, D. E., Radar Clutter in an Air Defense System. Part 1: Clutter Physics, Battelle Memorial Institute, Report No. RSIC-798 (January 1968).
41. Earing, D. G., and Smith, J. A., Target Signatures Analysis Center: Data Compilation, Institute of Science and Technology, University of Michigan, Report No. 7850-2-B (July 1966).
42. Hagn, G. H., The Direct Backscatter of High-Frequency Radio Waves from Land, Sea-Water, and Ice Surfaces, *IEEE Internat. Conv. Record* **15**, Part 2:150 (1967).

Chapter 10

IONIZED REGIONS

William D. Stuart

10.1. INTRODUCTION

This chapter discusses the radar cross section of ionized media or plasmas.[†] There are several excellent texts, among them References 1–18, which give derivations of the basic constitutive relations for plasmas. It is the purpose of Section 10.2 to define terms, to list in standard notation some of the more useful equations, and to interpret their graphical results. No attempt has been made in this chapter to assign historical priority to various formulations. Quasineutrality is assumed, i.e., the magnitude of the difference between the number of positive particles and the number of negative particles is very small compared to the number of atoms. The velocities of the particles are considered small with respect to the speed of light. This means, among other things, that the Luxemburg effect is not considered here. We consider only the weak-signal case. The general theory of anisotropic and gyrotropic media is not presented except for very brief discussions where it is especially pertinent. Cold plasmas (collision-free ionized gases) are discussed only briefly, since this type of plasma rarely occurs.

The basic constitutive relationships in an ionized gas are defined in the following section. The effective dielectric constant, conductivity, and other parameters of an ionized media are discussed. Plasma frequency, collision frequency, temperature, and other terms are defined and their effects discussed. The average-particle or hydrodynamic model and the kinetic or distribution-function model are presented and briefly examined. These relationships are applied to ionized regions which form planar surfaces, cylinders, spheres, and other shapes. Sections 10.2.2.1, 10.2.2.2, and 10.2.2.3 draw heavily upon the previous material of Chapters 7, 4, and 3, respectively. Section 10.2.2.4, which covers miscellaneous shapes, draws upon Chapters 5 and 6. Section 10.2.3 briefly discusses multiple scattering, including scattering from turbulent media and incoherent scatter from electrons in the presence of a magnetic field.

In Section 10.3, the results of Section 10.2 are applied to various radar targets of current interest, such as meteors, reentry wakes, rocket plumes,

[†] In this chapter, the word plasma is considered synonymous with ionized gas, regardless of the degree of ionization. Solid-state plasmas are not discussed.

rocket-launch effects on the ionosphere, satellite-induced ionospheric effects, auroras, nuclear explosions, and chemical seeding of the ionosphere. Many related subjects such as radar scattering from the sun, comet trails, other extraterrestrial plasmas, and natural ionospheric reflections must, unfortunately, be neglected. In Section 10.2.1.1.2, there is a brief discussion of ionospheric effects on the scattering matrix of a high-altitude target. This chapter also includes some brief discussion of how to discriminate between natural disturbances and man's perturbations of the ionosphere.

Plasma diagnostics has recently been the subject of two excellent books[18,19]; therefore this subject has not been covered here.

10.2. THEORY

The general material in this section is based largely on the following sources; the books of Ginzburg,[1] Ratcliffe,[2] Budden,[3] Brandstatter,[4] and Tischer,[5] and the many papers and reports of Bachynski and coworkers.[20–33]

10.2.1. Constitutive Relations

10.2.1.1. *Macroscopic Approach*

10.2.1.1.1. *Isotropic Media*

10.2.1.1.1.1. *Collision-Free Ionized Gas.* The field of electrical discharges in gases is very old. During the years 1816–1819, Faraday tried to find the properties of matter by extrapolating the changes of solid–liquid–gas to a fourth state. Sir William Crookes declared, in 1879, that he had found a fourth state, radiating matter, in his discharge tubes. Langmuir, in the early 1930's, began using the designation "plasma." Descriptively, the term simply denotes ionized gases. Depending on the degree of ionization, such gases exhibit similarities to metals, semiconductors, strong electrolytes, and ordinary gases. Assuming a quasineutral plasma, or

$$| N_+ - N_- | \ll N \tag{10.2-1}$$

and a plane electromagnetic wave propagating in the positive z direction and polarized in the x direction with the form

$$\mathbf{E}^i = \hat{\mathbf{x}} E_0 \exp(-i\omega t + ikz) \tag{10.2-2}$$

then k is the propagation constant and can be written as

$$ik = \alpha + i\beta \tag{10.2-3}$$

For free space, α is zero; thus $k_0 = \beta_0$. For an ionized gas so rarefied that collisions can be neglected, the force exerted on a charged particle is given by

$$\mathbf{F} = q(\mathbf{E} + \mathbf{v} \times \mathbf{B}) \tag{10.2-4}$$

where q is the charge of the particle, $\mathbf{v}$ is the velocity of the particle, and $\mathbf{B}$ is the magnetic flux density. In Eq. (10.2-4), any magnetization of the medium produced by the electric field and any polarization of the medium produced by the magnetic field is neglected. From Maxwell's equations and Eq. (10.2-2), the magnetic flux density is

$$\mathbf{B}^i = \hat{\mathbf{y}}(E_0/c) \exp(-i\omega t + ikz + i\pi/2) \tag{10.2-5}$$

and substitution of Eq. (10.2-5) into (10.2-4) yields

$$\mathbf{F} = qE_0 \exp(-i\omega t + ikz) [\hat{\mathbf{x}} + i(\mathbf{v}/c) \times \hat{\mathbf{y}}] + q(\mathbf{v} \times \mathbf{B}_{\text{ext}}) \tag{10.2-6}$$

where $\mathbf{B}_{\text{ext}}$ is any magnetic flux density generated external to the propagating electromagnetic wave. If the velocities of the particles in the ionized gas are small compared to the speed of light, the $\mathbf{v}/c$ term can be neglected. All the above assumptions upon the effects of $\mathbf{E}^i$, $\mathbf{B}^i$, and $\mathbf{v} \times \mathbf{B}^i$ constitute the "weak-signal case."

If for the present $\mathbf{B}_{\text{ext}}$ is assumed zero, the acceleration of the ionized particle is

$$\partial \mathbf{v}/\partial t = (q/m_q)\, \mathbf{E} \tag{10.2-7}$$

where m_q is the mass of the particle.

Since the charge to mass ratio of the electron is very much greater than that of the ion, it is common to assume for many applications that the ions are unaccelerated by the field, and that the ions therefore remain at their thermal velocities. The q/m_q in Eq. (10.2-7) then becomes e/m_e (1.759×10^{-11}), the charge to mass ratio of the electron.

Integrating Eq. (10.2-7), one obtains

$$\mathbf{v} = \int_0^t \mathbf{E}(e/m_e)\, dt = ie\mathbf{E}/\omega m_e \tag{10.2-8}$$

The exclusion of any initial velocity term in this equation has eliminated the discrete-particle nature of the ionized gas and tacitly treated it as a continuous medium. This is known as the macroscopic or hydrodynamical approach. The result is an induced current $\mathbf{J}$ in the ionized gas which oscillates harmonically with the same direction as the incident electric field. This current is given by

$$\mathbf{J} = N_e e\mathbf{v} = +i(N_e e^2/\omega m_e)\, \mathbf{E} = \sigma_c \mathbf{E} \tag{10.2-9}$$

where N_e is the number density of electrons and σ_c denotes the electrical conductivity of the plasma. It is seen that the conductivity is purely imaginary in this case. Substituting Eq. (10.2-9) into Maxwell's curl **H** equation results in

$$\nabla \times \mathbf{H} = \frac{iN_e e^2}{\omega m_e} \mathbf{E} - (i\omega\epsilon)\,\mathbf{E} \tag{10.2-10}$$

$$= -i\omega\epsilon_0 \left[\frac{\epsilon}{\epsilon_0} - \frac{N_e e^2}{\omega^2 m_e \epsilon_0}\right] \mathbf{E}$$

$$= -i\omega\epsilon_0\epsilon_r \mathbf{E}$$

Thus the effective macroscopic relative dielectric constant of a collisionless plasma is given by

$$\epsilon_r = \frac{\epsilon}{\epsilon_0} - \frac{N_e e^2}{\omega^2 m_e \epsilon_0} \tag{10.2-11}$$

or

$$\epsilon_r = \frac{\epsilon}{\epsilon_0} - \frac{\sigma_c}{i\omega\epsilon_0} \tag{10.2-12}$$

For most cases of practical interest, the medium with the free electrons has a permittivity of ϵ_0. Thus, from Eq. (10.2-11), the relative dielectric constant for a collision-free ionized gas with no external magnetic field is given by

$$\epsilon_r = 1 - \frac{N_e e^2}{\omega^2 m_e \epsilon_0} \approx 1 - \frac{81 N_e}{f^2} \tag{10.2-13}$$

where f is the frequency in cps and N_e has dimensions of electrons per cubic meter. The relative permittivity can be positive, zero, or negative. The frequency f_p such that $\epsilon_r = 0$ is called the plasma frequency. The angular plasma frequency, $2\pi f_p$, is denoted ω_p.

Setting $\epsilon_r = 0$ in Eq. (10.2-13), one obtains

$$\omega_p = e\sqrt{N_e/m_e\epsilon_0} \approx 56.458\sqrt{N_e} \tag{10.2-14}$$

or

$$f_p = (e/2\pi)\sqrt{N_e/m_e\epsilon_0} \approx 8.984\sqrt{N_e} \tag{10.2-15}$$

10.2.1.1.1.2. *Effects of Collisions*; *Hydrodynamic Approach.* A collision is any event that causes a momentum change. In most plasmas of interest for radar applications, many slight changes in momentum are more important than are fewer large changes. When the electrons "collide" with other electrons, ions, neutral atoms, or molecules during their motion, energy is coupled out of the incident wave, causing it to be attentuated.

This means ϵ_r will have an imaginary part. The simplest way to obtain a first approximation to the effects of collisions is to add a viscous friction term to the force equation, Eq. (10.2-4). This causes the acceleration equation, Eq. (10.2-7), to have an added term, with the result

$$d\mathbf{v}/dt = (e/m)\,\mathbf{E} + \mathbf{A}(t) \tag{10.2-16}$$

where $\mathbf{A}(t)$ is a stochastic force due to the particle collisions. In general, $\mathbf{A}(t)$ cannot be evaluated, but is spoken of in terms of its average, $\langle\mathbf{A}(t)\rangle$, which is given by

$$\langle\mathbf{A}(t)\rangle = -\nu_c \mathbf{v}_d \tag{10.2-17}$$

where ν_c is the measure of the average number of collisions an electron undergoes per unit time, and is called the collision frequency. The term $\mathbf{v}_d$ is the drift velocity. In general, ν_c is a function of $\mathbf{v}_d$, but it is common to assume as a first approximation that it is not. The collision frequency is an effective numerical term, and the determination of a value to use is extremely difficult. A knowledge of the effective scattering cross sections for the various types of particles is required. These cross sections are functions of the velocity of the particle, although as a first approximation this is neglected. The collision frequency is then found by summing the products of the cross sections of the different types of particles times their respective number per unit volume times the appropriate drift velocity. The drift velocity is again an average or effective value. What is needed is knowledge of the distribution of velocities of each of the different species of particle. The determination of the collision frequency is discussed further in Section 10.2.1.2, which deals with plasma kinetic theory.

Ignoring the difficulty of obtaining a valid number which represents the collision frequency, and assuming that the effects of collisions can be represented by a frictional force, then by substitution of (10.2-17) into (10.2-16) and using the harmonic-time-dependence assumption, one obtains

$$-i\omega\mathbf{v} = \frac{e}{m_e}\mathbf{E} - \nu_c\mathbf{v}$$

or

$$\mathbf{v} = \frac{(e/m_e)\,\mathbf{E}}{-i\omega + \nu_c} = \left(\frac{\nu_c + i\omega}{\omega^2 + \nu_c{}^2}\right)\frac{e}{m}\mathbf{E} \tag{10.2-18}$$

Substituting Eq. (10.2-18) into Eq. (10.2-9), one obtains

$$\mathbf{J} = \frac{N_e e^2}{m_e}\left(\frac{\nu_c + i\omega}{\omega^2 + \nu_c{}^2}\right)\mathbf{E} \tag{10.2-19}$$

or

$$\mathbf{J} = (\sigma_c' + i\sigma_c'')\,\mathbf{E} \tag{10.2-20}$$

where

$$\sigma_c' = \frac{N_e e^2}{m_e}\left(\frac{\nu_c}{\omega^2 + \nu_c^2}\right) \tag{10.2-21}$$

and

$$\sigma_c'' = \frac{N_e e^2}{m_e}\left(\frac{\omega}{\omega^2 + \nu_c^2}\right) \tag{10.2-22}$$

From Eq. (10.2-21) it is seen that $\sigma' \to 0$ as $\nu_c \to 0$, and σ'' goes to the form observed in Eq. (10.2-9).

More generally, one can write

$$\sigma_c' = \sum_{j,l} \frac{N_j q_j^2}{m_j} K_j' \frac{\nu_{jl}}{\omega^2 + \nu_{jl}^2} = \sum_j \sigma_j' \tag{10.2-23}$$

and

$$\sigma_c'' = \sum_{j,l} \frac{N_j q_j^2}{m_j} K_j'' \frac{\omega}{\omega^2 + \nu_{jl}^2} = \sum_j \frac{\omega}{\nu_{jl}} \frac{K_j''}{K_j'} \sigma_j' \tag{10.2-24}$$

The symbols j and l represent the types of particles and can be either the same or different. For a Maxwellian distribution, the K's for electrons as given by Berman[34] are shown in Table 10-1.

Table 10-1†

Conductivity Parameters for a Maxwellian Distribution

	Highly Ionized Gas		Weakly Ionized Gas	
ω/ν_c	K'	K''	K'	K''
0	1.13	1.51	1.95	4.59
0.05	1.13	1.50	1.92	4.51
0.1	1.12	1.48	1.86	4.34
0.2	1.03	1.40	1.65	3.79
0.5	1.02	1.19	1.07	2.30
1	0.94	1.07	0.72	1.41
2	0.95	0.99	0.62	1.05
4	0.98	1.00	0.73	0.97
6	0.99	1.00	0.73	0.97
10	1.00	1.00	0.82	0.98
35	1.00	1.00	0.99	1.00
∞	1.00	1.00	1.00	1.00

† From Berman[34].

Substituting Eq. (10.2-19) into Maxwell's curl **H** equation gives

$$\nabla \times \mathbf{H} = -i\omega\epsilon_0 \mathbf{E}\left[1 + i\frac{\omega_p{}^2\nu_c}{\omega(\nu_c{}^2 + \omega^2)} - \frac{\omega_p{}^2}{\omega^2 + \nu_c{}^2}\right] \tag{10.2-25}$$

$$\epsilon_r' = \text{Re}\,\epsilon_r = 1 - \frac{\omega_p{}^2}{\omega^2}\left[\frac{1}{1 + (\nu_c/\omega)^2}\right] \tag{10.2-26}$$

and

$$\epsilon_r'' = \text{Im}\,\epsilon_r = \left[\frac{\omega_p}{\omega}\right]^2\left[\frac{\nu_c/\omega}{1 + (\nu_c/\omega)^2}\right] \tag{10.2-27}$$

As ω goes to ∞, Re ϵ_r goes to 1 and Im ϵ_r goes to zero; therefore all interaction between the plasma and the wave disappears at sufficiently high frequencies.

Figure 10-1 is a nomogram designed by Plugge[35] relating N_e, ω_p, ϵ_r', ω, f, and ν_c based on Eq. (10.2-26). The dashed lines in Figure 10-1 illustrate its use. Let the radar frequency be 10 Gcps. Then, on the second scale from the right, the point 0.1 is selected. This fixes the scale factor a at 11. The value of ω is seen to be 0.628×10^{11}. If the number of collisions is 5×10^{10}, then one selects 0.5 on the right-hand scale. These two points determine a straight line which intersects the center line as indicated. If either N_e, ω_p, or ϵ_r' is known, the other two parameters are determined. For example, if $N_e = 1 \times 10^{18}$, one multiplies N_e by $10^{4-2(11)}$ or 10^{-18} and selects 1 on the

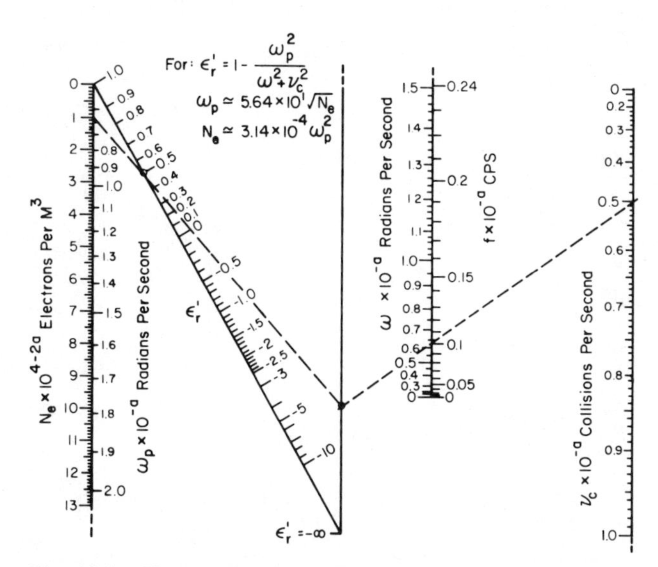

Fig. 10-1. Nomogram relating the real part of the relative permittivity of a plasma to the electron density, electron collision frequency, and radio frequency [after Plugge[35]].

left-hand side. This point, together with the previously chosen point on the center line, specifies the straight line intersecting ϵ_r' at 0.5.

The propagation constant k can be expressed as

$$k = (\omega^2 \mu_0 \epsilon_0 \epsilon_r)^{1/2} \tag{10.2-28}$$

or

$$ik = \alpha + i\beta = i\beta_0 \sqrt{\epsilon_r} = i\beta_0 m \tag{10.2-29}$$

where $\beta_0 = 2\pi/\lambda_0$ and m is the complex refractive index.† Thus from Eqs. (10.2-26) and (10.2-27), one finds that

$$\alpha + i\beta = i\beta_0 \left[1 - \frac{\omega_p{}^2}{\omega^2 + \nu_c{}^2} + i\frac{\nu_c}{\omega}\left(\frac{\omega_p{}^2}{\omega^2 + \nu_c{}^2}\right)\right]^{1/2} \tag{10.2-30}$$

Squaring both sides, one obtains

$$\alpha^2 - \beta^2 + 2i\alpha\beta = -\beta_0{}^2(\epsilon_r' + i\epsilon_r'') \tag{10.2-31}$$

Equating real and imaginary components gives

$$\alpha^2 - \beta^2 = -\beta_0{}^2\epsilon_r' \tag{10.2-32}$$

$$2\alpha\beta = -\beta_0{}^2\epsilon_r'' \tag{10.2-33}$$

Therefore

$$\begin{aligned} \alpha^2 &= \tfrac{1}{2}\beta_0{}^2\{-\epsilon_r' + [(\epsilon_r')^2 + (\epsilon_r'')^2]^{1/2}\} \\ &= \tfrac{1}{2}\beta_0{}^2(|\epsilon_r| - \epsilon_r') \end{aligned} \tag{10.2-34}$$

Then

$$\begin{aligned} \alpha &= -\beta_0\left(\frac{|\epsilon_r| - \epsilon_r'}{2}\right)^{1/2} \\ &= \frac{-\beta_0}{\sqrt{2}}\left(\frac{\omega^2 - \omega_p{}^2}{\omega^2 + \nu_c{}^2}\right)^{1/2}\left[\left(1 + \frac{\omega_p{}^4\nu_c{}^2}{\omega^2(\omega^2 - \omega_p{}^2)^2}\right)^{1/2} - 1\right]^{1/2} \end{aligned} \tag{10.2-35}$$

and, similarly,

$$\beta^2 = \tfrac{1}{2}\beta_0{}^2\{\epsilon_r' + [(\epsilon_r')^2 + (\epsilon_r'')^2]^{1/2}\} \tag{10.2-36}$$

or

$$\begin{aligned} \beta &= \beta_0\left(\frac{|\epsilon_r| + \epsilon_r'}{2}\right)^{1/2} \\ &= \frac{-\beta_0}{2}\left(\frac{\omega^2 - \omega_p{}^2}{\omega^2 + \nu_c{}^2}\right)^{1/2}\left[\left(1 + \frac{\omega_p{}^4\nu_c{}^2}{\omega^2(\omega^2 - \omega_p{}^2)^2}\right)^{1/2} + 1\right]^{1/2} \end{aligned} \tag{10.2-37}$$

† In the other chapters of this book, the quantity defined here as β_0 is designated k_0. The index of refraction is not normally denoted m in plasma papers. However, it will be denoted thus here in order to conform to the notation of the previous chapters.

The quantity given by the square root in the first equation of (10.2-36) is called the coefficient of extinction,

$$Q = -\alpha/\beta_0 \tag{10.2-38}$$

The quantity given by the square root in the first equation of Eq. (10.2-37) is the real part of the index of refraction, or

$$\text{Re}\, m = \beta/\beta_0 \tag{10.2-39}$$

Under the above assumptions, if an observing radar operates at a frequency such that ϵ_r is between zero and one, the ionized gas acts as a dielectric, and the results of the preceding chapters of this book hold. The plasma is termed underdense for this range of permittivity. For $\epsilon_r \leqslant 0$, $\omega_p \geqslant \omega$, the plasma is termed overdense. It is customary to divide discussions of ionized regions into overdense and underdense plasmas regardless of the effects of collisions. This division can sometimes be misleading. For $\epsilon_r \leqslant 0$, the lossless plasma acts as a waveguide beyond cutoff, in that only evanescent modes are present. This similarity gives rise to the commonly used approximation that an ionized gas with a nonpositive ϵ_r' reflects as a perfect conductor. Since for a perfect conductor ϵ_r'' goes to infinity, while for a plasma without collisions ϵ_r'' is zero, one might expect some differences in the scattering behavior. Even with the introduction of collisions, ϵ_r'' is not infinite. Neither is the ϵ_r' of metals less than unity. Some of the differences in scattering behavior for plasma and metallic spheres are discussed in Section 10.2.2.3. However, the perfect-conductor analogy is commonly used, and, if applied with care, will generally give good results. To this order of approximation, all of the results of the preceding chapters hold for ionized regions.

The principal effect of collisions is to introduce absorption. It is established practice to distinguish between deviative absorption and nondeviative absorption. If the refractive index in the medium differs appreciably from unity, the absorption is termed deviative. In nondeviative absorption, a ray path may deviate slightly from a straight line, but the absorption can be determined independently of the deviation of the ray path. If the assumption is made that ν_c is independent of velocity, manipulation of Eq. (10.2-29) shows that the nondeviative absorption per unit path length, α_n, is found from

$$\alpha_n = (\omega/c)\,\text{Im}\, m \tag{10.2-40}$$

Since, from Eq. (10.2-29), m is the square root of ϵ_r and since ϵ_r is assumed to be approximately one, m may be expanded in a series and all higher terms neglected. Eq. (10.2-40) then becomes

$$\alpha_n = \frac{-\omega_p{}^2}{2c}\,\frac{\nu_c}{\omega^2 + \nu_c{}^2} \tag{10.2-41}$$

Huxley[36] considered the case where ν_c is proportional to velocity, and obtained

$$\alpha_n = \frac{-\omega_p^2}{2c}\left\langle\frac{\nu_c}{\omega^2+\nu_c^2}\left(1+\frac{2}{3}\frac{\omega^2-\nu_c^2}{\omega^2+\nu_c^2}\right)\right\rangle \tag{10.2-42}$$

where the average is taken over the velocity distribution of the electrons. A common velocity distribution is the Maxwellian (see Section 10.2.1.2.). If a Maxwellian distribution is assumed[37], then

$$\alpha_n = (-25\omega_p^2/8c\nu_c)\, C_{5/2}\left(\frac{5\omega}{2\nu_c}\right) \tag{10.2-43}$$

where

$$C_p(X) = \frac{1}{p!}\int_0^\infty \frac{\exp(p-l)}{l^2+X^2}\,dl \tag{10.2-44}$$

The functions of Eq. (10.2-44) have been tabulated by Dingle *et al.*[38]

In reference 37, the relationship between the effective collision frequency of thermal electrons in air and the atmospheric pressure is given as $\nu_c = 2.3 \times 10^8 P$, where P is the pressure in millimeters of mercury, and this agrees with rocket measurements.[39]

Collisions and their effects are considered further in Section 10.2.1.2.

10.2.1.1.2. *Anisotropic Media.* The presence of a magnetic field has two major effects on radio-wave scattering from ionized media. First, it changes the medium. From Eq. (10.2-6), it is seen that a force is created by the interaction of moving electrons and the magnetic field. This force causes electrons to spiral along the field lines. Except for brief mention of such phenomena as the relation between reflection from auroras, the orthogonality of the geomagnetic field lines to the radio propagation path, and the creation of ionized regions at geomagnetic conjugate points by nuclear explosions, the effects on plasma dynamics of the magnetic field are beyond the scope of this book.

The second effect of the magnetic field is to change the electrical properties of the scattering medium. Neglecting the $\mathbf{v}/c$ term in Eq. (10.2-6) but including the $\mathbf{v} \times \mathbf{B}_{\text{ext}}$ term, assuming that E and v are harmonic variables, suppressing the "ext" subscript and the $\exp(-i\omega t)$ time dependence, and neglecting collisions, Hame and Stuart[40] obtained

$$v_x = \frac{(-im_e\omega/e)\,E_x + BE_y}{B^2 - (m_e\omega/e)^2} \tag{10.2-45}$$

$$v_y = \frac{(-im_e\omega/e)\,E_y - BE_x}{B^2 - (m_e\omega/e)^2} \tag{10.2-46}$$

and

$$v_z = eE_z/-im_e\omega \tag{10.2-47}$$

where the radio wave is propagating in the xz-plane and the magnetic field $\mathbf{B}$ is in the z direction. The propagation vector $\mathbf{k}^i$ lies at an angle θ from the z direction.

If the frequency is varied so that $(m_e\omega/e)$ approaches $|\mathbf{B}|$, v_x and v_y increase rapidly, and increase the probability of collisions to the point where, regardless of the value of collision frequency ν_c, it can no longer be neglected. Thus Eqs. (10.2-45), (10.2-46), and (10.2-47) become

$$v_x = \frac{[(-im_e\omega + m_e\nu_c)/e]\, E_x + BE_y}{B^2 - (m_e/e)^2\, (\omega^2 + \nu_c^2)} \tag{10.2-48}$$

$$v_y = \frac{[(-im_e\omega + m_e\nu_c)/e]\, E_y + BE_x}{B^2 - (m_e/e)(\omega^2 + \nu_c^2)} \tag{10.2-49}$$

$$v_z = eE_z/(-im_e\omega + \nu_c) \tag{10.2-50}$$

Collisions introduce further additional terms which are neglected in a first-order analysis. The angular frequency at which v_x and v_y would become infinite for a collisionless medium is called the angular gyromagnetic resonance frequency ω_H, or electron cyclotron frequency†; i.e.,

$$\omega_H = e\,|\,\mathbf{B}\,|/m_e \tag{10.2-51}$$

For ions, the angular gyromagnetic resonance frequency ω_i is given by

$$\omega_i = q_i\,|\,\mathbf{B}\,|/m_i \tag{10.2-52}$$

The value of ω_H is some 1836 times the value of ω_i for a hydrogen ion. If $\omega \gg \omega_i$, the effect of ω_i can thus be neglected.

Neglecting ω_i, Eq. (10.2-19) can be rewritten as‡

$$\begin{bmatrix} J_x \\ J_y \\ J_z \end{bmatrix} = \left(\frac{N_e e^2}{m_e}\right) \begin{bmatrix} +\dfrac{\nu_c - i\omega}{\omega_H^2 - (\omega + i\nu_c)^2} & +\dfrac{\omega_H}{\omega_H^2 - (\omega + i\nu_c)^2} & 0 \\ -\dfrac{\omega_H}{\omega_H^2 - (\omega + i\nu_c)^2} & +\dfrac{\nu_c - i\omega}{\omega_H^2 - (\omega + i\nu_c)^2} & 0 \\ 0 & 0 & \dfrac{1}{-i\omega + \nu_c} \end{bmatrix} \begin{bmatrix} E_x \\ E_y \\ E_z \end{bmatrix} \tag{10.2-53}$$

The off-diagonal terms in the matrix result in a current termed the Hall current. The current in any direction transverse to the direction of propagation has terms proportional to the fields in both transverse directions.

† The gyrofrequency should not be confused with the Larmor frequency, which is one-half the gyrofrequency.

‡ When the higher-order collision effects are considered, all the terms of (10.2-50) are nonzero.

It is seen from Eq. (10.2-53) that the conductivity is a dyad, and thus the permittivity is also a dyad, which can be written as

$$\tilde{\tilde{\epsilon}} = \begin{bmatrix} \epsilon_{11} & \epsilon_{12} & 0 \\ -\epsilon_{12} & \epsilon_{11} & 0 \\ 0 & 0 & \epsilon_{33} \end{bmatrix}$$

$$= \begin{bmatrix} \epsilon_0 \left\{ 1 + \dfrac{N_e e^2[1 - i(\nu_c/\omega)]}{\epsilon_0 m_e(\omega_H{}^2 - \omega^2 - \nu_c{}^2)} \right\} & +i \dfrac{N_e e^2 \omega_H}{m_e \omega(\omega_H{}^2 - \omega^2 - \nu_c{}^2)} & 0 \\ -i \dfrac{N_e e^2 \omega_H}{m_e \omega(\omega_H{}^2 - \omega^2 - \nu_c{}^2)} & \epsilon_0 \left\{ 1 + \dfrac{N_e e^2[1 - i(\nu_c/\omega)]}{\epsilon_0 m_e(\omega_H{}^2 - \omega^2 - \nu_c{}^2)} \right\} & 0 \\ 0 & 0 & \epsilon_0 \left\{ 1 - \dfrac{N_e e^2[1 - i(\nu_c/\omega)]}{\epsilon_0 m_e(\omega^2 + \nu_c{}^2)} \right\} \end{bmatrix} \quad (10.2\text{-}54)$$

The permittivity tensor components as seen by circularly polarized electromagnetic waves are found from

$$\epsilon_{\text{right}} = \epsilon_{11} + i\epsilon_{12} \quad (10.2\text{-}55)$$

$$\epsilon_{\text{left}} = \epsilon_{11} - i\epsilon_{12} \quad (10.2\text{-}56)$$

The permittivity matrix in this case has only zero terms off diagonal; ϵ_{left} appears in the upper left corner, ϵ_{right} in the center, and ϵ_{33} in the lower right corner.

For the case in which the magnetic field makes an angle θ with the direction of propagation, Hame and Stuart[40] give the square of the propagation constant as

$$k^2 = \omega^2 \mu_0 \frac{\left\{ \begin{aligned} (\epsilon_{11}^2 - \epsilon_{12}^2 - \epsilon_{11}\epsilon_{33}) \sin^2\theta + 2\epsilon_{11}\epsilon_{33} \mp [(\epsilon_{11}^2 - \epsilon_{12}^2 - \epsilon_{11}\epsilon_{33})^2 \sin^4\theta \\ + 4\epsilon_{33}^2\epsilon_{12}^2 \cos^2\theta]^{1/2} \end{aligned} \right\}}{2(\epsilon_{11} - \epsilon_{33}) \sin^2\theta + 2\epsilon_{33}} \quad (10.2\text{-}57)$$

For $\omega^2 \gg \omega_H{}^2$, $\epsilon_{11} \approx \epsilon_{33}$ and Eq. (10.2-57) reduces to

$$k^2 = \omega^2 \mu_0 \epsilon_{11} \left\{ 1 - \frac{\epsilon_{12}^2}{2\epsilon_{11}^2} \left[\sin^2\theta \mp \left(\sin^4\theta + \frac{4\epsilon_{11}^2}{\epsilon_{12}^2} \cos^2\theta \right)^{1/2} \right] \right\} \quad (10.2\text{-}58)$$

Propagation along the magnetic field lines is termed longitudinal propagation; propagation perpendicular to the magnetic field lines is termed transverse propagation. If θ is small, then the $\sin\theta$ terms in Eq. (10.2-58) can be neglected, giving the quasilongitudinal approximation†

$$k^2 = \omega^2\mu_0(\epsilon_{11} \pm \epsilon_{12}\cos\theta) \tag{10.2-59}$$

Near the transverse position, Eq. (10.2-58) may be approximated as

$$k^2 = \omega^2\mu_0\epsilon_{11}[1 - (\epsilon_{12}^2/2\epsilon_{11}^2)\sin^2\theta(1 \mp 1)] \tag{10.2-60}$$

or

$$k_1 = \omega\sqrt{\mu_0\epsilon_{11}} \tag{10.2-61}$$

and

$$k_2 = \omega\{\mu_0[\epsilon_{11} - (\epsilon_{12}^2/\epsilon_{11})\sin^2\theta]\}^{1/2} \tag{10.2-62}$$

The wave propagating at k_1 is known as the ordinary wave, and the wave propagating at k_2 as the extraordinary wave. For $\theta = 90°$, (10.2-61) and (10.2-62) then represent two orthogonally oriented, linearly polarized waves traveling at different velocities. The quasitransverse and quasilongitudinal approximations are discussed in more detail in References 1, 2, 17, and 18. The terms "ordinary" and "extraordinary" are not limited to the case of quasitransverse propagation, but are applied to the waves whose propagation constants reduce to the values given in the limit as θ goes to 90°. Evaluating Eq. (10.2-59) for keeping the square of the refractive index, we have

$$m^2 = 1 - \frac{(\omega_p/\omega)^2}{\left\{1 + i\dfrac{v_c}{\omega} - \dfrac{\frac{1}{2}(\omega_H/\omega)^2\sin^2\theta}{1-(\omega_p/\omega)^2 + i(v_c/\omega)} \pm \left[\dfrac{\frac{1}{4}(\omega_H/\omega)^4\sin^4\theta}{[1-(\omega_p/\omega)^2 + i(v_c/\omega)]^2} + \left(\dfrac{\omega_H}{\omega}\right)^2\cos^2\theta\right]^{1/2}\right\}} \tag{10.2-63}$$

which is known as the Appleton–Hartree equation. More generalized forms are available in the literature.[41,42] Ratcliffe[2] gives Booker's rules for determining which sign goes with the ordinary and which with the extraordinary wave. He states that the ordinary wave is represented by the plus sign where $(\omega_p/\omega)^2 < 1$; if $(\omega_p/\omega)^2 > 1$, the plus sign is used when $|(\omega_H\sin^2\theta)/(2\nu_c\cos\theta)| > 1$, and the minus sign when $(\omega_H\sin^2\theta)/(2\nu_c\cos\theta) < 1$. The extraordinary wave is, of course, represented by the opposite choice in each case. Kelso, on p. 131 of Reference 17, notes the rules on signs which must be followed if the expression is rearranged to remove a portion of the quantity under the square root sign.

† For the frequency range of 20 Mcps to 108 Mcps, Hame and Stuart state that this quasilongitudinal approximation is valid for a plasma with an electron density of 1.8×10^{12} electrons/m³ up to within a few degrees of the transverse position.

The presence of two values of the refractive index means that any linearly polarized wave will decompose into two elliptically polarized waves. These are circularly polarized in opposite senses for longitudinal propagation. In transverse propagation, the ordinary wave is polarized with the **E** vector parallel to the external magnetic field. The two waves are orthogonal only in a collisionless medium. The two rays propagate with different velocities and, in general, along different paths. The signal after passing through an ionized region consists of two signals traveling in different directions with a time delay between the signals. The birefringence-induced delay is inversely proportional to the cube of the frequency and directly portional to the magnetic field strength. A received wave packet is longer in duration and lower in peak amplitude than the same packet transmitted through a vacuum. Also, since the refractive index is frequency dependent, the different frequency components of a pulse will have their phase relationship changed. This chromatic aberration is inversely proportional to the square of the frequency.

Perhaps a more significant effect of the anisotropy is the Faraday rotation of the plane of polarization. The Faraday rotation is found by integrating one-half the difference between the phase constants of the two waves over the path of the waves, assuming the paths are identical. Under the quasilongitudinal approximation and neglecting collisions, Hame and Stuart,[(40)] as corrected by Potts,[(43)] derive the total angle of rotation ψ_F as

$$\psi_F = \sum_{j=1}^{\infty} \frac{K_j}{f^{2j}} \int_a^b N_e^{\,j} B \cos\theta \, ds \tag{10.2-64}$$

where

$$K_1 = e^3{}_0/8\pi^2 c\epsilon_0 m^2$$

$$K_2 = e^5{}_0/64\pi^4 c\epsilon_0{}^2 m^3$$

$N_e^{\,j}$ is the total number of electrons between a and b, raised to the jth power, and a, b are the limits of the ionized region along the ray path. Usually, the first term of the series gives sufficient accuracy. Other assumptions implicit in deriving Eq. (10.2-64) were that $\omega \gg \omega_H$, that $\omega \gg \omega_p$, and that the ionized medium is stratified with all boundaries perpendicular to the ray path.

For the quasitransverse case, similar assumptions give

$$\psi_F \approx \frac{e^4}{4c\epsilon_0 m^3 \omega^3} \int_a^b N_e B^2 \sin^2\theta \, ds \tag{10.2-65}$$

Considerably more rotation of polarization will occur in the quasilongitudinal case.

It should be noted that Faraday rotation effects are additive. That is, a radar signal traversing an ionized region twice has a Faraday rotation which is the sum of the rotations for each traversal. Millman[44] has shown for a radar return from a target at an altitude of 1000 km that, with longitudinal propagation, polarization rotation of more than 1 radian can occur during the daylight over most elevation angles at a frequency of 1000 Mcps.

It might be thought that Faraday rotation would greatly hinder the use of the radar scattering matrix discussed in Section 2.1.4. However, for frequencies high enough that path separation can be neglected, if there is no birefringence, the effect of Faraday rotation can always be removed from the measured scattering matrix of an arbitrary target.[45] For circular polarization the phase difference between the off-diagonal elements of the scattering matrix is equal to four times the one-way Faraday rotation angle. If birefringence is present, the propagation path can be calibrated by measuring the scattering matrix of a scatterer exhibiting four-way symmetry. This calibration could allow the calculation and removal of the effects of both Faraday rotation and birefringence. In general, the determinant of the polarization-scattering matrix and the sum of the diagonal elements (trace) of the power-scattering matrix are unaffected by the presence of an anisotropic medium.

The effects of collisions on the conductivity tensor in the presence of a static magnetic field are given in the next section. Inclusion of thermal pressure effects on a plasma consisting of electrons, ions, and neutrals also allows three longitudinal waves, as well as the two transverse waves. The three longitudinal waves can also exist in isotropic plasmas. The coupling between transverse and longitudinal waves, and the loss (or generation) of energy by these longitudinal waves, are briefly mentioned in the following section. Because of the complexity of the problem there are not many practical results for radar scattering from anisotropic regions. Brief discussions are given in the sections on planar and cylindrical scatterers (Sections 10.2.2.1 and 10.2.2.2). Much of the anisotropic theory has been concerned with ionospheric propagation or with propagation through plasmas. This is covered in References 2, 3, 16, 17, 46, and 47, and will not be discussed here.

10.2.1.2. *Microscopic Approach*

In Section 10.2.1.1.1.2, the effects of collisions were considered briefly. A more exact representation of these effects involves kinetic theory requiring statistical methods. An important tool in these methods is the distribution function, $f(\mathbf{r}, \mathbf{v}, t)$. The distribution function is a probability density function in six-dimensional phase space. The distribution function when multiplied by the total number of particles and integrated over a volume in configuration space and a volume in velocity space gives the number of particles which lie within the given spatial and velocity coordinates at a given time. The continuity equation in phase space for a weakly ionized gas is termed the Boltzmann equation. It represents the effects of external forces on the distribution function. Considering a nonrelativistic unbounded plasma, neglecting all motion other than translatory and assuming omnidirectional particle forces, one obtains a system of N coupled Boltzmann equations for a plasma consisting of N particles. The equations have the form

$$\frac{\partial f_n}{\partial t} + \mathbf{v} \cdot \nabla_r f_n + \frac{q_n}{m_n} (\mathbf{E}^i + \mathbf{v} \times \mathbf{B}) \cdot \nabla_v f_n = \sum_{j=1}^{N} \left(\frac{\partial f_n}{\partial t} \right)_{n,j} + \xi \qquad (10.2\text{-}66)$$

where $n, j = 1, 2, \ldots, N$; ∇_r and ∇_v denote gradients in configuration and velocity space; ξ represents the effects of ionization, recombination, attachment, excitation, and other inelastic processes; and $(\partial f_n/\partial t)_{n,j}$ represents the effects of all short-range particle interactions which may be considered elastic collisions. The subscripts n, j indicate collisions between the nth and the jth types of particle. The subscripts take on all values of n and j, including $n = j$. The term $(q_n/m_n)(\mathbf{v} \times \mathbf{B})$ is the Lorentz force. In this term, $\mathbf{B}$ is the external magnetic field. The Lorentz force caused by the magnetic field of the radar wave is neglected, since it is usually small in comparison to both the Coulomb force and the Lorentz force of the external field.

If no collisions and no inelastic processes occur, the terms on the right-hand side of Eq. (10.2-66) are zero, and the resulting equation is denoted the Vlasov equation.

Electron–electron interactions are difficult to evaluate. For low-temperature plasmas, these interactions are negligible. For high-temperature plasmas, they must be included. The best way to consider them seems to be not by means of the Boltzmann equation, but by the Fokker–Planck equation. The Fokker–Planck equation, neglecting higher-order terms, is

$$\frac{\partial f(\mathbf{r}, \mathbf{v}, t)}{\partial t} = - \sum_n \frac{\partial}{\partial v_n} f(\langle \Delta v_n \rangle) + \frac{1}{2} \sum_{n,j} \frac{\partial^2}{\partial v_n \, \partial v_j} f(\langle \Delta v_{n,j} \rangle) \qquad (10.2\text{-}67)$$

where Δv_n is a particular component of the vector change in velocity in the n, j coordinate system, and $\Delta v_{n,j}$ is a particular component of the dyad $\Delta \mathbf{v} \, \Delta \mathbf{v}$. The summations over n and over n, j take place over all three possible coordinates.

Equation (10.2-67) is valid if the number of electrons in the Debye sphere is large, say greater than one hundred. The Debye sphere has a radius equal to the Debye shielding length λ_D. The Debye length gives the magnitude of charge separation distance for which the resulting electrostatic energy equals the particle thermal energy. The surrounding plasma shields a charged particle so that the effects of its charge are not felt beyond a few Debye lengths. The Debye length is given by

$$\lambda_D = (\epsilon_0 K_B T / N_e e^2)^{1/2} \tag{10.2-68}$$

where K_B is Boltzmann's constant (1.38054×10^{-23} joules/deg). Therefore the number N_D of electrons in a Debye sphere is

$$N_D = \tfrac{4}{3}\pi \lambda_D{}^3 N_e \tag{10.2-69}$$

Equation (10.2-67) can be used to consider any small-deflection interaction, rather than merely electron–electron interactions. However, the assumptions necessary for the derivation of the equation are similar to those necessary for the derivation of the Boltzmann equation, and thus the results from the two equations are quite similar.

More advanced methods of treating the effects of collisions involve nonequilibrium statistical mechanics and quantum statistics. A good discussion of these methods is given in the book by Balescu.[48]

Returning to the Boltzmann equation, it is sometimes possible to integrate without specifying the exact nature of the distribution function. The results are a set of hydromagnetic equations (or Boltzmann transport or moment equations) in the microscopic variables of mass velocity, pressure, and current density.

The two most common distribution functions discussed in the literature are the Maxwellian and the Druyvesteyn. The Druyvesteyn is useful for low-temperature plasmas in the presence of large dc electric fields. An example of this is the plasma found in laboratory gas discharges. The Maxwellian distribution function is useful for high-temperature plasma in the presence of weak electric fields. An example of this is the plasma in shock fronts. The Maxwellian distribution function is given by

$$f_0(v) = (m_e/2\pi K_B T)^{3/2} \exp(-m_e v_c{}^2/2K_B T) \tag{10.2-70}$$

For the Maxwellian distribution, the collision frequency is independent of the velocity. For the Druyvesteyn distribution, the mean free path of the electron is constant.

In the presence of electric and magnetic fields, the most useful technique for solving the Boltzmann equation has involved expanding the distribution function in spherical harmonics. The zero-order term is the unperturbed distribution function (Maxwellian under proper circumstances). The higher-order terms correspond to various perturbing forces. The details of this expansion are available in References 4 and 5.

If only the zero-order and first-order terms in the expansion are used,

$$f(\mathbf{r}, \mathbf{v}, \mathbf{t}) = f_0(v^2) + f_1(\mathbf{r}, \mathbf{v}, t) \tag{10.2-71}$$

and the current density may be written as

$$\mathbf{J}(\mathbf{r}, t) = e \int \mathbf{v} f_1(\mathbf{r}, \mathbf{v}, t)\, d^3\mathbf{v} \tag{10.2-72}$$

Equation (10.2-72) neglects the contributions to the current from particles other than the electrons. In general, it should be written as a sum with each type of particle represented by its respective charge, velocity, and first-order distribution function.

Assuming a Maxwellian distribution, propagation in the xz-plane, and a static magnetic field along $\hat{\mathbf{z}}$, the components of the conductivity tensor for a warm plasma are

$$\sigma_{xx} = \frac{\epsilon_0 \omega_p{}^2}{\omega} \int_0^\infty \left[\cos \frac{\omega_H}{\omega} s - \frac{\omega^2 \zeta \sin^2\theta}{\omega_H{}^2} \left(1 - \cos^2 \frac{\omega_H}{\omega} s\right)\right] \exp[\Phi(s)]\, ds \tag{10.2-73}$$

$$\sigma_{yy} = \frac{\epsilon_0 \omega_p{}^2}{\omega} \int_0^\infty \left[\cos \frac{\omega_H}{\omega} s + \frac{\omega^2 \zeta \sin^2\theta}{\omega_H{}^2} \left(1 - \cos \frac{\omega_H}{\omega} s\right)^2\right] \exp[\Phi(s)]\, ds \tag{10.2-74}$$

$$\sigma_{zz} = \frac{\epsilon_0 \omega_p{}^2}{\omega} \int_0^\infty (1 - s^2 \zeta \sin^2\theta) \exp[\Phi(s)]\, ds \tag{10.2-75}$$

$$\sigma_{xy} = -\sigma_{yx} = \frac{\epsilon_0 \omega_p{}^2}{\omega} \int_0^\infty \left(-\sin \frac{\omega_H}{\omega} s\right)\left[1 - \frac{\omega^2 \zeta \sin^2\theta}{\omega_H{}^2} \left(1 - \cos \frac{\omega_H}{\omega} s\right)\right] \times \exp[\Phi(s)]\, ds \tag{10.2-76}$$

$$\sigma_{xz} = \sigma_{zx} = \frac{\epsilon_0 \omega_p{}^2}{\omega} \int_0^\infty \frac{\omega s \zeta \sin\theta \cos\theta}{\omega_H} \left(-\sin \frac{\omega_H}{\omega} s\right) \exp[\Phi(s)]\, ds \tag{10.2-77}$$

$$\sigma_{yz} = -\sigma_{zy} = \frac{\epsilon\ \omega_p{}^2}{\omega} \int_0^\infty \frac{\omega s \zeta \sin\theta \cos\theta}{\omega_H} \left(-1 + \cos \frac{\omega_H}{\omega} s\right) \exp[\Phi(s)]\, ds \tag{10.2-78}$$

where

$$\Phi(s) = i\left(1 + i\frac{\nu_c}{\omega}\right) s - \frac{\omega^2 \zeta \sin^2\theta}{\omega_H{}^2} \left(1 - \cos \frac{\omega_H}{\omega} s\right) - \frac{1}{2} s^2 \zeta \cos^2\theta$$

$$\zeta = m^2 (K_B T / m_e c^2) \tag{10.2-79}$$

m is the complex refractive index as before, and θ is the angle between the direction of propagation and the z direction.

Sitenko and Stepanov,[50] expanding to the first order in ζ, give the components of the permittivity tensor as†

$$\epsilon_{xx} = 1 - \left[\frac{\omega_p{}^2}{\omega} \frac{\omega + i\nu_c}{(\omega + i\nu_c)^2 - \omega_H{}^2}\right]\left\{1 + \frac{\zeta \omega^2 \cos^2\theta\, [(\omega + i\nu_c)^2 + 3\omega_H{}^2]}{[(\omega + i\nu_c)^2 - \omega_H{}^2]^2} + \frac{3\omega^2 \zeta \sin^2\theta}{(\omega + i\nu_c)^2 - 4\omega_H{}^2}\right\} \tag{10.2-80}$$

† The parameter ζ is directly proportional to the square of the ratio of the electron thermal velocity to the velocity of light. For the expansions of Eqs. (10.2-80) to (10.2-85) to be valid, $m^2 \gg 1$. This is the condition under which the warm-plasma results are significantly different from the cold-plasma case.[18]

$$\epsilon_{yy} = 1 - \left\{ \frac{\omega_H^2(\omega + i\nu_c)}{\omega[(\omega + i\nu_c)^2 - \omega_H^2]} \right\} \left\{ 1 + \frac{\zeta\omega^2 \cos^2\theta \, [(\omega + i\nu_c)^2 + 3\omega_H^2]}{[(\omega + i\nu_c)^2 - \omega_H^2]^2} + \frac{\omega^2\zeta \sin^2\theta \, [(\omega + i\nu_c)^2 + 8\omega_H^2]}{(\omega + i\nu_c)^2[(\omega + i\nu_c)^2 - 4\omega_H^2]} \right\} \tag{10.2-81}$$

$$\epsilon_{zz} = 1 - \left[\frac{\omega_p^2}{\omega(\omega + i\nu_c)} \right] \left[1 + \frac{3\omega^2\zeta \cos^2\theta}{(\omega + i\nu_c)^2} + \frac{\omega^2\zeta \sin^2\theta}{(\omega + i\nu_c)^2 - \omega_H^2} \right] \tag{10.2-82}$$

$$\epsilon_{xy} = -\epsilon_{yx} = i \frac{\omega_p^2\omega_H}{\omega(\omega + i\nu_c)^2 - \omega_H^2} \left\{ 1 + \frac{\zeta\omega^2 \cos^2\theta \, [3(\omega + i\nu_c)^2 + \omega_H^2]}{[(\omega + i\nu_c)^2 - \omega_H^2]^2} + \frac{6\omega^2\zeta \sin^2\theta}{(\omega + i\nu_c)^2 - 4\omega_H^2} \right\} \tag{10.2-83}$$

$$\epsilon_{xz} = \epsilon_{zx} = -\frac{(2\omega\omega_p^2\zeta \sin\theta \cos\theta)(\omega + i\nu_c)}{[(\omega + i\nu_c)^2 - \omega_H^2]^2} \tag{10.2-84}$$

$$\epsilon_{yz} = -\epsilon_{zy} = -i \frac{\zeta\omega\omega_p^2\omega_H \sin\theta \cos\theta \, [3(\omega + i\nu_c)^2 - \omega_H^2]}{(\omega + i\nu_c)^2[(\omega + i\nu_c)^2 - \omega_H^2]^2} \tag{10.2-85}$$

Several other expressions for the permittivity and conductivity for various special cases are available in the literature.(51–60) Reference 4 has an especially valuable collection of equations pertaining to various anisotropic cases. Most of the results are given in terms of the refractive index.

Brandstatter(4) chooses his coordinate system such that the permittivity tensor for a high-temperature plasma is given by

$$\tilde{\tilde{\epsilon}} = \begin{bmatrix} \epsilon_{11} & -i\epsilon_{12} & 0 \\ +i\epsilon_{21} & \epsilon_{22} & 0 \\ 0 & 0 & \epsilon_{33} \end{bmatrix} \tag{10.2-86}$$

He then defines

$$\epsilon_1 = \epsilon_{11} - i\epsilon_{12} \tag{10.2-87}$$

$$\epsilon_2 = \epsilon_{22} + i\epsilon_{12} = \epsilon_{22} - i\epsilon_{21} \tag{10.2-88}$$

$$\epsilon_3 = \epsilon_{33} \tag{10.2-89}$$

If ion motion, electron–electron collisions, and plasma oscillations are neglected and the distribution function is limited to the form of Eq. (10.2-71), then

$$\epsilon_1 = 1 - \frac{4\pi}{3} \frac{\omega_p^2}{N_e} \int \frac{f_1 v^4 \, dv}{\omega[\omega - (\omega_H + i\nu_c)]} \tag{10.2-90}$$

$$\epsilon_2 = 1 - \frac{4\pi}{3} \frac{\omega_p^2}{N_e} \int \frac{f_1 v^4 \, dv}{\omega[\omega + (\omega_H - i\nu_c)]} \tag{10.2-91}$$

$$\epsilon_3 = 1 - \frac{4\pi}{3} \frac{\omega_p^2}{N_e} \int \frac{f_1 v^4 \, dv}{\omega(\omega - i\nu_c)} \tag{10.2-92}$$

For ω sufficiently greater than the gyromagnetic frequency of the electrons, ω_H can be neglected, giving

$$\epsilon_1 = \epsilon_2 = \epsilon_3 = 1 - \frac{4\pi}{3} \frac{\omega_p{}^2}{N_e} \int \frac{f_1 v^4 \, dv}{\omega(\omega - i\nu_c)} \qquad (10.2\text{-}93)$$

Eq. (10.2-93) is restricted to frequencies much greater than

$$\sqrt{\tfrac{2}{3}}\,(eE^i/\sqrt{k_0 T_e m_i}) + (2m_e/m_i)$$

Ginsburg[1] compares the results of Eq. (10.2-93) with that obtained by macroscopic theory, i.e., Eqs. (10.2-26) and (10.2-27). For the case of weak electric fields, where $(E^i)^2 \ll 6K_B T(\omega^2 + \nu_c{}^2)(m_e{}^2/m_i)$, he indicates that the macroscopic theory does not give as good an approximation if $\omega \leqslant 5\nu_c$. As pointed out by Brandstatter[4], the permittivity at higher frequencies is essentially independent of which theory is used; thus, since the macroscopic theory is the simpler, it is preferred. The use of the macroscopic theory at low frequencies is not recommended. However, Brandstatter states that the only presently available results valid at low frequencies are those of Vilensky.[61] His equations are rather complex and will not be presented here.

The effects of a strong magnetic field have also been neglected in Eq. (10.2-93). If the radar frequency is near the gyrofrequency of the plasma, then the effects of the magnetic field will be more important in determining an effective permittivity.

It can also be shown[14,15] that at some radar frequencies and magnetic field strengths, the plasma has dielectric properties which permit the propagation of longitudinal waves, as discussed briefly in Section 10.2.1.1.2. In these plasma oscillations, the pressure is transmitted *via* the collective space-charge field instead of by individual particle collisions. A point charge moving through the plasma continuously excites plasma oscillations. On the other hand, the oscillations continuously lose energy because of the random motion of the particles. Either process can be dominant. That is, the plasma oscillations can either be damped or enhanced, depending upon conditions. The net effect of the wave–particle interactions is a diffusion of the distribution function in velocity space. The effects of the oscillations can be several orders of magnitude larger than the effects of ordinary collisions.

Thus the plasma is modeled by a dual medium, consisting of both discrete particles with some distribution of random velocities and a continuous fluid. This model currently appears to be the most promising for obtaining practical results in the near future.

10.2.2. Application of Classical Scattering and Diffraction Theory

Since a plasma is a collection of moving particles, one might expect that sharp boundaries would not exist. However, since many phenomena create a large number of free electrons in a short time, in a region with relatively small cross-sectional area, it has been customary to assume sharp discontinuities. For a review of current literature on the boundary problem, see References 27 and 28. The assumption that a region exists over which the usual electromagnetic boundary conditions for a dielectric or metallic material hold, plus the assumption that a permittivity can be defined, allows the use of the solutions developed in the previous chapters of this book. One may regard this section as giving special examples of the previous chapters.

10.2.2.1. *Planar Regions*

The simplest planar ionized region is a uniform semi-infinite region with a sharp boundary. This may be treated as a free-space–dielectric interface as in Chapter 7, and the results given there are valid if the values of effective dielectric parameters derived in Section 10.2.1.1.1 are used. The reflection coefficient is dependent upon the polarization of the incident field. For a plane wave incident at an angle θ from the normal to the surface (see Figure 10-2), and for the electric vector normal to the plane of incidence, the reflection coefficient may be written as $R_\perp = |R_\perp| e_\perp^{i\Psi}$, where

$$|R_\perp| = \sqrt{R_\perp R_\perp^*} = \left\{\frac{[\cos\theta_1 - (Q_2/k_0)]^2 + (P_2/k_0)^2}{[\cos\theta_1 + (Q_2/k_0)]^2 + (P_2/k_0)^2}\right\}^{1/2} \qquad (10.2\text{-}94)$$

and

$$\Psi_\perp = \tan^{-1}\left[\frac{2(P_2/k_0)\cos\theta_1}{\cos^2\theta_1 - (Q_2/k_0)^2 - (P_2/k_0)^2}\right] \qquad (10.2\text{-}95)$$

In Eqs. (10.2-94) and (10.2-95), the parameters P_2 and Q_2 are given by

$$P_2/k_0 = [\tfrac{1}{2}[(\epsilon' - \sin^2\theta_1)^2 + (\epsilon'')^2]^{1/2} - \tfrac{1}{2}(\epsilon' - \sin^2\theta_1)]^{1/2} \qquad (10.2\text{-}96)$$

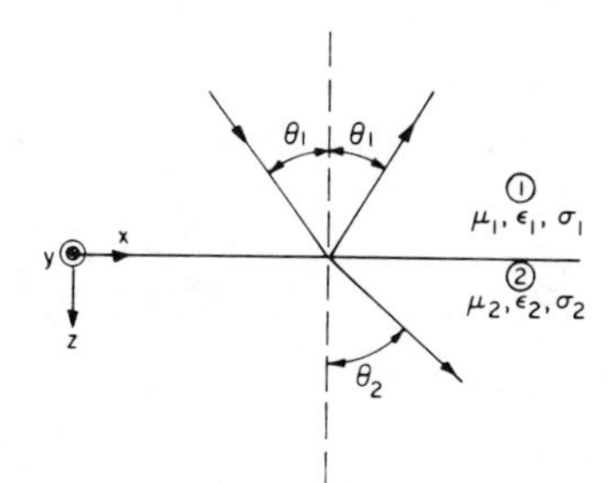

Fig. 10-2. Coordinates for incidence on an interface between free space and a semi-infinite plasma.

and

$$Q_2/k_0 = [\tfrac{1}{2}[(\epsilon' - \sin^2 \theta_1)^2 + (\epsilon'')^2]^{1/2} + \tfrac{1}{2}(\epsilon' - \sin^2 \theta_1)]^{1/2} \quad (10.2\text{-}97)$$

For the same situation but with the electric vector in the plane of incidence, the reflection coefficient is given by

$$|R_{\parallel}| = \sqrt{R_{\parallel} R_{\parallel}^*} = \left\{ \frac{[\epsilon_r' \cos \theta_1 - (Q_2/k_0)]^2 + [\epsilon_r'' \cos \theta_1 - (P_2/k_0)]^2}{[\epsilon_r' \cos \theta_1 + (Q_2/k_0)]^2 + [\epsilon_r'' \cos \theta_1 + (P_2/k_0)]^2} \right\}^{1/2} \quad (10.2\text{-}98)$$

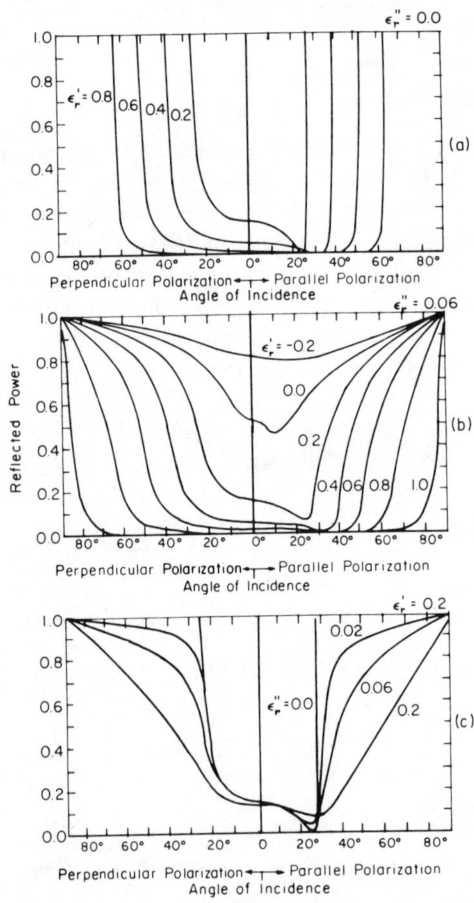

Fig. 10-3. Power reflected from various plasmas for a perpendicular-polarized and a parallel-polarized incident wave as a function of the angle of incidence [after Bachynski, Cloutier, and Graf[32]].

and

$$\Psi_{\parallel} = \tan^{-1}\left[\frac{2[\epsilon'(P_2/k_0) - \epsilon''(Q_2/k_0)\cos\theta_1]}{[(\epsilon')^2 + (\epsilon'')^2]\cos^2\theta_1 - (Q_2/k_0)^2 - (P_2/k_0)^2}\right] \qquad (10.2\text{-}99)$$

Figure 10-3 gives a plot of Eqs. (10.2-94) and (10.2-98) as a function of θ_1 for $\epsilon_r' = 0.2$, 0.4, 0.6, and 0.8. Figure 10-3(a) is plotted for $\epsilon_r'' = 0.0$, and Figure 10-3(b) is plotted for $\epsilon_r'' = 0.06$. In Figure 10-3(c), ϵ_r' is held constant at 0.2, and $|R_\perp|$ and $|R_\parallel|$ are plotted for $\epsilon_r'' = 0.0$, 0.02, 0.06, and 0.2. In all three cases, $|R_\perp|$ is plotted on the left and $|R_\parallel|$ on the right, with θ_1 decreasing to zero at the center. At $\theta_1 = 0°$, or normal incidence, $|R_\perp|$ must equal $|R_\parallel|$, as seen on the graphs and from the equations. For this case, Eqs. (10.2-94) and (10.2-98) may be rewritten as

$$|R| = \left\{\frac{[1 - (\beta/k_0)]^2 + (\alpha/k_0)^2}{[1 + (\beta/k_0)]^2 + (\alpha/k_0)^2}\right\}^{1/2} = \left|\frac{1-m}{1+m}\right| \qquad (10.2\text{-}100)$$

where α and β are defined in Eq. (10.2-30). Hochstim[62] has computed the values of $|R|$ given by Eq. (10.2-100) as a function of ω/ν_c for numerous values of ω_p/ν_c as shown in Figures 10-4–10-6. Plots of α/k_0 and β/k_0 for various ν_c/ω and ω_p/ω are available in References 22 and 23.

The familiar Brewster-angle phenomenon also holds for a plasma, in that no reflection occurs from a lossless plasma if the wave is polarized in the plane of incidence and is incident at the angle $\theta_1 = \sqrt{\epsilon_r'}$. For a lossy plasma, some reflection will always occur regardless of the angle of incidence or polarization, but the incident wave can never be totally reflected. For a lossless plasma, total reflection occurs for angles of incidence greater than the critical angle $\theta_c = \sin^{-1}\sqrt{\epsilon_r'}$. Substituting the value of ϵ_r' from Eq. (10.2-26) and performing a few straightforward manipulations gives the familiar relation for the critical frequency of ionospheric propagation:

$$f_c = f_p \sec\theta_1 = f_p \operatorname{cosec}(90° - \theta_1) \qquad (10.2\text{-}101)$$

The angle $(90° - \theta_1)$ is the angle often discussed in ionospheric propagation. Equation (10.2-101) states that for any θ_1, total reflection will occur from a lossless plasma for all frequencies below the critical frequency. The ionosphere, as well as most plasmas, is lossy, inhomogeneous, and anisotropic.

The general expressions for incidence at an arbitrary angle θ_1 on a stratified medium composed of M layers are quite involved. These are the equations of Section 7.1 with the appropriate values of ϵ_r' and ϵ_r'' inserted. The effective impedance of such a plasma medium is obtained by using the

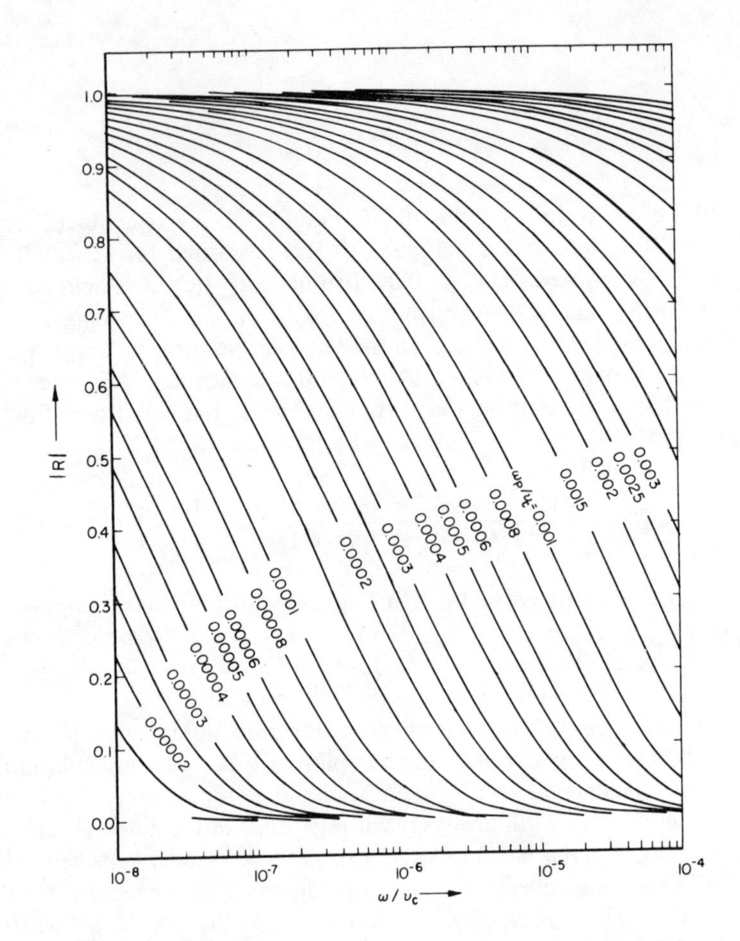

Fig. 10-4. Reflection coefficient of a homogeneous semi-infinite plasma versus ω/ν_c for values of ω_p/ν_c from 0.00002 to 0.003 [after Hochstim[62]].

proper values of permittivity in the impedance definitions of Eqs. (7.1-6) and (7.1-7), or Eqs. (7.1-15) and (7.1-16).

The simplest layered medium is a uniform plasma slab of permittivity ϵ_2 bounded on both sides by free space. While it is obvious that the reflections from this slab can be obtained from the formulas of Section 7.1, many authors have discussed this case for a variety of plasma parameters and it is therefore possible to observe the reflectivity in detail for various parameters and extend this case to anisotropic media.

For isotropic media, Eq. (7.1-42) can be written as

$$R = \frac{-i[z_2^2 - z_0^2] \tan \alpha_2}{2z_0 z_2 - i(z_2^2 + z_0^2) \tan \alpha_2} \tag{10.2-102}$$

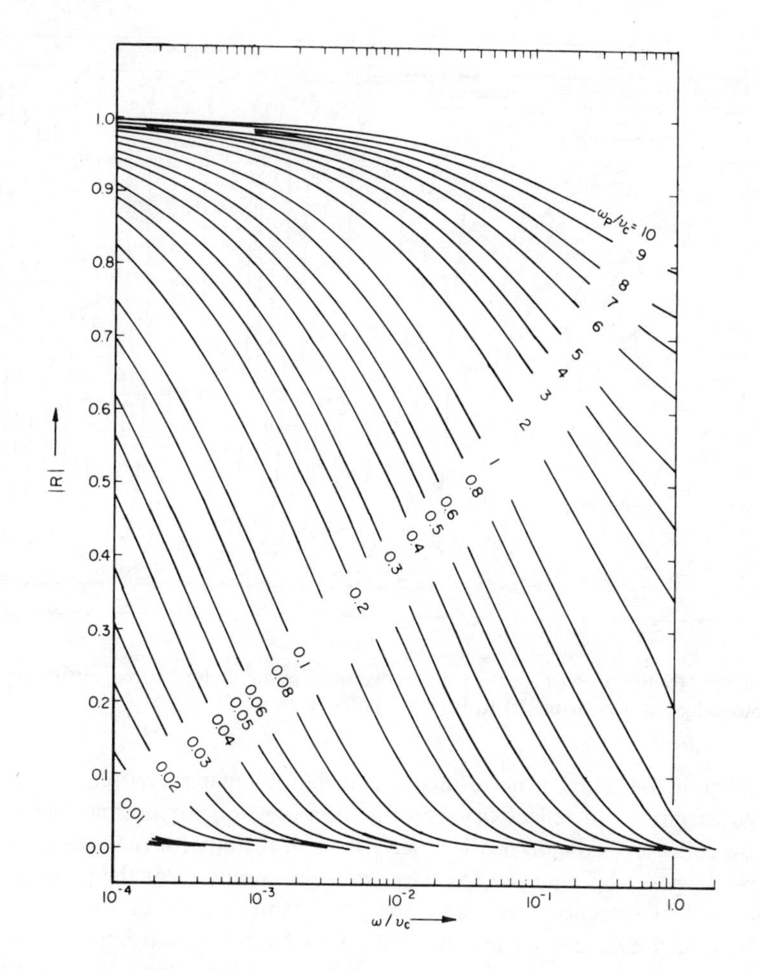

Fig. 10-5. Reflection coefficient of a homogeneous semi-infinite plasma versus ω/ν_c for values of ω_p/ν_c from 0.01 to 10 [after Hochstim[62]].

where

$$z_2 = [1 - (k_0/k_2)^2 \sin^2 \theta_1]^{1/2} (\mu_0/\epsilon_2)^{1/2} \quad \text{for parallel polarization} \qquad (10.2\text{-}103)$$

$$= \frac{(\mu_0/\epsilon_2)^{1/2}}{[1 - (k_0/k_2)^2 \sin^2 \theta_1]^{1/2}} \quad \text{for perpendicular polarization} \qquad (10.2\text{-}104)$$

and

$$k_2 = (\mu_0 \epsilon_2)^{1/2}, \qquad z_0 = (\mu_0/\epsilon_0)^{1/2} \qquad (10.2\text{-}105)$$

$$\alpha_2 = k_2 d[1 - (k_0/k_2)^2 \sin^2 \theta_1]^{1/2} \qquad (10.2\text{-}106)$$

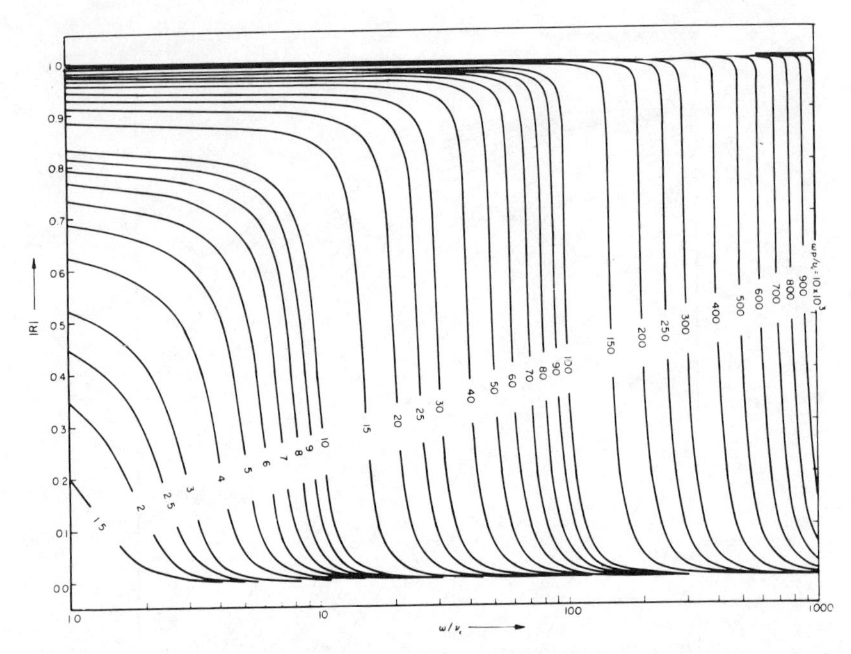

Fig. 10-6. Reflection coefficient of a homogeneous semi-infinite plasma versus ω/ν_c for various values of ω_p/ν_c from 1.5 to 10×10^3 [after Hochstim[62]].

Figure 10-7 shows the results of calculations made by French *et al.*[22] of the variations of reflectivity as a function of ν_c/ω_p for normal incidence. At low collision frequencies, the reflectivity is near unity for radar frequencies below the plasma frequency. A sharp transition occurs near the point where $\omega/\omega_p = 1$. The reflectivity suddenly drops and begins to oscillate. The oscillations are caused by internal reflections from the slab boundaries and are characteristic of a large range of slab thicknesses. Since the multiply-reflected signals are attenuated more rapidly as the collisions increase, the oscillations are damped out at the higher collision frequencies. Also, the average value of the reflectivity decreases as the collision frequency increases. This is illustrated in Figure 10-8. Figure 10-8(a) shows the reflectivity versus the normalized plasma frequency for various collision frequencies for a slab which is one free-space wavelength thick ($L = 1$). Figure 10-8(b) shows the phase angle of reflection. Figure 10-9 shows the variation for several collision frequencies for a slab $1.84\lambda_0$ thick. The corresponding phase angles are given in Figure 10-10.

Nicoll and Basu[63] also illustrate the effects of plasma thickness for $L = \frac{1}{2}$, 1, 2 free-space wavelengths. Figure 10-11 illustrates the reflection coefficient amplitude (a) and phase (b) from a loss-free plasma. Figure 10-12 gives the amplitude (a) and phase (b) for a plasma with a collision frequency

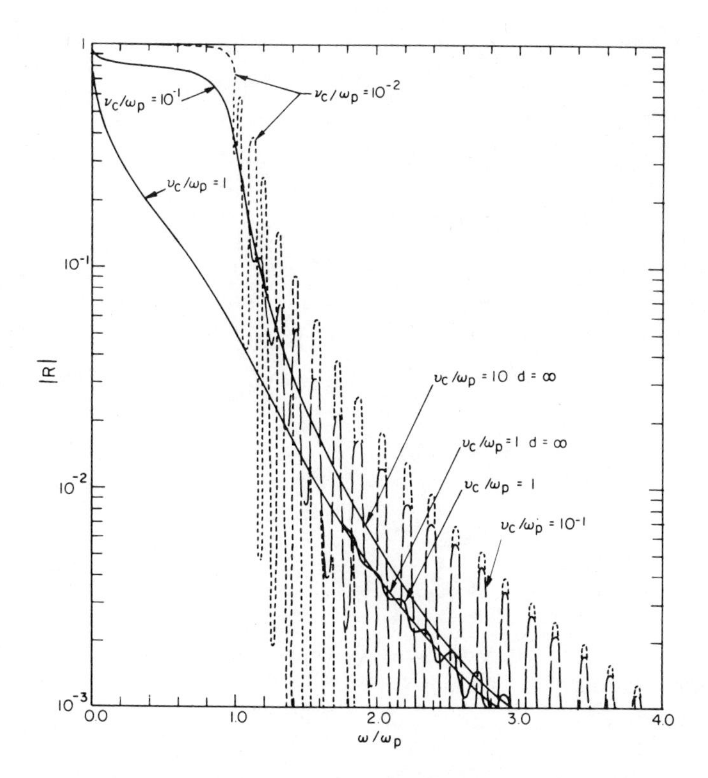

Fig. 10-7. Reflection coefficient of an isotropic plasma slab as a function of frequency for various electron collision frequencies. Slab thickness 2.5 λ_p [after French *et al.*[22]].

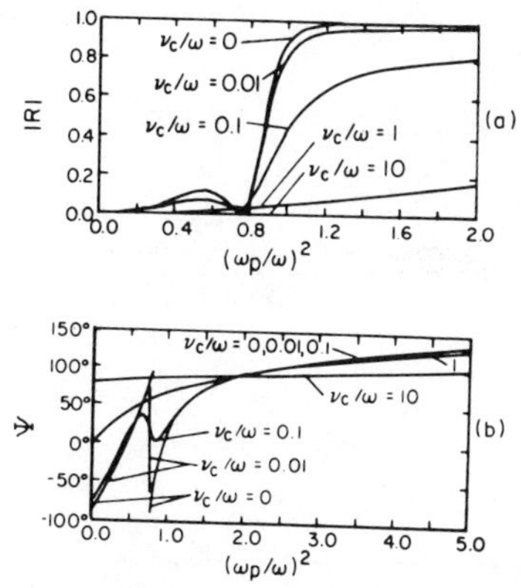

Fig. 10-8. Reflection characteristics of a one-wavelength-thick plasma slab versus ω_p/ω for several values of ν_c/ω, rectangular distribution of electron density [after Nicoll and Basu[63]].

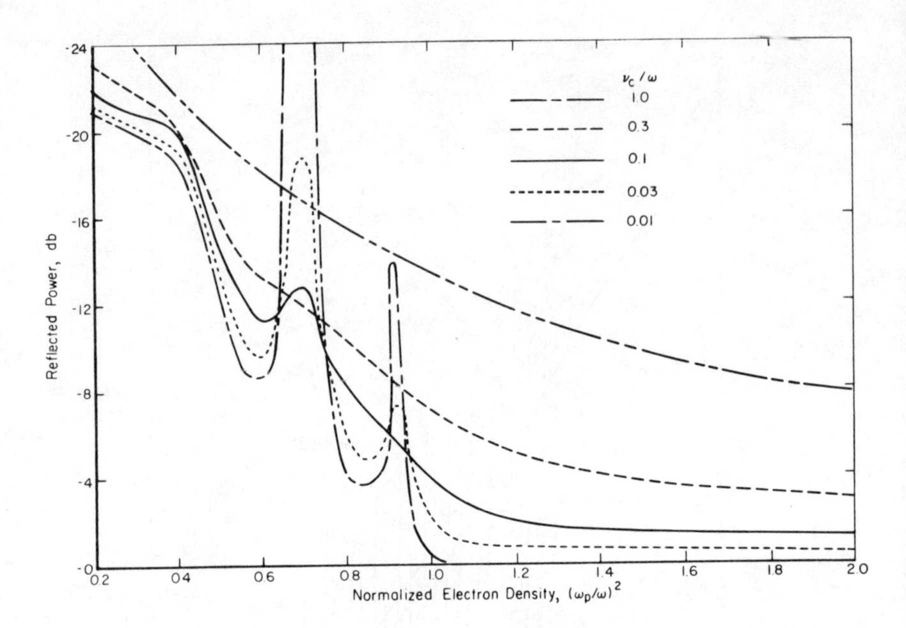

Fig. 10-9. Variation of reflected power with normalized electron density from a plasma slab of thickness $1.84\lambda_0$ bounded by free space [after Bachynski and Graf[24]].

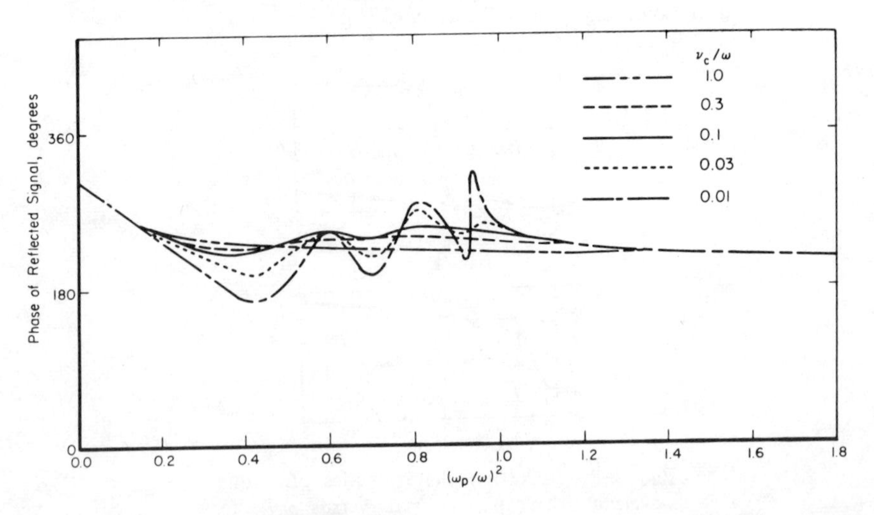

Fig. 10-10. Variation with normalized electron density of the phase of the reflected wave from a plasma slab of thickness $1.84\lambda_0$ bounded by free space [after Bachynski and Graf[24]].

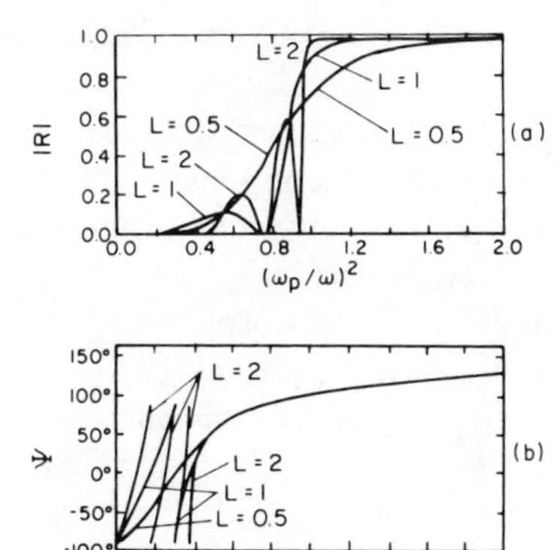

Fig. 10-11. Reflection characteristics of a collisionless plasma with a rectangular distribution of electron density [after Nicoll and Basu[63]].

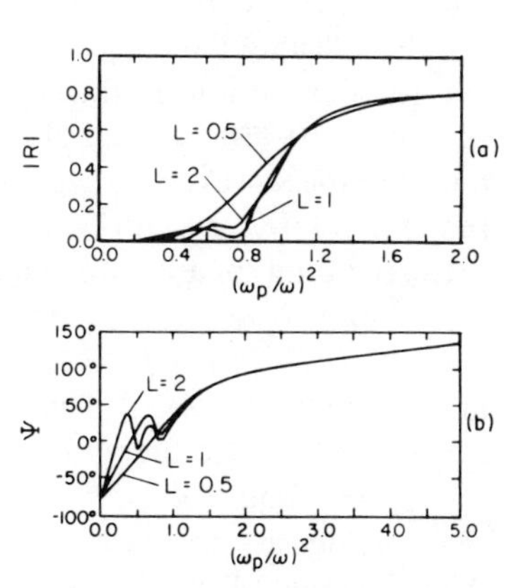

Fig. 10-12. Reflection characteristics of a plasma with a rectangular distribution of electron density for $\nu_c/\omega = 0.1$ [after Nicoll and Basu[63]].

one-tenth that of the observing frequency. For thin plasma, there is not sufficient phase change to allow cancellation. For thicker plasma, the cancellations and enhancements result in oscillations as the electron density changes.

For the case of normal incidence, if one is interested in the average effects of a layered slab and not the individual oscillations, the reflectivity may be calculated by assuming the power reflected at each interface is incoherent. The resulting expression is

$$\langle | R | \rangle \approx \frac{b[1 + (1 - 2b)\exp(-4\alpha l)]}{1 - b^2 \exp(-4\alpha l)} \tag{10.2-107}$$

where

$$b = |(1 - \sqrt{\epsilon})/(1 + \sqrt{\epsilon})|^2 \tag{10.2-108}$$

and l is the overall length measured in meters.

Inhomogeneous ionized media have been considered by many authors. Miller[64] has solved the wave equation for an electron density which increases exponentially into the medium, with solutions in the form of Bessel functions of complex argument and imaginary order. Klein *et al.*[65] have combined numerical techniques with the WKBJ method[3] to obtain the reflection coefficient for a variety of situations. The electron density changes only along the normal to the plasma surface. In the region of rapid variation of the index of refraction, the wave equation is solved by numerical integration. In the region of slow variation, the ray peak was calculated by geometrical optics, and the absorption was obtained by the WKBJ method. The solutions were then matched. Only the short-wavelength case was considered. For small values of ν_c/ω, the amount of energy reflected is small except for sharp density discontinuities. For $\nu_c/\omega \geqslant 10$, the reflection coefficient is almost independent of ν_c/ω.

Nicoll and Basu[63] have also considered the reflections from electron density distributions in the shapes of half of a cosine wave and of a triangle. In Figure 10-13(a) the amplitude of the reflection coefficient is given for a one-free-space-wavelength slab with each of these two distributions as a function of $(\omega_p/\omega)^2$ for various values of ν_c/ω. Figure 10-13(b) gives the corresponding phases. In Figure 10-14, $|R|$ and Ψ are plotted against $(\omega_p/\omega)^2$ for $L = \frac{1}{2}$, 1, 2 for a lossless plasma, and in Figure 10-15 for a plasma with $\nu_c/\omega = 0.1$. The curves no longer have the oscillations seen in the previous curves. This is because of the gradual boundary between free space and the highly ionized region.

Albini and Jahn[66] have performed calculations for a slab where electron density increases linearly. The solution is given in terms of the Airy function. Penico[67] obtained a similar solution in terms of Hankel functions. Golant[68] has considered a reflection from a linearly increasing electron gradient for the case where $\omega \gg \nu_c$ and found a solution in terms of Bessel functions of orders $\pm\frac{1}{3}$. Albini and Jahn[66] calculated numerical

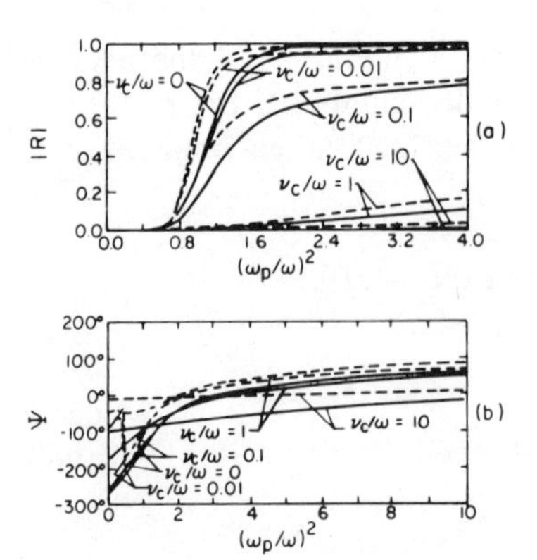

Fig. 10-13. Reflection characteristics of a plasma slab of one-wavelength thickness for fixed values of ν_c/ω; solid lines: triangular distribution of electron density; dotted lines: half-cosine distribution [after Nicoll and Basu[63]].

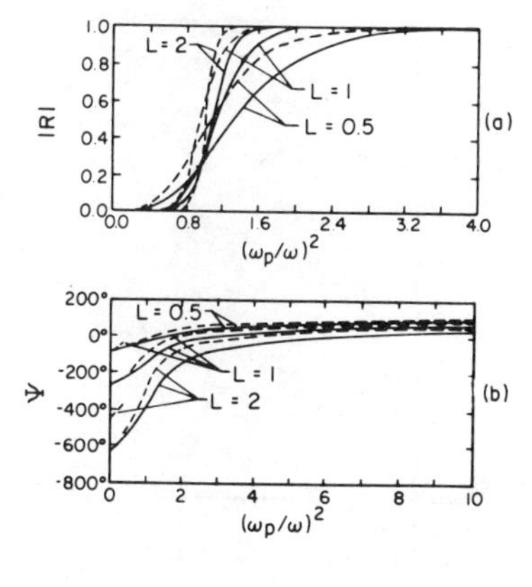

Fig. 10-14. Reflection characteristics for a collisionless plasma; solid lines: triangular distribution of electron density; dotted lines: half-cosine distribution [after Nicoll and Basu[63]].

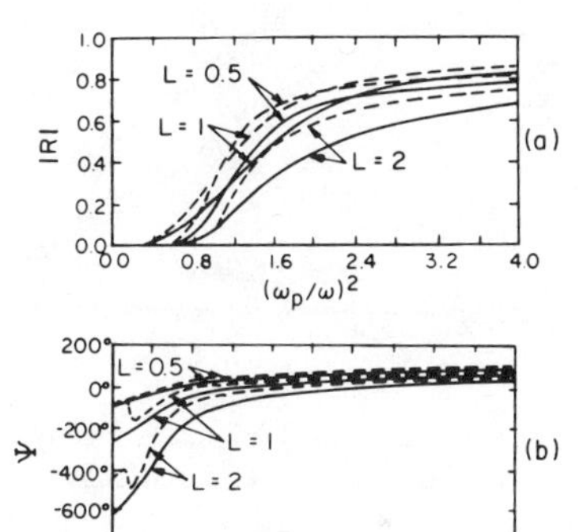

Fig. 10-15. Reflection characteristics for a plasma with $\nu_c/\omega = 0.1$; solid lines: triangular distribution of electron density; dotted lines: half-cosine distribution [after Nicoll and Basu[63]].

results for a slab with an electron density which is uniform in the center and varies linearly at both edges, i.e., a trapezoidal electron-density distribution. The overall width of the slab is z_t. The length of each linear range is z_0. Thus if the origin is placed at the front edge of the slab, the coordinates of the uniform portion of the slab are z_0 and $z_t - z_0$. The results of this calculation are shown in Figure 10-16. In Figure 10-16(a) and (b), the amplitudes of the reflected signal and of the transmitted wave are shown as a function of $(z_t - z_0)/\lambda_0$ for various values of z_0/λ_0. In Figure 10-16(a), the permittivity is chosen to be 0.25. In Figure 10-16(b), the permittivity is chosen to be $0.24 + i0.10$. The corresponding phase shifts are shown in Figures 10-16(c) and (d), respectively.

The more gradual the ramp, the less the signal is reflected from the plasma. The positions of the maxima and minima of both the reflected and transmitted signals depend on the total electron density or the physical length of the central uniform plasma slab, and not on the actual physical dimensions of the slab-and-ramp combination until the ramp dimensions become quite

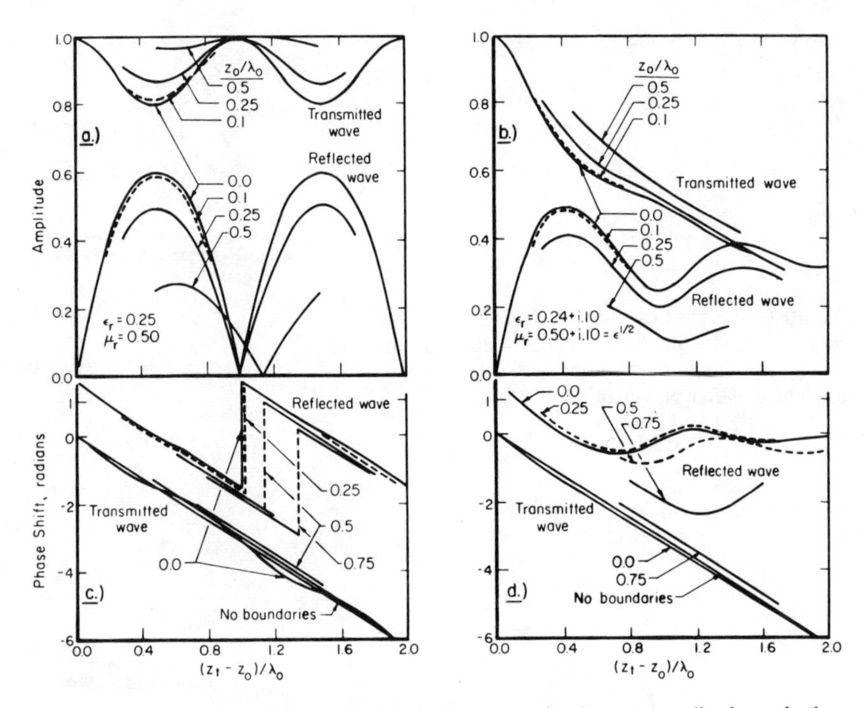

Fig. 10-16. Effect of spatial distribution of electron density on amplitude and phase shift of transmitted and reflected waves [after Bachynski and Graf[24]].

significant ($z_0/\lambda_0 \geqslant 0.50$). In the presence of losses, the signals become less sensitive to the shape of the boundaries.

Albini and Jahn[66] compute the total phase shift Φ_T of the transmitted wave upon passing through a plasma slab of thickness z_t, since this includes the free-space path as well. The form of the phase change introduced by the plasma, $\Delta\Phi$, was obtained by subtracting the phase change in a path length z_t in free space from Φ_T.[261] Thus

$$\Delta\Phi = 2\pi(z_t/\lambda_0) - |\Phi_T| \tag{10.2-109}$$

The phase shift $\Delta\Phi$ is plotted in Figure 10-16(c) and (d). It is very nearly the same as the phase shift for an unbounded plasma. The effect of boundaries is to make the phase undulate slightly about the no-boundary value. The influence of losses does not introduce any significant modifications.

For the phase of the reflected signal, the reflections are considered to occur from the slab as if the boundary was located at the midpoint between the points where the plasma starts and where the maximum electron density has been reached. This is again replacing the slab by an equivalent (in total electrons) uniform slab of the maximum density. Therefore $(2\pi/\lambda_0)\, z_0$ is added to the phase of the reflected wave, Φ_R, as calculated by Albini and Jahn. Thus

$$\Delta\Phi_R = \Phi_R + (2\pi/\lambda_0)\, z_0 \qquad (10.2\text{-}110)$$

As is shown, only at values of $z_0/\lambda_0 > 0.5$ does the character of the reflected phase depart notably from that of a uniform slab whose electron density is the same as the maximum of the trapezoidal electron distribution and which contains the same total number of electrons as the trapezoidal slab.

The reflection from a layer of isotropic plasma, which has an electron density variation with distance that has two maxima separated by a nonzero minimum, cannot be solved in terms of a simple set of known wave functions. Rydbeck and Rönnäng[69] have considered the problem in principle. A precise evaluation of the amplitude and phase would require numerical evaluation by a computer.

Bein[70] has recently considered the inhomogeneous slab by dividing it into many adjacent homogeneous slabs. He found good agreement between his method and other approximate and exact solutions for several electron-density and collision-frequency distributions.

Yeh[71] has considered the reflection coefficient of moving, cold, semi-infinite plasma slabs. He found that when the ionized medium moves parallel to the interface, the reflected fields for perpendicular polarization are independent of the movement of the medium. However, when the plasma moves perpendicular to the interface, the reflection coefficient is a rather complicated function of the angle of incidence, velocity, and plasma parameters. For normal incidence on semi-infinite plasma where $\omega_p/\omega > 1$, the reflection coefficient is always greater than unity if the half-space is moving toward the incident wave. That is, the reflected energy is greater than the incident energy.

A considerable number of other planar-plasma region solutions have been found. These include a simple exact solution of the exponential plasma profile by Taylor.[72] Taylor has also solved the linear ramp and parabolic ramp,[73] and a special case of the inverse parabolic ramp.[74] Klein extended earlier work (of Reference 65) to include numerical and WKBJ solutions of both exponential and inverse parabolic profiles adjacent to a conducting wall,[75] a Gaussian plasma slab profile,[76] and the case where radiation originates within a dielectric medium which is in contact with the inside of an exponential-profile plasma sheet.[77] Yen[78] has numerically determined approximate solutions for a modified exponential ramp which starts at zero electron density and asymptotically reaches a maximum. The reflection of pulses from plasma have been considered by Schmitt[79,80] and Case.[81] Most of these calculations have been made on the assumption of constant electron-collision frequency. This assumption is often reasonable in very weakly ionized plasmas but, in general, changes in collision frequency, as well as changes in electron density, affect the permittivity and thus change the reflectivity.

The classic anisotropic layer is the Epstein layer.[82] In this layer, there is a parabolic variation of the permittivity. Consider a layer such that the z axis is directed into the layer. The x direction is chosen such that

$$\mathbf{B} \cdot \hat{\mathbf{x}} = 0 \tag{10.2-111}$$

where $\mathbf{B}$ is the external magnetic field. The coordinate system is located in such a manner that $z = 0$ occurs in the middle of a layer of thickness l. The electron-density variation is such that

$$\omega_p{}^2 = \omega_{p,\max}^2 \, [1 - (2z/l)^2] \tag{10.2-112}$$

where $\omega_{p,\max}$ is the maximum angular plasma frequency of the layer. The relative permittivity of the layer for a plane wave incident normally on the layer is

$$\epsilon_r = 1 - \frac{\omega}{\omega_{p,\max}^2} [\omega^2 - 2\omega\omega_H + \omega_H{}^2 + \nu_c{}^2]^{1/2} \times \left[1 - \left(\frac{z}{l/2}\right)^2\right] \exp\left(\arctan \frac{\nu_c}{\omega - \omega_H}\right) \tag{10.2-113}$$

The reflection coefficient for this case, as given by Rydbeck,[83] is, for $u < 1$

$$|\, R \,| = \left[\frac{1 + u^2 + 2u \cos \xi}{1 + u^2 - 2u \cos \xi}\right]^\eta \exp\left[(a_1 \cos \xi)(1 - u^2) \arctan \left(\sin \xi \frac{2u}{1 - u^2}\right)\right] \tag{10.2-114}$$

$$\eta = \tfrac{1}{2} a_1 (1 + u^2) \sin \xi$$

$$u^2 = (\omega/\omega_{p,\max}^2)[\omega^2 - 2\omega_H\omega + \omega_H^2 + \nu_c{}^2]^{1/2} \tag{10.2-115}$$

and

$$\xi = \tfrac{1}{2} \arctan[\nu_c/(\omega - \omega_H)] \tag{10.2-116}$$

For $u > 1$, the exponential factor in Eq. (10.2-114) is replaced by

$$\exp\left[(a_1 \cos \xi)(u^2 - 1) \arctan \left(\sin \xi \frac{2u}{u^2 - 1}\right) - (\pi a_1 \cos \xi)(u^2 - 1)\right] \tag{10.2-117}$$

For $\nu_c = 0$, Eq. (10.2-114) predicts that $|\, R \,| \approx 1$, but, for the case where $u > 1$, using the modification to Eq. (10.2-114) given by Eq. (10.2-117) results in

$$|\, R \,| \approx \exp[-a_1 \pi (u^2 - 1)] \tag{10.2-118}$$

which indicates that there is still reflection for $\omega > \omega_p$.

For the case of low losses, where ξ is approximately equal to $\frac{1}{2}\nu_c/(\omega - \omega_H)$, then the reflection coefficients are given by

$$| R | \approx \left[\frac{1+u}{1-u}\right]^{-\zeta} \exp\left[\frac{\nu_c}{\omega - \omega_H} - a_1(u^2 - 1)\right] \qquad (10.2\text{-}119)$$

for $u < 1$, $\nu_c^2/(\omega - \omega_H)^2 \ll 1$; and

$$| R | \approx \left[\frac{u+1}{u-1}\right]^{-\zeta} \exp\left[\frac{\nu_c a_1 u}{\omega - \omega_H} - a_1\pi(u^2 - 1)\right] \qquad (10.2\text{-}120)$$

for $u > 1$, $\nu_c^2/(\omega - \omega_H)^2 \ll 1$. In both equations, $\zeta = \nu_c a_1(1 + u^2)/2(\omega - \omega_H)$. For the case where the external field is zero, setting $\omega_H = 0$ in Eqs. (10.2-113)–(10.2-120) gives the isotropic results.

Additional discussion of planar anisotropic media is available in the literature. See References 1–4, 14–17, and 80–85 for examples.

10.2.2.2. *Cylindrical Regions*

10.2.2.2.1. *Isotropic Circular Cylinders.* Numerous solutions for scattering from dielectric cylinders and dielectric-coated metallic cylinders are given in Chapter 4. These are valid for plasma and plasma-coated cylinders provided the plasma can be represented at each point on a macroscopic basis by a value of permittivity (for a discussion of when this representation is not valid, see Reference 86.) This section will review briefly some special forms pertinent primarily for plasma cylinders. Effects of turbulence are considered in Section 10.2.3. Some applications in which cylindrical geometry is important are discussed in Section 10.3, especially 10.3.1 (meteors) and 10.3.2 (auroras). The coordinate system used is illustrated in Figure 10-17 for normal incidence.

The modified geometrical-optics approximation developed at the Ohio State University (see Section 4.1.3.4) has been applied to this problem by Swarner and co-workers[87–92] and by Thomas.[93] Defining the radar echo width as

$$\sigma^c = \lim_{r\to\infty} 2\pi r \mid E^s \mid^2 / \mid E^i \mid^2 \qquad (10.2\text{-}121)$$

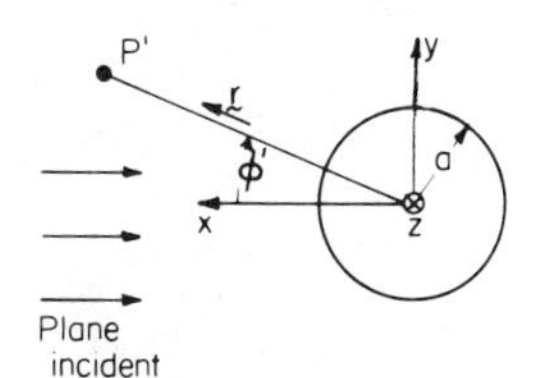

Fig. 10-17. Scattering geometry for a circular plasma cylinder, with axis perpendicular to the page.

they write for backscatter

$$\sigma^c = \pi a \left\{ R_{12} + T_{12}T_{12}R_{21} \left(\frac{\sqrt{\epsilon_r}}{2-\sqrt{\epsilon_r}} \right)^{1/2} \exp \left[-i \left(4k_0 a - \frac{\pi}{2} \right) \right] \right\}^2 \qquad (10.2\text{-}122)$$

where R_{12} is the reflection coefficient at normal incidence for a wave striking the cylindrical boundary from the outside, T_{12} is the transmission coefficient at normal incidence for a wave entering the cylinder, R_{21} is the reflection coefficient at normal incidence for a wave striking the cylindrical boundary from the inside, T_{21} is the transmission coefficient at normal incidence for a wave leaving the cylinder. The approximation of Eq. (10.2-122) is polarization-insensitive. Figure 10-18 compares the results of Eq. (10.2-122) with the exact solution from an infinite plasma dielectric cylinder with relative permittivity equal to $\frac{1}{2}$ and $\frac{3}{4}$. It is seen that the approximate values agree surprisingly well with the exact results.[92,337]

Peters and Green[94] consider a long, thin, metallic cylinder coated by plasma as an antenna which receives and reradiates energy. This reradiated energy can be thought of as reflected from the mismatch which exists at fictitious antenna terminals. The echo area of the antenna is given by

$$\sigma(\theta, \phi) = \gamma_r^2 [G^2(\theta, \phi)/4\pi] \lambda_0^2 \qquad (10.2\text{-}123)$$

where γ_r is the voltage reflection coefficient at the antenna terminals and $G(\theta, \phi)$ is the antenna gain. Peters[95] uses as a value of $G(\theta, \phi)$

$$G = 7L/\lambda_0 \qquad (10.2\text{-}124)$$

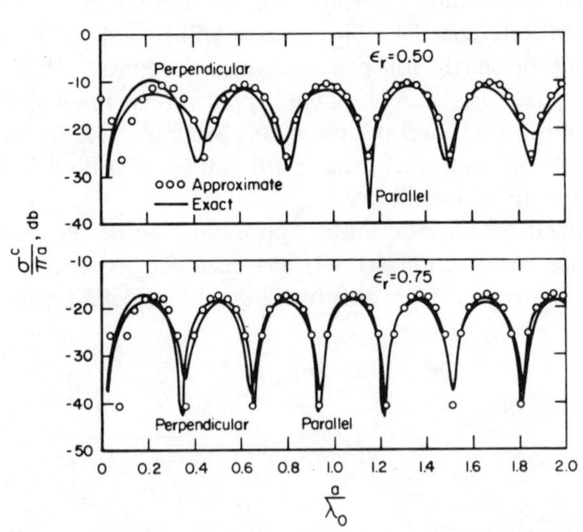

Fig. 10-18. Comparison of exact and approximate backscatter echo widths of an infinite circular plasma cylinder with relative permittivity $\epsilon_r = 0.50, 0.75$ [after Peters *et al.*[92]].

where L is the length of the cylinder. He points out that Eq. (10.2-124) is approximate since it requires that a uniform amplitude distribution be present along the cylinder, which is not true here. He also approximates γ_r by the Fresnel reflection coefficient, i.e.,

$$\gamma_r^2 = \left[\frac{\sqrt{\epsilon_r} - 1}{\sqrt{\epsilon_r} + 1}\right]^2 \tag{10.2-125}$$

where ϵ_r is given by Eq. (10.2-13).

Tang[96] has considered the problem of scattering from plasma columns in a one-component compressible electron gas without a magnetic field. Using Cohen's[97] set of linearized plasma equations, he decomposed the vector field into three components, each of which is represented by a scalar potential function. Electron–electron interactions are considered only through the introduction of an isotropic pressure term. General expressions for the cross sections for both polarizations are given, including the effects of the excitation of longitudinal oscillations. These expressions are very complex and are not included here. When the conductivity of the plasma column is allowed to become infinite and the permittivity of the exterior plasma is allowed to go to that of free space, Tang's equations reduce to those obtained in scattering from a thin metallic wire.

Gildenburg and Kondrat'ev[98] studied the influences of the thermal motion of electrons on the diffraction characteristics of a plasma cylinder. Using the hydrodynamic approximation (see Section 10.2.1.1.1.2), they find that all the problems which are solved by the method of separation of variables in the case of a cold plasma may be generalized without much difficulty for a hot plasma.

The problem of plasma resonance is discussed by Herlofson,[99] who has shown that a uniform plasma cylinder has a natural mode of oscillation at the frequency $\omega_p/\sqrt{2}$ for an incident field with the electric vector parallel to the axis. Fejer[100] considers this resonance, and solves the problem of scattering from a homogeneous cylinder analytically. He obtains numerical solutions for two cases in which the plasma is radially inhomogeneous. The thermal motion of the electrons is taken into account approximately by a scalar pressure term. The results of the calculations show that resonant scattering takes place at several frequencies. At the lowest resonant frequency, the thermal motion of the electrons plays little part, and the plasma behaves essentially as a dielectric. Thermal motion plays a prominent part in the resonances which occur at the higher frequencies; these are caused by standing "plasma waves" (electron acoustic waves). If the plasma cylinder is homogeneous and its diameter is much greater than the Debye length, these higher resonant frequencies are rather closely spaced. (Fejer obtains resonance at $\omega/\omega_p \approx 1.0037, 1.0098, 1.0187, 1.0303,\ldots$ besides the resonance at $\omega/\omega_p = 0.707$.) The separations between the resonant frequencies increase, however, if the electron density is taken to be substantially greater on the axis of the cylinder than at the boundary. Fejer states that the results of the computations for such an inhomogeneous cylinder appear to be in qualitative agreement with data obtained from experiments on the resonant scattering of microwaves by a gas discharge column.

An extensive set of scattering curves for plasma cylinders, which have electron-density profiles expressed as polynomials, have been calculated by Ridder and Peterson.[101,102] The Born approximation has been used by Midzuno[103] and by Ruquist[104] to calculate the cross section of an electron-density distribution, which is arbitrary except for the requirement that the plasma frequency must be much less than the radar frequency.

Axial variations of the permittivity make the variation of the scattered field three dimensional. Brysk[105] has used the Born approximation to compute scattering for low-density meteor trails with axially varying permittivity.

10.2.2.2.2. *Anisotropic Cylinders.* Wait[106] and Platzman and Ozaki[107] have considered the case of a homogeneous plasma cylinder in the presence of a uniform, axial, static magnetic field. When the incident field is polarized such that the incident magnetic field is along the axis of the cylinder, the static field effects the scattering; for the other polarization, the static magnetic field has no effect. In the remainder of this section, the incident magnetic field will be assumed to be parallel to the cylinder axis. Platzmann and Ozaki numerically evaluated their equations and found a pronounced resonance at $\phi = \pi/2$ for ω near (but less than) the cyclotron frequency ω_H. Adachi[108] demonstrated that this theory predicts that the bistatic pattern is antisymmetrical, i.e., $H_z^s(\phi) = H_z^s(-\phi)$ for $0 < \phi < \pi$. Messiaen and Vandenplas[109] experimentally find reasonably good agreement with this theory. The modified geometrical-optics approximation has also been applied to this problem. Wait[106] presented a solution for a plasma cylinder consisting of concentric layers with different properties.

Messiaen and Vandenplas[110] have studied, both theoretically and experimentally, the existence of a scattered E_z field, even for normal incidence. This phenomenon represents a depolarization which does not occur for normal dielectric cylinders. This scattered E_z field exists even when the axial magnetic field is zero. The mechanism responsible is the drift velocity. The scattered field has a sin ϕ dependence when the incident field has a cos ϕ dependence. When an axial magnetic field is present, a more complicated spectrum is obtained.

Wilhelmsson[111] has calculated the backscatter cross section of a homogeneous cylinder in a uniform, axial magnetic field. He found a very sharp high peak in an otherwise almost constant cross section versus angle of incidence. The position of the peak depends upon the plasma frequency, and the height of the peak is a function of both the plasma frequency and the size of the cylinder. He presents curves illustrating this effect for $\omega_H = 4\omega$.

Dnestrovskiy and Kostomarov[112] consider a plasma cylinder with a diffuse boundary. The unperturbed electron-distribution function is assumed to be Maxwellian. The mean Larmor radius [$(1/\omega_H)\sqrt{2T/m_e}$, where T is the electron temperature] is assumed to be much less than both the radius of the cylinder and the free-space wavelength. Using the known solution for the diffraction of a dielectric cylinder, they reduced the diffraction problem to an integral equation which is solved by successive approximations. They report that their technique is valid for any shape of plasma cylinder, provided the solution of the corresponding dielectric cylinder is known.

Chen and Cheng[113,114] have considered the scattering by an infinitely long, anisotropic, plasma-coated conducting circular cylinder. Their formulation contains the scattering from an anisotropic plasma column and from an isotropic plasma-coated conducting cylinder as special cases. Rusch and Yeh[115,116] have obtained numerical results for normal incidence on an infinite cylinder of radius b coated with inhomogeneous and anisotropic plasma of thickness $a - b$. A parabolic electron-density profile is assumed. The backscattering cross section per unit length is given by

$$\sigma^c = \frac{4}{k_0}\left[\sum_{n=-\infty}^{\infty} B_n \exp(-in\pi/2)\right]^2 \tag{10.2-126}$$

The far-field pattern of the scattered wave is proportional to

$$\xi(\phi) \sim \left|\sum_{n=-\infty}^{\infty} B_n \exp[-in(\phi - \pi/2)]\right| \tag{10.2-127}$$

where B_n is given by

$$B_n = \frac{(-i^n)}{\Delta}\left\{P_n(k_0a, k_0b)\, J_n'(k_0a) + \frac{\epsilon_0 J_n(k_0a)}{[\epsilon_2^2(a) - \epsilon_1^2(a)]} \times \left[\frac{\epsilon_2(a)\, n}{k_0a} P_n(k_0a, k_0b) + \epsilon_1(a)\, S_n(k_0a, k_0b)\right]\right\} \tag{10.2-128}$$

with

$$\Delta = \frac{-\epsilon_0 H_n^{(2)}(k_0a)}{[\epsilon_2^2(a) - \epsilon_1^2(a)]}\left[\frac{\epsilon_2(a)\, n}{k_0a} P_n(k_0a, k_0b) + \epsilon_1(a)\, S_n(k_0a, k_0b)\right] - P_n(k_0a, k_0b)\, H_n^{(2)\prime}(k_0a) \tag{10.2-129}$$

and

$$P_n(k_0a, k_0b) = R_n^{(1)}(k_0a) - Q_n(k_0b)\, R_n^{(2)}(k_0a) \tag{10.2-130}$$

$$S_n(k_0a, k_0b) = R_n^{(1)\prime}(k_0a) - Q_n(k_0b)\, R_n^{(2)\prime}(k_0a) \tag{10.2-131}$$

$$Q_n(k_0b) = \frac{[\epsilon_2(b)/\epsilon_1(b)](n/k_0b)\, R_n^{(1)}(k_0b) + R_n^{(1)\prime}(k_0b)}{[\epsilon_2(b)/\epsilon_1(b)](n/k_0b)\, R_n^{(2)}(k_0b) + R_n^{(2)\prime}(k_0b)} \tag{10.2-132}$$

$J_n(k_0r)$ and $H_n^{(2)}(k_0r)$ are, respectively, the Bessel function and the Hankel function of the second kind. The prime denotes the derivative of the function with respect to its argument. $R_n^{(1),(2)}(k_0r)$ are the two independent solutions of

$$\frac{d^2R(r)}{dr^2} + \frac{1}{rg(r)}\frac{d}{dr}[rg(r)]\frac{dR(r)}{dr} + \frac{1}{g(r)}\left\{\omega^2\mu_0\epsilon_0 \pm \frac{n}{r}\frac{d}{dr}\left[\frac{\epsilon_2(r)}{\epsilon_1(r)}\, g(r)\right] - \frac{n^2}{r^2}\, g(r)\right\} R(r) = 0 \tag{10.2-133}$$

where

$$g(r) = \frac{\epsilon_0\epsilon_1(r)}{\epsilon_1{}^2(r) - \epsilon_2{}^2(r)} \tag{10.2-134}$$

$$\epsilon_1(r) = \epsilon_0\left[1 - \frac{\omega_p{}^2(r)/\omega^2}{1 - (\omega_H{}^2/\omega^2)}\right] \tag{10.2-135}$$

$$\epsilon_2(r) = -\epsilon_0\left[\frac{\omega_p{}^2(r)\,\omega_H}{\omega(\omega^2 - \omega_H{}^2)}\right] \tag{10.2-136}$$

$$\omega_p{}^2(r) = \frac{e^2A_1}{m_e\epsilon_0}\left[1 - \frac{r^2}{A_2}\right] \tag{10.2-137}$$

and A_1 and A_2 are arbitrary constants. Numerous curves of bistatic scattering for a variety of cylinder sizes and sheath thicknesses are presented by Rusch and Yeh.[115,116] The inner metallic cylinder dominates the backscattering cross section except near the upper hybrid resonance frequency found from

$$\omega^2 = \omega_H{}^2 + \omega_p{}^2 \tag{10.2-138}$$

The results are plotted in normalized parameters, but the authors suggest that a typical curve could represent a cylinder with a total (plasma plus metallic cylinder) radius a_0 of 10 cm, $\lambda_0 = 21.3$ cm (1430 Mcps), magnetic flux density of 510 gauss, and an axial electron density of 0.2×10^{16} electrons/m^3. Seshadri[117] has also presented numerical results for a perfectly-conducting circular cylinder embedded in a gyrotropic medium, with both gyrotropic axis and the magnetic vector of the incident plane wave parallel to the axis of the cylinder. Calculations of backscatter cross section as a function of cylinder size and of bistatic cross section for various cylinder sizes were presented for the case when the gyrofrequency ω_H is equal to the plasma frequency ω_p.

10.2.2.3. *Spherical Regions*

10.2.2.3.1. *Plasma Spheres*

10.2.2.3.1.1. *Homogeneous Spheres.* Various cases of scattering from dielectric spheres are discussed in Chapter 3. Exact solutions for the scattering from dielectric spheres are given by Eqs. (3.3-3), (3.3-4), and (3.1-6)–(3.1-9), which are valid for all values of permittivity. The exact solution has the disadvantage that the effects of varying a parameter, such as collision frequency, electron density, or radar frequency, are not easily seen. Numerous authors have given a variety of approximate solutions for plasmas with various material parameters.

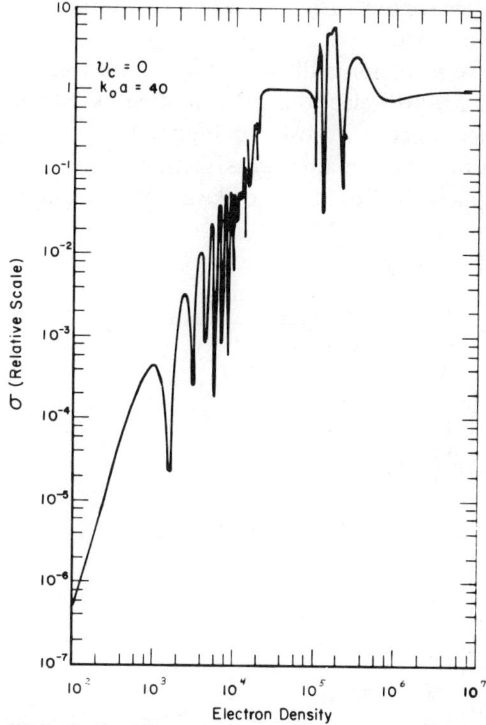

Fig. 10-19. The relative radar cross section as a function of electron density for a collisionless homogeneous plasma sphere with $k_0a = 40$. The incident frequency of 1272.36 Mcps corresponds to a critical density of 2.0082×10^4 meter^{-3} [after Wyatt[119]].

Wyatt[118,119] has considered scattering from large homogeneous plasma spheres. He examined the concept that a large overdense ionized volume may be treated as a perfect reflector. He applied the general equations of Chapter 3, Section 3.3 to the case of a homogeneous sphere of radius a, and programed the resulting expressions. For a sphere of radius $1.5m$ and k_0a of 40, the radar cross section as a function of density within the range 10^2–10^7 electrons/m^3, is shown in Figure 10-19. For a value of k_0a of 40, the critical density where the observing frequency is equal to the plasma frequency is 2.0082×10^4 electrons/m^3. "Anomalous" scattering occurs for densities from about four to about $k_0a/2$ times the critical density. A detailed plot of the cross-section behavior in this region is given in Figure 10-20. This behavior is caused by interference effects between the backscattered wave and tightly bound surface waves. These surface waves have small attenuation around the sphere as compared to creeping waves. This phenomenon is similar to that reported by Probert-Jones[120] on the scattering of microwaves by ice spheres. Thus a plasma has characteristics of a dielectric even when it is overdense.

The surface-wave behavior is illustrated in Figure 10-21. The cross section for a density of 1.5×10^5 electrons/m^3 is plotted as a function of k_0a. This density was chosen since, as shown in Figure 10-19, it is in the region of large deviation from the metallic cross section. It is seen from the figure that the plasma sphere does not behave like a metallic sphere for this situation until k_0a is in the vicinity of 200.

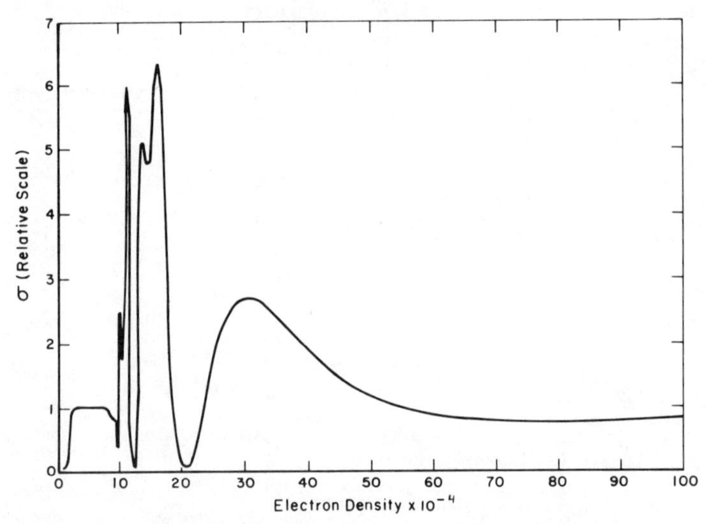

Fig. 10-20. Expanded portion of Figure 10-19 in the vicinity of anomalous behavior [after Wyatt[119]].

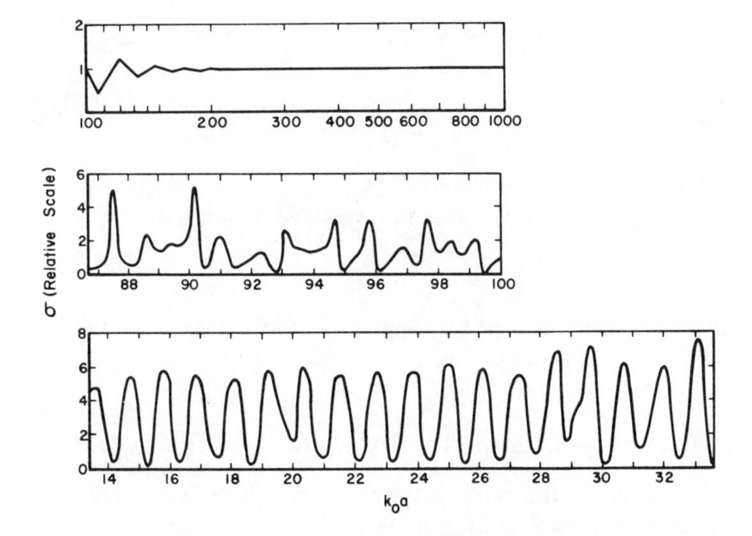

Fig. 10-21. The relative radar cross section of a collisionless homogeneous plasma sphere as a function of k_0a over several ranges at an incident frequency of 1272.36 Mcps. The electron density of 1.5×10^5 meter^{-3} corresponds to a region of considerable anomaly in Figures 10-19 and 10-20 [after Wyatt[119]].

10.2.2.3.1.2. *Inhomogeneous Spheres.* Mar'in[121] has considered scattering from three cases of spherically-shaped plasma regions and solved them by the method of geometrical optics. For a hyperbolic distribution of electrons, the permittivity is given by

$$\epsilon_r = 1 - (a_0^2/r^2) \tag{10.2-139}$$

where a_0 is the radius for which $\epsilon_r = 0$. The backscatter cross section of the plasma sphere at radius a_0 is given as

$$\sigma = (4/\pi^2)\, \sigma_{p.c.} \tag{10.2-140}$$

where $\sigma_{p.c.}$ is the backscatter cross section of a perfectly conducting sphere of radius a_0.

For a parabolic distribution of electrons, the permittivity is given by

$$\begin{aligned} \epsilon_r &= 1 - b^2[1 - (r^2/a_n^2)], \qquad & r \leqslant a_n \\ &= 1 & r > a_n \end{aligned} \tag{10.2-141}$$

where b is an arbitrary constant such that $1 - b^2 < 0$, and a_n is the radius where $\epsilon_r = 1$ (free space). The backscatter cross section for the plasma sphere of radius a_n is given by

$$\sigma = \frac{\pi a_n^2}{1 + [1/(b^2 - 1)\, b]} \tag{10.2-142}$$

In the limit when $b^2 \to \infty$, the effective cross section goes to πa_n^2.

For an exponential variation of electron density, the permittivity is given by

$$\epsilon_r = 1 - \exp[-b^2(r^2 - a_0^2)] \tag{10.2-143}$$

where b is an arbitrary constant and a_0 is the radius where $\epsilon_r = 0$. The backscatter cross section is given by

$$\sigma = \frac{\sigma_{p.c.}}{\pi^2[(a_0/a_t) + \{\operatorname{arctanh} \sqrt{1 - \exp[-b^2(a_t^2 - a_0^2)]}\}/2b^2 a_0^2]} \tag{10.2-144}$$

with a_t the value of the radius at which the sphere is assumed bounded, and $\sigma_{p.c.}$ the cross section of a perfectly-conducting sphere of radius a_0. The exponential plasma sphere can focus the reflected energy if

$$b^2 a_0^2 \geqslant \frac{\operatorname{arctanh}\{1 - \exp[-b(a_t^2 - a_0^2)]\}^{1/2}}{2[1 - (a_0/a_t)]} \tag{10.2-145}$$

Margulies and Scarf[122] compare the exact solution for scattering from an expanding plasma sphere having an electron density proportional to $1/r^2$ with the scalar solution and the iterated Born approximation. It is found that the exact solution and its scalar approximation can differ by many orders of magnitude, as well as by the appearance of fine structure which may be attributed to diffraction of transverse waves, or to interference between the surface wave and a wave which is scattered geometrically. It is shown that an appropriate Born approximation, however, can give reasonably good results over a small range of frequency, size, and electron density (i.e., $\omega_p < c/a_t$).

Bisbing[123] has considered scattering from a sphere that has an electron-density distribution which decreases exponentially with radial distance. It is a diffuse obstacle which blends into the ambient medium at large distances. He formulates the problem rigorously and numerically integrates the appropriate differential equations on a digital computer. The permittivity is given by

$$\epsilon_r = 1 - \frac{\omega_p^2(0) \exp(-r/b)}{\omega(\omega + i\nu_c)} \tag{10.2-146}$$

where $\omega_p(0)$ is the plasma frequency at the center of the sphere, r is the distance from the center of the sphere, and b is an arbitrary constant. He compares the numerical results with those obtained from a Born approximation given by

$$\sigma = 4\pi \mid g(\theta)|^2(1 - \sin^2\theta \cos^2\theta) \tag{10.2-147}$$

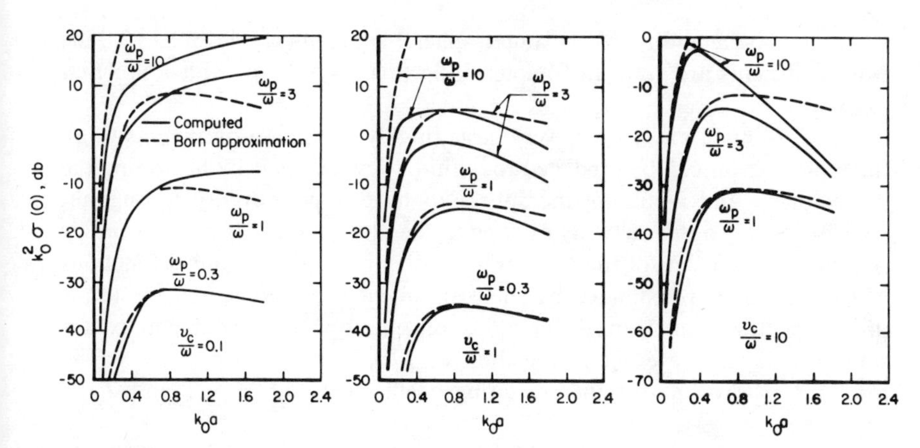

Fig. 10-22. Normalized backscattering cross section of exponentially inhomogeneous plasma spheres [after Bisbing[123]].

where

$$g(\theta) = \frac{k_0}{2\sin(\theta/2)} \int_0^\infty r[1 - \epsilon(r)] \sin\left(2k_0 r \sin\frac{\theta}{2}\right) dr \qquad (10.2\text{-}148)$$

Evaluating Eq. (10.2-147) for the permittivity of Eq. (10.2-146) gives

$$\sigma = \frac{16\pi k_0^4 b^6 \omega_p^4(0)[1 - \sin^2\theta \cos^2\theta]}{\omega(\omega^2 + \nu_c^2)[(1 + 2k_0 b \sin \frac{1}{2}\theta)^2]^4} \qquad (10.2\text{-}149)$$

The results are illustrated in Figure 10-22. This approximation is fairly good at small values of $k_0 b$, and when the plasma is underdense, i.e., $\omega_p^4 < \omega(\omega^2 + \nu_c^2)$.

10.2.2.3.2. *Plasma-Coated Metallic Spheres*

10.2.2.3.2.1. *General Expressions.* Solutions for the scattering from dielectric-covered metallic spheres are given in Chapter 3, Section 3.4. These are also valid for metallic spheres coated with isotropic plasma. Combining Eqs. (3.1-12) and (3.1-13) enables the backscatter cross section to be determined by

$$\sigma(0) = (4\pi/k_0^2) \left| \sum_{n=1}^{\infty} (-i)^{n-1} [n(n+1)/2](A_n + iB_n) \right|^2 \qquad (10.2\text{-}150)$$

where A_n and B_n are the Mie coefficients. For a perfectly conducting metallic sphere of radius a_1 coated with a plasma of thickness $a_0 - a_1$, the Mie coefficients are given by Eqs. (3.4-1), (3.4-2), (3.4-8), and (3.4-9). If the center sphere is not perfectly conducting, Eqs. (3.4-4)–(3.4-7) are used with (3.4-1) and (3.4-2).

For spheres of such size that the product of the free-space wave number

k_0, and a_0, the radius of the outer sphere, is small (say, less than $\frac{1}{2}$), the Mie coefficients are given in Chapter 3, Section 3.4.1.2.1 provided also that $|k_1| a_0 < \frac{1}{2}$.

If a_0 is small and $|k_1|$ is very large (highly overdense plasma coating), the cross section can be predicted to within a few per cent by neglecting the inner sphere and assuming the sphere to have a permittivity throughout which is equal to that actually existing in the plasma shell. If more accuracy is required, the terms of the Mie series involving Bessel or Hankel functions of $k_0 a_0$ can be approximated by small-argument expansions, and the terms involving Bessel or Hankel functions of $k_1 a_0$ by large-argument expansions.[124] These are exact in the limit of $k_0 a_0 \ll 1$ and $|k_1| a_0 \gg 1$. The resulting expressions for the scattering coefficients with $n = 1$ and 2 are

$$A_1 = -\frac{i}{2}(k_0 a_0)^3 \left[1 - \frac{3}{1 + k_0 \sqrt{\epsilon_r}\, a_0 \cot \tau}\right] \tag{10.2-151}$$

$$A_2 = \frac{(k_0 a_0)^3}{27} \left[1 - \frac{\frac{5}{2}}{1 + \frac{1}{2} k_0 \sqrt{\epsilon_r}\, a_0 \cot \tau}\right] \tag{10.2-152}$$

$$B_1 = (k_0 a_0)^3 \left[1 - \frac{\frac{3}{2}}{1 - (\sqrt{\epsilon_r}/k_0 a_0) \cot \tau}\right] \tag{10.2-153}$$

$$B_2 = \frac{i}{18}(k_0 a_0)^5 \left[1 - \frac{\frac{5}{3}}{1 - (\sqrt{\epsilon_r}/2k_0 a_0) \cot \tau}\right] \tag{10.2-154}$$

where $\tau = k_0 \sqrt{\epsilon_r}(a_0 - a_1)$. By writing the coefficients in this manner, one sees that they are related to the scattering coefficients for a metallic sphere minus a complicated correction term that compensates for the fact that part of the sphere is plasma. Since $\sqrt{\epsilon_r}$ is a complex number, the resulting coefficients may be either larger or smaller than the coefficients for a perfectly-conducting sphere of radius b.

The main contribution to the backscattering cross section of Eq. (10.2-150) comes from the term with $n = 1$. Reevaluating Eqs. (10.2-151) and (10.2-153), more exactly one finds that

$$A_1 + iB_1 = -\frac{3i}{2}(k_0 a_0)^3 \left\{ 1 + \frac{\left(\frac{a_0 - a_1}{a_1}\right)\cos\tau - \left(\frac{1 + k_0^2 \epsilon_r a_0 a_1}{k_0 \sqrt{\epsilon_r}\, a_1}\right)\sin\tau}{k_0^2 \epsilon_r a_0^2 \cos\tau + k_0 \sqrt{\epsilon_r}\,\frac{a_0}{a_1}\sin\tau} - \frac{1}{1 + \epsilon_r \left[-1 + \dfrac{k_0^3 \epsilon_r^{3/2} a_0^2 a_1 \cos\tau + k_0^2 \epsilon_r a_0^2 (1 - k_0^2 \epsilon_r a_1^2)\sin\tau}{\left\{\begin{array}{l}(1 + k_0^2 \epsilon_r a_0 a_1)(k_0 \sqrt{\epsilon_r})(a_0 - a_1)\cos\tau \\ + [k_0^2 \epsilon_r (a^2_0 - a_0 a_1 + a^2_1) - 1 - k_0^4 \epsilon_r^2 a_0^2 a_1^2]\sin\tau\end{array}\right\}}\right]} \right\} \tag{10.2-155}$$

Equation (10.2-155) is valid for spheres small with respect to the free-space wavelength. In using Eqs. (10.2-151)–(10.2-154), caution should be taken to avoid letting a_0 or $a_0 - a_1$ shrink to zero, since these equations require τ to be large.

10.2.2.3.2.2. *High-Frequency Limit.* Numerous authors[88–94] have considered the scattering from a plasma-coated metallic sphere by various approximation techniques. In the simplest technique, the metallic sphere of radius a and a plasma coating of thickness $b - a$ and relative permittivity ϵ_r is replaced by a metallic sphere of radius b. However, the cross section σ_m of the metallic sphere is multiplied by the square of the Fresnel reflection coefficient for normal incidence on a semi-infinite plasma of relative permittivity ϵ_r. The resulting cross section is

$$\sigma = \sigma_m[(1 - \epsilon_r)/(1 + \epsilon_r)]^2 \tag{10.2-156}$$

As discussed in Section 2.2.2.3, this is only valid when the Leontovich boundary conditions apply at the surface of the sphere.

A second technique considers the defocusing by the plasma, and replaces the configuration with a metallic sphere of radius equal to $a\sqrt{\epsilon_r}$. This is only valid for a very underdense, nearly-lossless plasma. Under these limitations, however, it gives good agreement with the exact solution.

A technique which gives good results over a wide range of inner- and outer-sphere conditions involves ray-tracing through the ionized shell. Although this technique is a high-frequency approximation, it has been shown to work nearly to the Rayleigh region.[88–93] The technique is discussed in detail in Section 3.3.2.2. The apparent disagreement at the higher frequencies, shown in the figures of Reference 88 obtained by using this technique, was caused by the use of an improper caustic phase shifts.[337] As shown in Section 3.3.2.2, when properly executed, this technique yielded very good agreement at the higher frequencies and good agreement at the lower frequencies. One advantage of this technique is that it can be usefully applied to nonconcentric dielectric-clad spheres.[93]

Musal[125] has given a physical-optics approximation for finding the radar cross section of a metallic object covered by a dielectric coating whose thickness is small compared to the radius of curvature of the metal surface. (This does not mean that the thickness is necessarily small compared to the wavelength of the incident electromagnetic wave.) For an axisymmetric plasma coating, the backscatter from a sphere is given by

$$\sigma = \pi k_0^2 a^4 \left| \int_{\pi/2}^{\pi} [R_{\parallel}(\pi - \theta) - R_{\perp}(\pi - \theta)] \sin\theta \cos\theta \times \exp[-i2k_0(a + \Delta)\cos\theta]\, d\theta \right|^2 \tag{10.2-157}$$

In the above equation $R_{\parallel}(\pi - \theta)$ and $R_{\perp}(\pi - \theta)$ are the reflection coefficients for a slab of thickness Δ for parallel polarization and for perpendicular polarization, respectively, as defined by Eq. (8.3-17). [The proper values of z for use in Eq. (8.3-17) are determined by Eqs. (7.1-6) and (7.1-7)]; a is the radius of the metallic sphere; and Δ is the thickness of the plasma coating. This approximation has also been formulated for the cone and the sphere-cone, as discussed in the following section.

10.2.2.4. *Other Regions*

As mentioned in the previous section, Musal[125] has formulated a physical-optics approximation for scattering from dielectric-covered metal objects. The coating must be thin with respect to the curvature of the surface. However, he has applied it to scattering from a plasma-coated cone with an axisymmetrical coating. The physical-optics approximation, strictly speaking, is not valid for a cone because of the discontinuity at the tip. However, as shown in Chapter 6, it does give good results for axial-incidence scattering from an infinite cone. Musal's expression for backscatter from a flat-backed cone viewed nose-on is

$$\sigma = \pi k_0^2 \tan^4\alpha_c \left\{ \int_0^{h_c} [R_{\parallel}(\tfrac{1}{2}\pi - \alpha_c) - R_{\perp}(\tfrac{1}{2}\pi - \alpha_c)]\, x \times \exp[+i2k_0(x - \Delta \sin \alpha_c)]\, dx \right\}^2 \qquad (10.2\text{-}158)$$

where α_c is the cone half-angle, h_c is the height of the cone, Δ is the thickness of the plasma coating, x is the distance along the cone axis, and $R_{\parallel}(\frac{1}{2}\pi - \alpha_c)$ and $R_{\perp}(\frac{1}{2}\pi - \alpha_c)$ are the reflection coefficients for parallel and perpendicular polarizations, respectively. The reflection coefficients are found from Eq. (8.3-17) by choosing the proper parameters from Eqs. (7.1-6) or (7.1-7). Allowing ϵ_r to go to unity in the Fresnel reflection coefficients reduces Eq. (10.2-158) to the ordinary physical-optics expression, as discussed in Section 6.3.1.2.2.1.1. For an axisymmetric dielectric coating on a metal cone with a spherical nose viewed nose-on, Musal obtains

$$\begin{aligned}\sigma = \pi k_0^2 \bigg| a \int_0^d \left[R_{\parallel}\left[\frac{\pi}{2} - \sin^{-1}\left(1 - \frac{x}{a}\right)\right] - R_{\perp}\left[\frac{\pi}{2} - \sin^{-1}\left(1 - \frac{x}{a}\right)\right]\right] \\ \times \left(1 - \frac{x}{a}\right) \exp\left[-i2k_0(a + h_c)\left(1 - \frac{x}{a}\right)\right] dx \\ + \tan^2\left(\frac{\alpha}{2}\right) \int_d^{d+h_c} \left[R_{\parallel}\left(\frac{\pi}{2} - \alpha_c\right) - R_{\perp}\left(\frac{\pi}{2} - \alpha_c\right)\right](x - d) \\ \times \exp[+2ik_0(x - a - \Delta \sin \alpha_c)]\, dx \bigg|^2 \qquad (10.2\text{-}159)\end{aligned}$$

where a is the nose radius and d is the depth of the spherical nose segment. Equation (10.2-159) reduces to (10.2-158) or (10.2-157) if the radius of the sphere or the height of the cone, respectively, is allowed to shrink to zero.

Mar'in[126] has also considered conical plasmas. He has considered scattering from conical trails formed by point sources of ionization moving through the atmosphere. The point of the cone is placed at the origin, and the trail extends along the $+z$ axis. He used geometrical optics for a plane wave incident on the stationary ionized cone whose electron concentration varied as r^{-2} along the axis of the trail and as $(\alpha_c^2 - \theta^2)$ across the trail. The distance from the apex of the cone is r, and α_c is the coordinate defining the boundary of the cone (the half-cone angle). The electron concentration is then a maximum radially along the axis of the cone and a maximum axially at the apex.

Backscattering occurs in this formulation if

$$\exp\left[\frac{(\pi - 2\theta_1)\, b_m}{R_0 \sin(\theta - \alpha_c)}\right] = \frac{\alpha_c b_m + R_0 \sin(\theta - \alpha_c)}{\alpha_c b_m - R_0 \sin(\theta - \alpha_c)} \tag{10.2-160}$$

where θ_1 is the angle between the radius vector and the ray at the point of incidence;

$$b_m^2 = \frac{4\pi e^2 N_e}{m_e(\omega^2 + \nu_c^2)} \frac{r_0^2}{\alpha_c^2} \tag{10.2-161}$$

r_0 is the distance to the point on axis where the electron density N_e is known or specified; and R_0 is the distance from the apex to the point where the ray first strikes the cone.

For $\alpha_c \ll \pi$, the backscatter cross section is given by

$$\sigma = \frac{2\pi R_0^2 \sin 2\theta_1}{\sin^3\theta} \left[(\pi - 2\theta_1)\cos\theta_1 + \sin\theta_1 + \frac{2R_0^2 \alpha_c \cos\theta_1 \sin^2(\theta - \alpha_c)}{\alpha_c^2 b_m^2 - R_0^2 \sin^2(\theta - \alpha_c)}\right]^{-1}$$
$$\times \left[\frac{1}{2} + \sin^2\theta_1 + \frac{2R_0^2 \sin(\theta_1 - \alpha_c)\sin(\theta - \alpha_c)}{\alpha_c^2[b_m^2 \alpha_c - R_0^2 \sin^2(\theta - \alpha_c)]}\right]^{-1} \tag{10.2-162}$$

When $b_m \to \infty$, the trail backscatters at the surface, $\theta - \alpha_c$, since $\sin\theta_1 \to \pi/2$. If b_m decreases, the rays are not backscattered from the trail for $\alpha_c^2 b_m^2 - R_0^2 \sin^2(\theta - \alpha_c) \leqslant 0$.

Mar'in[127] solves by ray-tracing the backscattering from an ionized paraboloid of revolution in which the electron concentration depends on two of the coordinates in a parabolic system of coordinates. He gives expressions for nose-on and broadside aspects, and a general expression which, with the application of differential geometry, could give values for other aspect angles. Mar'in[128] also considers the ray paths, effective reflecting surface, and absorption in an ionized toroid. Åström[129] has studied the scattering from elliptical cylinders. Numerous theoretical papers on objects embedded in anisotropic plasma are available in the literature (see, e.g., References 130–137).

Moorcroft,[138] in attempting to model patches of auroral ionization, has considered scattering from prolate-spheroidal ionized volumes. He terms the scattering body a Gaussoid. The distribution Moorcroft chooses is

$$N = N_0 \exp\left[-\frac{x^2}{a^2} - \frac{y^2}{a^2} - \frac{z^2}{b^2}\right] \tag{10.2-163}$$

where x, y, z are the spatial coordinates measured in meters, a is the semimajor axis, and b is the semiminor axis. The spheroidal portion is defined by the surface on which the permittivity goes to zero. Refraction by the ionized region outside the spheroid is taken into account by means of a correlation factor which alters the scattering radius of an electron. This correlation factor is a function which varies between zero and one. The backscatter cross section is then

$$\sigma = \left[\arcsin\sqrt{\frac{N_c}{N_0}}\right]^2 \left[\frac{b^2 \ln(N_0/N_c)}{4[\cos^2\theta + (b/a)^2 \sin^2\theta]^2}\right] \tag{10.2-164}$$

where N_0 is the electron density at the center of the distribution, N_c is the critical electron density $= (m_e\epsilon_0/e^2)\,\omega_p{}^2$, and θ is the angle that the z axis makes with the incident wavefront. For backscattering from this prolate-spheroidal Gaussian distribution for the underdense case, Moorcroft obtains

$$\sigma = (\mu_0{}^2 e^4/16\pi^2 m_e{}^2)\, N_0{}^2 a^4 b^2 \pi^3 \exp[-2k_0{}^2(a^2\cos^2\theta + b^2\sin^2\theta)] \tag{10.2-165}$$

Papas[139] has considered the case of dense plasma spheres embedded in less-dense planar plasma. The equivalent permittivity of the plasma is derived with the spheres present, and the planar plasma is then treated as a homogeneous region with this effective permittivity. The spheres are assumed to have radii such that $a \ll \lambda_0$ and to have permittivity equal to ϵ_1. The ambient ionized gas has a permittivity equal to ϵ_2. It is assumed there are N_s spheres per unit volume. Values of the equivalent permittivity are obtained by two different methods. By consideration of the dipole moment or polarization of each individual sphere without the presence of the others, Papas obtains an equivalent permittivity ϵ_{eq} which is given by

$$\epsilon_{eq}/\epsilon_2 = 1 + 4\pi[(\epsilon_1 - \epsilon_2)/(\epsilon + 2\epsilon_2)]\, a^3 N_s \tag{10.2-166}$$

By consideration of the amount of charge, neglecting the inhomogeneous configuration, he obtains

$$\epsilon_{eq}/\epsilon_2 = 1 + 4\pi[(\epsilon_1 - \epsilon_2)/3\epsilon_2]\, a^3 N_s \tag{10.2-167}$$

If ϵ_1 is approximately equal to ϵ_2, the values given by Eqs. (10.2-166) and (10.2-167) do not differ greatly. When ϵ_1 and ϵ_2 differ greatly, edge effects are significant. If the polarization due to the arrangement of the spheres cannot be calculated, the best results are obtained by assuming the total charge per unit volume is distributed homogeneously.

Kalmykova and Kurilko[140] have considered the case of a right-angle wedge, filled with anisotropic dielectric, in contact with a perfectly-conducting right-angle wedge. They show that the diffraction of a plane wave at the junction of these wedges is given by the solution of a singular integral equation with a Cauchy kernel, which, because of the special nature of the problem, reduces to a Fredholm equation which can be solved numerically.

The method of modified geometrical optics developed at the Ohio State University (see Sections 3.3.2.2 and 4.1.3.4.1) has also been extended to gyrotropic bodies.[141–143]

10.2.3. Application of Multiple Scattering

Multiple scattering is used here in the general sense, i.e., the effects of the interaction of an incident electromagnetic wave with many objects. The various scattering particles of an ionized gas are the many objects. Much of the application of multiple scattering has been for ionospheric phenomena, but applies to any type of ionized region.

10.2.3.1. *Incoherent Scatter*

When a radar signal traverses an ionized region where the maximum plasma frequency is well below the operating radar frequency, the free electrons exhibit a weak scattering of the incident energy. In determining the value of such scattered energy, it was first assumed[144] that the free electrons would scatter incoherently, i.e., each electron would move about in a magnetic-field-free plasma independently of the motion of the other electrons and of the ions. An electron absorbs and reradiates as a short dipole the amount of energy in the incident wave which occupies an area denoted as the differential Thomson cross section σ_e. The Thomson cross section was used by J. J. Thomson to calculate the scattering of X rays by the loosely bound electrons in the atoms of a crystal lattice. The cross section of an electron in square meters is given by

$$\sigma_e = (\mu_0{}^2 e^4 \sin^2 \gamma)/4\pi m_e{}^2 \approx 0.9978 \times 10^{-28} \sin^2 \gamma \qquad (10.2\text{-}168)$$

where γ is a function of the scattering and polarization angles,†

$$\cos \gamma = \sin \theta \cos (\Psi - \Phi) \qquad (10.2\text{-}169)$$

† The incident field is assumed to be propagating in the positive z direction. The E-field is then parallel to the xy-plane, with Ψ the angle between the incident electric field vector and the x-axis.

It should be noted here that the power scattered is not a function of frequency. From the microscopic analysis in the previous section, it was found that as the frequency increases toward infinity, the effects of the presence of the plasma disappears; in this equation, the effect of the electron remains constant with frequency. The differential Thomson scattering from ions can be found from Eq. (10.2-168) by replacing e^4/m_e^2 with q_i^4/m_i^2, where q_i and m_i are the charge and the mass of the ion, respectively. It is seen that most of the scattering would arise from free electrons, since their Thomson-scattering cross section is much greater than that of the ions.

Early experimental results[145] on the scattering of radar signals from the ionosphere demonstrated that each electron did not move independently of the other particles. The Coulomb interactions needed to be considered. Several authors[146–159] thoroughly investigated the problem of incoherent scatter and found that for the limiting case of long wavelength, the scattering cross section per free electron is related to the classical Thomson differential cross section by the approximate formula:

$$\sigma = \sigma_e \frac{1}{[1 + (T_e/T_i)]} \tag{10.2-170}$$

as long as

$$\lambda_0 \ll 4\pi\lambda_D$$

where T_e and T_i are the electron and ion kinetic temperatures, respectively, and λ_D is the Debye shielding distance. This correction factor is caused by the ions. In general, there are two effects, one an ionic correction factor, given more exactly by

$$\{[4k_0^2\lambda_D^2 + 1][4k^2\lambda_D^2 + 1 + (T_e/T_i)]\}^{-1} \tag{10.2-171}$$

which reduces to the term in brackets in Eq. (10.2-170) for the limitation given. In addition, there is also a component to the scattered power from the electron interactions. This gives a correction factor of $1 - [4k_0^2\lambda_D^2 + 1]^{-1}$, which is near zero for $\lambda_0 \ll 4\pi\lambda_D$.

The total scattered power can thus be considered to arise from two components. One component has a frequency spectrum spread around the transmitted frequency in a band corresponding to the thermal motions of the electrons. The other component has a frequency spectrum spread around the transmitted frequency in a band corresponding to the thermal motions of the ions. Bowles[145], Laaspere[152], and Farley *et al.*[147] have predicted that when the angle of incidence is within a degree or two of normal to the geomagnetic field lines, the continuous echo spectrum will become a line spectrum spaced by the ion gyrofrequency.

The correction factors effectively modify the Thomson cross section by a factor of $\frac{1}{2}$. Thus the backscattering cross section of a unit volume of tenuous ionized gas is, approximately,

$$\sigma = \tfrac{1}{2}N_e\sigma_e(\pi/2) \tag{10.2-172}$$

10.2.3.2. *Plasma Instabilities and Oscillations*

In many of the incoherent scatter experiments,[160] a weak scatter was noted which is many times stronger than the incoherent scatter, but still weak in the Born approximation sense. Farley[161] and Dougherty and Farley[162] developed a theory of a two-stream plasma instability, including the effects of the magnetic field as well as the collisions of the charged particles with neutrals. They find that the plasma instabilities give a periodic order to the random motions. The order greatly enhances the scattering. A sufficient condition for the spontaneous generation of this type of instability is that the velocity of the electrons relative to the ions be somewhat greater than the acoustic velocity of the medium. The instabilities travel at the acoustic velocity, which in the ionosphere at the magnetic equator is 360 m/sec. The instabilities are related to the electrojet current. These instabilities help explain both equatorial scatter and auroral scatter. For a further discussion of effects in auroral regions, see Section 10.3.2.4.

The strength of the scattered signal is closely associated with the strength of the electrojet current. In addition, the jet current must generally be greater than a certain threshold value if strong echoes are to be obtained. The echoes are aspect-sensitive. The direction of the radar beam must be nearly perpendicular to the magnetic field. The frequency spectrum of the echoes when the radar is pointing off-perpendicular usually consists of a fairly sharp peak at a discrete Doppler shift. The magnitude of the shift is roughly independent of the angle between the beam and the jet current; the sign depends on the direction of the current. Echoes from normal incidence are weaker than those from oblique angles and have a more variable, often diffuse, frequency spectrum. Farley[161] has suggested that the scattering is due to the disturbance of electron number density resulting from the two-stream instability that occurs when there is a sufficient relative drift of the electrons through the ions. By extending the theory of this instability to allow for the presence of both a magnetic field and collisions with neutral molecules, Farley showed that the threshold effect is accounted for by the fact that the strength of the electrojet is proportional to the relative ion–electron drift velocity, and it is really this velocity that has a threshold value. The aspect sensitivity arises from the fact that only waves propagating very nearly perpendicular to the magnetic field are unstable. All other waves are heavily damped. The quantitative agreement between the predicted and observed aspect sensitivity is good. The discrete Doppler shift corresponds to the speed of the plasma waves in the medium. The echoes come only from those waves whose wave vector is parallel to the direction of the radar beam. At oblique angles, the speed of these waves is simply the acoustic velocity. The main peak in the observed spectrum does indeed have a Doppler shift in agreement with this prediction, but there is a substantial scattered signal over a band of frequencies. Dougherty and Farley[162] have shown that some secondary features exhibited by the scattering can be explained by an extension of the theory to include nonlinear coupling between the unstable plasma waves. Such coupling produces further wavelike disturbances that can also scatter the incident waves.

Scattering from plasma oscillations has been the subject of numerous other investigations too voluminous to detail here. See References 163–167 for examples. References 4, 11, 14, and 15 also give useful background information.

10.2.3.3. *Plasma Turbulence*

Turbulence is closely related to plasma oscillations and instabilities. Generally, it develops as a result of an instability of an initial laminar state. In order to visualize the transition from the laminar to the turbulent state it is often convenient to examine the behavior of the plasma while changing a parameter, the change of which results in a loss of stability. Such a parameter is called an equivalent Reynolds number. For small equivalent Reynolds numbers the plasma is stable; at some higher number a normal mode of the plasma configuration is excited; at still higher equivalent Reynolds numbers more modes are excited. Unless the plasma system is such that the modes are harmonically related and synchronized, a turbulent state is created. The equivalent Reynolds number is similar to the Reynolds number used in neutral hydrodynamics, but differs in that it may be expressed in terms of a relative drift motion occurring between the positive ions and the electrons of the ionized gas. Once the critical Reynolds number is exceeded and turbulence has formed the scattering characteristics of the plasma change considerably. For example, Akhiezer[167] has shown that the interaction between transverse electromagnetic waves and longitudinal acoustic waves is much stronger for turbulent plasma than for equilibrium or quasiequilibrium plasma.

The classic theory of scattering from turbulence is that of Booker and Gordon.[168] When the scattering volume is much larger than λ_0^3, and when the electron density is such that the plasma frequency is less than the wave frequency, turbulent irregularities in electron density responsible for the backscattering are those with wave vector equal to $2k_0$, where k_0 is the wave vector of the incident electromagnetic wave. With an exponential correlation, they obtain

$$\sigma = \frac{2}{\pi} \frac{k_0^4 \langle | \Delta\epsilon/\epsilon |^2 \rangle \sin^2\beta_2}{[1 + 4k_0^2 l^2 \sin^2(\theta/2)]^2} \tag{10.2-173}$$

where $\langle | \Delta\epsilon/\epsilon |^2 \rangle$ is the average of the square of the ratio of the change of permittivity to the mean permittivity of the ionized region under observation,

l is the correlation length, θ is the angle between $\hat{\mathbf{k}}^i$ and $\hat{\mathbf{k}}^s$, and β_2 is the angle between the direction of the incident electric field and $\hat{\mathbf{k}}^s$.

From this expression, it is seen that (1) for $4\pi l/\lambda_0 \gg 1$, the scattering is mostly in the forward direction; (2) for $4\pi l/\lambda_0 \ll 1$, the scattering is independent of θ. (i.e., large blob sizes are important for forward scattering and small blob sizes are important for backscattering.)

Turbulence causes irregularities of electron density by several mechanisms: through the pressure fluctuations associated with the turbulence, through the randomization of a mean gradient in electron density and through the randomization of a mean gradient in electron temperature. The pressure fluctuations would cause scattering even if there were no mean gradient of electron density or temperature. However, for meteor trails, the randomization of the mean gradient of density is thought to be the predominant mechanism of scatter.

Significant backscattering from the wake of a high-speed reentry body takes place when the wake flow is turbulent. The turbulence scales in the longitudinal and transverse directions are expected to be different, and the intensity of the turbulence varies throughout the wake (decaying because of dissipation). When the turbulent Reynolds number is high enough, there exists a range of small eddies (high wave numbers) for which the turbulence is essentially isotropic. If the eddies responsible for the scattering are in this range, the theory of locally isotropic turbulence may be used. When the length of the incoming electromagnetic wave is comparable to or smaller than the wake diameter, the larger eddies are the chief sources of scattering. The effect of nonisotropy may become significant, e.g., the scattering cross section may become aspect-angle dependent.

A prominent feature of turbulent wake flows is the phenomenon of intermittency.[169,170] The cause of the flow intermittency is the production of a convoluted boundary surface between the turbulent and nonturbulent fluid by the large eddies (of fluid motion). Thus across the wake an intermittency factor is introduced to indicate statistically the fraction of the time that the flow is found to be turbulent. Intermittency of the turbulent flow will consequently introduce, at a fixed point, fluctuations of electron density in addition to those present in the entirely turbulent fluid. Yen[169,170] has shown that turbulence intermittency will introduce additional electron-density fluctuations and thus increase the radar cross section. The intermittency reduces the aspect-angle dependence and makes the wavelength dependence irregular. Ichikawa[171] has studied the dynamical effect of turbulent fluctuation on the dielectric properties of a weakly turbulent plasma. The turbulent fluctuation gives rise to appreciable modification of the dielectric properties in the low-frequency regions at $f_0 = 2f_p$, and perhaps at $f_0 = f_p$.

For underdense plasma, the mean radar-scattering cross section has been obtained by Tatarski[172] using the Born and geometrical-optics approximations:

$$\sigma = (2\pi)^3 \sigma_e V \langle N_e^2 \rangle \, \Phi[2k_0 \sin(\theta/2)] \tag{10.2-174}$$

where σ_e is the Thomson-scattering cross section for the electron, V the volume of the ionized region examined by the radar, θ the angle between the incident propagation vector from the transmitting antenna and the vector from the scattering volume to the observation point, $\langle N_e^2 \rangle$ the mean square electron density, and $\Phi(x)$ the three-dimensional power spectral-density function of the electron fluctuations.

For an exponential correlation function, Booker and Gordon[168] give

$$\Phi(x) = l^3(1 + x^2 l^2)^{-2} \tag{10.2-175}$$

where l is the scale of the turbulence. For the Kolmogorov spectrum of electron density fluctuations, the correlation function is

$$\Phi(x) = l^3(1 + x^2 l^2)^{-1-(1/6)} \tag{10.2-176}$$

which is practically the same.

Macrakis[173] has introduced fluctuations in Maxwell's equations describing a space-time dispersive plasma by considering the Boltzmann–Vlasov equations with an external field. The cross section is derived by evaluating the imaginary part of the correction to the transverse permittivity due to fluctuation scattering. The cross section describes the total incoherent scattering from density fluctuations in the plasma.

Ruffine and deWolf[174] have calculated the steady-state incoherent cross-polarized scattering of electromagnetic waves by a turbulent, tenuous plasma using the second Born approximation. When the dimensions of the scattering volume are much greater than the wavelength of the radiation and the correlation length of the plasma, the principal result is that the main contribution to the cross section comes from terms corresponding to the first and second scatterings occurring in different correlation cells. Thus the effect of statistical averaging is solely that of determining the magnitude of the scattered amplitude of each cell.

Salpeter and Treiman[175] and Albini[176] have considered turbulent scatter from a cylindrical plasma. Salpeter and Treiman first approximate the cylinder by a slab, discuss the effects of a variety of parameters, and then apply the results to a cylinder. Albini considers highly underdense plasma cylinders that are cylindrically symmetric, but axially and radially inhomogeneous. The cross section of turbulent plasma cylinders is discussed further in Section 10.4.4.3.

10.3. APPLICATIONS

This section applies the theories of the previous section to various important plasmas. The amount of space allocated to each topic is not a reflection of the opinion of the author as to its importance, but rather is an indication of the amount of published material available.

The discussions have arbitrarily been broken into four portions. First comes a discussion of the effects of entrance into the atmosphere of natural objects—meteors. This is followed by a section on auroras caused by the entrance of particles or waves into the atmosphere. Nuclear explosions, which can also create auroras, are discussed next. While a nuclear explosion affects the ionosphere mainly, the nuclear burst is also of interest to a radar observing a region of the burst at any altitude. Therefore it is considered separate from the last section on man-made ionospheric disturbances. The section on man-made ionospheric disturbances is divided into chemical releases in the ionosphere, effects near rocket launch, interaction of satellites and rockets in midcourse with the ionosphere, and reentry phenomena.

10.3.1. Meteors

10.3.1.1. *Introduction*

When a meteor enters the atmosphere, it collides with atoms and molecules of the atmosphere. This heats the meteor, causing it to melt and boil away. The rate of evaporation increases as the meteor falls to lower altitudes, until it reaches a peak value, and then suddenly decreases to zero. The velocities of the evaporating atoms are very high relative to the ambient atmosphere. These fast-moving atoms lose their kinetic energy in three ways:

(a) By elastic collisions with the air molecules, which raises the temperature of the air.

(b) By inelastic collisions, which raise an outermost electron of an air atom to a higher-energy state. (The return of the electron to the ground state accounts for the optical radiation seen from meteors.)

(c) By inelastic ionization collisions. (The outermost electron is freed from an atom, resulting in the creation of positively-charged ions and free electrons.)

The primary use of this meteor-induced ionization is for the bistatic reflection of brief communication signals. However, the long columns of ionized particles from the billions of meteors that enter the atmosphere every day can serve several other purposes. The columns give indications of the constituents of the upper atmosphere and their behavior. They act as clutter or interference to radars observing other targets. These columns also provide an application for some of the scattering theory of the preceding sections.

There are some differences in the literature concerning the meaning of certain commonly used words. In this book, "meteors" is the common name for all natural (as opposed to artificial or man-made) bodies entering the atmosphere from space. The majority of the larger meteors are moving at speeds from 20 to 70 m/sec. The earth's escape velocity is approximately 10 km/sec. The sum of this and the solar-system escape velocity is about 72 km/sec. Very few meteors have been detected with apparent velocities higher than 72 km/sec. "Meteorites" are those remnants which reach the ground. "Micrometeors" are particles smaller than 0.03 cm equivalent radius. Their mass per volume of space apparently exceeds that of meteors of all other sizes taken together. Micrometeors reach the earth's atmosphere most frequently at sizes of around 10^{-3} cm and speeds of the order of 12 km/sec.

Table 10-2. Order-of-Magnitude Estimates of the Properties of Sporadic Meteors†

Behavior	Mass (g)	Visual Magnitude	Radius	Number at this mass, or greater, swept up by the earth each day	Electron line density (electrons per meter of trail length)
Particles pass through the atmosphere and fall to the ground	10^4	−12.5	8 cm	10	—
Particles totally disintegrate in the upper atmosphere	10^3	−10.0	4 cm	10^2	—
	10^2	−7.5	2 cm	10^3	—
	10	−5.0	0.8 cm	10^4	10^{18}
	1	−2.5	0.4 cm	10^5	10^{17}
	10^{-1}	0.0	0.2 cm	10^6	10^{16}
	10^{-2}	2.5	0.08 cm	10^7	10^{15}
	10^{-3}	5.0	0.04 cm	10^8	10^{14}
	10^{-4}	7.5	0.02 cm	10^9	10^{13}
	10^{-5}	10.0	80 μ	10^{10}	10^{12}
Approximate limit of radar measurements	10^{-6}	12.5	40 μ	10^{11}	10^{11}
	10^{-7}	15.0	20 μ	10^{12}	10^{10}
	10^{-8}	17.5	8 μ	—	—
Micrometeorites (Particles float down unchanged by atmospheric collisions)	10^{-9}	20.0	4 μ	Total for this group estimated as high as 10^{20}	Practically none
	10^{-10}	22.5	2 μ		
	10^{-11}	25.0	0.8 μ		
	10^{-12}	27.5	0.4 μ		
Particles removed from the solar system by radiation pressure	10^{-13}	30	0.2 μ	—	—

† From Sugar(177).

At this size and velocity, they dissipate most of their energy through radiation and reach the earth with most of their original mass. Thus, they can be called "micrometeorites." Some of the common characteristics of typical meteors of each size are listed in Table 10-2 (from Reference 177), which is based on data from Manning and Eshleman[178].

"Meteroid" is a term used to refer to meteors in space or to the solid or liquid core of the meteor. Here the term "kernel" is used for the meteor core. Some meteors do not possess a kernel and are termed "dustballs." They collapse under their aerodynamic drag and are thus converted into a dust cloud. "Artificial meteors" are high-speed pellets produced by the explosion of shaped charges.

There are two categories of meteors—sporadic and shower meteors. The sporadic meteors have an almost random distribution of velocities with respect to the earth. More sporadic meteors occur near dawn than near sunset, since the earth overtakes and meets more meteors than meteors overtake the earth. Discussions of the biannual and annual variations of both sporadic and shower meteors are readily available in the literature, and will not be included here. A "meteor shower" consists of a group of particles moving with a small range of velocities in a stream in a well-defined orbit. When they are incident on the atmosphere, they are in nearly parallel trajectories, and so appear to radiate from a fixed point on the celestrial sphere known as the "radiant point of the shower."

The "coma" is a mixture of the jet of vaporized atoms which emerges from the meteor and the ambient air. The constituents of the coma have an appreciable portion of the original kinetic energy of forward motion. The main dissipation of the kinetic energy usually occurs in the coma. The impact radiation occurring here is the main source of luminosity for visual meteors.

The "wake train" or "ionized trail" is the ionized column in which the translational velocity of the coma has decelerated to well below the mean molecular velocity of the surrounding atmosphere. It is the convention of this book to refer to "trails" for meteors and to "wakes" for reentry vehicles and missiles. The trail expands because of turbulence and gas

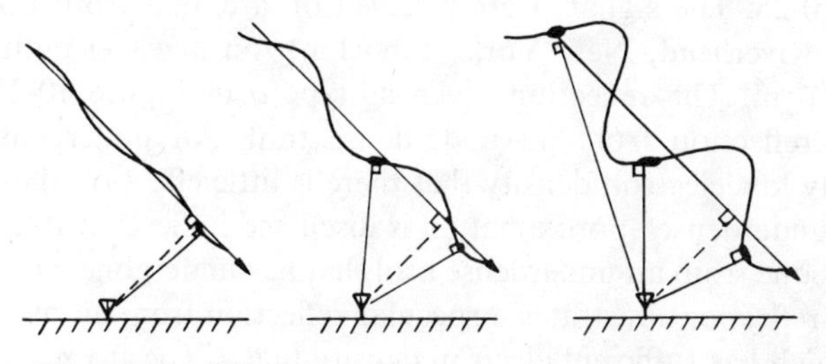

Fig. 10-23. Distortion of meteor trails resulting in glint.

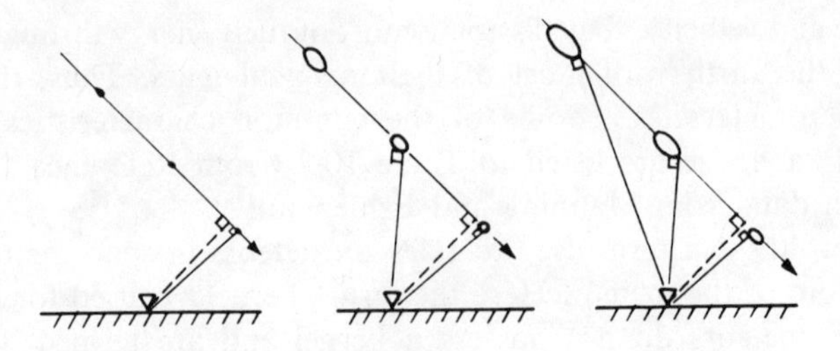

Fig. 10-24. Increase in number of specular reflection points as blobs increase in size.

diffusion. It is a long, very narrow paraboloid, which is generally closely approximated by a cylinder for radar purposes. However, the column does not always remain cylindrical. Wind shears in the upper atmosphere often tend to distort the trail increasingly as time progresses. Figure 10-23 illustrates the distortion of a trail and shows the formation of additional specular reflecting points. This is known as "glint theory," and has been discussed by many authors. Manning[179] has computed the theoretical delay in the appearance time of reflectors for these additional points. McKinley[180] compares the theoretical curve with the experimental data[181] and finds good agreement.

It must be noted that not all meteors create ionization uniformly along the trail. The less-dense portions soon become underdense, and the primary returns are from the expanding overdense blobs, as illustrated in Figure 10-24.

From geometrical-optics theory, the backscattering cross section of both the curved glint segment and the blob are nearly the same in magnitude and the same in form: $\pi\rho_1\rho_2$, where ρ_1 and ρ_2 are the radii of curvature at the point perpendicular to the radar line of sight. Practically all the echo is received from the surface area near this point. The cross section is larger at a longer wavelength, since the radius to the point where the dielectric constant equals zero, "the critical radius," is proportional to wavelength, as discussed in Section 10.2.

Typical bistatically-received signals from meteor trails are illustrated in Figure 10-25. The signals were received on a circuit from Long Branch, Illinois, to Riverhead, New York (1460 km), on a wavelength of slightly more than 6 m. The reflection given as type *a* in Figure 10-25 illustrates a specular reflection from an underdense trail. An underdense trail has a sufficiently low electron density that there is little effect on the transmitted wave. The underdense approximation is discussed in Section 10.2.2. A type-*b* reflection is one from an underdense trail that has undergone wind distortion. The type-*c* reflection illustrates a specular reflection from an overdense trail, i.e., one which has sufficient electron density that ϵ_2' goes to zero. The type-*d*

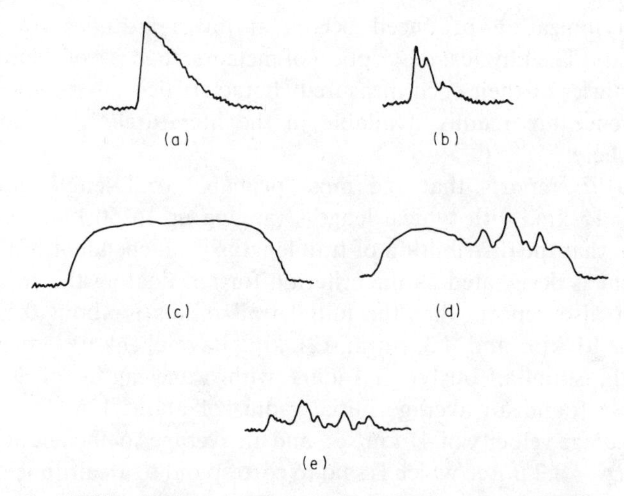

Fig. 10-25. Typical signal strength versus time plots of radio reflections from meteor trails: (a) through a specular reflection, underdense trail; (b) underdense trail with wind distortion; (c) overdense trail; (d) overdense trail with wind distortion; (e) nonspecular reflection, overdense trail with wind distortion [after Reference 182].

reflection is typical of specular reflection from an overdense trail with wind distortion, and type *e* represents nonspecular reflection for an overdense trail with wind distortion. Types *a* and *b* usually lasted less than 2 sec, while types *c* and *d* lasted about 30 sec or more. Table 10-3 gives the distribution by types of 2007 bursts received during the fall and winter of 1958–1959.[182]

Table 10-3. Frequency of Occurrence of Various Meteor-Trail Reflection Types

Type†	Number of Bursts	Per cent of Total
a	1264	63.0
b	25	1.2
c	8	0.4
d	443	22.1
e	267	13.3
Totals	2007	100.0

† See Figure 10-25.

The observed characteristics of the meteors vary with the initial mass and velocity of the meteors. For a given mass, the higher-velocity meteors produce observable trails at higher altitudes. For a given velocity, the maximum ionization produced occurs at lower altitudes for objects of greater mass. The physical description of meteors, theories of their formation, and summaries of their such measured characteristics as velocities and rates of occurrence are readily available in the literature(183–189), and are not discussed here.

Sugar(177) reports that the most probable trail length for sporadic meteors is 15 km, with typical lengths ranging up to 50 km. Eshleman(190) has shown that the distribution of trail lengths is independent of the electron density that is designated as the criterion for terminating the trail.

Sugar also reports that the initial trail radius is about 0.55 m at an altitude of 81 km, and 4.35 m at 121 km. Bayrachenko(191) measured the initial radii simultaneously on radars with wavelengths of 9.59 m and 6.49 m. He found an average initial radius of about 1 m for 50 meteors with an average velocity of 41 km/sec, and an average coefficient of ambipolar diffusion $D = 8.2$ m/sec which is said to correspond to an altitude of 96.8 km.

10.3.1.2. "*Head*" *Echoes*; *Echoes from Terminal Aspects*

A very short-duration radar echo moving with the velocity of the meteor is sometimes seen as meteors approach the observer and as meteors recede from the observer. These brief returns are termed "head" echoes. As the meteor approaches, it can be treated as a spheroidal reflector. It has the reflecting characteristics of a metallic object at the lower frequencies and a lossy dielectric-coated object at higher frequencies. As the aspect changes, more and more of the trail is observed, and the echo from the front is combined with echoes from more than one Fresnel zone, and thus disappears as a separate entity. Typical shapes of such meteor signal returns for an incident pulsed signal is illustrated in Figure 10-26. It shows the typical Fresnel integral variation. The figure is drawn for a meteor whose

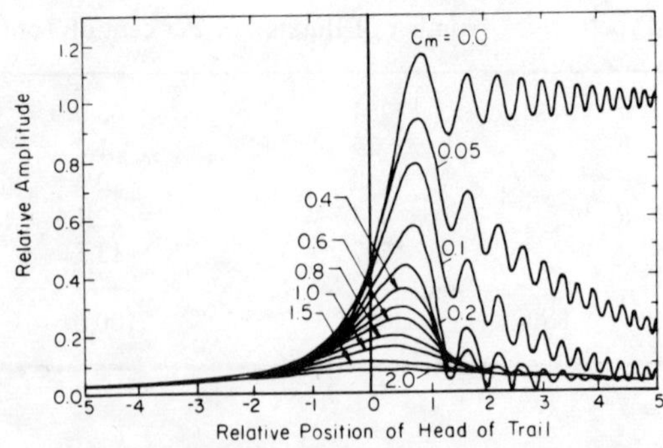

Fig. 10-26. Relative amplitude of meteor echo as a function of position at head of trail; C_m is proportional to diffusion coefficient divided by velocity [see Eq. (10.3-1)]. [After Loewenthal.(192)]

trail presents a broadside aspect to the radar. The size and shape of the curve will vary with the aspect presented. The return for a CW radar would have the oscillations at both ends of the return signal. The parameter C_m determines the effect of diffusion and finite velocity. It is equal to the distance a meteor travels in the decay time divided by π times the length of the central Fresnel zone. This is evaluated by Loewenthal[192] to be

$$C_m = 2(D/v_m)\sqrt{K_0^3(r + r')/2\pi} \tag{10.3-1}$$

where D is the ambipolar diffusion coefficient in square meters per second, v_m is the velocity of the meteor in meters per second, and r and r' are the distances to the transmitter and receiver, respectively. Greenhow and Neufield[193] state that D varies from 1 m^2/s at 85 km height to 142 m^2/s at 115 km.

It has been suggested that the "head" echo for the frontal case is merely the initial rise of Figure 10-26 at a near-frontal aspect; however, the head echo sometimes appears larger than would be predicted theoretically and disappears suddenly. It is as if the meteor were a large dense ball of ionization. These echos can be treated by assuming spherical, spheroidal, ellipisoidal, parabolodial, or conical shapes and using the standard theory discussed in previous chapters. The Soviet engineer Mar'in[126] has considered scattering from conical and parabolodial trails formed by a point source of ionization moving through the atmosphere. These are discussed in Section 10.2.2.4.

As a meteor recedes, the head echo is sometimes explained as a lack of cancellation from different Fresnel zones due to a discontinuity at the front of the meteor. Kaiser[185] gives the relation between the amplitude A_h of the head echo to the amplitude A_t of a trail echo at the same range and same electron line density (electrons per meter of trail length) as

$$(A_h/A_t) = (1/4\pi)\sqrt{\lambda_0/d} \tag{10.3-2}$$

where d is the difference between the range to the head of the meteor and the range to the specular reflection point. Kaiser also states that the magnitude of the head echo is proportional to the fourth root of the electron line density for high line densities. If the line density is less than about 10^{12} electrons/cm, then the head echo is seldom observed; but at these densities, the magnitude is directly proportional to the electron line density.

It might appear that the meteor trail acting as a traveling-wave antenna could account for some of the head echoes, especially in a receding case; however, using the criteria of Peters[95] discussed in detail in Section 10.2.2.2.1 we find the trails do not support the traveling-wave mode.

Evans[194] has recently published the results of a unique experiment in which meteors are observed traveling radially toward a narrow-beam, high-power radar operating at a wavelength of 68 cm. The beamwidth of the antenna pattern precludes observation of a meteor whose aspect changed by more than 2°. Because of equipment limitations, precise observation of the radar backscattering echo area was not possible; however, the majority of the cross sections were between 0.1 and 0.25 m^2, which is 10 or more decibels below the average cross sections of meteors viewed from the sides by Evans.[195] The mean mass of the meteor was estimated to be in the range 10^{-2}–10^{-3} g, corresponding to a visual magnitude of +5.

10.3.1.3. *Underdense Trails*

One assumption commonly made is that the trail is an infinitely long right circular cylinder of electrons whose density is low enough so that the incident wave passes through the trail without undergoing a large modification. This is the underdense approximation discussed in Section 10.2.2. The diameter of the cylinder is assumed to be very small with respect to the wavelength. The energy backscattered by each electron is summed with the correct phase relationships. It is then found that the principal backscatterer is the first Fresnel zone of the trail. That is a region centered at the point where the trail is perpendicular to direction of propagation from a monostatic radar and is of length $\sqrt{2\lambda r}$. This gives an initial variation similar to the Fresnel diffraction pattern of a straight edge, as mentioned in the previous section and as illustrated in Figure 10-26. It also gives a cross section which is related to the distance r from the radar, which is not normally expected. After the Fresnel oscillations have died, the cross section is normally estimated for either high frequencies or low frequencies as discussed below.

10.3.1.3.1. *Long-Wavelength Reflection from Low-Density Trails.* For a line of electrons (cylinder of initial radius zero) expanding so that the distribution of electrons across the trail is Gaussian, the backscattering cross section may be written as

$$\sigma(0) = \frac{\mu_0{}^2 e^4 N_m{}^2 r \lambda_0}{32\pi^2 m_e{}^2} \exp\left(-\frac{32\pi^2 D}{\lambda_0{}^2}\, t\right) \tag{10.3-3}$$

where N_m is the electron line density of the trail in electrons per meter and t is the time measured for the formation of the trail in seconds.

As seen from Eq. (10.3-3), the signal amplitude falls to $1/e$ of its initial value when the trail radius reaches $\lambda_0/2\pi$. The drop in amplitude is caused by the phase difference increasing from electrons on opposite sides.

Since the effect of a finite initial radius is equivalent to a shift of the time scale for a Gaussian electron distribution, the initial radius may be considered nonzero by multiplying the right-hand side of Eq. (10.3-3) by an additional attenuation factor, $\exp(-8\pi^2 r_i^2/\lambda_0^2)$, where r_i is the terminated radius of the trail in meters. This gives

$$\sigma(0) = \frac{\mu_0^2 e^4 N_m^2 r \lambda_0}{32\pi^2 m_e^2} \exp\left(-\frac{32\pi^2 D}{\lambda_0^2} t - 8\pi^2 \frac{r_i^2}{\lambda_0^2}\right) \qquad (10.3\text{-}4)$$

Discussion of bistatic effects can be found in the literature.(188)

10.3.1.3.2. *Short-Wavelength Reflection from Low-Density Trails.* The time of formation of a trail cannot always be considered as negligible; sometimes the formation appears to be a steady state. The time of formation of the trail can be considered as the time of traversal of one-half the first Fresnel zone. By equating this time to the echo duration until decay to $1/e$ amplitude, one obtains an expression for the transition wavelength λ_T:

$$\lambda_T^3 = 128\pi^4 D^2 r / v_m^2 \qquad (10.3\text{-}5)$$

Eshleman(188) states that λ_T varies between 0.1 to 100 m, with the most probable λ_T near 2.9 m for backscatter and from 0.5 to 1.6 m for forward scatter.

The effective length of the trail is found by multiplying the velocity by the time it takes the trail to diffuse to $1/e$ of its initial value. The backscatter cross section is then proportional to the square of this length, with the result being

$$\sigma(0) = \frac{\mu_0^2 e^4 \lambda_0^4 N_m^2 v_m^2}{4096\pi^6 m_e^2 D^2} \exp\left(-\frac{8\pi^2 r_0^2}{\lambda_0^2}\right) \qquad (10.3\text{-}6)$$

where r_0 is the initial radius of the trail in meters. The complete equation for the bistatic case is available in the literature.(188)

It should be noted that the cross section varies as λ_0^4, instead of λ_0 as in the long-wavelength case. Thus as λ_0 decreases from a large value, the cross section at first decreases linearly, and then, for short wavelength, the cross section decreases quadratically, which explains the lack of meteor scatter at the higher frequencies.

10.3.1.4. *Overdense Trails*

If the meteor creates a large amount of ionization, the underdense assumption that the incident wave passes through the trail essentially unmodified is not valid. The simplest assumption made is that the meteor trail may be represented by a cylindrical metallic reflector. Any ionization outside the effective radius is assumed to have no effect. The radius of the trail expands with time, and the electron density falls until the underdense approximation holds. However, the radius is usually quite large by this time, and the resulting backscattered signal is small. Again two cases must be considered: long wavelengths and short wavelengths. It should be remembered that it requires a larger electron density to be overdense at the higher frequencies.

10.3.1.4.1. *Long-Wavelength Reflections from High-Density Trails.* If the distance R between the radar operating at a long wavelength and the overdense trail is very much larger than the effective radius, the backscattering cross section may then be shown to be

$$\sigma(0) = \frac{r}{4}\,[4Dt\ln(\mu_0 e^2 N_m \lambda_0^2/16\pi^3 D m_e t)]^{1/2} \tag{10.3-7}$$

The quantity under the square root sign is the effective radius of an expanding Gaussian trail as found by Hines and Forsyth.[196]

As the time t increases, the logarithm goes to zero at time t_f, which is given by

$$t_f = 16\pi^3 m_e D/\mu_0 e^2 N_m \lambda_0^2 \tag{10.3-8}$$

At times greater than t_f, the relation for underdense trails, Eq. (10.3-3), is said to be applicable.

The forward-scatter case has been considered by Hines and Forsyth[196] and by Manning,[197] and is not discussed here.

10.3.1.4.2. *Short-Wavelength Reflections from High-Density Trails.* Hawkins and Winter[198] have considered the case of high-frequency reflections from overdense trails under the conditions of severe diffusion. They found that the overdense ionized cloud forms a prolate spheroid, with the meteor and its trail along the major axis. They neglect the initial diameter of the trail and the effect of fragmentation of the meteors. The resulting radar cross section is given by

$$\sigma(0) = \lambda_0^4 v_m^2 N_m^2 \sigma_e/16^2\pi^4 D^2 \tag{10.3-9}$$

where σ_e is the Thomson-backscatter cross section of an electron, as discussed in Section 10.2. Equation (10.3-9) is not valid for incidence along the axis of the trail. Equation (10.3-9) is identical to the equation for scattering for underdense trails as obtained by Hawkins[199] for the case of severe diffusion.

Eshleman[188] gives the maximum cross section in the backscatter case as

$$\sigma(0) = \lambda_0^4 v_m^2 N_m^{5/2} r_e^{5/2}/32\pi^4 D^2\sqrt{2.718} \tag{10.3-10}$$

where r_e is the classical radius of an electron and is equal to $(\mu_0 e^2)/(4\pi m_e)$. Equation (10.3-10) varies as $N_m^{5/2}$ and Eq. (10.3-9) varies as N_m^2. The two expressions are equal when N_m equals 10^{14} electrons/meter. Equation (10.3-10) gives a larger cross section for all higher electron densities.

The echo duration as given by Eshleman is

$$\tau = 2\pi^2 D r/\lambda_0 v_m^2 N_m r_e \tag{10.3-11}$$

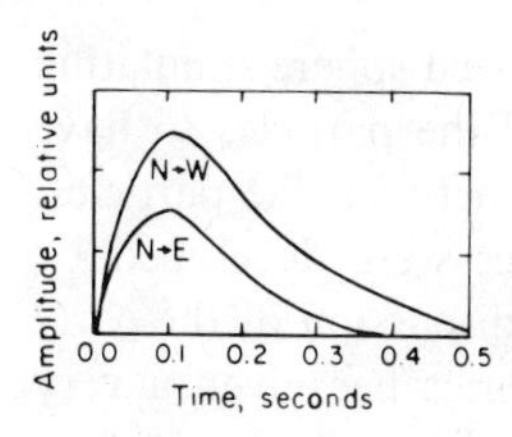

Fig. 10-27. Polarization effect at 23.1 Mcps. Presence of a received echo with a north-to-east rotation occurs when the transverse reflection coefficient differs from the longitudinal [after Manning and Eshleman[178]].

10.3.1.5. *Other Aspects of Meteor Reflections*

The above discussion neglects such factors as the effects of low-density ionization surrounding an overdense cylinder. Manning[200] has shown that the refractive effects of this ionization can weaken the return signal considerably more than predicted by the overdense assumption.

Also neglected are polarization effects. For trails whose dielectric constant is near minus one, a plasma resonance phenomenon[184] may occur when the incident electric field has a polarization which is perpendicular to the axis of the trail. The plasma oscillations can double the reflection coefficient, or quadruple the reflected power. Figure 10-27 illustrates the effect on circularly-polarized 23.1-Mcps signal returns. The transmission had a north-to-west polarization rotation. A received signal with a north-to-east polarization rotation occurs when the transverse reflection coefficient differs from the longitudinal.

Lagutin and Kashcheyev,[201] operating at 36.9 Mcps, found that over a four day period, 50 % of the meteors observed exhibited some polarization effects.

Watkins[202] has found experimentally that returns much stronger than expected could be explained by enhancement of echoes from meteor trails parallel to the geomagnetic field. However, Kaiser[185] and Francey[203] both conclude theoretically that the geomagnetic field should have only an insignificant effect on the diffusion coefficient.

10.3.2. Auroras

10.3.2.1. *Introduction*

The beautiful lights which illuminate the northern sky have long intrigued man. These visual auroras are termed aurora borealis or northern lights. The aurora in the southern sky are termed aurora australis or southern lights. When speaking of both or either, they are called aurora polaris. The idea prevalent in the nineteenth century was that the aurora was a gigantic electrical discharge in the atmosphere—somewhat akin to a lightning discharge. The beginning of the modern approach to auroral physics was the work of Birkeland, who fired electrons at a magnetized sphere simulating the earth. Stormer developed a theory considering all the particles to have the same electrical charge and neglecting interactions between the particles. In the following years, a variety of alternate theories were developed by Chapman, Ferraro, Alfven, Martyn, and others. For discussion of the early theories, see References 204–209. None of these theories has given a very satisfactory explanation or description of most of the observed phenomena. The discovery of the Van Allen radiation belts of the earth opened a new era in auroral study.

The cause of aurora seems to be the solar wind. The phenomenon of the interaction of the solar wind with the magnetosphere is not yet well understood. This interaction eventually causes particles from the Van Allen belt to precipitate into the upper atmosphere. If one classifies auroras according to their excitation mechanism, there appear to be at least four types: the polar-glow aurora, high red arcs, electron-excited aurora, and proton-excited aurora.[210] The polar-glow aurora may have little relation to the events in the magnetosphere which cause the ordinary aurora. This aurora is manifested as a weak glow over the entire polar cap and seems to be caused by protons with energies of the order of millions of electron volts. The high red arc is not well understood. The various auroras excited by fast-moving electrons exhibit similar optical spectra, but vary considerably in geometric forms, time of appearance, and fluctuations with time. The proton-excited aurora has different optical behavior and a different precipitation zone than the electron-excited aurora, although they may be partially caused by electrons also. They appear to lie south of the bright electron aurora early at night and to lie north in the early morning hours. Around midnight, the two types appear in the same regions. A more detailed discussion of the location of auroras is given in the following section.

The optical auroras are divided by their appearance into several types. The conventional polar-aurora type (electron- and proton-excited) are subdivided into several more types. Definitions, descriptions, and nomenclature for these are available in the literature.[209–211] There seem to be two general categories, those with rayed structure and those without it. Significantly, the rays lie along the geomagnetic field. The alignment of auroras along the geomagnetic field is quite important to radar observation, as is discussed in Section 10.3.2.4. Since the optical luminosity of auroras is attributed in part to recombination following ionization, it is reasonable to assume that some radio effects should be observable when a visual aurora is present. These radio effects are not simply related. The correlation of optical and radio auroras is discussed in Section 10.3.2.3.

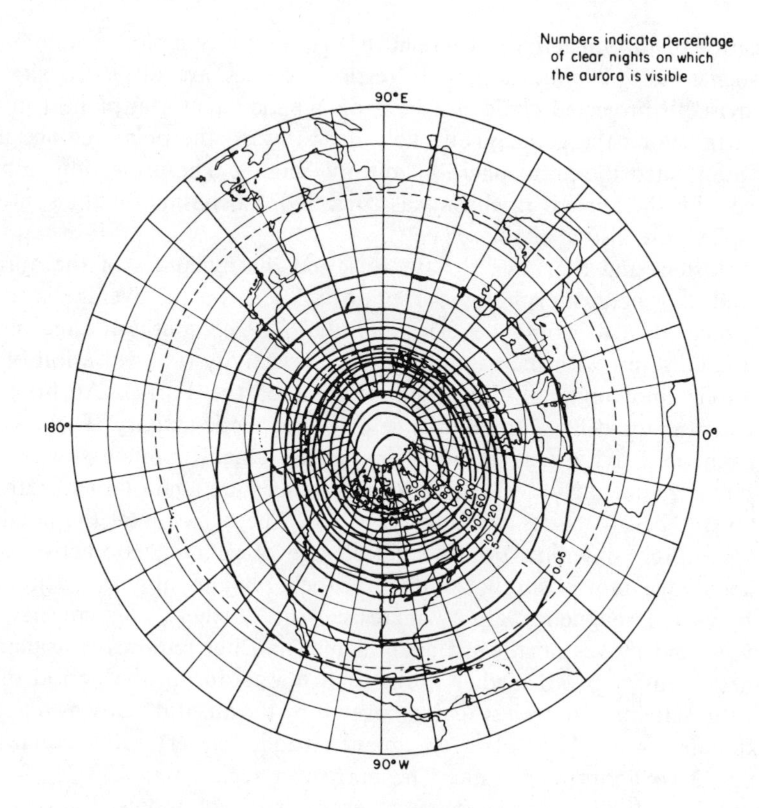

Fig. 10-28. Auroral isochasms [after Vestine(212)].

10.3.2.2. *Geographic and Temporal Variations*

In 1881, Fritz† drew a map showing equal occurrences of visual auroral sightings. These lines of the equal likelihood are called isochasms. The region around the isochasm of maximum occurrence is termed the Fritz zone in many articles. Vestine,(212) in 1944, revised, updated, and improved Fritz's work. Vestine's curves for the northern hemisphere are shown in Figure 10-28. The numbers indicate the percentage of clear nights on which an aurora is visible somewhere in the sky from the given geographic location. These isochasms were drawn after years of observations. All the diurnal, seasonal, sun-spot cycle, and magnetic-storm variations are smoothed out. There has also been considerable discussion about the lack of correlation between visible and radar aurora. As shown in the next section, there is often a fairly good correlation. Therefore Figure 10-28 can be used as an average

† Fritz, H., *Das Polarlicht*, F. A. Brockhaus, Leipzig (1881). See Reference 204 for a reproduction of his map.

indication of auroral activity. There have been some recent plottings[210,213–216] of isochasms which indicate that Vestine's curves are displaced slightly. Hultqvist[217] projected circles in the geomagnetic equatorial plane (outside the earth) along the geomagnetic field lines toward the polar regions until they intersected the atmosphere at auroral heights. The projections almost agreed with the auroral isochasms as corrected. Isochasms for the southern hemisphere are also available.[218,219]

Feldstein and Starkov[216] have reviewed the dynamics of the auroral belt and its general morphology. The auroral zone on the average is found to be located at a geomagnetic latitude of 67°. The southern edge moves toward the south with increased geomagnetic activity (i.e., variation of the magnitude and angle of the magnetic field of the Earth). At first, the movement is rapid for a small change of geomagnetic activity. Then, as the geomagnetic activity increases further, the movement is much slower. For still larger geomagnetic disturbances, the changes are again larger, with the edge of the belt reaching as far south as 63°. The behavior of the northern edge is entirely different. For a small change of geomagnetic activity, the northern edge moves south, so that the belt remains almost constant in width; as the magnetic activity increases, the northern edge reverses its direction and moves northward again somewhat, but requires considerable magnetic activity to go farther north than it was during the period of no magnetic activity. The widest region covered by the auroral belt was reported as extending from 63° to 74° geomagnetic latitude for very active conditions during 02–06 hours local time. The narrowest region was 73°–74° for no activity from 06–10 hours. The northernmost extent reported was about 77.5° for no activity from 10–14 hours. During magnetically calm periods, auroras were always found at 70° geomagnetic latitude in all the data examined. The details of the correlation between the movements and the magnetic activity depend upon the geomagnetic index used. A further discussion of this correlation is found in Reference 216.

Visual auroras have been reported at heights from 70 to 1100 km, but most occur at a height of 90–120 km. Hultqvist and Egeland[220] summarized radar observations of auroral height, at frequencies of 40 Mcps–1300 Mcps. Heights from 50 to 190 km have been reported. Some of the measurements need modification for tropospheric and ionospheric refractions. The measurements at the highest and lowest frequencies agree fairly well. The mean heights reported for each of the measurements lie in the 90–100 km range. Hultqvist also reported that the majority of auroral absorption takes place below 90 km altitude.

The number of auroras, both visual and radar, vary directly with the sun-spot cycle. It has been suggested by Kelley *et al.*[221] that the auroras related to ionospheric storms would occur most frequently about two years after the sun-spot peak. Krasovskij[222] states that there is a definite tendency

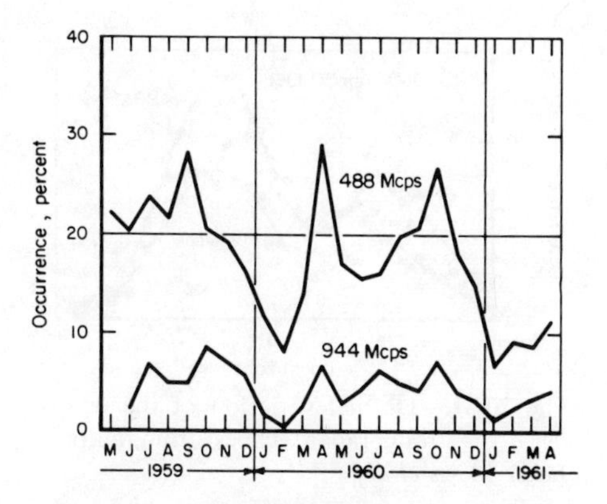

Fig. 10-29. Monthly variation of the per cent of a day that a radar echo from an aurora occurred [after Blevis *et al.*(224)].

for auroras to reoccur after about 27 days (i.e., the period of solar rotation), particularly during years of minimum solar activity.

Bullough and Kaiser(223) reported a maximum of activity at about the equinoxes, with a minimum occurring at the solstices. To a large extent, Blevis *et al.*,(224) support this observation. Blevis *et al.* report a deep minimum occurring in about February, with a shallower, but broad minimum in the summer, as shown in Figure 10-29. This figure shows the variation of occurrences during the period from May 1959 to April 1961 as measured at 488 Mcps and 944 Mcps. The day was divided into one hundred intervals. An occurrence was defined as an interval in which the presence of a radar echo was detected. Blevis *et al.* took backscatter observations from May 1959 to April 1961 at Ottawa on two similar radar systems

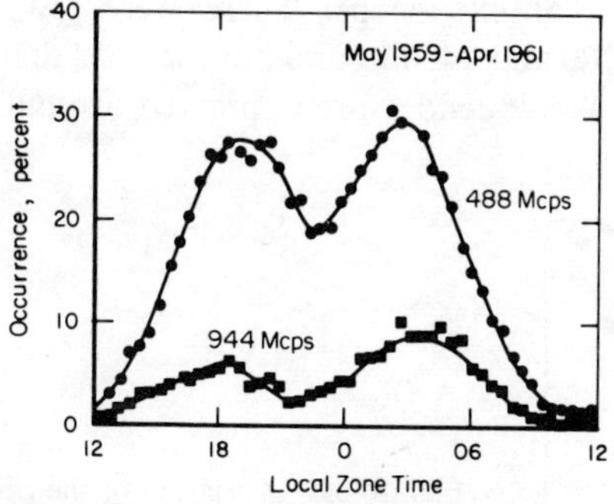

Fig. 10-30. Diurnal variation of the occurrence of radar echoes from auroras [after Blevis *et al.*(224)].

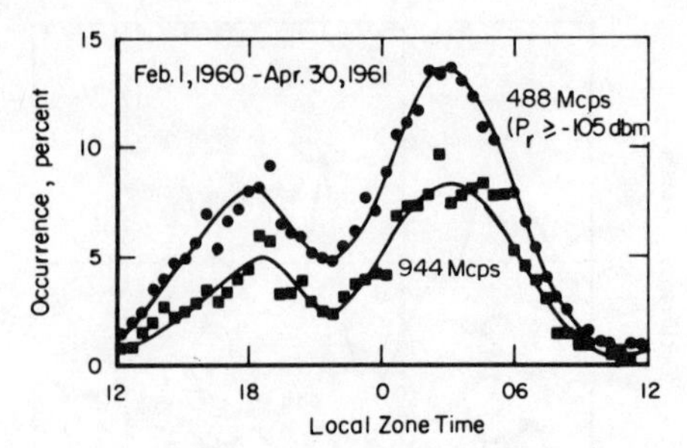

Fig. 10-31. Diurnal variation of the occurrence of strong radar echoes from auroras [after Blevis *et al.*[224]].

operating at 488 and 944 Mcps. Narrow-beam transmitting and receiving antennas were fixed in position to intercept the ionosphere at auroral heights where the radar ray path strikes the magnetic field lines at normal incidence. This gives specular reflection from geomagnetic-field-aligned irregularities. This was about 700 km from Ottawa at an altitude of 150 km, a latitude of 51.0°N, a longitude of 80.0°W, and a geomagnetic latitude of 62.2°N.

Figure 10-30 presents the diurnal variation of echo occurrence for the period May 1959–April 1961. It is seen that the two curves are quite similar, with a much lower percentage of occurrence of echoes at the higher frequency. When only the stronger signals at the lower frequency are plotted along with all the signals at the higher frequency, as is done in Figure 10-31 for the period February 1, 1960 to April 30, 1961, the agreement is much closer. The echoes shown for 488 Mcps all have received signal powers of -105 dbm or stronger, with a transmitter peak power of 5 kW and antenna gain of 36 db. The 488-Mcps curve no longer has equal peaks; it now has a clear maximum after midnight. Also, the peaks have shifted away from midnight slightly. There was a definite tendency for echoes to be received at 944 Mcps whenever strong echoes were received at 488 Mcps. There were few echoes received at 488 Mcps between 0700 and 1400 hours, but the ones that were received were generally all strong. Weak echoes predominated around 2000 hours.

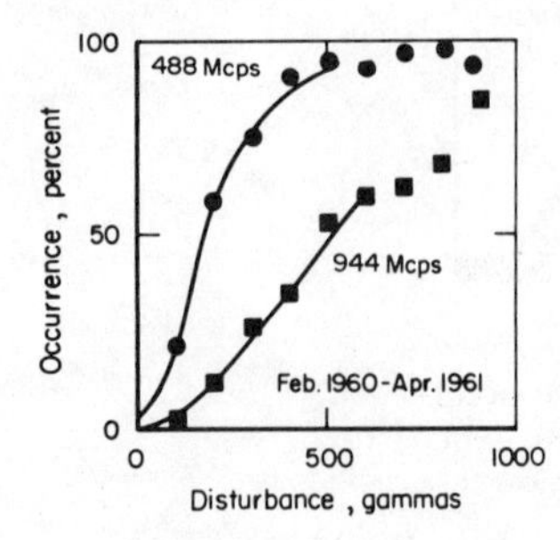

Fig. 10-32. Variation of the per cent of occurrence of auroral echoes as a function of the strength of the geomagnetic field disturbance at the surface [after Blevis *et al.*[224]].

Blevis *et al.*[224] also considered the variation of the occurrence of radar echoes with geomagnetic-field variations. A recording flux-gate magnetometer was installed at Moosonee, Ontario (latitude 51.3°N., longitude 80.6°W.). This location was chosen since it was almost directly beneath the region from which most auroral specular echoes were expected. For the data from February 1960 through April 1961, the per cent occurrence of radar echoes was compared against the absolute value of the magnetic disturbance in units of gamma (1 gamma = 10^{-5} gauss). As illustrated in Figure 10-32, the probability of an echo at 488 Mcps rises rapidly with increasing magnetic disturbance. For disturbances greater than about 400 gammas, the occurrences lie in the 90–100% bracket. At 944 Mcps, the average per cent occurrence increases more slowly with increasing disturbance. A "knee" in the data is seen in Figure 10-32 near 500 gammas. Blevis *et al.* also noted that the nature of the dependence on magnetic disturbance is affected by the time of day.

Additional summaries of the variation of auroral occurrences with sun-spot cycle, seasons, time of day, magnetic activity, and other variables, as well as summaries of the locations of aurora, appear in References 205–211, and 225–228.

10.3.2.3. *Correlation of Optical and Radio Auroras*

In many of the early auroral papers, it was felt that there was little relationship between the optical and radio aurora. Radar reflections were received when no auroras were visible, and auroras which were visible produced no echoes. In experiments at College, Alaska,[228] which is under the maximum visual auroral zone, it was noted that most of the radar echoes came from the north. Bates *et al.*[229] note that two main factors have tended to confuse the optical–radio comparison. First, the VHF and UHF radars observe only the auroras which occur in a limited volume of the ionosphere. Direct backscatter occurs only when the radar propagation path intersects the geomagnetic field lines at nearly right angles. Second, refraction changes the ray path of HF radars in such a manner that HF echoes can occur over a large range of angles.

Bates *et al.*[229] reported that the position of the aurora measured in College, Alaska during twelve days in December 1964 with radar and with all-sky cameras corresponded closely. The major discrepancies occurred in the following situations: when the aurora was overhead; when an aurora well to the north occurred together with a strong overhead aurora; when a weak, patchy aurora was present. The first discrepancy was explained by difficulties in data processing; because of difficulties in identifying the source of the returns, all returns of ranges of less than 300 km were ignored. The second discrepancy was attributed to the local aurora being accompanied by *D*-region absorption and sporadic *E*-ionization, which prevented returns from the more distant aurora. The third discrepancy was expected, since patchy auroras are generally of low optical intensity, indicating a low production rate of *E*-region ionization. For this case, only weak backscatter echoes at 4–8 Mcps would be expected. At these frequencies, many different echoes occur, so that data processing is difficult.

Kelly(230) reported measurements made by the Prince Albert Radar Laboratory of the Defense Research Board of Canada, located in central Saskatchewan about 500 km south of the auroral-zone maximum. A section of the northern sky was scanned with a 448-Mcps radar and a photometer, both with beamwidths of about 2.2°. This gave a resolution of 40 km perpendicular to the beam at the heights of most-frequent auroral occurrence. An interference filter with a half-width of 28 Å was used to select the oxygen green line at 5577 Å. A stationary camera photographed the region being scanned to permit identification of the visual auroral form. Good spatial and temporal correlation between radar and visual observation was found when the visual form was at high enough elevation so that atmospheric extinction did not preclude its detection, when the sky was clear, when background light was low enough to permit visual observations, and when the radar aspect angle was close to normal.

The correlation between the visual and radio auroras, and the mechanism of their formation, will be better understood as the result of present coordinated measurement programs involving both ground and satellite-based equipment. It should be noted that in laboratory observations, Hyde and Coleman(231) have shown that optically-dark regions can be strong scatterers. The particular case cited is that of the dark-space regions of a glow-discharge plasma.

10.3.2.4. *Radar Reflections*

Auroral reflections were first noted by amateur radio operators, who communicated by means of signals reflected bistatically from the aurora. A peculiar modulation was impressed on the signals by the auroral reflection. The sound created has been variously described as "hissing," "garbling," "fluttering," "rumbling," etc. Fading is common, but on some occasions the signal is very strong.

The early theories of auroral reflection considered the aurora as another ionospheric layer and assumed critical reflection from a large planar surface. This type of assumption requires too large an electron density to explain the observed returns. Herlofson(232) considered a planar aurora, but suggested that the returned signal was caused by underdense reflection from strong density gradients in fairly weak ionization. This theory requires an electron density about three powers of ten less than that needed for critical reflections. An example of the measurements that have been made on the aurora is that reported by Hultqvist and Egeland.(220) They found a measured

CW radar cross section of 37 km^2 for $\lambda_0 = 7.3$ m. This was calculated from strong echoes observed with a transmitting and a receiving antenna each of which had a 3 db gain. If the aurora is compared to specular reflection from a large, flat metal sheet at the distance of the auroral scatterers, the reflection coefficient of a strong aurora would be between 10^{-5} and 10^{-3} between 40 Mcps and 120 Mcps. Several references contain useful summaries of experimental data.[214,223,225,226]

The fading and angular distribution of the observed echoes indicated that reflection, either critical or partial, from plasma surfaces could not explain the observations. Various other models have been proposed. Because of the observed aspect sensitivity, most require irregularities in ionization aligned along the geomagnetic field. It was noticed that echoes could be obtained only if a radar beam was pointed in the direction of geomagnetic pole. This was not too surprising, since the observers were at middle latitudes, and the aurora, as seen in Section 10.3.2.2, occurs almost concentrically about the geomagnetic pole. However, Dyce[228] observed the same phenomenon at Point Barrow, Alaska, even though visual auroras were occurring overhead and to the south. The explanation of this phenomenon had already been advanced by Chapman.[233] He proposed that auroral ionization is aligned along the field lines. He predicted that the echoes from auroral scatterers at a height of 100 km could be observed most easily by radars located where the magnetic dip is of the order of 70–75°, since the radar beam could then easily be orthogonal to the field lines at auroral heights. Chapman also indicated that echoes would not be received from any points where the dip angles are in excess of 80°. This has been disproved experimentally. His aspect criterion was a bit too sharp, and he neglected refraction along the ray path. However, his pioneering work, which has been extended by several authors,[214,220] forms the basis for current practice.

Echoes have been classified as diffuse or discrete depending upon their characteristics on a radar PPI scope. Diffuse echoes have rapid fluctuations in amplitude and a large spread in range. The range spread is sometimes several hundred kilometers. Discrete echoes have relatively precise ranges, intense amplitudes, and range spreads which vary from a few kilometers to a few tens of kilometers, depending on the pulse length of the observing radar. Booker[234] has pointed out that there is a great deal of difference in the usage of these terms. For example, echoes observed in England and divided into discrete and diffuse classes there would probably all be classified as discrete by the observers in Alaska. The weak, diffuse echo covering a large part of the sky over Alaska has probably not been observed in England.

The mechanism of radar reflection from auroras has been the subject of much discussion. Several authors[223,235,236] have considered the reflections to be from an overdense medium. This type of reflection is probably important at low radar frequencies. Other authors[232] have emphasized the mechanism of partial reflection from small irregularities in electron density. Estimates of the size of the irregularities range from a fraction of a meter[237] to 100 m.[238] The ionization is considered to be elongated along the geomagnetic-field lines, but estimates of this elongation have ranged from four-to-one[239] to forty-to-one.[237] Lyon[240] compared the simultaneous signals received on two scatter circuits of quite different geometry, but viewing a common scattering volume. The receivers were located at London, Ontario (latitude 43°N., longitude 81°W.). One transmitter was located at Great Whale River (55.3°N., 77.7°W.) and operated at 40.36 Mcps, and the other transmitter was located at Ottawa (45.5°N., 75.6°W.) and operated at 40.35 Mcps. The Great-Whale-River-to-London circuit was a forward-scatter path. The Ottawa-to-London circuit was a bistatic path, with the angle being about 50° from backscatter. All the antennas were identical five-element Yagis. Isotropic scatterers were detected at a height of about 85 km. Scatterers anisotropic with an axial ratio of at least ten occurred at a height of 100 km upward. The occurrence of different types of scattered signal simultaneously on the two circuits shows that the scatterers may exist simultaneously at both levels. The type of signal that can be seen on any given circuit depends markedly on the geometry. The theory of bistatic scattering is not yet well formulated.

At present, three theories of auroral scattering find much use. They are those of Booker[234,237] and of Moorcroft,[138] and the plasma-wave theory of Farley[161] and others (discussed in Section 10.2.3). Booker[237] suggested that the radar auroral echoes originated from columns of ionization fairly restricted in length and aligned along the geomagnetic field. Small-diameter columns with lengths of the order of a few tens of meters are necessary to explain the observed return. The mean electron density of the columns corresponds to a plasma frequency of around 10 Mcps, with irregularities having a mean-square fractional deviation of electron density of the order of 10^{-3}. The irregularities are assumed to have a Gaussian autocorrelation function and to be symmetrical about the geomagnetic field, with an axial correlation distance of 10 m and a transverse correlation distance of 0.1 m or less. Booker[237] suggested that the irregularities originate from turbulence. The use of the Gaussian autocorrelation function was dictated more by mathematical necessity than by the belief that this was the proper function. The choice of a correct autocorrelation function has eluded all investigators.

Booker[237] obtained the following relation for backscatter from field-aligned irregularities, provided that the direction of incidence is not parallel to the geomagnetic field lines:

$$\sigma(0) = \frac{2^{3/2}\pi^{7/2}}{\lambda_p^4} \left\langle \left(\frac{\Delta N}{N}\right)^2 \right\rangle T_B^2 L_B \exp\left(-\frac{8\pi^2 T_B^2}{\lambda_0^2}\right) \times \exp\left[-\frac{8\pi^2}{\lambda_0^2}(L_B^2 - T_B^2)\sin^2\Psi\right] \tag{10.3-12}$$

where λ_p is the plasma wavelength (the free-space wavelength corresponding to the plasma frequency, (i.e., $\lambda_p = c/f_p$); $\langle(\Delta N/N)^2\rangle$ is the mean-square fractional deviation of electron density; T_B is the correlation distance transverse to the axis of symmetry; L_B is the correlation distance along the axis of symmetry; and ψ is the angle between the direction of wave travel and the normal to the longitudinal axis of irregularity. In the derivation of Eq. (10.3-12), a Gaussian autocorrelation function was assumed for $(\Delta N/N)$, and the cross section was normalized to a unit volume and unit solid angle. Since $L_B \gg T_B$, the T_B term in the second exponential is usually neglected; sin Ψ is also often approximated by Ψ.

For the 20–100 Mcps frequency range, Booker concluded that $2\pi L_B$ was about 40 m. Peterson[241] states that from 100 to 800 Mcps, measurements of aspect sensitivity by the Stanford Research Institute have indicated a value of $2\pi L_B = 20$ m. Peterson suggests that 6×10^{-4} could be a reasonable value for $\langle(\Delta N/N)^2\rangle$ for the highest frequency range. Booker had suggested 3×10^{-7}. Booker suggested 0.16 m for T_B; Peterson suggests 0.1 m for the 100–800 Mcps range.

Booker[234] deduces from the observations of Leadabrand *et al.*[242] that the scattering properties of auroral ionization vary with height in the region between 80 and 130 km. One suggestion is that the scale of irregularities decreases with height.

Booker[234] reports that a number of observers have found that the polarization of auroral echoes agrees closely with the transmitted polarization. Hultqvist and Egeland[220] state that depolarization has been observed occasionally up to 500 Mcps, but, in general, the echoes at frequencies above 100 Mcps favor the transmitted plane of polarization. The depolarization usually occurs in brief, abrupt time intervals.

Moorecroft[138] attempted to combine the mechanism of overdense reflection with weak scattering. He considered the medium to consist of an assembly of individual scatterers. The individual scatterers are spheroidal with a Gaussian distribution of electron density. These are termed Gaussoids, and are discussed in Section 10.2.2.4. Moorcroft considered a number of possible sizes and assemblies of these. Experimental evidence in support of this model of ionization has been given by Moorcroft[243] and by Lyon and Forsyth.[244]

The simplest combination of Gaussoids is one in which the scattering coefficient per unit volume is the scattering coefficient of an individual scatterer times the number of scatters per unit volume, n_g. Thus for an assembly of weak Gaussoids, n_g times Eq. (10.2-165) gives

$$\sigma = \frac{\mu_0^2 e^4}{16\pi^2 m_e^2} N_0^2 a^4 b^2 \pi^3 n_g \exp[-2k_0^2(a^2 \cos^2\theta + b^2 \sin^2\theta)] \qquad (10.3\text{-}13)$$

where a is the semimajor axis and b is semiminor axis of the Gaussoid.

Moorcroft applies Booker's technique by taking the autocorrelation function of a Gaussian electron density as given by Eq. (10.2-175) and showing that the two approaches give results that are essentially similar. Moorcroft's model is not unique in terms of Booker's autocorrelation function, in that an infinite number of other physical models would give the same autocorrelation function.

Moorcroft also considers assemblies of Gaussoids of different sizes but all having the same peak electron density. The size distribution has the form

$$n(a) = n_0 a^p \exp(-pa^2/2a_{\max}^2) \tag{10.3-14}$$

where p and n_0 are arbitrary constants with $a_{\max}$, the most likely value of the semimajor axis a. This distribution has a maximum value at $a_{\max}$, falling to zero at both $a = 0$ and $a = \infty$. As p approaches infinity, the distribution becomes a very narrow peak about $a = a_{\max}$. It thus approaches the case of an assembly of identical scatterers. The cross section is found by integrating the product of the size distribution $n(a)$ and the cross section of the individual scatterers as a function of a. The result for $\lambda_p/\lambda_0 > 1$ is

$$\sigma = \frac{N_T b(2\pi)^4 A_p}{\lambda_p{}^4 N_0[1 + (16\pi^2 a_{\max}^2/p\lambda_0{}^2)(\cos^2\theta + (b^2/a_{\max}^2)\sin^2\theta)]^{(p+7)/2}} \tag{10.3-15}$$

For $\lambda_p/\lambda_0 < 1$, the cross section is found from

$$\sigma = \frac{a_{\max} N_T b f^2 \lambda_p{}^2 B_p \ln(\lambda_0{}^2/\lambda_p{}^2)}{\lambda_0{}^2 N_0[a_{\max}^2 \cos^2(\theta + \delta) + b^2 \sin^2(\theta + \delta)]^2} \tag{10.3-16}$$

where

$$A_p = \frac{\Gamma[\frac{1}{2}(p + 7)]}{2^{5/2}\pi^{1/2}p^{3/2}\Gamma[\frac{1}{2}(p + 4)]} \tag{10.3-17}$$

$$B_p = \frac{p^{1/2}\Gamma[\frac{1}{2}(p + 3)]}{2^{5/2}\pi^{3/2}\Gamma[\frac{1}{2}(p + 4)]} \tag{10.3-18}$$

and $\Gamma[x]$ is the Gamma function of x.[124] The quantity N_T is the total electron density in the volume seen by the radar. The angle δ is determined by ray-tracing through the refractive medium exterior to the critical reflecting surface. Without detailed ray tracing, all that can be said about δ is that it is zero when θ is 0° or 90°. Moorcroft[243] found b/a to be between 5 and 10 and p to be between 1 and 5.

Moorcroft also observes that for waves penetrating a scatterer there is a force on the electrons which is nearly equal across the scatterer. The resulting electrostatic fields can increase the amplitude of the reflected wave to as much as twice the value predicted by the theory of Booker for polarization transverse to the cylinder axis. For polarization parallel to the cylinder axis, there is no increase. For most practical applications, this factor of two or less is neglected.

Echoes from auroras are usually Doppler-shifted and Doppler-broadened. The Doppler shift is sometimes as much as several hundred cycles. Positive Doppler shifts are usually associated with visually homogeneous forms, and negative shifts with rayed forms. The broadening could be caused by the rapidly-varying amplitude of the echoes, which sometimes broadens the frequency of a CW signal by several hundred cycles.

Bullough and Kaiser[223] reported on auroral observations made during a study of meteor reflections from June 1945 to December 1953. They pointed their radar beam 50° west of the magnetic meridian at Jodrell Bank, England. They deduced the motion to be westward in the early evening, changing to eastward between 2100 and 2200 hours local time. The mean velocity changed from 600 m/sec westward at 1500 hours to 600 m/sec eastward at 0500 hours. There was an absence of echoes between about 0600 and 1300 hours.

Blevis *et al.*[224] noted some Doppler shifts that remained quite steady or varied only slowly over periods of the order of several hours. On other occasions, the Doppler shift of the reflected signals fluctuated rapidly between large positive and negative values. On the average, echoes with negative Doppler shifts occurred normally between noon and midnight,

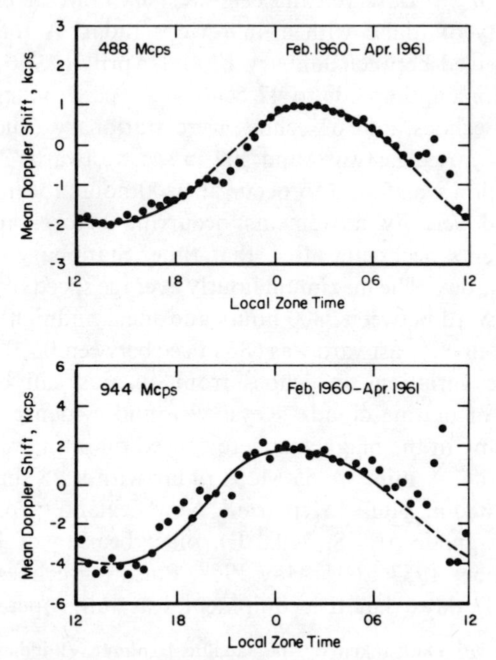

Fig. 10-33. Doppler shift of auroral radar echoes as a function of the time of day [after Blevis *et al.*[224]].

while echoes with positive shifts reached their peak in occurrence after midnight. The diurnal variation of the near Doppler shift is shown in Figure 10-33. Figure 10-33(a) gives the data for 488 Mcps, and Figure 10-33(b) for 944 Mcps. Also shown are empirical sinusoidal curves. For 488 Mcps, the Doppler shift Δf is empirically fitted by

$$\Delta f + 0.5 = 1.5 \sin[(2\pi/24)(t + 3.5)] \tag{10.3-19}$$

where the time t is measured in hours. The relationship for 944 Mcps is given by

$$\Delta f + 1.0 = 3.0 \sin[(2\pi/24)(t + 4.5)] \tag{10.3-20}$$

This is double the curve of the lower frequency, and shifted one hour earlier in time.

Assuming that the motions causing the Doppler shift are east–west along the parallels of geomagnetic latitude, one obtains a maximum apparent mean westward velocity of 1.2 km/sec for the reflecting surface and a maximum mean eastward velocity of 0.6 km/sec, relative to a point on the surface of the earth below the echoing region. After correction for the rotational velocity of the earth's surface, a mean velocity in either direction of 0.9 km/sec is determined.

Wilkins *et al.*[245] observed the east–west motions of the aurora north of the University of Idaho with a 46.8-Mcps radar. A total of 284 radar echoes were received between January 17 and April 1, 1966. It was possible to assign directions and speeds to 47 echoes. A speed but no direction was assigned to 174 echoes, and 63 echoes were stationary. The average radial velocity was 414 m/sec eastward and 502 m/sec westward. A reversal in the direction of motion was found to occur at 0200 hours local time, with 83 % of the westward velocity movements occurring before, and 82 % of the easterly movements occurring after, that time. Stationary echoes occurred at any time of the day. The maximum hourly average speed of movement was 1210 m/sec westward between 2300 hours and local midnight. The maximum hourly average velocity eastward was 685 m/sec between 0600 and 0700 hours.

Not all the variations in echoes from auroras can be attributed to simple motions of plasma clouds. Keys[246] found evidence that hydromagnetic waves arising in the magnetosphere caused pulsating radar echoes with a period range of 1–9 min.† A 55-Mcps radar with a 25° half-power beamwidth was operated at Bluff, Invercargill, New Zealand (geographic 46°37′S, 168°49′E; geomagnetic 51.8°S, 249.3°E), on a bearing of 175° geographic from 22 September 1957 to 15 May 1958. Pulsating echoes were observed on 77 of the 237 days that the equipment was fully operational. Normal

† *Note added in proof:* Data taken by the satellite Explorer VI indicate hydromagnetic wave activity with a period-range of 100 to 500 seconds. See Judge, D. L., and Coleman, P. J., Jr., Observations of Low-Frequency Hydromagnetic Wires in the Distant Geomagnetic Field: Explorer VI, *J. Geophys. Res.* **67**:5071 (1962).

nonpulsating echoes were observed on 203 days. (Both types of echo were observed on some days.) The pulsations exhibited a diurnal variation in occurrence, periods being dependent on latitude, and onset times dependent on magnetic activity.

Lyon and Forsyth(244) published a sequence of individual, simultaneous radar observations of the aurora taken from 42 to 104 Mcps. Moorcroft(247) analyzed the results and found that twelve of the curves of cross section versus frequency might be explained by weak scattering, nine others might be explained with the introduction of critical reflection, four more might be explained with the addition of absorption, and the remaining seven must be explained in terms of other effects. Therefore, if the scattering medium is assumed to consist of a random assembly of different scatterers, critical reflection ($\omega \leqslant \omega_p$) must be introduced to explain the observed frequency dependence. The standard objection to critical reflection is the requirement of high electron densities. Three of the curves imply densities as high as 10^{14} electrons/m^3, which is critical reflection for approximately 90 Mcps. This high an electron density has not been found naturally present in the ionosphere. However, since this density would not have to exist over a very large volume, it is not impossible for it to exist. A more serious objection is that the type of reflection process has to change rapidly with time. The curves were taken in a 2-hr period. The first two curves require critical reflection between 82 and 104 Mcps, the next two are explained by weak scattering, the next two by critical reflection between 60 and 82 Mcps, and so on. Moorcroft states that all of the observations can be explained in terms of a wave-generating process giving ordered structure to the ionization. A logical candidate is a two-stream instability, similar to that associated with the equatorial electrojet. For both the equatorial electrojet and the radio aurora, the direction of the radar beam must be nearly perpendicular to the direction of the geomagnetic field. Both have a similar type of Doppler behavior. The scattering is due to the disturbance of electron density resulting from a two-stream plasma instability, as analyzed by Farley(161) and discussed in Section 10.2.3. This instability occurs when there is a sufficient relative drift of electrons through the ions. The plasma waves generated by the instability grow in amplitude until they are limited by nonlinear effects. The aspect sensitivity arises from the fact that the only unstable waves are those propagating very nearly perpendicular to the geomagnetic field. In his laboratory simulation of aurora, Coleman(248) noted similar scattering from two-stream instabilities. He observed the change in scattering cross section as the mean electron drift was made to exceed a critical drift level. It was demonstrated that when the critical level was exceeded, the scattering cross section increased over five orders of magnitude above that predicted for Thomson scattering [see Eq. (10.2-172)]. The scattering spectrum observed under these conditions corresponded to the modulation envelope predicted for a continuum of growing ion acoustic waves. The Doppler spectra measured was quite similar to that observed in auroral radar observations. It should be pointed out that Leadabrand[249] has found that the spectral characteristics of 400 and 800 Mcps auroral radar data obtained in Scotland are not completely described by the Farley theory alone.

In summary, the reflections in the HF band can perhaps be largely explained by partial reflections from many small overdense regions, with some underdense regions also present to create some refraction and absorption. At UHF, use of the model of Moorcroft, consisting of underdense Gaussoids, will usually give good monostatic results. At VHF, however, the phenomena are more complicated. Moorcroft's model of some overdense and some underdense Gaussoids will sometimes predict the proper amount of echo. The HF model is sometimes valid, especially in the lower portion of the VHF band. There are other times when a wave-generating process, such as a two-stream plasma instability, is the only satisfactory model. It is this latter model in which there is the most current interest and the most hope of developing an explanation of auroral reflections. The models of Moorcroft and Booker fail to explain the following auroral effects:

(1) The existence of a threshold electrojet current level below which auroral scatter becomes negligible.

(2) Observed Doppler shifts that do not follow the laws of directed charge motion, but are predictable from the concept of longitudinal acoustic plasma waves in the medium.

(3) Apparent turbulent "blob" velocities that greatly exceed estimated wind velocities of the ionosphere (360 m/sec or about 1000 mph).

(4) Relaxation of aspect sensitivity with higher electrojet current levels.

(5) Increase in the extent of spectral content with electrojet current level.

(6) Enhanced radio noise levels when the critical electrojet current level is exceeded.

(7) Enhanced field-sensitive cross-section levels that again depend upon a critical current level.

The two-stream plasma instability gives an analytical basis for the plausibility of all seven of these auroral effects. Coleman[248] observed these same effects in the laboratory. However, no adequate computation of the structure function of ion acoustic wave turbulence currently exists. Farley's work does not completely predict the scattering spectrum since it is based on a linear theory. Since both the spatial scales and the time scales of auroral fluctuations are small, the true magnetic aspect dependence of the aurora is probably masked in field experiments. Also, other phenomena may occur simultaneously with these instabilities. Bowles *et al.*[160] have observed that other wave modes were possible and recent evidence[250] indicates that gravity waves could also be present, accounting for a very-low-velocity wave motion in the disturbance.

10.3.3. Nuclear Explosions

A nuclear explosion produces many effects upon radar signals. It creates a plasma cloud which can reflect, scatter, absorb, and refract the radar signal; it generates noise; and it perturbs the ionosphere and geomagnetic field. Nuclear explosions have produced aurora at the magnetic conjugate points† of explosions. The ionization is produced, either directly or indirectly, by the gamma rays and neutrons of the initial nuclear radiation, by the beta particles and gamma rays of the residual nuclear radiation, and by the X rays and ultraviolet light present. Hence an expanding ionized cloud is created. Since plasmas interact with a magnetic field, the geomagnetic field is distorted.

Glasstone[251] calculates that if 10 % of the energy of a megaton weapon were expended in ionizing the atmosphere, an amount of electrons approximately equal to the number of free electrons in the entire ionosphere would be created (about 10^{32} free electrons). He divides nuclear explosions into types according to the height of the explosion. The height ranges are roughly as follows: surface and subsurface; below 15 km; 15–60 km; 60–100 km; and above 100 km. Surface and subsurface bursts introduce clouds of dust, debris, and water droplets or ice crystals into the atmosphere. Most of the particles are so small as to effect only X-band or shorter-wavelength radiation. The clouds may exist for minutes or hours, depending upon the weapon yield, the type of surface on or under which it was detonated, and the meteorological conditions at the time of the explosion.

For explosions below 15 km altitude, considerable ionization is produced in the fireball, but most of the ionizing radiation will not penetrate more than a few hundred yards in the atmosphere, and the free electrons created will attach themselves to particles in the atmosphere almost immediately. Therefore, unless the radar is examining the fireball itself or trying to look at something immediately behind the fireball, the nuclear explosion will have little effect upon radio wave propagation. As time progresses, the fireball rises. For nuclear explosions of the size of a megaton or more, the fireball may rise to an altitude of more than 30 km. At these altitudes, the gamma rays from fission products and weapon debris are no longer absorbed in the immediate vicinity of the fireball. The result is abnormally high electron densities in the D-region of the ionosphere, causing attenuation, refraction, clutter, and, at the lower frequencies, reflection for several hours.

† Magnetic conjugate points have the same geomagnetic longitude, and geomagnetic latitude values differing only in sign.

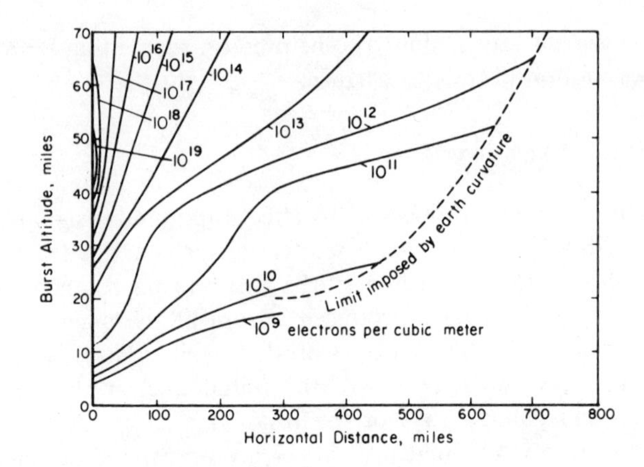

Fig. 10-34. Calculated electron density at an altitude of 45 miles produced by a 1-megaton fission-yield explosion at various burst heights as a function of horizontal distance [after Glasstone[251]].

If the explosion occurs initially in the 15–60-km altitude range, gamma rays, neutrons, and, to a lesser degree, thermal X rays—all initial radiation—will ionize the *D*-region. The density of ionization is predicted to be a maximum at an altitude of 70 km. Figure 10-34 shows the theoretical electron density increase, at an altitude of 70 km caused by a 1-megaton fission-yield explosion as a function of burst altitude and horizontal distance. The electrons freed by the initial radiation will remain free for a few minutes. However, beta particles and gamma rays from the rising fireball will free more electrons; thus the effects may last for hours.

For bursts in the 60–100-km altitude range, the horizontal distance over which the effects of the initial gamma rays and neutrons acts is limited only by the horizon distance of the earth. Another factor causing the larger amounts of ionization at the higher altitudes, observed in Figure 10-34, is the effect of thermal X rays. The X rays possess as much as two-thirds of the energy of the explosion. At the lower altitudes, they are heavily attenuated by the atmosphere; while at this higher altitude, they cause ionization for large distances.

Approximate values of the distances in which one half the ionizing radiation is absorbed in the atmosphere are given in Table 10-4. The values are given to illustrate the magnitudes of the distances, not to give precise values. The distances are for horizontal paths. The half distance would be larger for an upward path and smaller for a downward path. For the beta rays, it is assumed that the magnetic field and the beta-ray path are both horizontal.

The beta particles actually follow a spiral path along the geomagnetic

Table 10-4. Effect of Atmospheric Absorption on Ionizing Radiation

Ionization Cause	Half-Distance		Altitude at Which
	Sea Level	100-km Altitude (km)	100 km > Half-Distance (km)
Gamma rays	120 m	400,000	50
Neutrons	120 m	400,000	50
Beta rays	3 m	10,000	75
X rays	3 cm	100	100
Ultraviolet rays	0.16 cm	5	125
Debris particles	0 03 cm	1	140

field lines, with part traveling toward the magnetic conjugate point in one hemisphere and part toward the conjugate point in the other hemisphere. At about 70 km in altitude in both hemispheres, ionized clouds about 15 km thick are created. Some of these clouds have been observed to contain auroras.

For explosions above 100 km altitude, the various radiations penetrate downward to the *D*-region, creating ionization over very large volumes. The interaction between the electrons and ions and the geomagnetic field is especially important. There are effects at the conjugate points, and there is interaction between the expanding debris and the excluded geomagnetic field lines causing the expansion to stop. Glasstone[251] estimates that the expansion would continue to a radius of 600 miles for a 1-megaton bomb and to 60 miles for a 1-kiloton bomb. If the explosion occurs at a height of the order of a few hundred miles, the expansion downward of ionized weapon residue is stopped by the magnetic field at an altitude of about 140 km.

For all high-altitude detonations, daytime explosions produce larger amounts of longer-lasting ionization than night-time explosions. Some of the recombination which occurs at night, however, could result in the formation of negative ions which separate again at sunrise.[252,253]

An extensive bibliography of the published effects of high-altitude explosions has been given by Saha and Mahajan.[254] They observed ionospheric effects over Delhi, India following three high-altitude nuclear explosions over central Asia by the Soviet Union on October 22, October 28, and November 1, 1962, and two high-altitude explosions over Johnston Island by the United States on October 26 and November 1, 1962. The ionospheric phenomena, noted by a variety of means, were discussed in a previous paper[255] which reported observations of ionospheric changes caused by 24 lower-atmospheric nuclear explosions by the USSR over Novaya Zemlya in the period August–December 1962. The effects of both types of explosions occurred mainly in the *D*- and *F*2-regions of the ionosphere. The height range affected in the high-altitude explosions appeared higher than for the lower-altitude explosions. For the Novaya Zemlya explosions, the ionization enhancement in the *D*-region was more localized to heights lower than the effective height of the radio noise absorption during a sudden ionospheric disturbance caused by a solar flare. The *D*-region effects of the lower explosions had a sizeable delay over the effects of high-altitude explosions. The Central Asian detonations appeared to cause increases in the heights of maximum ionization of the *F*-layer simultaneously with decreases in the ionization levels at various heights. No significant changes in the *F*-layer height or shape were noted for the Johnston Island explosion, except for a general increase in the ionization at all heights. For the Novaya Zemlya explosions, the *F*-layer was distorted considerably and the layer peak brought down significantly. From ionospheric radar soundings it was found that there was enhancement of electron density at all levels above 200 km. The maximum effect in the *F*2-layer was found to occur around 240 km, which is somewhat below the *F*2 peak. There was also considerable enhancement in the topside of the ionosphere. Velocity computation for the *F*2-layer disturbance gave an average speed of 550 m/sec, which is lower than the acoustic velocity for the *F*-region (700 m/sec at 240 km). A wide scatter of points was noted around the line $\Delta N/N = 4M$, where $\Delta N/N$ is the ratio of enhancement of maximum electron density to average expected electron density, and M is the explosion strength in megatons.

Since the effects of nuclear explosions on the ionosphere cover such an extensive volume, for many applications the techniques discussed for planar media in Section 10.2.2.1 and in Chapter 7 are sufficient. For radar returns from the ion cloud induced by the nuclear explosion, or for objects directly behind the ion cloud, the techniques discussed in Sections 10.2.2.3 and 10.2.2.4 for calculating the cross section of plasma spheres and bodies embedded in plasma, could be useful. The attenuation and refraction of the signal by the ion cloud can be very significant. For an ionized cloud layer, Crain and Booker[256] give an approximate relationship for the attenuation ΔA (in db) as

$$\Delta A \approx \frac{\csc \theta'}{f^2} \Delta N_m T^3 \times 10^6 \tag{10.3-21}$$

where N_m is the maximum electron density in electrons/m³, T is measured in keV, and f is measured in cps. The additional phase shift $\Delta\psi$ introduced by the ionized cloud is given by

$$\Delta\psi \approx \frac{2\pi}{\lambda_0} \frac{\csc \theta'}{f^2} \Delta N_m \tag{10.3-22}$$

The value of ΔN is directly proportional to device yield. A value of 10^9 electrons/m^3 is said to correspond to a device with about 15 kilotons of X-ray yield exploded in space at a distance of 100,000 nautical miles in a direction near the zenith. For the effects caused by the aurora at the geomagnetic conjugate point of the explosion, the techniques discussed in Section 10.3.2 are useful.

10.3.4. Man-Made Ionospheric Disturbances

In the previous section, it was shown that nuclear explosions, besides creating a highly-ionized fireball, also had considerable effect on the ionosphere. This section considers other ionized regions artificially created in the ionosphere. The first section considers chemical explosions in the ionosphere; these have a variety of uses, including simulating some of the effects of meteors, auroras, nuclear explosions, and rocket launches. The phenomena occurring in the ionosphere near a rocket launch are considered next. The last two sections consider the remainder of the trajectory, the midcourse or orbital portion of the flight of a vehicle, and the reentry of the vehicle.

10.3.4.1. *Chemical Releases*

Marmo *et al.*[257] reported the creation of an artificial electron cloud with an initial average density of about 4×10^{12} electrons/m^3 by the release of 18 pounds (275 moles) of nitric oxide gas at an altitude of 95 km in March 1956. The cloud was observed by various radars for more than 10 min. Other tests were carried out at Holloman Air Force Base, New Mexico, in 1957 and 1958.[258] The releases were at various altitudes from 69 to 121 km. Negative results were obtained at 69 km with the release of 28 moles of cesium. It was felt that this altitude was too low for the generation of a long-lived, high-density, solar-induced artificial electron cloud. A release of 27 moles of cesium at 82 km created a detectable cloud.

Gallagher and Villard[259] report the analysis of the radar cross section of a release on November 19, 1957. They assumed a model of the cloud which has an electron density that goes from 10% of the maximum density to 90% in a distance a. The cloud has a thickness b between its 50% density points. The backscatter radar cross section for coherent scattering is given by

$$\sigma = N_T^2 \sigma_e (\lambda_0/2\pi b)^2 \sin^2(2\pi b/\lambda_0) \exp(-8\pi a^2/\lambda_0^2) \qquad (10.3\text{-}23)$$

where N_T is the total electron content of the cloud and σ_e is given by Eq. (10.2-168). As the cloud expands, the cross section predicted by Eq. (10.3-23) oscillates from zero to a maximum value of

$$\sigma = N_T^2 \sigma_e [2/\pi(2n + 1)]^2 \exp(-8na^2/\lambda_0) \qquad (10.3\text{-}24)$$

where b has a value of $\frac{1}{4}(2n + 1)\,\lambda_0$, $n = 0, 1, 2,\ldots$. The number n increases as the cloud size increases. This equation ceases to be valid after n becomes equal to 4 or 5. Another reasonable model for obtaining the cross section of the cloud suggested by Gallagher and Villard is a perfectly reflecting oblate spheroid. The cross section predicted by this model can be obtained by the equations of Section 5.1.3.

Gallagher and Barnes[260] have presented many of the early measurements of radar cross section. They found that the radar cross section increased with frequency until a certain critical value was reached; it was then almost independent of frequency until a second critical value was reached. At higher frequencies, there was a rapid decrease with frequency. These critical frequencies were in the neighborhood of 10–20 Mcps. The amplitude of the returned signals generally followed a Rayleigh or some higher-order Rice distribution. It is suggested that the clouds were composed of multiple, independent reflecting centers. In most of the clouds, signals could propagate internally. In two clouds, the high density of scatterers made the clouds overdense at 26 Mcps. They found that much of their data of cross section versus frequency could be approximated on a log–log scale by straight lines. The cross section can be expressed in three different frequency regions as arbitrary constants A_1, A_2, A_3 times the number of irregularities, times λ^{-4}, λ^0, or λ^{+4}. Assigning the wavelengths λ_1 and λ_2 to the critical frequencies discussed above, then

$$\sigma \approx A_1 n \lambda_0^{-4} \qquad \text{for} \quad \lambda_0 > \lambda_1 \tag{10.3-25}$$

This is the expression for Rayleigh scatter from n small dielectric spheres of radius a, where $\lambda_1 = 10a$. For $\lambda_1 > \lambda_0 > \lambda_2$,

$$\sigma \approx A_2 n \tag{10.3-26}$$

which is the sum of cross sections from n large dielectric spheres. For $\lambda_1 > \lambda_0$, the expression is

$$\sigma \approx A_3 n \lambda_0{}^4 \tag{10.3-27}$$

which is a form of the Booker–Gordon theory for isotropic turbulance (see Section 10.2.3.3).

The value of n is a function of altitude of release, and the time after burst as well as the amount and type of chemical used. Gallagher and Barnes[260] have tabulated estimates of n and other burst parameters for ten releases. Values varied from 4×10^3 to 10^6 at the time of burst. One burst was estimated as having 4.4×10^4 irregularities immediately after the burst, 7.7×10^5 at 65 seconds after the burst, 2.09×10^5 at 180 seconds, and 4×10^4 at 720 seconds.

The idealistic nature of these models was pointed out by the authors. From these models and their measured cross sections (some of which went as high as 10 km^2), they estimated a total electron yield of 10^{20} electrons, which was less than expected. The fading rates of the returned signals were reported as comparable to those observed for long-enduring meteors and for normal-mode ionospheric propagation, and were considerably less than were observed in auroral observations. This work was part of the Project Firefly sponsored by the US Air Force Cambridge Research Laboratories.

The "Firefly" program,[(261)] conducted from 1959 through 1963, also involved the formation of an electron-depleted region in the *F*-layer of the ionosphere. The objectives were to study the nature of reflected radio-frequency waves from the perturbation and to determine the rate of return of the region to normal. In one of the 27 successful launchings during 1962, 22 kg (about 10^{26} molecules) of sulfur hexafluoride (known to attach thermal electrons with a high rate constant) were carried aloft aboard an Aerobee vehicle launched near mid-day from Eglin AFB, Florida. The SF_6 was released at an altitude of 222 km, 80 km south of the launch point and the ionosonde. Based upon the electron-density profile determined for the time of release, and upon an assumed Gaussian density distribution of the SF_6 immediately after release, contours of electron density form an electron-depleted hole in the *F*-region. The electron density near the center is less than 10% of the ambient, which was near 7×10^{11} electrons/m^3. Both ionograms and multiple fixed-frequency radar returns from the region show a disturbance which died out in about $\frac{1}{2}$ hr. Tracings of the ionograms are available in the papers of Wright.[(262,263)] The ionograms are clearly of the kind described by Booker[(264)] as the typical echo pattern from a region of reduced electron density. Booker's ionograms are discussed in the following section (10.3.4.2).

Rosenberg *et al.*[(265)] reported on a series of cesium releases denoted Project RED LAMP. The detonations took place near 120 km altitude in the vicinity of 78°W longitude, 27.3°N latitude, and between 0130 and 0200 EST on successive nights. The explosive weighed 13 kg and released 65×10^{13} ergs of energy. The radar cross section of the first release was measured as 5.5×10^5 m^2 at 24 Mcps, and 5.2×10^5 m^2 at 15 Mcps, 0.25 sec after the release; both measurements were taken at Orlando, Florida. Using a spherical scatterer as a model, the calculated radii are 419 m and 407 m, respectively. This agrees quite well with radar Doppler data. On the second test, radar cross sections measured 0.25 sec after release were 0.86×10^5 m^2 at 15 Mcps, 0.81×10^5 m^2 at 24 Mcps, and 3.8×10^5 m^2 at 7 Mcps as observed at Jupiter, Florida.

The fully-expanded radius of these gaseous clouds is said to be 600 m, which is reached after 0.8 sec, with electron densities of 10^{13} to 10^{14} electrons/m^3. At 20 msec, the radii are 150 m, and the electron densities are 10^{15} to 10^{16} electrons/m^3. The expanding cloud has a highly overdense surface at these frequencies. The surface is rough, perhaps resembling a pin cushion. It may act as a diffuse scatterer. Both receding and approaching Doppler were observed.

Föppl *et al.*[266] reported on experiments performed with the metals barium and strontium in the Sahara and in Sardinia in 1964. The ionization proceeded in two steps with characteristic times of 5 sec and 100 sec. The increase in the diameter of the clouds was the same as predicted by purely molecular diffusion. The rate of increase in the central intensity of the clouds at sunrise was greater than the rate of decrease at sunset. Atmospheric wind velocities of 50 to 130 m/sec were measured. The motion of ion clouds perpendicular to the geomagnetic field indicated the presence of electric fields. The initial expansion velocity for explosive mixtures was greater than the mean thermal velocity of the atoms at a temperature of about 3000°K by about a factor of four.

In summary, a variety of chemicals have been introduced into the ionosphere, both reducing and increasing the electron concentration. The radar effects are best summarized by the remarks made by Golomb *et al.*[267] at the end of the 1963 tests. They summarize the causes of signal return as (i) perfect or partial reflection from a wholly overdense volume; (ii) reflective scattering from a large number of overdense irregularities; and (iii) Booker–Gordon-type scattering from underdense irregularities. For type (i) scattering, the cross section is proportional to the physical area of the cloud, and no short-period fading is observed. For type (ii) scattering, the cross section is proportional to the number of irregularities and their scale, and considerable short-period fading is observed. For type (iii) scattering, the cross section is proportional to the mean-square departure of the electron density from the mean and the scale size of the turbulent irregularities, and fading is observed. For electron clouds in the 100-km altitude region, all three types of scattering were observed simultaneously. After tens of minutes in sunlight releases, and several minutes in night releases, only type (iii) remains.

10.3.4.2. *Effects near Rocket Launch*

The radar cross section of a flaming rocket is not well documented in the literature. The flame behind the rocket is a highly-ionized tapered cylinder;[268] thus the techniques discussed in Chapter 4 and in Section 10.2.2.2 apply.

Several disturbances are associated with the passage of a rocket up through the ionosphere. One such disturbance, confirmed by both observational evidence and theoretical analyses, is that powered-missile transit through the *F*-region of the ionosphere produces a spatially extensive, electron-deficient hole which can persist as long as an hour. Such an electron deficient region is usually referred to as an "ionospheric hole" or "duct." The deficiency in electron density appears to the radar as if it were a dielectric

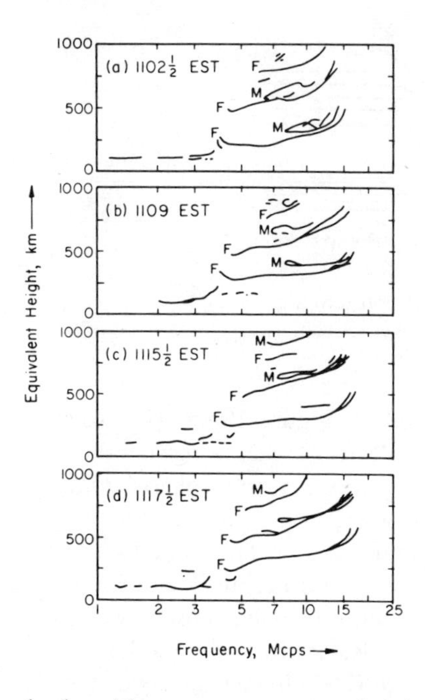

Fig. 10-35. Ionograms taken on Grand Bahama Islands after the firing of Vanguard II [after Booker[264]].

body with an apparent permittivity ϵ_a such that

$$\epsilon_a/\epsilon_0 = \epsilon_h/\epsilon_i \qquad (10.3\text{-}28)$$

where ϵ_h is the permittivity of the hole and ϵ_i is the permittivity of the ambient ionosphere. The radar cross section of the hole can then be found by means of the standard techniques for dielectric bodies.

Sketches of ionograms recorded by the US Army Signal Radio Propagation Agency at Grand Bahama Island in conjunction with the firing from Cape Kennedy of Vanguard II are shown in Figure 10-35. The *F* indicates a normal *F*-region echo, and the letter *M* an unusual echo associated with the passage of the missile. As is normal with ionograms, traces caused by multiple bounces between the earth and the ionosphere are seen at equivalent heights of 500 km and up. These are not actual discontinuities at these heights, but delayed returns from the discontinuities at the lower equivalent height. The secondary (and higher-order) echoes tend to magnify small irregularities so that after 11 : 15½ EST, the primary missile-produced echo disappeared, but the anomolous echoes were still identifiable until after 11 : 30 EST.

Booker[264] states that these ionograms are similar to those obtained after a number of missile firings, and hypothesizes that the missile exhaust gases "punched a hole" in the *F*-region. He interprets the contours of constant ionization density, originally quasihorizontal and with an upward positive gradient (to the *F*-region peak), as being deformed so that they roughly

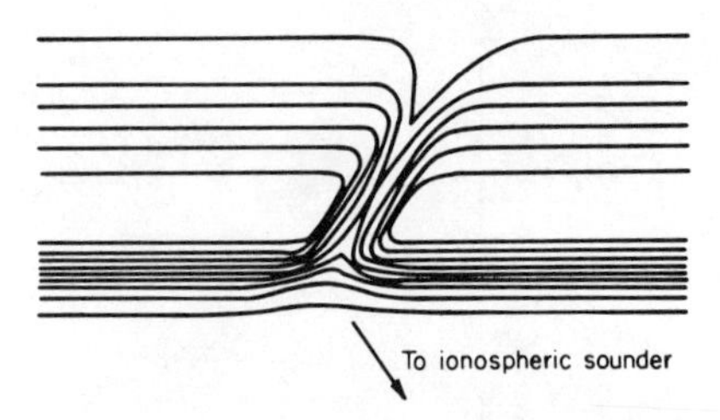

Fig. 10-36. Contours of constant ionization density after a rocket launch, as hypothesized by Booker.[264]

parallel the axis of the hole, as illustrated in Figure 10-36. He observes that at sufficiently low frequencies, the tilting of the lower contours is not enough to provide a vertical-incidence echo to the ionospheric sounder; with an increase in frequency, normal incidence onto the far side of the hole would occur, thus providing the initial point of an anomalous echo. With further increase in frequency, the echo would become a double trace, as illustrated in Figure 10-37. Successive multiple reflections between the earth and the ionosphere can result in the pulse approaching the hole from a direction nearer to the vertical. Thus, successively higher multiples should appear at successively lower frequencies, as observed in Figure 10-35. Also, as observed, the primary echo should disappear first as the hole fills up.

Booker suggests that although the exhaust gases are highly ionized, they would be hot, and therefore would rise. They would be replaced by relatively un-ionized air from below. Diffusion of electrons to fill this hole would be inhibited by the geomagnetic field, so that the principal process tending to fill the hole would be reionization by solar radiation. The time observed for the disappearance of indications of the hole's existence was of the same order of magnitude as the time the *F*-region requires to recover from a solar eclipse.

As an alternative to Booker's rising-gas theory, Barnes[269] suggests that the exhaust gas, upon emergence from the nozzles into the near-vacuum of the *F*-layer, expands immediately by a factor of more than 10^7 until it reaches ambient pressure. It cools in the process, and he states that its electron density could decrease from roughly 10^{16} to 10^9 electrons/m^3. The ion density of the *F*-layer being typically above 10^{10} electrons/m^3 in the daytime, the expanded exhaust constitutes an electron-deficient region.

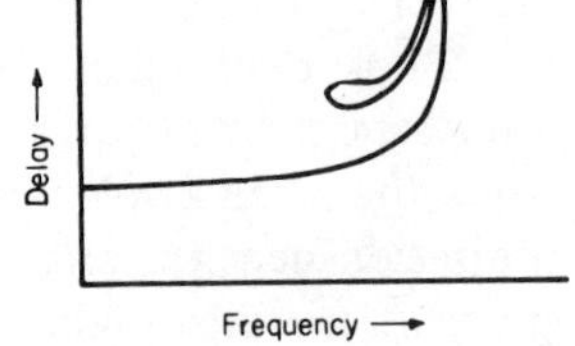

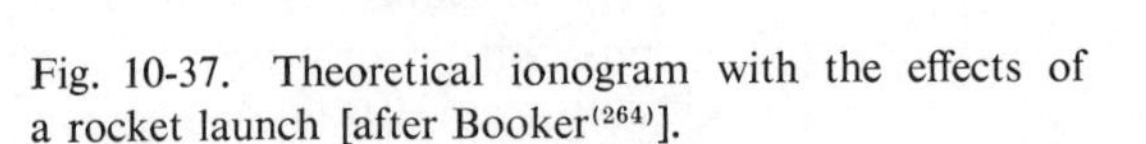
Fig. 10-37. Theoretical ionogram with the effects of a rocket launch [after Booker[264]].

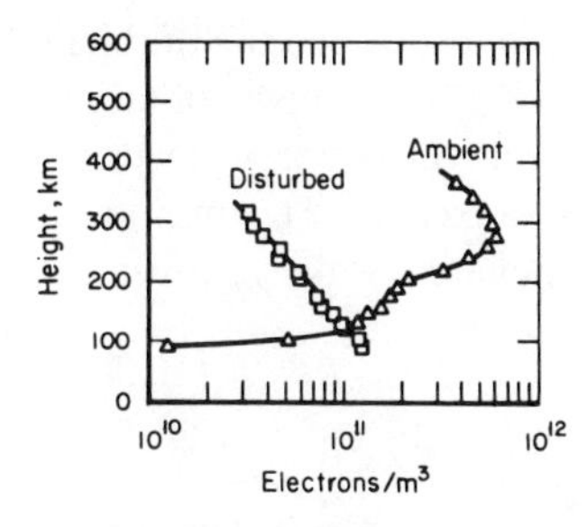

Fig. 10-38. Comparison of the electron density in a rocket plume with the ambient ionosphere as determined by Faraday rotation [after Stone *et al.*[270]].

Lincoln Laboratory personnel[270] measured the Faraday rotation of signals passing through the length of a rocket plume for five Atlas missile firings on the Atlantic Missile Range. It was determined that the trajectory was such that the axis of the receiving antenna could be aligned in approximate coincidence with the aft extension of the roll axis of the missile during the part of the flight between altitudes of about 100 and 400 km. The amount of Faraday rotation, as discussed in Section 10.2.1.1.2, varies directly with the integrated electron density along the path. Results indicate, as shown in Figure 10-38, that the electron density is reduced below the ambient by a factor of 10–20, reaching values of the order of 3×10^{10} electrons/m^3 at altitudes above 300 km. They also indicate electron densities greater than ambient in the low *E*-region.

The above missiles were liquid-fueled. Jackson *et al.*[271] report on effects observed for a solid-fuel Scout vehicle. Faraday-rotation measurements at 73.6 Mcps showed an electron-density enhancement at third-stage ignition, an apparent reduction of electron densities along the remainder of the third-stage burning path (120–220 km), and similar but less-pronounced effects for the smaller fourth stage, which ignited just below the electron density peak of the F_2-region. Ionograms obtained during the flight reportedly show oblique echoes similar to those observed by Booker.

Data[272] obtained both from rocket-borne probes and from propagation experiments[272] indicate that outgassing of unburned fuel after burnout decreases the electron density in the rocket vicinity by a factor of five from the ambient; in this case, about 5×10^{11} electrons/m^3 in the *F*-region for liquid-fuel Aerobee rockets launched from Eglin AFB, Florida, and White Sands, New Mexico. Tailpipe observations of pulse delay for seven Aerobee rocket flights into the *E*-region at Holloman Missile Test Center, New Mexico gave similar results.

Peters[95] has analyzed the possible echo area from an evacuated wake created by a satellite moving through the ionosphere. He treats the wake as a dielectric rod antenna. His results may have some application to returns from rocket exhausts interacting with the ionosphere, if his various conditions are met. The transmitter and receiver must be in such a position that the rocket is moving either toward or away from both of them. The frequency range for which significant echo areas are obtained is just above the ambient plasma frequency f_p (measured in cps); i.e., $f_p < f < 1.8f_p$. This is obtained by the requirement that the dielectric constant relative to the ionosphere be sufficiently large that the dielectric rod mode is excited. The minimum permissible diameter (d_{min}, in meters) of the evacuated wake is given by

$$d_{min} = 1.5 \times 10^8/f_p \qquad (10.3\text{-}29)$$

The echo area is given by

$$\sigma_{max}(0) = 3.9\gamma^2L^2 \qquad (10.3\text{-}30)$$

where L is the length of the wake in meters, and γ is the voltage reflection coefficient ($\gamma_{max} = 1$), a function of the relative dielectric constant and the rod diameter. Assuming that $\gamma = \gamma_{max}$ in Eq. (10.3-30) perhaps overestimates the echo area. This equation, derived by assuming that the Hansen and Woodyard antenna condition is satisfied, requires a uniform amplitude distribution be set up in the wake, which is not probable. The actual echo area might be two orders of magnitude below that predicted by this equation.

10.3.4.3. *Satellite–Ionosphere Interactions*

10.3.4.3.1. *Experimental Observations*

10.3.4.3.1.1. *Bistatic Observations.* In early 1958, Kraus[273] and Kraus and Albus[274] of the Radio Observatory, the Ohio State University, reported receiving bursts of 20-Mcps signals that correlated with the passages of the artificial earth satellites 1957 Alpha and 1957 Beta, arousing a great deal of interest. Despite a large amount of effort that has gone into the study of these phenomena, very little is known about the causes and nature of this Kraus effect. Kraus monitored signals emanating from WWV, the time service and standard-frequency radio station located near Washington, D.C., a distance of about 330 miles from Columbus, Ohio. Under these conditions, any temporary patch of ionization, such as a meteor trail, auroral activity, or any ionization associated with a satellite, would give an increase in signal. A meteor-induced burst usually lasts less than a few seconds. The bursts recorded by Kraus lasted several minutes. From the pattern of the bursts, he concluded that the breakup of Sputnik I (1957 Alpha) was a gradual affair, with an initial breakup in December followed by a more complete fragmentation in January.[275] To explain the WWV bursts and an asymmetrical variation of the signals received from the satellite, he suggested that a rather large ion cloud was accompanying the satellite. Kraus and Dreese[276] calculated that the effective cross section of the ionospheric reflecting region was a minimum of 600 m². The earlier

Sputnik observations were all made while the satellites were below an altitude of 500 statute miles. However, Kraus *et al.*[277] soon noticed bursts that correlated with passages of Explorer I and III at altitudes as high as 1000 miles.

Flambard and Reyssat[278] in France and Dieminger[279] in Germany also reported observations that appeared to be the result of satellite–ionosphere interactions. However, in an analysis of the WWV signals received at the University of Illinois on a frequency of 20 Mcps, no indications were found of a column of electrons associated with the USSR 1957 satellites, which were at high altitudes during the Illinois experiments.[280]

An investigation[281] at the Antenna Laboratory of the Ohio State University concluded that no significant correlation existed between the 20-Mcps WWV bursts and the transits of the satellites 1958 Delta 1 and 1958 Delta 2 (Sputnik booster and spacecraft). It also was shown that the irregularities in received signal from satellite transmissions occurred most often when the satellite was to the north of Columbus, regardless of the direction of travel.[282–284]

At the Ohio State University Antenna Laboratory, WWV and other transmitters were monitored on a variety of frequencies from 5 to 50 Mcps from August 1958 to September 1960. Several bursts of WWV signals observed at 10, 15, and 20 Mcps appeared to be related to satellite passages, but other bursts were received that did not appear to be correlated. There also were occasions when no bursts were received under conditions similar to those when bursts did appear.[285,286]

The frequency spectra of WWV (10-Mcps, 15-Mcps, and 20-Mcps) signals over a bandwidth of ± 2000 cps centered on the carrier frequencies were filmed nearly continuously with a spectrum analyzer from December 1959 to May 1960. During this period, only seven pronounced indications of Doppler shift were recorded. All of these occurred when a satellite was near Columbus. Two types of Doppler effects were expected; one proportional to the mean velocity of the reflecting region, and the other a frequency-broadening effect due to turbulence. The theoretical value of the Doppler shift was compared with the measured values. In all cases, the shifts had some resemblance, but were not the same. In general, throughout this period, whenever WWV amplitude bursts were noted in the charts of received signal strength, there was little evidence of coherent Doppler shifts on the filmed data. Hame and Stuart concluded that sometimes there seemed to be bursts related to satellite passages, but that the bursts had very little value for deducing orbit parameters. With the large number of objects in orbit, they had practically no value even to indicate the presence of an object in orbit.[283,284,287]

At the Ohio State University Radio Observatory, correlations of bursts with satellite passages were being observed consistently. WWV signal enhancements were occurring in precursor bursts 15 to 20 min before the near approach of artificial satellites.[288–290] The receiver was tuned to a frequency slightly above that of the transmitter, on the hypothesis that the bursts were reflections from clouds of particles driven forward at high velocities by the satellite. Sudden drops in the received levels of the signals transmitted by satellites were observed at the Radio Observatory on several occasions.[290–293] These often occurred simultaneously with WWV bursts, perhaps indicating an ion cloud between the satellite and Columbus, Ohio dense enough to reflect WWV and to prevent the satellite signal from reaching Columbus.

Crone[294] outlined a simplified mechanism describing how such a cloud might be formed by simple collisions of the satellite body with the gas or plasma particles. He pointed out that under idealized conditions, one might expect the velocities of such particles to be from one to three (or perhaps more) times the velocity of the satellite. He calculated from some specific radar-echo events that the dimensions of a cylinder which would give returns of the magnitude observed would be a minimum of 10 m in radius and a minimum of 1 km in length. The dimensions would increase by an order of magnitude for stronger bursts.

When the apparent velocity of the clouds causing the bursts were calculated, it was found that the velocities were between 30 and 150 km/sec.[289] Bursts at 15 Mcps were reported[295] which showed reflected signal with a Doppler spread of at least 5 kcps into both upper and lower side frequencies. Such Doppler shifts imply electron velocities of at least 50 km/sec. Later, some radar returns[296] indicated objects or disturbances originating near a satellite and traveling away from it at velocities up to 100 km/sec. This is 14 times the satellite velocity. Some of these appeared to move along the geomagnetic-field lines, indicating trapped electron clouds. Other evidence indicated that some of the high-velocity disturbances traveled across the field lines.

Another interesting effect observed[291] was that of WWV 20-Mcps enhancements, which apparently occurred upon the passage of Sputnik III through the auroral region in the southern hemisphere after passing near Columbus on its northern excursion. Many of the signal enhancements occurred when the satellite was in or near the lower extremities of the outer Van Allen belt or in the auroral zone.[296] The between-hemisphere effects appeared to occur when Sputnik III was in the region of the northern Van Allen and auroral zones. A survey of a series of Sputnik III passes indicated a marked correlation between precursor bursts and the passage of the satellite through the Van Allen outer belt. Correlation of CW bursts following the near-approach times of Sputnik III and passage through the belt region in a northerly direction was also observed. Many of the deletions in the signals from the satellite occurred while the satellite was in similar locations.[291–293,297]

It was suggested that many of these effects were caused by interaction of a satellite with particle concentrations in the Van Allen outer belt,[296] and that the occurrence of a given event may therefore depend strongly on conditions within the belt in the region of a satellite path. In order to gain more complete data on events such as these, Herriman[298] undertook studies of the radiation belt. Telemetry of the corpuscular-radiation counting rates from 1959 Iota I (Explorer VII) were received at the Radio Observatory. The telemetry of the best five or six 10–20-min passes was recorded nearly every day during the periods of March 30 through September 14 and October 28 to December 19, 1960. The results indicated strong correlation of the counting rates observed in the belt and the structure of the belt with magnetic activity, but the relationship is not a simple one.

10.3.4.3.1.2. *Monostatic Observations.* During the period November 14, 1960 to December 15, 1960, the Ohio State University Radio Observatory received 41 unusual radar returns from their 17.5-Mcps radar. Hame[299] analyzed this data and found that six of the events agreed in time and range with the passage of particular satellites. The possibility that the other events could have been caused by a more distant satellite was not eliminated.

The Jodrell Bank staff[300] used the 80-m steerable radar telescope to investigate fading characteristics of echoes from Sputnik II. On three occasions, this group found scattering areas from 280 m^2 to 500 m^2 and suggested that the possible reason was a reflection from the satellite antennas. The major portion of the record was normal, and they believed no significant amounts of ionization were present.

The Russian scientist Al'pert reviewed[301] the western work on satellite-related anomalies and suggested that the quantitative data given by the Jodrell Bank staff led to the opposite conclusion. The satellite was a sphere 29 cm in radius. Al'pert assumed that the antenna structure did not noticeably influence the scattering, and calculated that the echo area was approximately 0.41 m^2 at 120 Mcps and 0.03 m^2 at 36 Mcps. The English measured approximately 0.4 m^2 at both frequencies. The factor of 13 at the lower frequency was attributed by Al'pert to inhomogeneous plasma formations in the vicinity of the satellite. He concluded that the ionization was not in a form similar to a meteor trail.

Tiuri[302] made observations in Finland with a 98.82-Mcps CW radar from August 1959 to April 1960. He observed echoes from ion trails when Sputnik III had traveled close to the latitude 65°N in an approximately west to east direction. The satellite varied in altitude from 200 to 600 km. The echoes were usually detected when auroral ionization was present, and when spread-*F* phenomena had occurred in the ionosphere near the observation site. The echoes had an average duration of 2.5 sec. The best returns had an average cross section of 1000 m^2, while the satellite was at an altitude of 600 km and at a distance of 700 km from the receiver. The reflections appeared to be specular in nature. Observations were made from 15 min before to 15 min after the nearest approach. Echoes delayed 8–10 min from the nearest approach were not observed. The distribution of echoes ± 0.7 min from the transit was such that the probability that the observed number would occur within this period from a random process was less than 0.0005. From his echoes, he estimated that they might be caused by a cylinder with an electron density of 10^{13} electrons/m with a diameter of a few meters.

In the U.S., radar searches for effects in the ionosphere in the immediate vicinity of Sputnik III and Echo I were conducted by Croft and Villard[303,304] and by Croft[305] at Stanford University. They had two high-frequency (10–30 Mcps) radar sets with a peak power of 20 or 50 kW and a bandwidth of 3 kcps. Four different antennas with gains of 8–30 db were used. The beamwidth was comparatively broad, both in azimuth and elevation. Both direct reflections from the vicinity of the vehicle, and perturbations within the *F*-layer sufficient to alter the structure and appearance of ground backscatter mirrored by the layer, were sought. On at least six occasions, the radar sensitivity was such that an object an order of magnitude larger than the satellite should have been detectable. No observed returns could be attributed to direct reflection. However, many layer anomalies were found to occur at locations below orbiting vehicles at times close to the time of the satellite passage. However, study of these anomalies and comparison with statistical and other characteristics of natural changes did not provide any basis for believing that any of the anomalies were caused by the satellite. Also, no apparent correlation was found between the time of day or the geomagnetic-field instability and the success of finding anomalies. If it is assumed that anomalies are induced by satellites through some mechanism involving the geomagnetic field, it might be expected that the degree of success in the detection of the anomalies would be a function of antenna heading. No such correlation was found. If ion clouds caused by the satellite were separated from the satellite by over 500 km, they would not have been detected in this experiment.

Lootens[306,307] reported that the Ballistics Research Laboraties, with the support of the Advanced Research Projects Agency, operated a Doppler satellite-tracking system with a 108-Mcps transmitter in Fort Sill, Oklahoma, and receiving stations at White Sands Missile Range, New Mexico and at Forrest City, Arkansas. Seventeen out of sixty-seven satellite passes recorded by the DOPLOC center antenna revealed a constant-frequency reflection either preceding or following the Doppler signal reflected from the satellite. These constant-frequency reflections, termed "flats," appeared on the analog records as a horizontal line close to or equal to the bias frequency. This indicated zero velocity. This could have been caused either by a large ionized mass moving through the antenna beam at a very low velocity, or by a stationary ionized mass having a lifetime equal to the duration of the observed signal. On nine of the records, the flat produced a stronger signal level than the corresponding satellite reflection.

It was concluded that the dimensions of the reflecting mass causing the flats are comparable to or larger than those of the satellite, and that the flats may be observed almost as frequently associated with satellites at low altitudes as at high altitudes. (The DOPLOC system recorded flats associated with satellites at altitudes from 110 miles to 600 miles.) Many flats were recorded that were not associated with a satellite.

Tiuri made a statistical analysis(308) of some of the older data of Kraus. For an interval of 20 min before to 10 min after a passage, a probability of less than 0.0001 % that the distribution of bursts was due to a random phenomenon was found. For 10-min intervals arbitrarily chosen throughout the day, without regard to satellite position, it was reported that the enhancement per interval gave a histogram with a probability of 72 % that it was due to a random phenomenon. This approach was applied to Looten's data, with similar results. Two different research groups identifying effects on two different satellites gave similar results.

In a 20-min interval, the satellite travels 2000 km. However, both Looten's work(307) and Tiuri's work(302) in Finland suggested that effects occur at relatively short ranges from the satellite (within a few hundred kilometers). These effects tended to occur when the satellite was south of the observer. The magnetic dip angle of the earth's field at both observing locations is about 70°. If the ionization was along a magnetic-field line through the satellite and below it (i.e., at a magnetic subsatellite point), it would be to the north of the satellite at both observing locations, and hence in the antenna beam when the satellite was to the south, as observed. Accordingly, these observations suggest the possibility that the satellite may produce a disturbance which propagates down along magnetic-field lines to a point in the beam. If, e.g., the disturbance were a magnetohydrodynamic wave, a reflection would be expected at the base of the ionosphere, where there is a transition in conductivity of the medium from a large value above to a small value below. However, some entirely different mechanism may be responsible for the effect.

10.3.4.3.2. *Theoretical Considerations.* The theoretical question of the interaction of a space vehicle or missile with the ionosphere is exceedingly complicated and has received considerable attention. The satellite might lose portions of itself and thus form a trail.(309) It becomes charged as it moves through the ionosphere. The motion of a charged body in the ionosphere is an interesting theoretical problem and has been attacked by many persons.

The possibility of extraordinary electromagnetic waves being caused by the interaction between the charged body and the gyrating charged particles was considered by Chopra and Singer.[310] The generation of a collective disturbance of the charged particles was given a preliminary examination by Kraus and Watson[311] and Grielfinger.[312] Chopra[313] described three kinds of waves generated: plasma oscillations in the wake of the body, extraordinary electromagnetic waves, and magnetohydrodynamic waves.

Calculations of the echo area of the satellite plus an accompanying ion cloud have been carried out at the University of Michigan[314–316] and at the Ohio State University.[87–95] The theoretical effects of the plasma are not enough, in general, to explain the large echoes sometimes observed. Singer and Walker[317,318] have proposed an ion cloud caused by the satellite, which detaches itself and moves on a separate path. Hame and Stuart[286] briefly summarized some of the possible effects by stating that the satellite and any accompanying ion cloud and shock waves could disturb the ionosphere, scatter electrons from high-density regions into lower-density regions, and perhaps, destroy local clouds and create new ones. The possibility of detectable reflections from the various clouds would depend upon the following: frequency, cross-sectional area of the vehicle, angle of look, relative velocity, and electron density (this varies with height, time, and solar activity; the variability of solar activity may be the reason correlated disturbances are sometimes noticed, while many passes appear to have little effect). For clouds not traveling as an integral unit with the vehicle, there are several possibilities. There might be an increase in ionization level traveling at greater velocity or lesser velocity than the satellite, or remaining stationary with respect to the atmosphere along the satellite track. Alternately, or perhaps even concurrently, there might be a duct of low ionization level traveling at a greater velocity or a lesser velocity than the satellite, or remaining stationary with respect to the atmosphere.

The classic theoretical work is that of Gurevich,[319] Al'pert and Pitayevskii,[320] and Pitayevskii and Klesin,[321] who have advanced a model which involves the sweeping away of ions by the vehicle. The velocity of the vehicle far exceeds the ion thermal speed. A hole in the electron distribution directly behind the body is developed, since the electrons remain roughly within a Debye length from the ions. The depletion in electrons represents a discontinuity in permittivity, and scatters radar waves. They obtain large cross sections when the radar is directly under the wake, but very small cross sections when the radar is otherwise located.

Shea[322] and Rand and Albini[323] have extended and improved the work of the Russians. Rand and Albini include Fresnel effects, and do not assume that the electrons are in Boltzmann equilibrium in the space-charge electrostatic field when the vehicle is moving parallel to the geomagnetic field.

For the radar cross section of a vehicle moving parallel to the geomagnetic field, Rand and Albini obtain

$$\sigma = \frac{\lambda_0 r}{32}\left(\frac{\omega_p}{c}\right)^4 R_w{}^4 \exp\left(-\frac{32\pi^2 D_0 L}{\lambda_0{}^2 V_s}\right) \tag{10.3-31}$$

where R_w is the initial radius of the depleted region, D_0 the diffusion coefficient corresponding to ambient density, L the distance behind the vehicle as a parameter, and V_s the velocity of the vehicle. As a numerical example, they choose $r = 200 \times 10^3$ m, $N_e = 10^{12}$ electrons/m^3, $R_w = 10^{-1}$ m, $D_0 = 0.54$ m^2/sec, $V_s = 7 \times 10^3$ m/sec, $\lambda_0 = 20$ m, and $L = 10^4$ m, resulting in a cross section of 84 m^2.

For a vehicle moving obliquely to the geomagnetic field, they obtain

$$\sigma = \frac{r\lambda_0{}^3 R_w{}^4 \omega_p{}^4 \omega_H{}^2}{102.4\pi^3 c^4 V_i{}^2}\left[1 + \frac{16\pi^2 V_i{}^4 L^4 \sin^4\xi}{10^4 \lambda_0{}^2 r^2 V_s{}^4}\right]^{-1/2} \tag{10.3-32}$$

where ω_H is the gyrofrequency of the principal ionic species, V_i the ion thermal speed, ξ the angle between the direction of motion and the geomagnetic field, and V_s the velocity of the satellite. They assume the same numerical values as before and $\xi = 30°$, a magnetic field strength of 0.2 gauss, $V_i{}^2 = (8.31 \times 10^4/M)(\text{m/sec})^2$ ($T = 1000°$), where M is the atomic weight of the principal ionic species ($M = 15$). The resulting cross section is 1.4 m^2.

10.3.4.3.3. *Discussion.* The effects produced by a satellite on the nearby ionosphere seem to be best explained by References 319–323.† However, the theory does not explain the rather large cross section observed in several of the experiments, nor does it explain the apparently satellite-correlated reflections from areas of the ionosphere widely separated from any satellite. None of the theories presently available appear to explain these phenomena satisfactorily. Extensions of the various wave theories, and/or the detached cloud theory of Singer and Walker, are perhaps needed to explain the observations completely. Unfortunately, none of the scale-model experiments, such as reported by Clayden and Hurdle,[(324)] can yet simulate a large portion of the ionosphere.

10.3.4.4. *Reentry Effects*

10.3.4.4.1. *Introduction.* A phenomenon which in some respects is similar to the entry of meteors is the reentry of missiles and space vehicles. Much of the terminology used to describe meteor trails is also used in

† *Note added in proof:* A recent paper by R. K. Gupta, On the Evaluation of Scattered Power and Radar Cross Section of a Space Vehicle in Warm Plasmas, *Intern. J. Electron.* **25**:449 (1968), presents equations for the radar cross section of a thin cylinder moving in a plasma without a magnetic field, including contributions from electroacoustic waves generated by the vehicle. The analysis is based on a linearized hydrodynamic approach.

studies of reentry phenomena. The reentry vehicle is, of course, larger than most meteors. Whereas the ionized trail for most meteors is largely created by ablation of the meteor body, the reentry vehicle creates its ionization partly by ablation and partly by the hypersonic shock wave formed by the body. The shock wave causes a deceleration of the air flow around the vehicle. This region between the vehicle and the shock wave is an inviscid flow region, and has temperatures sufficiently high that dissociation and ionization of gas molecules occur. This region is called the plasma sheath, and has been the subject of extensive investigations both for its effects upon radar observations of reentry and for its effects upon communications with reentry vehicles. The treatments of radar observables can in general be divided into two classes: those treatments in which the radar is looking at the target in the frontal aspect, similar to the "head" echoes in meteor trail observations; and those in which the radar is concerned primarily with the wake, usually near broadside. The first case is considered in the following section, and the latter case is considered in Section 10.3.4.4.3.

10.3.4.4.2. *Frontal Scattering from Plasma Sheaths.* As an object descends into the atmosphere surrounding a planet, it finds the atmosphere growing more dense as the altitude above the surface decreases. Above 300,000 feet in the earth's atmosphere, the atmospheric density is so low that the aerodynamics of the object are not of significance. Altitudes greater than 300,000 feet above the surface of the earth are referred to as exoatmospheric, and altitudes below 300,000 feet are called endoatmospheric. At endoatmospheric heights, an object traveling at hypersonic speeds, such as a meteor, reentering space vehicle, or missile, causes a system of shock waves to form. These shock waves cause the formation of high-temperature regions around the body. As the temperature continues to increase, portions of the surface of the body may ablate and ionize. The ionization is not uniformly distributed, but is a function of the temperature and the presence of ablation products. Details as to the relevant aerothermochemical phenomena are available in books by Martin[325] and by Loh.[326]

The plasma sheath at the front of the reentry vehicle scatters radio waves at the surface of the sheath and at the permittivity discontinuities present in the sheath. This ionized region refracts the incident signal away from the reentry vehicle and, because of the high collision frequency, attenuates both the incident and reflected signals. Figure 10-39 illustrates the change in cross section as a function of altitude as given by Gunar and Mennella.[327] In altitude region 1, there is little effect on the observed free-space radar cross section. As the vehicle reaches region 2, the sheath refracts significant energy away from the vehicle. In region 3, the electron density of the sheath has increased such that $\omega_p > \omega$, and the surface of the sheath reflects most of the energy. By the time the vehicle reaches

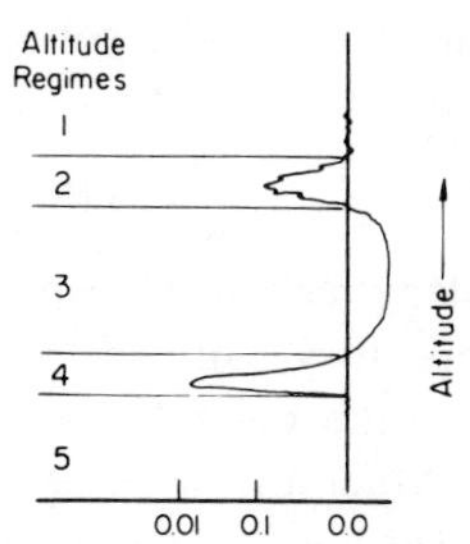

Fig. 10-39. Plasma modification of the nose-on radar cross section of a reentering vehicle [after Gunnar and Mennella(327)].

region 4, the atmosphere has slowed the vehicle; thus the electron density and plasma frequency drop, allowing the signal to penetrate the sheath. The collision frequency constantly increases from regions 1 through 4, so that in region 4, the sheath absorbs much of the incident signal. At lower altitudes, the speed has decreased to the point where ionization no longer occurs, and the sheath disappears. Gunnar and Mennella(327) note that the upper-altitude radar cross section reduction is expected to occur between the altitudes of 250,000 and 150,000 feet for a *C*-band radar under usual reentry conditions. The cross section could theoretically be reduced to zero, but they report that the maximum reduction that had been observed was about 20 db. The lower-altitude reduction occurs in the region between about 70,000 feet to 50,000 feet. The exact shape of the curve of Figure 10-39 is a function of the shape and surface composition of the reentry vehicle, the trajectory, and the free-space wavelength of the observing radar. In fact, some of the altitude regions may not be present in all cases.

This type of variation has been predicted theoretically by Musal and co-workers.(328–331) They assumed that the reentry vehicle was a perfectly-conducting metallic sphere and that the plasma sheath was a homogeneous coating with an electron density and a collision frequency equal to that present in the stagnation region at the front of the sheath. The theoretical equations useful for this model are available in Section 10.2.2.3.2 [Eqs. (10.2-50)–(10.2-54)]. Measurements(330) of small projectiles fired at hypersonic velocities in a laboratory ballistic range showed that the reduction in cross section was much greater than that predicted by a homogeneous layer. This reduction was attributed partially to absorption, but largely to diffraction by gradients in the plasma permittivity parallel to the body surface. Musal *et al.*(331) state that this effect can produce large reductions in the radar cross section with plasma layers as thin as one-tenth the free-space radar wavelength, with or without losses in the plasma, provided the plasma frequency at the stagnation point is greater than the radar frequency. Good agreement has been shown(331) between measurements and theory which takes this gradient into account.

10.3.4.4.3. *Scattering from Wakes.* The length of the wake behind a reentry vehicle as seen by a backscatter radar can be considered to be

$$W_L = (T_R - T_t)\, c/(2 \cos \theta) \tag{10.3-33}$$

where T_R is the length of the received pulse, T_t is the length of the transmitted pulse, and the positive z axis is along the vehicle trajectory, with the aspect angle θ between the $\mathbf{k}^i$ vector and the z axis. The wake behind a reentry vehicle is in many respects similar to a meteor trail. Much of the discussion in Sections 10.3.1.3 and 10.3.1.4 is applicable to wake problems. Pippert and Edelberg[(332)] have reported on a program conducted by the MIT Lincoln Laboratory at Wallops Island, Virginia which is essentially an artificial meteor program. Many laboratory experiments involving hypervelocity particles are also quite similar.[(333)] Lin[(334)] reported observations of five clearly separated ionized trails associated with the reentry of the MA-6 Mercury capsule (which carried John Glenn in his orbital flight) on February 20, 1962. The most prominent trail was identified with the wake of the main capsule. It was visible for a duration of 20 sec, starting near 230,000 feet (70 km), and was relatively insensitive to aspect angle. The other four trails were considered to be caused by fragments of the disintegrating retrorocket package. The echoes from these four trails showed only short-duration glints of the order of 1 sec. The glints were similar to those received from smooth, rapidly-decaying trails of small meteors. The overall characteristics, but not all the fine structure, of the returned signal for the capsule wake is said to be accounted for by a combination of signals. One arises from a strongly-reflecting front segment of the ionized wake, which appeared to be electromagnetically smooth to the approximately 10-m free-space wavelength employed. Another is caused by diffuse scattering from a turbulent rear segment of the wake, which appeared rough for this wavelength. The strong signal from the smooth frontal portion was located approximately 14,000 feet (4.2 km) behind the capsule itself. The apparent velocity of this portion of the wake agreed well with the velocity of the capsule. The distance between the capsule and the "bright" portion of the wake did not change with time in any significant manner during the observation except for some short-period fluctuations. This specularly-reflecting portion of the wake is termed the laminar portion, and can be treated as a plasma cylinder by the methods of Section 10.2.2.2.

Analysis of the turbulent portion of the wake requires the more difficult formulations of Section 10.2.3. If there are perturbations of the order of a wavelength in electron-density contours near the wake boundary, the wake will have the properties of a rough surface. The Booker–Gordon[(168)] model discussed in Section 10.2.3 consists of closely-packed, approximately spherical irregularities varying in size and electron density (but remaining underdense). The irregularities are randomly moving and have a characteristic length l. The deviation in dielectric constant is small compared to ϵ_0. Combining the Booker–Gordon model of radio-wave scattering with the Lees–Hromas[335] aerothermodynamical model for the wake of a blunt-nosed reentry vehicle, gives, for $\lambda_0 \ll 4\pi l$, a backscatter cross section of

$$\sigma = a_0^2 L_t / 64 l \langle \epsilon_r^2 \rangle \tag{10.3-34}$$

where a_0 is the radius of the wake for which the electron density has decayed to 1/10 the axial concentration, and L_t is the longitudinal distance along the wake for which the electron density has decayed to $(1/e)$ times its peak value.

Equation (10.3-34) gives backscatter cross sections which are independent of wavelength for $\lambda_0 \ll 2\pi l$, independent of polarization, and relatively insensitive to θ for $15° < \theta < 75°$.

For frequencies sufficiently high, and antenna beamwidths sufficiently narrow that the wake can be approximated by a planar surface, the rough-surface formulations presented in Chapter 9 can be used for turbulent-wake cross-section predictions. The overdense wake is modeled by a conducting plane with various types of irregularities, depending upon the formulation used. The underdense wake would require the use of the formulations for scattering from a rough dielectric. Inclusion of the effects of curvature of the wake so as to extend the results to lower frequencies is also possible.

The instantaneous radar cross section, based on the instantaneous returned signal intensity, is termed by Lin[334] the equivalent isotropic scattering cross section, σ_E. The maximum value of σ_E observed for the wake was 1.6×10^6 m^2. The value of σ_E varied continually, with variations by a factor of one hundred or more in about $\frac{1}{4}$ sec. Lin estimates an electron line density in the neighborhood of 2×10^5 electrons/m in the "bright" spot. At the beginning of the turbulent portion of the wake, the width of the wake is estimated at about 30 m. He found that there was apparently a large population of irregularities or eddies with characteristic dimensions lying in the range of $0.8 \lesssim l \lesssim 3$ m, and that the root-mean-square electron-density fluctuation associated with these eddies must have been close to the value of 10^{13} electrons/m^3.

Kalinin[336] took the data from Lin's article and interpreted it by means of a diffusion model of a meteor trail. He found good agreement between the theoretical and experimental data.

Figure 10-40 illustrates measured reentry cross sections at L-band for three separate tests of similar vehicles. The altitudes covered a range from about 180,000 feet to about 50,000 feet. The aspect angles varied from 87° to 104°. The marked increase in rapid signal amplitude fluctuations, starting slightly below 170,000 feet and continuing until around 85,000 feet, could be attributed to reflections from a turbulent wake. The sudden decrease in signal level near 85,000 feet is probably related to absorption

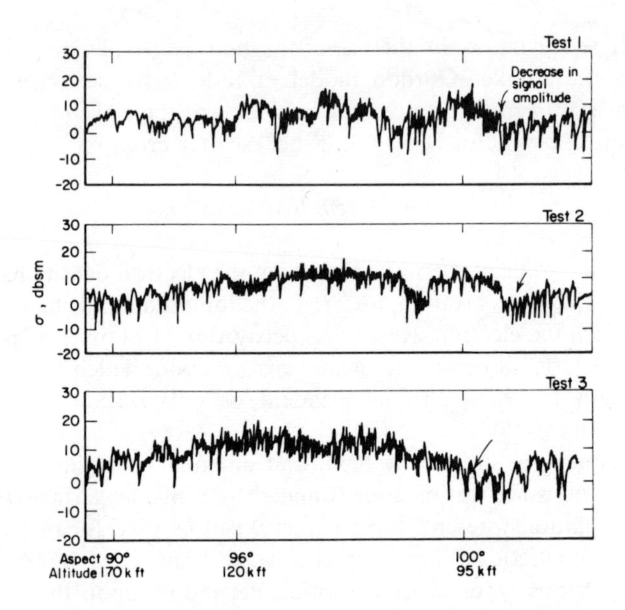

Fig. 10-40. Measured radar cross section of a typical reentering vehicle [after Gunar and Mennella[327]].

of the signal at low altitudes, as discussed for frontal echoes in the previous section. At still lower altitudes, the bare-body cross section is observed.

10.4. REFERENCES

1. Ginzburg, V. L., *Propagation of Electromagnetic Waves in Plasma*, Gordon and Breach Science Publishers, New York (1961).
2. Ratcliffe, J. A., *The Magneto-Ionic Theory and Its Applications to the Ionosphere*, Cambridge University Press, London (1959).
3. Budden, K. G., *Radio Waves in the Ionosphere*, Cambridge University Press, London (1961).
4. Brandstatter, J. J., *An Introduction to Waves, Rays and Radiation in Plasma Media*, McGraw-Hill Book Co., New York (1963).
5. Tischer, F. J., *Basic Theory of Space Communications*, D. Van Nostrand Co., Princeton, New Jersey (1965).
6. Arzimovich, L. A., *Elementary Plasma Physics*, Blaisdell Publishing Co., New York (1965).
7. Chandrasekhar, S., *Plasma Physics*, The University of Chicago Press, Chicago, Illinois (1960).
8. Spitzer, L., Jr., *Physics of Fully Ionized Gases*, John Wiley (Interscience), New York (1956).
9. Delcroix, J. L., *Introduction to the Theory of Ionized Gases*, John Wiley (Interscience), New York (1960).
10. Delcroix, J. L., *Plasma Physics*, John Wiley & Sons, New York (1965).

11. Linhart, J. G., *Plasma Physics*, North-Holland Publishing Co., Amsterdam (1960).
12. Kunkel, W. B., *Plasma Physics in Theory and Application*, McGraw-Hill Book Co., New York (1966).
13. Longmire, C. L., *Elementary Plasma Physics*, John Wiley (Interscience), New York (1963).
14. Allis, W. P., Buchsbaum, S. J., and Bers, A., *Waves in Anisotropic Plasmas*, MIT Press, Cambridge, Massachusetts (1963).
15. Stix, T. H., *The Theory of Plasma Waves*, McGraw-Hill Book Co., New York (1962).
16. Al'pert, Ya. L., *Radio Wave Propagation and the Ionosphere*, Consultants Bureau, New York (1963).
17. Kelso, J. M., *Radio Ray Propagation in the Ionosphere*, McGraw-Hill Book Co., New York (1964).
18. Heald, M. A., and Wharton, C. B., *Plasma Diagnostics with Microwaves*, John Wiley & Sons, New York (1965).
19. Huddlestone, R. H., and Leonard, S. L., *Plasma Diagnostic Techniques*, Academic Press, New York (1965).
20. Shkarofsky, I. P., Johnston, T. W., and Bachynski, M. P., *The Particle Kinetics of Plasmas*, Addison-Wesley Publishing Co., Reading, Massachusetts (1966).
21. Bachynski, M. P., Johnston, T. W., and Shkarofsky, I. P., Electromagnetic Properties of High-Temperature Air, *Proc. IRE* **48**:347 (1960).
22. French, I. P., Cloutier, G. G., and Bachynski, M. P., The Absorptivity Spectrum of a Uniform, Anisotropic Plasma Slab, *Can. J. Phys.* **39**: 1273 (1961).
23. Graf, K. A., and Bachynski, M. P., Electromagnetic Waves in a Bounded, Anisotropic Plasma, *Can. J. Phys.* **40**:887 (1962).
24. Bachynski, M. P., and Graf, K. A., Electromagnetic Properties of Finite Plasma, *RCA Rev.* **XXV**:3 (1964).
25. Bachynski, M. P., and Gibbs, B. W., Propagation in Plasmas along the Magnetic Field—I. Circular Polarization, *Phys. Fluids* **9**:520 (1966).
26. Bachynski, M. P., and Gibbs, B. W., Propagation in Plasmas along the Magnetic Field—II. Linear Polarization, *Phys. Fluids* **9**:532 (1966).
27. Bachynski, M. P., Sources in Plasma, *RCA Rev.* **XXVIII**:111 (1967).
28. Bachynski, M. P., and Cloutier, G. G., Communications in the Presence of Plasma Media. *In* Scala, S. M., Harrison, A. C., and Rogers, M. (Eds.), *Dynamics of Manned Lifting Planetory Entry*, John Wiley & Sons, New York (1963), p. 206.
29. Graf, K., and Bachynski, M. P., Transmission and Reflection of Electromagnetic Waves at a Plasma Boundary for Arbitrary Angles of Incidence, RCA Victor Co., Ltd., AFCRL-378 (March 1961), AF 19(604)-7291, AD 260 305.
30. Cloutier, G. G., Bachynski, M. P., and Graf, K. A., Antenna Properties in the Presence of Ionized Media, RCA Victor Co., Ltd., AFCRL-62-191 (March 1962), AF 19(604)-7334.
31. Graf, K. A., and Bachynski, M. P., Existing Solutions for Interaction of Electromagnetic Waves with Plasma of Simple Geometries, RCA, DAMP Technical Monograph No. 62-07 (June 1962), DA 36-034-ORD-3144RD.
32. Bachynski, M. P., Cloutier, G. G., and Graf, K. A., Microwave Measurements of Finite Plasmas, RCA Victor Co., Ltd., AFCRL-63-161 (1963), AF19(604)-7334.
33. Bachynski, M. P., and Gibbs, B. W., Electromagnetic Wave Propagation in Anisotropic Plasmas along the Direction of a Static Magnetic Field, RCA Victor Co., Ltd., AFCRL-65-84 (1964), AF19(628)-4071.
34. Berman, M., A Study of the Interaction of Electromagnetic Waves with the Plasma Surrounding a Re-Entry Vehicle, Chrysler Corporation, Missile Division, Report No. SAE-TN-17-62, BSD-TDR-62-57 (April 1962), AF 04(694)-25, AD 283 543.
35. Plugge, R. J., Nomographs for Plasma EM Parameters, Antenna Laboratory, Department of Electrical Engineering, the Ohio State University, Columbus, Ohio (June 1963).
36. Huxley, L. G. H., Free Path Formulae for the Electric Conductivity of a Weakly Ionized Gas in the Presence of a Uniform and Constant Magnetic Field and a Sinusoidal Electric Field, *Australian J. Phys.* **10**:240 (1957).

37. Valley, S. L., *Handbook of Geophysics and Space Environments*, Air Force Cambridge Research Laboratories, Office of Aerospace Research, United States Air Force (1965), pp. 12–17.
38. Dingle, R. B., Arndt, D., and Roy, S. K., Semiconductor Integrals, *Appl. Sci. Res.* **6B**:158 (1957).
39. Kane, J. A., Arctic Measurements of Electron Collision Frequencies in the *D* Region of the Ionosphere, *J. Geophys. Res.* **64**:133 (1959).
40. Hame, T. G., and Stuart, W. D., The Polarization Rotation of Radio Transmissions from Artificial Earth-Satellites and the Electron Density in the Upper Ionosphere, Ohio State University, Report No. 889-5 (August 1959), AF 33 (616)-6137.
41. Sen, H. K., and Wyller, A. A., On the Generalization of the Appleton–Hartree Magneto-Ionic Formulas, *J. Geophys. Res.* **65**:3931 (1960).
42. Shkarofsky, I. P., Generalized Appleton–Hartree Equation for Any Degree of Ionization and Application to the Ionosphere, *Proc. IRE* **49**:1857 (1961).
43. Potts, B. C., Ionospheric Studies Using Polarization Rotation of Satellite Radio Signals, Ohio State University, Scientific Report No. 11, Report No. 1116-16, AFCRL-62-155 (January 1962) AF 19 (604)-7270.
44. Millman, G. H., A Survey of Tropospheric, Ionospheric, and Extra-Terrestrial Effects on Radio Propagation between the Earth and Space Vehicles, General Electric Company, Report TIS R66EMH1. (Also presented as a paper at the NATO–AGARD Symposium on Propagation Factors in Space Communications, Rome, Italy, September 21–25, 1965.)
45. Bickel, S. H., and Bates, R. H. T., Effects of Magneto-Ionic Propagation on the Polarization Scattering Matrix, *Proc. IEEE* **53**:1089 (1965).
46. Booker, H. G., Studies on Propagation in the Ionosphere—Part I, An Outline of the Magneto-Ionic Theory, Cornell University, Technical Report No. 1 (March 1950), W36-039-sc-44518.
47. Deschamps, G. A., and Weeks, W. L., Charts for Computing the Refractive Indexes of a Magneto-Ionic Medium, *IRE Trans. Antennas Propagation* **AP-10**:305 (1962).
48. Balescu, R., *Statistical Mechanics of Charged Particles*, John Wiley (Interscience), New York (1963), p. 69.
49. Mower, L., Conductivity of Warm Plasma, *Phys. Rev.* **116**:16 (1959).
50. Sitenko, A. G., and Stepanov, K. N., On the Oscillations of an Electron Plasma in a Magnetic Field, *Soviet Phys.—JETP* **4**:512 (1957).
51. Caldirola, P., De Barbieri, O., and Maroli, C., Electromagnetic Wave Propagation in a Weakly Ionized Plasma. *In* Wilsted, H. D. (Ed.), *Fundamental Studies of Ions and Plasmas*, Vol. 2, NATO Advisory Group for Aerospace Research and Development, Paris, (1965), p. 587.
52. Silin, V. P., and Shister, A. R., Theory of Transverse Diffusion and Static and High-Frequency Conductivity of a Plasma in a Strong Magnetic Field, *Soviet Phys.—JETP* **22**:139 (1966).
53. Shkarofsky, I. P., Dielectric Tensor in Vlasov Plasmas near Cyclotron Harmonics, *Phys. Fluids* **9**:561 (1966).
54. Whitmer, R. F., and Herrmann, G. F., Effects of a Velocity-Dependent Collision Frequency on Wave–Plasma Interactions, *Phys. Fluids* **9**:768 (1966).
55. Rybin, V. V., Comments on the Electric Conductivity of the Upper Atmosphere, *Geomagnetism and Aeronomy* **VI**:314 (1966).
56. Spitzer, L., Jr., and Harm, R., Transport Phenomena in a Completely Ionized Gas, *Phys. Rev.* **89**(5):977 (1953).
57. Allis, W. P., Motions of Ions and Electrons. *In* Flugge, S. (Ed.), *Handbuch der Physik*, Vol. XXI, Springer-Verlag, Berlin (1956), p. 383.
58. Buchsbaum, S. J., Waves in Uniform, Magnetoactive Plasmas. *In* Futerman, W. I. (Ed.), *Propagation and Instabilities in Plasmas*, Stanford University Press, Stanford, California (1963), p. 1.
59. Stepanov, K. N., Kinetic Theory of Magnetohydrodynamic Waves, *Soviet Phys.—JETP* **7**:892 (1958).

60. Drummond, J. E., Microwave Propagation in Hot Magneto-Plasmas, *Phys. Rev.* **112**(5):1460 (1958).
61. Vilensky, I. M., The Influence of the Earth's Magnetic Field on the Interaction of Radio Waves in the Ionosphere, *Zh. Eksperim. i. Teor. Fiz.* **26**(1):42 (1954).
62. Hochstim, A. R., Charts for Microwave Propagation through a Plasma, General Dynamics/Astronautics, Report No. AE62-0858, ARPA Order No. 39-62 (December 1962), DA-04-495-ORD-3383.
63. Nicoll, G. R., and Basu, J., *Reflection and Transmission of an Electromagnetic Wave by a Gaseous Plasma*, The Institution of Electrical Engineers, Monograph No. 498E, London (January 1962).
64. Miller, G. F., III, Propagation of Electromagnetic Waves in a Nonuniformly Ionized Medium, *Phys. Fluids* **5**:899 (1962).
65. Klein, M., Greyber, H., King, J. I. F., and Brueckner, K., Interaction of a Non-Uniform Plasma with Microwave Radiation, Space Sciences Laboratory, General Electric Company, Report No. TIS-R59SD467 (November 1959).
66. Albini, F. A., and Jahn, R. G., Reflection and Transmission of Electromagnetic Waves at Electron Density Gradients, *J. Appl. Phys.* **32**:75 (1961).
67. Penico, A. J., Propagation of Electromagnetic Waves in a Plasma with an Inhomogeneous Electron Density, *In*: *Proc. Symposium on the Plasma Sheath*, Vol. I, AFCRC-TR-60-108(1) (December 1959).
68. Golant, V. E., Microwave Plasma Diagnostic Techniques, *Soviet Phys.—Tech. Phys.* **5**:1197 (1961).
69. Rydbeck, O. E. H., and Rönnäng, B., Wave Propagation in a Double Hump, Isotropic Ionized Medium, *In:* Rydbeck, O. E. H. (Ed.), *Propagation of Low and High Power Radio Waves in Ionized Media*, Rydbeck Research Associates, Annual Summary Report No. 1 (1966), AF 61(052)-864, AD 637 665.
70. Bein, G. P., Plane Wave Transmission and Reflection Coefficients for Lossy Inhomogeneous Plasmas, *IEEE Trans. Antennas Propagation* **AP-14**:511 (1966).
71. Yeh, C., Reflection and Transmission of Electromagnetic Waves by a Moving Plasma Medium, *J. Appl. Phys.* **37**:3079 (1966).
72. Taylor, L. S., Electromagnetic Propagation in an Exponential Ionization Density, *IRE Trans. Antennas Propagation* **AP-9**:483 (1961).
73. Taylor, L. S., RF Reflectance of Plasma Sheaths, *Proc. IRE* **49**:1831 (1961).
74. Taylor, L. S., Reflection of a TE Wave from an Inverse Parabolic Ionization Density, *IRE Trans. Antennas Propagation* **AP-9**:582 (1961).
75. Klein, M. M., Interaction of Microwave Radiation with a Non-Uniform Plasma Sheath Adjacent to a Conducting Wall, General Electric Company, Report No. TIS-61SD118 (August 1961).
76. Klein, M. M., Calculation of the Interaction of a Non-Uniform Ionized Wake with Microwave Radiation, General Electric Company, Report No. TIS-61SD116 (August 1961).
77. Klein, M. M., Interaction of Microwave Radiation from a Re-Entry Vehicle with the Surrounding Plasma, General Electric Company, Report No. TIS-61SD117 (August 1961).
78. Yen, K. T., Microwave Reflection by Non-Uniform Plasma with Exponential Electron Distribution, Space Sciences Laboratory, General Electric Company, Report No. TIS-R62SD47, Report No. RADC-TDR-62-441 (June 1962).
79. Schmitt, H. J., Plasma Diagnostics with Short Electromagnetic Pulses, Harvard University, Report No. NR-371-015 (May 1964), Nonr-1866(26).
80. Schmitt, H. J., Pulse Dispersion in a Gyrotropic Plasma, Sperry Rand Research Center, Report No. SRRC-RR-65-16 (January 1965), AF 19(628)-4183.
81. Case, C. T., On Transient Wave Propagation in a Plasma, Air Force Cambridge Research Laboratories, Physical Sciences Research Paper No. 141, Report No. AFCRL-65-702 (September 1965).
82. Epstein, P. S., Reflection of Waves in an Inhomogeneous Absorbing Medium, *Proc. Natl. Acad. Sci.* (*U.S.*) **16**:629 (1930)

83. Rydbeck, O. E. H., On the Propagation of Radio Waves, *Trans. Chalmers Univ. Technol. Gothenburg*, No. 34 (1944).
84. Unz, H., Oblique Incidence on Plane Boundary between Two General Gyrotropic Plasma Media, *J. Math. Phys.* **6**:1813 (November 1965).
85. Unz, H., Oblique Wave Propagation in a Compressible General Magnetoplasma, *Appl. Sci. Res.* **16**:105 (1966).
86. Landau, L. D., and Lifshitz, E. M., *Electrodynamics of Continuous Media*, Pergamon Press, New York (1960), p. 247.
87. Swarner, W. G., *Radar Cross Sections of Dielectric or Plasma Coated Conducting Bodies*, Ph.D. Dissertation, Ohio State University (1962); University Microfilms, Inc., Ann Arbor, Michigan (1963).
88. Swarner, W. G., and Peters, L., Jr., Radar Cross Sections of Dielectric or Plasma Coated Conducting Spheres and Circular Cylinders, *IEEE Trans. Antennas Propagarion* **AP-11**:558 (1963).
89. Peters, L., Jr., Kawano, T., and Swarner, W. G., Approximations for Dielectric or Plasma Scatterers, *Proc. IEEE* **53**:882 (1965).
90. Peters, L., Jr., and Swarner, W. G., Coherent Scattering of a Metallic Body in the Presence of an Ionized Shell. *In* Singer, S. F. (Ed.), *Interactions of Space Vehicles with an Ionized Atmosphere*, Pergamon Press, New York (1956), p. 447.
91. Swarner, W. G., Radar Cross Section of Dielectric or Plasma Coated Conducting Bodies, Antenna Laboratory, Department of Electrical Engineering, Ohio State University, Report 1116-21 (August 1962), AF 19 (604)-7270, AFCRL-62-764, AD 286855.
92. Peters, L., Jr., Swarner, W. G., and Thomas, D. T., Further Studies of the Radar Cross Section of Plasma Clad Bodies. *In* Rotman, H., Moore, K., and Papa, R. (Eds.), *Electromagnetic Aspects of Hypersonic Flight* (2nd Symposium on the Plasma Sheath), Spartan Books, Baltimore, Maryland (1964), p. 320.
93. Thomas, D. T., Scattering by Plasma and Dielectric Bodies, Antenna Laboratory Ohio State University, Report 1116-20 (August 1962), AF 19 (604)-7270, AFCRL-62-735; Ph.D. Dissertation, Ohio State University (1962).
94. Peters, L., Jr., and Green, R. B., Echo Area of a Plasma-Coated Sphere. *In* Rotman, W., and Meltz, G. (Eds.), *Electromagnetic Effects of Reentry*, Pergamon Press (1961), p. 133.
95. Peters, L., Jr., Echo Area from Satellite Wakes Acting as Dielectric Rod Antennas, Antenna Laboratory, Ohio State University, Report 1116-3, AFCRL 89, (January 1961) AF 19 (604)-7270.
96. Tang, C. H., Scattering by Finite Objects in Compressible Plasma, *Nuovo Cimento* **XLVI B**:93 (1966).
97. Cohen, M. H., Radiation in a Plasma—I. Cerenkov Effect, *Phys. Rev.* **123**:711 (1961); II. Equivalent Sources, *Phys. Rev.* **126**:389 (1962); III. Metal Boundaries, *Phys. Rev.* **126**:398 (1962).
98. Gildenburg, V. B., and Kondrat'ev, I. G., Diffraction of Electromagnetic Waves by a Bounded Plasma in the Presence of Spatial Dispersion, *Radio Eng. Electron.* (*USSR*) (*English Transl.*) **10**:560 (1965).
99. Herlofson, N., Plasma Resonance in Ionospheric Irregularities, *Arkiv Fysik* **3**:247 (1951).
100. Fejer, J. A., Scattering of Electromagnetic Waves by a Plasma Cylinder, *Phys. Fluids* **7**:439 (1964).
101. Ridder, C. M., and Peterson, L. G., Scattering from a Homogeneous Plasma Cylinder of Infinite Length, MIT Lincoln Laboratory, Report 312G-4 (1962).
102. Ridder, C. M., and Edelberg, S., Scattering from Plasma Cylinders with Radial Variations in Electron Density. *In* Rotman, H., Moore, K., Papa, R. (Eds.), *Electromagnetic Aspects of Hypersonic Flight* (2nd Symposium on the Plasma Sheath), Spartan Books, Baltimore, Maryland (1964), p. 286.
103. Midzuno, Y., Scattering of Microwaves from a Cylindrical Plasma in the Born Approximation, *J. Phys. Soc. Japan* **16**:971, 1403 (1961).
104. Ruquist, R. D., A Diagnostic Theory for Cylindrical Plasmas Based on the Born Approximation, Harvard University, AFCRL-65-87 (December 1964), AF 19 (628)-2046.

105. Brysk, H., Electromagnetic Scattering by Low Density Meteor Trails, *J. Geophys. Res.* **63**:693 (1958).
106. Wait, J. R., Some Boundary Value Problems Involving Plasma Media, *J. Res. Natl. Bur. Std.* (*U.S.*), *B* **65B**:137 (1961).
107. Platzman, P. M., and Ozaki, H. T., Scattering of Electromagnetic Waves from an Infinitely Long Magnetized Cylindrical Plasma, *J. Appl. Phys.* **31**:1957 (1960).
108. Adach., S., Scattering Pattern of a Plane Wave From a Magneto-Plasma Cylinder, *IRE Trans. Antennas Propagation* **AP-10**:352 (1962).
109. Messiaen, A. M., and Vandenplas, P. E., Experiment and Theory of Plasma Cylinders in the Presence of a Magnetic Field, *Phys. Letters* **2**:192 (1962).
110. Messiaen, A. M., and Vandenplas, P. E., High-Frequency Effect Due to the Axial Drift Velocity of a Plasma Column, *Phys. Rev.* **149**:131 (1966).
111. Wilhelmsson, K. H. B., Interaction between an Obliquely Incident Plane Electromagnetic Wave and an Electron Beam in the Presence of a Static Magnetic Field of Arbitrary Strength, *J. Res. Natl. Bur. Std.* (*U.S.*) **66D**:439 (1962).
112. Dnestrovskiy, Yu. N., and Kostomarov, D. P., Diffraction of a Plane Electromagnetic Wave by a Circular Plasma Cylinder, *Radio Eng. Electron.* (*USSR*) (*English transl.*) **8**:359 (1963).
113. Chen, H. C., and Cheng, D. K., Scattering of Electromagnetic Waves by an Anisotropic Plasma-Coated Conducting Cylinder, *IEEE Trans. Antennas Propagation* **AP-12**:348 (1964).
114. Chen, H. C., and Cheng, D. K., Scattered Electromagnetic and Plasma Waves of a Conducting Cylinder Coated with a Compressible Plasma, *Appl. Sci. Res.* **B11**:442 (1965).
115. Yeh, C., and Rusch, W. V. T., Interaction of Microwaves with an Inhomogeneous and Anistropic Plasma Column, *J. Appl. Phys.* **36**:2302 (1965).
116. Rusch, W. V. T., and Yeh, C., Scattering by an Infinite Cylinder Coated with an Inhomogeneous and Anistropic Plasma Sheath, *IEEE Trans. Antennas Propagation* **AP-15**:452 (1967).
117. Seshadri, S. R., Scattering by a Perfectly Conducting Cylinder in a Gyroelectric Medium: Numerical Results, *Can. J. Phys.* **42**:860 (1964).
118. Wyatt, J., Electromagnetic Scattering by Finite Dense Plasmas, *J. Appl. Phys.* **36**:3875 (1965).
119. Wyatt, P. J., Improved Surface-Wave Description for Scattering from Overdense Plasmas, *J. Appl. Phys.* **37**:3641 (1966).
120. Probert-Jones, J. R., Surface Waves Associated with the Back-Scattering of Microwave Radiation by Large Ice Spheres. *In* Kerker, M. (Ed.), *Electromagnetic Scattering*, Pergamon Press, London (1963), p. 237.
121. Mar'in, N. P., On the Effective Reflecting Area of a Spherically Shaped Ionized Region, *Radio Eng. Electron. USSR* (*English transl.*) **10**:193 (1965).
122. Margulies, R. S., and Scarf, F. L., Electromagnetic Scattering from a Spherical Nonuniform Medium—Part II: The Radar Cross Section of a Flare, *IEEE Trans. Antennas Propagation* **AP-12**:91 (1964).
123. Bisbing, P. E., Electromagnetic Scattering by an Exponentially Inhomogeneous Plasma Sphere, *IEEE Trans. Antennas Propagation* **AP-14**:219 (1966).
124. Abramowitz, M., and Stegun, I. A. (Eds.), *Handbook of Mathematical Functions with Formulas, Graphs, and Mathematical Tables*, National Bureau of Standards, Applied Mathematics Series No. 55 (1964).
125. Musal, H. M., Jr., Electromagnetic Wave Scattering from Dielectric-Coated Metal Bodies, GM Defense Research Laboratories, Santa Barbara, California, CTN64-05 (August 1964), DA-04-495 ORD-3567 (Z), ARPA Order No. 347-63.
126. Mar'in N. P., Scattering of Rays of Electromagnetic Waves by a Conical Ionized Trail, *Geomagnetism and Aeronomy* **5**:198 (1965).
127. Mar'in, N. P., Scattering of Electromagnetic Waves by an Ionized Trace in the Shape of a Paraboloid of Revolution, *Radio Eng. Electron. USSR* (*English transl.*) **10**:1516 (October 1965).

128. Mar'in, N. P., Trajectories of Electromagnetic-Wave Rays in an Ionized Toroid and Its Effective Reflecting Surface, *Geomagnetism and Aeronomy*, **5**:437 (1965).
129. Åström, E., Electromagnetic Behavior of a Plasma, *Arkiv Fysik* **19**:163 (1961).
130. Seshadri, S. R., Scattering by a Narrow, Perfectly Conducting Infinite Strip in a Gyrotropic Medium, Harvard University, Technical Report No. 380, (October 1962), Nonr-1866(32), NR-371-016.
131. Seshadri, S. R., and Rajagopal, A. K., Diffraction by a Perfectly Conducting, Semi-Infinite Screen in an Anisotropic Plasma, Harvard University, Technical Report No. 376 (October 1962), Nonr 1866(32), NR 371-016.
132. Seshadri, S. R., Morris, I. L., and Mailloux, R. J., Scattering by a Perfectly Conducting Cylinder in a Compressible Plasma, Harvard University, Technical Report No. 409 (April 1963), Nonr-1866(32).
133. Przezdziecki, S., Diffraction by a Half-Plane Perpendicular to the Distinguished Axis of a Uniaxially Anisotropic Medium, *J. Appl. Phys.* **37**:2768 (1966).
134. Jull, E. V., Diffraction By a Wide Aperture in an Anisotropic Medium, *Can. J. Phys.* **44**:713 (1966).
135. Zitron, R., Diffraction by a Slit of Intermediate Width in an Anisotropic Plasma, *Can. J. Phys.* **44**:1067 (1966).
136. Rulf, B., Diffraction by a Half-Plane in a Uniaxial Medium, *SIAM J. Appl. Math.* **15**(1):120 (1967).
137. Rulf, B., Diffraction and Scattering of Electromagnetic Waves in Anisotropic Media, Microwave Research Institute, Polytechnic Institute of Brooklyn, Report No. PIBMRI-1262-65 (April 1965), AF 19(628)-2357.
138. Moorcroft, D. R., Models of Auroral Ionization, Part I. Auroral Ionization Models and Their Radio-Reflection Characteristics, *Can. J. Phys.* **39**:695 (1961).
139. Papas, C. H., On the Index of Refraction of Spatially Periodic Plasma. *In* Nilsson, N. R. (Ed.), *Proc. 4th International Conference of Ionization Phenomena in Gases*, North-Holland Publishing Co., Amsterdam (1959), p. IIIC-718.
140. Kalmykova, S. S., and Kurilko, V. I., The Diffraction of Surface Waves by a Perfectly Conducting Wedge, *Soviet Phys.—Doklady* **9**:145 (1964).
141. Lee, W. C. Y., Peters, L., Jr., and Walter, C. H., Electromagnetic Scattering by Gyrotropic Cylinders with Axial Magnetic Fields, Antenna Laboratory, Ohio State University, Report No. 1116-48, AFCRL-64-717 (May 1964), AF 19(604)-7270, AD 606095.
142. Lee, W. C. Y., Peters, L., Jr., and Walter, C. H., Electromagnetic Scattering by Gyrotropic Cylinders with Axial Magnetic Fields, *Radio Science J. Res. NBS/USNC-URSI* **69D**(2):227 (1965).
143. Lee, W. C. Y., Peters, L., Jr., and Walter, C. H., Geometrical Optics for Gyrotropic Bodies, *Radio Science J. Res. NBS/USNC-URSI* **69D**(3):349 (1965).
144. Gordon, W. E., Incoherent Scattering of Radio Waves by Free Electrons, with Applications to Space Exploration, *Proc. IRE* **46**:1824 (1958).
145. Bowles, K. L., Incoherent Scattering by Free Electrons as a Technique for Studying the Ionosphere and Exosphere: Some Observations and Theoretical Considerations, *J. Res. Natl. Bur. Std.* (*U.S.*) **65D**:1 (1961).
146. Dougherty, J. P., and Farley, D. T., A Theory of Incoherent Scattering of Radio Waves, *Proc. Roy. Soc.* **A259**:79 (1960).
147. Farley, D. T., Dougherty, J. P., and Barron, D. W., A Theory of Incoherent Scattering of Radio Waves: Part 2, *Proc. Roy. Soc.* **A263**:238 (1961).
148. Dougherty, J. P., and Farley, D. T., Jr., A Theory of Incoherent Scattering of Radio Waves by a Plasma—3. Scattering in a Partially Ionized Gas, *J. Geophys. Res.* **68**:5473 (1963).
149. Farley, D. T., A Theory of Incoherent Scattering of Radio Waves by a Plasma—4: The Effect of Unequal Ion and Electron Temperatures, *J. Geophys. Res.* **71**:4091 (1966).
150. Fejer, J. A., Radio-Wave Scattering by an Ionized Gas in Thermal Equilibrium, *J. Geophys. Res.* **65**:2635 (1960).
151. Fejer, J. A., Scattering of Radio Waves by an Ionized Gas in Thermal Equilibrium in the Presence of a Uniform Magnetic Field, *Can. J. Phys.* **39**:716 (1961).

152. Laaspere, T., The Effect of a Magnetic Field on the Spectrum of Incoherent Scatter, *J. Geophys. Res.* **65**:3955 (1960).
153. Salpeter, E. E., Electron Density Fluctuations in a Plasma, *Phys. Rev.* **120**:1528 (1960).
154. Salpeter, E. E., Plasma Density Fluctuations in Magnetic Field, *Phys. Rev.* **122**:1663 (1961).
155. Hagfors, T., Density Fluctuations in a Plasma in a Magnetic Field, with Application to the Ionosphere, *J. Geophys. Res.* **66**:1699 (1961).
156. Rosenbluth, M. N., and Rostoker, N., Scattering of Electromagnetic Waves by a Nonequilibrium Plasma, *Phys. Fluids* **5**:776 (1962).
157. Bowles, K. L., Ochs, G. R., and Green, J. L., On the Absolute Intensity of Incoherent Scatter Echoes from the Ionosphere, *J. Res. Natl. Bur. Std.* (*U.S.*) **66D**(4):395 (1962).
158. Farley, D. T., Jr., The Effect of Coulomb Collisions on Incoherent Scattering of Radio Waves by a Plasma, *J. Geophys. Res.* **69**:197 (1964).
159. Moorcroft, D. R., On the Determination of Temperature and Ionic Composition by Electron Backscattering from the Ionosphere and Magnetosphere, *J. Geophys. Res.* **69**:955 (1964).
160. Bowles, K. L., Balsley, B. B., and Cohen, R., Field-Aligned *E*-Region Irregularities Identified with Acoustic Plasma Waves, *J. Geophys. Res.* **68**:2485 (1963).
161. Farley, D. T., Jr., A Plasma Instability Resulting in Field-Aligned Irregularities in the Ionosphere, *J. Geophys. Res.* **68**:6083 (1963).
162. Dougherty, J. P., and Farley, D. T., Ionospheric *E*-Region Irregularities Produced by Nonlinear Coupling of Unstable Plasma Waves, *J. Geophys. Res.* **72**:895 (1967).
163. Arunasalam, V., and Brown, S. C., Microwave Scattering Due to Acoustic-Ion–Plasma-Wave Instability, *Phys. Rev.* **140**(2A):A471 (1965).
164. Perepelkin, N. F., Raman Scattering of Microwaves by Plasma Oscillations, *JETP Letters* **3**:165 (1966).
165. Dong, N. Q., Interaction of Electromagnetic Waves with Quantum and Classical Plasmas—Correlation Effects, *Phys. Rev.* **148**(1):148 (1966).
166. Ivanov, A. A., and Ryutov, D. D., Scattering of Electromagnetic Waves by Plasma Oscillations in a Plane Plasma Layer, *Soviet Phys.—JETP* **21**:913 (1965).
167. Akhiezer, I. A., Scattering and Conversion of Electromagnetic Waves in a Turbulent Plasma, *Soviet Phys.—JETP* **21**:774 (1965).
168. Booker, H. G., and Gordon, W. E., A Theory of Radio Scattering in the Troposphere, *Proc. IRE* **38**:401 (1950).
169. Yen, K. T., Some Aspects of Turbulent Scattering of Electromagnetic Waves by Hypersonic Wake Flows, General Electric Company, Missile and Space Division, Report No. R63SD58 (December 1963).
170. Yen, K. T., Effect of Turbulence Intermittency on the Scattering of Electromagnetic Waves by Underdense Plasmas, *AIAA Journal* **4**(1):154 (1966).
171. Ichikawa, Y. H., Dielectric Properties of a Weakly Turbulent Plasma, *Phys. Fluids* **9**:111 (1966).
172. Tatarski, V. I., *Wave Propagation In a Turbulent Medium*, McGraw-Hill Book Co., New York (1961), p. 157.
173. Macrakis, M. S., Fluctuation Scattering and Absorption Cross Section in a Space Dispersive Plasma, *Phys. Fluids* **9**:1124 (1966).
174. Ruffine, R. S., and DeWolf, D. A., Cross-Polarized Electromagnetic Backscatter from Turbulent Plasmas, *J. Geophys. Res.* **70**:4313 (1965).
175. Salpeter, E. E., and Treiman, S. B., Backscatter of Electromagnetic Radiation from a Turbulent Plasma, *J. Geophys. Res.* **5**:869 (1964).
176. Albini, F. A., Scattering and Absorption of Plane Waves by Cylindrically Symmetrical Underdense Zones, *AIAA Journal* **2**:524 (1964).
177. Sugar, George R., Radio Propagation by Reflection from Meteor Trials, *Proc. IEEE* **52**(2):116 (1964).
178. Manning, L. A., and Eshleman, V. R., Meteors in the Ionosphere, *Proc. IRE* **47**(2):186 (1959).
179. Manning, L. A., Air Motions and the Fading, Diversity, and Aspect Sensitivity of Meteoric Echoes, *J. Geophys. Res.* **64**:1415 (1959).

180. McKinley, D. W. R., *Meteor Science and Engineering*, McGraw-Hill Book Co., New York (1961), p. 220.
181. McKinley, D. W. R., and Millman, P. M., A Phenomenological Theory of Radar Echoes From Meteors, *Proc. IRE* **37**:364 (1949).
182. Meteor Facsimile via Meteor Trail Propagation, RCA Laboratories, Radio Corporation of America, Princeton, New Jersey, AFCRL-TR-60-190 (August 1960), AF 19 (604)-4102, AD 248 723.
183. Lovell, A. C. B., Prentice, J. P. M., Porter, J. G., *et al.*, Meteors, Comets, and Meteoric Ionization, *In*: *Rept. Progr. Phys.* **XI**:389 (1947), The Physical Society, London.
184. Kaiser, T. R., and Closs, R. L., Theory of Radio Reflections from Meteor Trails: I, *The Philosophical Magazine, 7th Ser.* **43**(336):1 (1952).
185. Kaiser, T. R., Radio Echo Studies of Meteor Ionization, *Adv. Phys.* **II**:495 (1953).
186. Lovell, A. C. B., Geophysical Aspects of Meteors. *In* Flugge, S. (Ed.), *Handbuch Der Physik*, Springer-Verlag, Berlin (1957), p. 427.
187. Davies, J. G., Radio Observation of Meteors. *In* Martin, L. (Ed.), *Advances in Electronics and Electron Physics* (1957), pp. 95–128.
188. Eshleman, V. R., Meteor Scatter. *In* Menzel, D. H. (Ed.), *The Radio Noise Spectrum*, Harvard University Press, Cambridge, Massachusetts (1960), p. 19. [See also Eshleman, V. R., Meteor Scatter, Stanford Electronics Laboraties, Stanford University, Scientific Report No. 4, Contract AF19 (604)-2193 (Air Force Cambridge Research Center, AFCRC-TN-58-370) (August 1958), AD 160819.]
189. Hawkins, G. S., A Radio Survey of Sporadic Meteor Radiants, *Monthly Notices, Royal Astron. Soc.* **116**:92 (1956).
190. Eshleman, V. R., The Theoretical Length Distribution of Ionized Meteor Trails, *J. Atmospheric Terrest. Phys.* **10**:57 (1957).
191. Bayrachenko, I. V., Measurements of the Initial Radii of Ionized Meteor Trails from Simultaneous Observations of Radio Meteors on Two Wavelengths, *Geomagnetism and Aeronomy* **5**(3):353 (1965).
192. Loewenthal, M., On Meteor Echoes from Underdense Trails at Very High Frequencies, Lincoln Laboratory, Massachusetts Institute of Technology, Lexington, Massachusetts, Technical Report No. 132 (December 1956), AD 137 608.
193. Greenhow, J. S., and Neufeld, E. L., The Diffusion of Ionized Meteor Trails in the Upper Atmosphere, *J. Atmospheric Terrest. Phys.* **6**:133 (1955).
194. Evans, J. V., Radar Observations of Meteor Deceleration, *J. Geophys. Res.* **71**:171 (1966).
195. Evans, J. V., Radio-Echo Studies of Meteors at 68-Centimeter Wavelength, *J. Geophys. Res.* **70**:5395 (1965).
196. Hines, C. O., and Forsyth, P. A., The Forward Scattering of Radio Waves from Overdense Meteor Trails, *Can. J. Phys.* **35**(9):1033 (1957).
197. Manning, L. A., Oblique Echoes From Over-Dense Meteor Trails, *J. Atmospheric Terrest. Phys.* **14**:82 (1959).
198. Hawkins, G. S., and Winter, D. F., Radar Echoes from Overdense Meteor Trails under Conditions of Severe Diffusion, *Proc. IRE* **45**:1290 (1957).
199. Hawkins, G. S., Radar Echoes from Meteor Trails under Conditions of Severe Diffusion, *Proc. IRE* **44**:1192 (1956).
200. Manning, L. A., The Strength of Meteoric Echoes from Dense Columns, *J. Atmospheric Terrest. Phys.* **4**:219 (1953).
201. Lagutin, M. F., and Kashcheyev, B. L., Polarization Effect on Radio Signals Scattered by Meteor Trails, *Radio Eng. Electron. USSR* (*English transl.*) **9**(8):1227 (1964).
202. Watkins, C. D., Influence of the Geomagnetic Field on Meteor Trains, *Nature* **206**(4988):1027 (1965).
203. Francey, J. L. A., Diffusion of Meteor Trails on the Earth's Magnetic Field, *Australian J. Phys.* **17**:315 (1964).
204. Harang, L., *The Aurorae*, John Wiley & Sons, New York (1951).
205. Störmer, C., *The Polar Aurora*, Oxford Press, London (1955).
206. Dalgarno, A., and Armstrong, E. B. (Eds.), *The Airglow and the Aurorae*, Pergamon Press, London (1957).

207. Ratcliffe, J. R. (Ed.), *Physics of the Upper Atmosphere*, Academic Press, New York (1960).
208. Chamberlain, J. W., *Physics of the Aurora and Airglow*, Academic Press, New York (1961).
209. Ferraro, V. C. A., The Aurorae, *Adv. Phys.* **2**:265 (1953).
210. Walt, M. (Ed.), *Auroral Phenomena*, Stanford University Press, Stanford, California (1965).
211. Maehlum, B. (Ed.), *High Latitude Particles and the Ionosphere*, Logos Press, London, and Academic Press, New York (1965).
212. Vestine, E. H., The Geographic Incidence of Aurora and Magnetic Disturbance, Northern Hemisphere, *Terrestrial Magnetism and Atmospheric Electricity* **49**(2):77 (1944).
213. Isayev, S. I., The Geographical Distribution of Auroras and the Earth's Radiation Belts, *Geomagnetism and Aeronomy* **2**:552 (1962).
214. Feldstein, Y. I., Peculiarities in the Auroral Distribution and Magnetic Disturbance Distribution in High Latitudes Caused by the Asymmetrical Form of the Magnetosphere, *Planet. Space Sci.* **14**:121 (1966).
215. Starkov, G. V., and Fel'dshteyn, Ya. I., Change in the Boundaries of the Oval Auroral Zone, *Geomagnetism and Aeronomy* **7**:48 (1967).
216. Feldstein, Y. I., and Starkov, G. V., Dynamics of Auroral Belt and Polar Geomagnetic Disturbances, *Planetary Space Sci.* **15**(2):209 (1967).
217. Hultqvist, B., Circular Symmetry in the Geomagnetic Plane for Auroral Phenomena, *Planetary Space Sci.* **8**:142 (1961).
218. Fel'dshteyn, Ya. I., and Shevnina, N. F., The Position of the Auroral Zone in the Southern Hemisphere, *Geomagnetism and Aeronomy* **2**:240 (1962).
219. Fel'dshteyn, Ya. I., and Solomatina, E. K., Auroras in the Southern Hemisphere, *Geomagnetism and Aeronomy* **1**:475 (1961).
220. Hultqvist, B., and Egeland, A., Radio Aurora, *Space Science Reviews* **3**:27 (1964).
221. Kelley, L. C., Perlman, S., Russell, W. J., Jr., and Stuart, W. D., Tentative Evaluation of Transmission Factors for Space Vehicle Communication, U.S. Army Signal Corps, Radio Propagation Agency (September 1958).
222. Krasovskij, V. I., Polar Auroras, *Space Science Reviews* **3**:232 (1964).
223. Bullough, K., and Kaiser, T. R., Radio Reflections from Aurora, Parts I and II, *J. Atmospheric Terrest. Phys.* **5**:189 (Part I), 198 (Part II) (1954).
224. Blevis, B. C., Day, J. W. B., and Roscoe, O. S., The Occurrence and Characteristics of Radar Auroral Echoes at 488 and 944 Mc, *Can. J. Phys.* **41**:1359 (1963).
225. Bagaryatskii, B. A., Radar Reflections from Aurora, *Soviet Phys.—Uspekhi* **4**:70 (1961).
226. Bierfeld, Ya. G., Radar Reflections From Aurora, *Izv. Akad. Nauk SSSR, Ser. Geofiz.* **12**:1248 (1960) [English translation: *Bull. Acad. Sci. USSR: Geophysics Series* (April 1961)].
227. Stringer, W. J., Belon, A. E., and Akasofu, S. I., The Latitude of Auroral Activity during Periods of Zero and Very Weak Magnetic Disturbances, *J. Atmospheric Terrest. Phys.* **27**:1039 (1965).
228. Dyce, R. B., Auroral Echoes Observed North of the Auroral Zone, *J. Geophys. Res.* **60**(3):317 (1955).
229. Bates, H. F., Belon, A. E., Romick, G. J., and Stringer, W. J., On the Correlation of Optical and Radio Aurora, *J. Atmospheric Terrest. Phys.* **28**:439 (1966).
230. Kelly, P. E., The Association between Radio and Visual Aurora, *Can. J. Phys.* **43**:1167 (1965).
231. Hyde, G., and Coleman, J., Microwave Properties of Plasma, *RCA Engineer* **8**:40 (1963).
232. Herlofson, N., Interpretation of Radio Echoes from Polar Auroras, *Nature* **160**:867 (1947).
233. Chapman, S., The Geometry of Radio Reflections from Aurorae, *J. Atmospheric Terrest. Phys.* **3**:1 (1952).
234. Booker, H. G., Radar Studies of the Aurora. *In* Ratcliffe, J. A. (Ed.), *Physics of the Upper Atmosphere*, Academic Press, New York (1960), p. 355.

235. Collins, C., and Forsyth, P. A., A Bistatic Radio Investigation of Auroral Ionization, *J. Atmospheric Terrest. Phys.* **13**:315 (1959).
236. Lovell, A. C. B., Clegg, J. A., Ellyett, C. D., Radio Echoes from the Aurora Borealis, *Nature* **160**:372 (1947).
237. Booker, H. G., A Theory of Scattering by Nonisotropic Irregularities with Application to Radar Reflections from the Aurora, *J. Atmospheric Terrest. Phys.* **8**:204 (1956).
238. Aspinall, A., and Hawkins, G. S., Radio-Echo Reflections from the Aurora Borealis, *J. Brit. Astron. Assoc.* **60**:130 (1950).
239. Forsyth, P. A., and Vogan, E. L., The Frequency of Radio Reflection from Aurora, *J. Atmospheric Terrest. Phys.* **10**:215 (1957).
240. Lyon, G. F., The Anisotropy of Ionospheric Irregularities Deduced from VHF Scatter Measurements, *J. Atmospheric Terrest. Phys.* **27**:1213 (1965).
241. Peterson, A. M., The Aurora and Radio Wave Propagation. *In* Menzel, D. H. (Ed.), *The Radio Noise Spectrum*, Harvard University Press, Cambridge, Massachusetts (1960), p. 7.
242. Leadabrand, R. L., Dolphin, L., and Peterson, A. M., *IRE Trans. Antennas Propagation* **AP-7**:127 (1959).
243. Moorcroft, D. R., Models of Auroral Ionization, Part II: Applications to Radio Observations of Aurora, *Can. J. Phys.* **39**:677 (1961).
244. Lyon, G. F., and Forsyth, P. A., Radio-Auroral Reflection Mechanisms, *Can. J. Phys.* **40**:749 (1962).
245. Wilkins, L. K., Assendrop, J. D., and Kim, J. S., E–W Motion of Auroral as Observed at Moscow, Idaho, *J. Atmospheric Terrest. Phys.* **29**:311 (1967).
246. Keys, J. G., Pulsating Auroral Radar Echoes and Their Possible Hydromagnetic Association, *J. Atmospheric Terrest. Phys.* **27**:385 (1965).
247. Moorcroft, D. R., The Interpretation of the Frequency Dependence of Radio Aurora, *Planetary Space Sci.* **14**:275 (1966).
248. Coleman, J. T., The Control of Ion Acoustic Wave Instabilities in a Gaseous Discharge, *J. Appl. Phys.* **38**:5 (1967).
249. Leadabrand, R. L., A Comparison of Radar Auroral Reflection Data with Acoustic Wave Theory, *Radio Science J. Res. NBS/USNC* **69D**:959 (1965).
250. Sterling, D. L., *F*-Region Travelling Disturbances at the Magnetic Equator, Paper 3.8-3 at 1967 URSI Spring Meeting, Ottawa, Canada.
251. Glasstone, S., *The Effects of Nuclear Weapons*, United States Atomic Energy Commission, Washington, D.C. (1962), p. 502.
252. Singer, S. F., Nuclear Explosions in Space, *Missiles Rockets* **5**(March 30):33 (Part I); **5**(April 6):36 (Part II); **5**(April 13):21 (Part III); **5**(April 20):26 (Part IV) (1959).
253. Kompaneets, A. S., Radio Emission From an Atomic Explosion, *Soviet Phys.—JETP* **35**:1538 (1958).
254. Saha, A. K., and Mahajan, K. K., Ionospheric Effects of Nuclear Explosions: High Altitude Explosions, *Indian J. Pure Appl. Phys.* **4**:117 (1966).
255. Saha, A. K., and Mahajan, K. K., Ionospheric Effects of Nuclear Explosions: Low Altitude Explosions, *Indian J. Pure Appl. Phys.* **3**:433 (1965).
256. Crain, C. M., and Booker, H. G., The Effects of Nuclear Bursts in Space on the Propagation of High Frequency Radio Waves Between Separated Earth Terminals, *J. Geophys. Res.* **63**:2159 (1963).
257. Marmo, F. F., Pressman, J., Aschenbrand, L. M., Jursa, A., and Zelikoff, M., Formation of an Artificial Ion Cloud; Photoionization of NO by Solar Lyman Alpha at 95 km, *J. Chem. Phys.* **25**:187 (1956).
258. Marmo, F. F., Aschenbrand, L. M., and Pressman, J., Artificial Electron Clouds—I. Summary Report on the Creation of Artificial Electron Clouds in the Upper Atmosphere, *Planetary Space Sci.* **1**:227 (1959).
259. Gallagher, P. B., and Villard, Jr., O. G., Radio Reflections from Artificially Produced Electron Clouds: Smokepuff II, Stanford Electronics Laboratories, Stanford University, Stanford, California (December 1958), AF19(604)-2075, AD 230 690.
260. Gallagher, P. B., and Barnes, R. A., Radio-Frequency Backscatter of Artificial Electron Clouds, *J. Geophys. Res.* **68**:2987 (1963).

261. Rosenberg, N. W. (Ed.), *Project Firefly 1962–1963*, Environmental Research Paper No. 15, AF Cambridge Research Laboratories, USAF, L. G. Hanscom Field, Bedford, Massachusetts.
262. Wright, J. W., Ionosonde Observations of Artificially Produced Electron Clouds Firely 1960, Boulder Laboratories, Technical Note No. 135, PB 161636 US Department of Commerce, National Bureau of Standards (April 1962).
263. Wright, J. W., Ionosonde Studies of Some Chemical Releases in the Ionosphere, *J. Res. Natl. Bur. Std.* (*U.S.*), *Part D*, *Radio Science* **68D**:189 (1964).
264. Booker, H. G., A Local Reduction of *F*-Region Ionization Due to Missile Transit, *J. Geophys. Res.* **66**:1073 (1961).
265. Rosenberg, N. W., Conley, T. D., Curley, S. R., and Lorentzen, A. H., Studies of Detonations at Ionospheric Altitudes under Project Red Lamp, AF Cambridge Research Laboratories, L. G. Hanscom Field, Bedford, Massachusetts, AFCRL-65-356 (May 1965).
266. Föppl, F., Haerendel, G., Haser, L., *et al.*, Artificial Strontium and Barium Clouds in the Upper Atmosphere, *Planetary Space Sc.* **15**:357 (1967).
267. Golomb, D., Rosenberg, N. W., Wright, J. W., and Barnes, R. A., Formation of an Electron Depleted Region in the Ionosphere by Chemical Releases. *In: Space Research*, Vol. IV, John Wiley & Sons, New York (1963), p. 389.
268. Rashad, A. R. M., Rocket Exhaust Scattering Effects on Incident Microwave Propagation Characteristics, *IEEE Trans. Aerospace Electronic Systems* **AES-2**:425 (1966).
269. Barnes, C., Jr., Comment on the Paper by Henry G. Booker, "A Local Reduction of *F*-Region Ionization Due to Missile Transit," *J. Geophys. Res.* **66**:2589 (1961).
270. Stone, M. L., Bird, L. E., and Balser, M., A Faraday Rotation Measurement on the Ionospheric Perturbation Produced by a Burning Rocket, *J. Geophys. Res.* **69**:971 (1964).
271. Jackson, J. E., Whale, H. A., and Bauer, S. J., Local Ionospheric Disturbance Created by a Burning Rocket, *J. Geophys. Res.* **67**:2059 (1962).
272. Pfister, W., and Ulwick, J. C., Effects of Rocket Outgassing on RF Experiments, *J. Res. Natl. Bur. Std.* (*U.S.*) *Part D*, *Radio Science* **69D**:1219 (1965).
273. Kraus, J. D., Detection of Sputnik I and II by CW Reflection, *Proc. IRE* **46**:611 (1958).
274. Kraus, J. D., The Last Days of Sputnik I, *Proc. IRE* **46**:612 (1958).
275. Kraus, J. D., and Albus, J. S., A Note on Some Signal Characteristics of Sputnik I, *Proc. IRE* **46**:610 (1958).
276. Kraus, J. D., and Dreese, E. E., Sputnik I's Last Days in Orbit, *Proc. IRE* **46**:1580 (1958).
277. Kraus, J. D., Higgy, R. C., and Albus, J. S., Observations of the U.S. Satellites, Explorers I and III, by CW Reflection, *Proc. IRE* **46**:1534 (1958).
278. Flambard, A., and Reyssat, M., Ondes Electromagnétiques et Satellites: Echos des Trainées Ionisées de Satellites en H. H. *In: Avionics Research: Satellites and Problems of Long Range Detection and Tracking*, Pergamon Press, New York (1960), p. 101. See also Electromagnetic Waves and Satellites: Echoes from Ionized Trails of Satellites at High Frequency, *L'Onde Elect.* **38**:830 (1958).
279. Dieminger, W., Ground Scatter by Ionospheric Radar. *In: Avionics Research: Satellites and Problems of Long Range Detection and Tracking*, Pergamon Press, New York (1960), p. 29.
280. Hendricks, C. D., Jr., Swenson, G. W., Jr., and Schorn, R. A., Radio Reflections From Satellite Produced Ion Clouds, *Proc. IRE* **46**:1763 (1958).
281. Techniques for Echo Area Determination, Antenna Laboratory, Ohio State University, Report 792-5 (February 1959), Contract AF 33(616)-5398.
282. Hame, T. G., and Kennaugh, E. M., Recordings of Transmissions from the Satellite 1958 δ2 at the Antenna Laboratory, The Ohio State University, *Proc. IRE* **47**:991 (1959).
283. Hame, T. G., and Stuart, W. D., Detection of Satellites by Their Influence on the Ionosphere, paper presented at the URSI–IRE Meeting, December 12-15, 1960, Boulder, Colorado.

284. Hame, T. F., and Stuart, W. D., The Detection of Artificial Satellites by Their Influences on the Ionosphere. *In* Singer, S. F. (Ed.), *Interactions of Space Vehicles with an Ionized Atmosphere*, Pergamon Press, New York (1965), p. 373.
285. Hame, T. G., On the Detection of Artificial Earth Satellites, Antenna Laboratory, Ohio State University Research Foundation, Report 889-7 (October 1959), AF 33(616)-6137.
286. Hame, T. G., and Stuart, W. D., Reflection Characteristics of High Velocity Aerial Targets, Antenna Laboratory, Ohio State University Research Foundation Report 889-8 (November 1959), AF 33(616)-6137.
287. Stuart, W. D., and Potts, B. C., Experimental Program to Determine Properties of Ionization Which is Produced by Space Vehicles, Antenna Laboratory, Ohio State University, Report 1108-6 (June 1961), AF 19(604)-7274, AFCRL-751.
288. Kraus, J. D., and Higgy, R. C., Evidence of Ionization Induced by Sputnik III, Ohio State University Radio Observatory Report No. 10, Report 884-2 (March 1959), DA-33-019-ORD-2867.
289. Kraus, J. D., Higgy, R. C., Crone, W. R., and Scheer, D. J., Early Detection of Artificial Earth Satellites Using CW Reflection and Reception at Side Frequencies, Ohio State University Radio Observatory Report No. 13, Report 884-4 (April 1959), DA-33-019-ORD-2867.
290. Kraus, J. D., Higgy, R. C., Scheer, D. J., and Crone, W. R., Observations of Ionization Induced by Artificial Earth Satellites, *Nature* **185**:520 (1960).
291. Kraus, J. D., Evidence of Satellite-Induced Ionization Effects between Hemispheres, *Proc. IRE* **48**:1913 (1960). See also Ohio State University Radio Observatory, Report No. 17, Report 884-8 (March 1960), DA 33-019-ORD-2867.
292. Kraus, J. D., and Higgy, R. C., The Relation of the Satellite Ionization Phenomenon to the Radiation Belts, *Proc. IRE* **48**:2027 (1960).
293. Kraus, J. D., and Herriman, A. G., Further Evidence of Satellite Induced Ionization by CW-Reflection, Ohio State University Radio Observatory, Report No. 884-11 (April 1961), DA-33-019-ORD-2867.
294. Crone, W. R., Some Experimental and Theoretical Considerations of the Satellite Ionization Phenomenon, Master's Thesis, Ohio State University, Columbus, Ohio (1960).
295. Kraus, J. D., and Crone, W. R., Doppler Effects Associated with CW-Reflection Observations of Satellite-Induced Ionization, Ohio State University Radio Observatory, Report No. 16, Report 884-7 (July, 1959), DA-33-019-ORD-2867.
296. Kraus, J. D., and Higgy, R. C., Some Radar Observations of Satellite-Related Ionization, Ohio State University Research Foundation, Report No. 884-10 (March 1961), DA-33-019-ORD-2867.
297. Kraus, J. D., Higgy, R. C., and Crone, W. R., The Satellite Ionization Phenomenon, *Proc. IRE* **48**:672 (1960).
298. Herriman, A. G., Studies of the Earth's Outer Radiation Belt from Observations of the U.S. Artificial Satellite Explorer VII Made at the Ohio State University Radio Observatory, Master's Thesis, Ohio State University, Columbus, Ohio (1961).
299. Hame, T. G., The Analysis of HF Radar Observations during the Period November 15, 1960 to December 15, 1960, Radio Observatory and Antenna Laboratory, Department of Electrical Engineering, The Ohio State University, Report 884-12 (March 1961), DA-33-019-ORD-2867.
300. Staff of Jodrell Bank Experimental Station, Radar Observations of the Russian Earth Satellite, *Proc. Roy. Soc. (London), Series A*, **248**:25 (1958); **250**:367 (1959).
301. Al'pert, Ya. L., Investigation of the Ionosphere and Interplanetary Gas, *Soviet Phys.—Uspekhi* **3**:479 (1961).
302. Tiuri, M., Investigations of Radio Reflections From Satellite-Produced Ion Trials Using 100Mc CW Radar, The State Institute for Technical Research, Finland, Publication 59, 1960, Doctoral Thesis, UDC 621.396.9:550.389.2.
303. Croft, T. A., and Villard, Jr., O. G., An HF-Radar Search for the Effects of Earth Satellites upon the Ionosphere. *In* Singer, S. F. (Ed.), *Interactions of Space Vehicles with an Ionized Atmosphere*, Pergamon Press, New York (1965), p. 389.

304. Croft, T. A., and Villard, O. G., Jr., An HF-Radar Search for Possible Effects of Earth Satellites upon the Upper Atmosphere, *J. Geophys. Res.* **66**:3109 (1961).
305. Croft, T. A., An HF-Radar Search for the Effects of Earth Satellites upon the Ionosphere, Stanford Electronics Laboratories, Stanford University, Stanford California, Technical Report No. 24, Project No. 087090 (March 1961), Contract Nonr-22533.
306. Lootens, H. T., DOPLOC Observations of Reflection Cross Sections of Satellites, Memorandum Report No. 1330, AD 259123, ARPA Satellite Force Series, No. 22, Ballistic Research Laboratories, Aberdeen Proving Ground, Maryland (March 1961).
307. Lootens, H. T., Satellite-Induced Ionization Observed with the DOPLOC System, Memorandum Report No. 1362, ARPA Satellite Force Series No. 23, Ballistic Research Laboratories, Aberdeen Proving Ground, Maryland (1961).
308. Kraus, J. D., and Tiuri, M. E., Observations of Satellite Related Ionization Effects between 1958 and 1960, *Proc. IRE* **50**:2076 (1962).
309. Medved, D. B., On the Formation of Satellite Electron Sheaths Resulting from Secondary Emission and Photoeffects. *In* Singer, S. F. (Ed.), *Interactions of Space Vehicles with an Ionized Atmosphere*, Pergamon Press, New York (1965), p. 305.
310. Chopra, K. P., and Singer, S. F., Drag of a Sphere Moving In a Conducting Fluid In the Presence of a Magnetic Field, University of Maryland, Physics Department Technical Report 97 (1958); also in *Heat Transfer and Fluid Mechanics Institute* (University of California, Berkeley, California, June 19, 20, 21, 1958) Stanford University Press, Stanford, California (1958), p. 166.
311. Kraus, L., and Watson, K. M., Plasma Motions Induced By Satellites in the Ionosphere, *Phys. Fluids* **1**:480 (1958); see also Convair Report ZPH-016 (May 1958).
312. Greifinger, P., Induced Oscillations in a Rarefied Plasma in a Magnetic Field. *In* Masson, D. J. (Ed.), *Aerodynamics of the Upper Atmosphere*, Rand Corporation, Santa Monica, California, Report R-399 (1959).
313. Chopra, K. P., Interactions of Rapidly-Moving Bodies in Terrestrial Atmosphere, University of Southern California Engineering Center, USEC Report 56-212, AFOSR TN-60-398 (March 1960), AD 236347.
314. Chen, K. M., Disturbance Due to a Satellite in a Plasma Medium and Its Effect on Radar Return. *In* Singer, S. F. (Ed.), *Interactions of Space Vehicles with an Ionized Atmosphere*, Pergamon Press, New York (1965), p. 465.
315. Chen, K. M., Studies in Radar Cross Section XLIII—Plasma Sheath Surrounding a Conducting Spherical Satellite and the Effect on Radar Cross Section, The University of Michigan Radiation Laboratory, Report No. 2764-6T (October 1960), DA 36-039 sc-75041, ARPA Order No. 120-60.
316. Dolph, C. L., and Weil, H., On the Change in Radar Cross Section of a Spherical Satellite Caused by a Plasma Sheath. *In* Rotman, W., and Meltz, G. (Eds.), *Electromagnetic Effects of Reentry*, Pergamon Press, London (1961), p. 123.
317. Walker, E. H., and Singer, S. F., Wake of a Charged Body Moving in a Plasma, *Bull. Am. Phys. Soc.* **5**:234 (1960). See also Wake of Charged Body in an Ionized Gas, *Bull. Am. Phys. Soc.* **5**:47 (1960).
318. Singer, S. F., and Walker, E. H., Plasma Compression Effects Produced by Space Vehicles in a Magneto-Ionic Medium. *In* Singer, S. F. (Ed.), *Interactions of Space Vehicles with an Ionized Atmosphere*, Pergamon Press, New York (1965), p. 483.
319. Gurevich, A. V., Perturbations in the Ionosphere Caused by a Moving Body, *Akad. Nauk SSSR*, *Institut Zemnogo Magnetizma Ionosfery i rasprostraneniya Radiovoln*, *Trudy* **17**(27), *Rasprostraneniye Radiovoln i ionosfere*, Moscow (1960), p. 173. [English translation appears in *Planetary Space Sci.* **9**:321 (1962) and in *ARS Journal* **32**:1161 (1962). Reviewers Comment: Francis S. Johnson, p. 1167.]
320. Al'pert, Ya. L., and Pitayevskii, L. P., Scattering of Electromagnetic Waves by Perturbations Caused in a Plasma by a Rapidly Moving Body, *Geomagnetism and Aeronomy* **1**:709 (1961).

321. Pitayevskii, L. P., and Klesin, V. Z., On the Problem of Perturbations Occurring during the Motion of a Body in a Plasma, *Zh. Eksperim. i Teor. Fiz.* **40**:271 (1961) [English transl.: *Soviet Phys.—JETP* **13**:185 (1961).]
322. Shea, J. J., Collisionless Plasma Flow around a Conducting Sphere in a Magnetic Field. *In* Brundin, C. L. (Ed.), *Rarefied Gas Dynamics, Advances in Applied Mechanics*, Vol. II, Academic Press, New York (1967), p. 1671.
323. Rand, S., and Albini, F., Radar Return from Vehicles in the Ionosphere, *AIAA Journal* **5**:1174 (1967).
324. Clayden, W. A., and Hurdle, C. V., An Experimental Study of Plasma-Vehicle Interaction. *In* Brundin, C. L. (Ed.), *Rarefied Gas Dynamics*, Vol. II, *Advances in Applied Mechanics*, Academic Press, New York (1967), p. 1717.
325. Martin, J. J., *Atmospheric Reentry*, Prentice-Hall, Englewood Cliffs, New Jersey (1966).
326. Loh, W. H. T., *Dynamics and Thermodynamics at Planetary Entry*, Prentice-Hall, Englewood Cliffs, New Jersey (1963).
327. Gunar, M., and Mennella, R., Signature Studies for a Reentry System. *In: Proceedings of the Second Space Congress—New Dimensions in Space Technology*, Canaveral Council of Technical Societies (April 1965), p. 505.
328. Musal, H. M., Jr., The Plasma Sheath Effect on the Scattering Cross Section of a Hypervelocity Sphere—Part One, Bendix Aviation Corporation, Research Note 1, ARPA Order No. 39 (December 1959).
329. Musal, H. M., Jr., The Plasma Sheath Effect on Radar Cross Section of a Hypersonic Sphere as Predicted by Lossless Geometrical Optics, Bendix Aviation Corporation, Research Note 1, ARPA Order No. 39 (March 1960).
330. Robillard, P. E., Musal, H. M., Jr., and Primich, R. I., Millimeter Radar Techniques for Studying Plasma Effects Associated with Hypersonic Velocity Projectiles, General Motors Corporation, Report No. TR 63-217B, ARPA Order No. 347-63 (July 1963), DA-04-495-ORD-3567(Z). (Presented as a paper at the Millimeter and Submillimeter Wavelength Conference at Orlando, Florida in January, 1963.)
331. Musal, H. M., Jr., Primich, R. I., Blore, W. E., and Robillard, P. E., Millimeter Radar Instrumentation for Studying Plasma Effects Associated with Hypersonic Flight, General Motors Corporation, Report No. TR 64-02J, ARPA Order No. 347-63 (August 1964), DA-04-495-ORD-3567(Z). (Paper presented at the First International Congress on Instrumentation in Aerospace Simulation Facilities, 28–29 September 1964 in Paris, France.)
332. Pippert, G. P., and Edelberg, S., The Electrical Properties of the Air Around a Reentering Body, Institute Aerospace Sciences, Reprint 61-40 (1961).
333. Primich, R. I., and Hayami, R. A., The Application of the Focussed Fabry–Perot Resonator to Plasma Diagnostics, *IEEE Trans. on Microwave Theory Techniques* **MTT-12**:33 (1964).
334. Shao-Chi Lin, Radio Echoes from a Manned Satellite during Reentry, *J. Geophys. Res.* **67**:3851 (1962).
335. Lees, L., and Hromas, L., Turbulent Diffusion in the Wake of a Blunt-Nosed Body at Hypersonic Speeds, *J. Aeronautical Sciences* **29**:976 (1962).
336. Kalinin, Yu. K., Use of the Diffusion Model of a Meteor Trail for the Interpretation of Data on Radio-Wave Scattering by the Reentering Capsule of the American MA-6 Satellite, *Geomagnetism and Aeronomy* **5**:220 (1965).
337. Barrick, D. E., A Note on Scattering from Dielectric Bodies by the Modified Geometrical-Optics Method, *IEEE Trans. Antennas Propagation* **AP-16**(2):275–277 (March 1968).

Chapter 11

RADAR CROSS-SECTION MEASUREMENTS

C. K. Krichbaum

11.1. INTRODUCTION

The scattering cross section of an object, as defined in Section 2.1.1, is given by

$$\sigma(\theta, \phi, \theta', \phi') = \lim_{R\to\infty} 4\pi r^2 \frac{|\mathbf{E}^s(\theta', \phi')|^2}{|\mathbf{E}^i(\theta, \phi)|^2} \tag{11.1-1}$$

where θ', ϕ' define the direction in which the scattered energy is measured, θ, ϕ define the direction from which the scatterer was illuminated, and σ is the "differential" bistatic cross section of the object. If σ is integrated over 4π radians at a large distance from the scatterer, the total scattering cross section is obtained. For $\theta = \theta'$ and $\phi = \phi'$, the backscattering cross section is obtained. This chapter is concerned with the measurement of the scattering cross section of various objects, and the factors which must be considered when such measurements are made.

11.2. COORDINATE SYSTEMS

Two coordinate systems are generally required to describe the scattering properties of a body and to define the orientation of the body with respect to a reference frame. If the body has a natural axis of symmetry, consider a body-fixed coordinate system with the z' axis along the body axis of symmetry and the x' and y' axes rigidly fixed to the body. Directions of illumination and reflection of electromagnetic energy can be described by the co-latitude angle θ and longitude angle ϕ. The polarization state of a plane wave incident or scattered at a direction (θ, ϕ) can be described by two transverse components of the electric field, along $\hat{\boldsymbol{\theta}}$ and along $\hat{\boldsymbol{\phi}}$. When static scattering measurements are made, the illuminating antenna is generally fixed, and the target is rotated to obtain data for a set of viewing directions. The body position can be considered the result of three successive rotations of the body from a coordinate system determined by three planes intersecting at the body center of mass. This inertial coordinate system is determined by an initial or reference position of the body-fixed system (x, y, z). Body

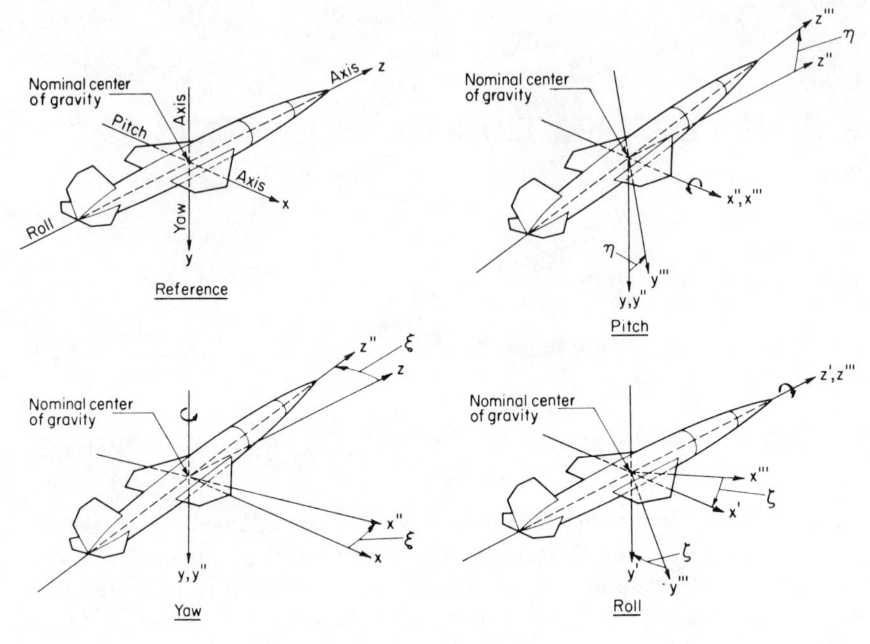

Fig. 11-1. Coordinate rotations.

motions relative to the reference system (x, y, z) can be described by three angles and associated maneuvers, as illustrated in Figure 11-1:

(1) *Yaw angle* ξ. Clockwise rotation viewed from below the body (about the y axis).

(2) *Pitch angle* η. Clockwise rotation viewed from the left side of the body (about the x'' axis).

(3) *Roll angle* ζ. Clockwise rotation viewed from the rear of the body (about the z''' axis).

The transformations between unit vectors in the primed and unprimed systems, resulting from successive yaw, pitch, and roll rotations, are

$$\begin{bmatrix} \hat{\mathbf{x}}' \\ \hat{\mathbf{y}}' \\ \hat{\mathbf{z}}' \end{bmatrix} = \begin{bmatrix} \cos\zeta\cos\xi - \sin\zeta\sin\eta\sin\xi & \sin\zeta\cos\eta & \cos\zeta\sin\xi + \sin\zeta\sin\eta\cos\xi \\ -\sin\zeta\cos\xi - \cos\zeta\sin\eta\sin\xi & \cos\zeta\cos\eta & -\sin\zeta\sin\xi + \cos\zeta\sin\eta\cos\xi \\ -\cos\eta\sin\xi & -\sin\eta & \cos\eta\cos\xi \end{bmatrix} \begin{bmatrix} \hat{\mathbf{x}} \\ \hat{\mathbf{y}} \\ \hat{\mathbf{z}} \end{bmatrix} \qquad (11.2\text{-}1)$$

and

$$
\begin{bmatrix} \hat{\mathbf{x}} \\ \hat{\mathbf{y}} \\ \hat{\mathbf{z}} \end{bmatrix} = \begin{bmatrix} \cos\zeta \cos\xi - \sin\zeta \sin\eta \sin\xi & -\sin\zeta \cos\xi - \cos\zeta \sin\eta \sin\xi & -\cos\eta \sin\xi \\ \sin\zeta \cos\eta & \cos\zeta \cos\eta & -\sin\eta \\ \cos\zeta \sin\xi + \sin\zeta \sin\eta \cos\xi & -\sin\zeta \sin\xi + \cos\zeta \sin\eta \cos\xi & \cos\eta \cos\xi \end{bmatrix} \begin{bmatrix} \hat{\mathbf{x}}' \\ \hat{\mathbf{y}}' \\ \hat{\mathbf{z}}' \end{bmatrix} \tag{11.2-2}
$$

Scattering data for all directions of illumination can be obtained by fixing either the pitch angle or the roll angle and by varying the remaining two angles. Equations describing the transformation of angles and of polarization for such procedures will be given below.

11.2.1. Roll-Angle System

Data are obtained with transmitting and receiving antennas on the z axis, and $0 \leqslant \xi \leqslant 180°$, $\eta \equiv 0°$, $0 \leqslant \zeta \leqslant 360°$. The incident field is

$$
\mathbf{E}^i = \hat{\mathbf{x}} E_x + \hat{\mathbf{y}} E_y \tag{11.2-3}
$$

The parallel component of the incident field is E_x, and the perpendicular component is E_y. The direction of illumination relative to the target coordinates is given by

$$
\theta' = \xi, \qquad \text{and} \qquad \phi' = -\zeta \tag{11.2-4}
$$

11.2.2. Pitch-Angle System

Data are obtained with transmitting and receiving antennas on the z axis, and $0 \leqslant \xi \leqslant 360°$, $-90° \leqslant \eta \leqslant 90°$, $\zeta \equiv 0°$. The incident field is

$$
\mathbf{E}^i = \hat{\mathbf{x}} E_x + \hat{\mathbf{y}} E_y \tag{11.2-5}
$$

$\hat{\boldsymbol{\phi}}$ is a unit vector perpendicular to the plane determined by $\hat{\mathbf{z}}'$ and $\hat{\mathbf{z}}$, or

$$
\hat{\boldsymbol{\phi}}' = \frac{\hat{\mathbf{z}}' \times \hat{\mathbf{z}}}{|\,\hat{\mathbf{z}}' \times \hat{\mathbf{z}}\,|} = \frac{-\sin\eta}{\sqrt{\sin^2\eta + \cos^2\eta \sin^2\xi}}\,\hat{\mathbf{x}} + \frac{\cos\eta \sin\xi}{\sqrt{\sin^2\eta + \cos^2\eta \sin^2\xi}}\,\hat{\mathbf{y}} \tag{11.2-6}
$$

$\hat{\boldsymbol{\theta}}'$ is a unit vector in the plane determined by $\hat{\mathbf{z}}'$ and $\hat{\mathbf{z}}$, and is perpendicular to $\hat{\mathbf{z}}$, or

$$
\hat{\boldsymbol{\theta}}' = -\frac{\hat{\mathbf{z}} \times \hat{\boldsymbol{\phi}}}{|\,\hat{\mathbf{z}} \times \hat{\boldsymbol{\phi}}\,|} = \frac{-\cos\eta \sin\xi}{\sqrt{\sin^2\eta + \cos^2\eta \sin^2\xi}}\,\hat{\mathbf{x}} + \frac{\sin\eta}{\sqrt{\sin^2\eta + \cos^2\eta \sin^2\xi}}\,\hat{\mathbf{y}} \tag{11.2-7}
$$

The parallel component of the incident field is

$$\hat{\boldsymbol{\theta}}' \cdot \mathbf{E}^i = E_x \frac{\cos\eta \sin\eta}{\sqrt{\sin^2\eta + \cos^2\eta \sin^2\xi}} + E_y \frac{\sin\eta}{\sqrt{\sin^2\eta + \cos^2\eta \sin^2\xi}} \tag{11.2-8}$$

The perpendicular component of the incident field is

$$\hat{\boldsymbol{\phi}}' \cdot \mathbf{E}^i = -E_x \frac{\sin\eta}{\sqrt{\sin^2\eta + \cos^2\eta \sin^2\xi}} + E_y \frac{\cos\eta \sin\xi}{\sqrt{\sin^2\eta + \cos^2\eta \sin^2\xi}} \tag{11.2-9}$$

The polarization angle (angle between the x axis and the projection of the z' axis onto the xy-plane) is

$$\Upsilon = \tan^{-1}(\tan\eta/\sin\xi) \tag{11.2-10}$$

The direction of illumination relative to the target is given by

$$\theta' = \cos^{-1}(\cos\eta\cos\xi) \tag{11.2-11}$$

and

$$\phi' = \tan^{-1}(\sin\eta/\tan\xi) \tag{11.2-12}$$

11.3. MEASUREMENT OF THE SCATTERING MATRIX

The radar range equation, as defined in Section 2.1.1, is

$$P_r = P_t G_t G_r \lambda_0^2 \sigma/(4\pi)^3 r_t^2 r_r^2 \tag{11.3-1}$$

where σ is the far-field scattering cross section of the target for the polarizations transmitted and received by a particular radar. The polarizations used in radar range measurements will, in general, be different from those needed for calculations. A set of parameters which describes the polarization properties of scatterers and permits calculation of cross section amplitudes for arbitrary sensor polarizations have been formalized by Sinclair[(1)] and by Kennaugh.[(2)] These scattering matrix parameters (see Section 2.1.4) are defined by a matrix equation relating the polarization of the incident electric field, scattering properties of the target, coordinate transformations between target-oriented and radar-oriented reference frames, and polarization of the measured scattered field, or

$$V_r = (\eta_0 I_t/4\lambda_0 r^2 \sqrt{\pi})\, h_r U^{-1} S U h_t \tag{11.3-2}$$

where V_r is the received voltage at the antenna terminals; I_t the current at the antenna terminals; η_0 the intrinsic impedance of the medium; λ_0 the wavelength; r the slant range to the target; $h_r = [h_{\theta_r}, h_{\phi_r}]$ the receiving antenna vector height; $S = \begin{bmatrix} a_{11} & a_{12} \\ a_{21} & a_{22} \end{bmatrix}$, the scattering matrix of the target;

$U = \begin{bmatrix} \cos\Upsilon & \sin\Upsilon \\ -\sin\Upsilon & \cos\Upsilon \end{bmatrix}$ the transformation matrix, with Υ the polarization angle; and $h_t = \begin{bmatrix} h_{\theta_t} \\ h_{\phi_t} \end{bmatrix}$ the transmitting antenna vector height.

The two polarizations most commonly used in reflectivity measurements are parallel (horizontal) polarization, for which the electric field is oriented in the plane of the model symmetry axis or longest dimension, and perpendicular (vertical) polarization, for which the electric field is oriented perpendicular to this plane.† The cross section is related to the matrix elements by

$$\sigma_{\parallel} = 4\pi r^2 \mid a_{11} \mid^2 \tag{11.3-3a}$$

$$\sigma_{\perp} = 4\pi r^2 \mid a_{22} \mid^2 \tag{11.3-3b}$$

The matrix elements are complex quantities containing both amplitude and phase terms[1] (e.g. $a_{11} = \mid a_{11} \mid e^{i\phi_{11}}$, etc.); consequently, S contains eight unknown quantities. For certain conditions, this number can be reduced.‡ If the same antenna is used for transmitting and receiving,[2]

$$a_{12} = a_{21} \tag{11.3-4}$$

If the target possesses a plane of symmetry, and the matrix elements are referenced to this plane (i.e., orthogonal polarizations are either parallel to this plane or perpendicular to it),[3]

$$a_{12} = 0 \tag{11.3-5}$$

In addition, the absolute phase of the return is unimportant for many applications, so that one of the elements can be considered real. Thus, for the monostatic case, three amplitudes and two relative phases are often sufficient to define S. For the symmetrical case described, two amplitudes and one relative phase are sufficient.

The most direct method of measuring the elements of the scattering matrix, and also the most difficult method to implement, is the absolute-phase method. A minimum number of transmissions are required; two orthogonal linearly polarized fields are transmitted in succession, and the amplitudes and absolute phases of both scattered fields are measured. The elements of a linear scattering matrix can be directly determined by means of the measurements listed in Table 11-1.

If the coherent scattered fields from two or more targets are to be determined by scattering matrix calculation, the matrix parameters must be

† The terms horizontal and vertical apply to these polarization states only when the model symmetry axis, or longest dimension, lies in a horizontal plane.

‡ For the remainder of this section, only monostatic measurements will be considered.

Table 11-1. Matrix Determination by Measurement of Amplitude and Absolute Phase

Transmitted Polarization	Received Polarization	Measured Parameters†
Parallel	Parallel	$\lvert a_{11} \rvert, \phi_{11}$
Parallel	Perpendicular	$\lvert a_{21} \rvert, \phi_{21}$
Perpendicular	Perpendicular	$\lvert a_{22} \rvert, \phi_{22}$
Perpendicular	Parallel	$\lvert a_{12} \rvert, \phi_{12}$

† $\phi_{ij} = \text{Arg } a_{ij}$ for $i, j = 1, 2$.

determined by the absolute-phase method. If only scattered powers are required, sufficient matrix information can be determined for the monostatic case by the relative-phase method, which requires simultaneous reception of two orthogonal polarizations and a measurement of the relative phase between these two polarizations. An example of the measurements required by the relative phase method is listed in Table 11-2.

The amplitudes and relative phases of the scattering matrix elements can also be determined by amplitude measurements alone. The errors encountered in phase measurements are eliminated; however, a larger number of measurements are required, and relative phase terms can only be obtained from algebraic expressions involving several measured amplitude terms. This can lead to a large cumulative error in the calculated relative phase, due to the measurement error in each amplitude term. For symmetrical bodies which possess diagonal scattering matrices, and for matrix solutions involving the more common transmitter–receiver polarization combinations, the algebraic expressions involving only amplitude terms are inherently no more inaccurate than those expressions involving amplitude and phase terms. However, for those bodies having matrices with cross-polarized terms which

Table 11-2. Matrix Determination by Measurement of Amplitude and Relative Phase—Monostatic Case

Transmitted Polarization	Received Polarizations	Measured Parameters	Calculated Parameters
$a_{12} = a_{21} \neq 0$			
Parallel	Parallel and perpendicular	$\lvert a_{11} \rvert, \lvert a_{21} \rvert, (\phi_{21} - \phi_{11})$	None
Perpendicular	Parallel and perpendicular	$\lvert a_{22} \rvert, \lvert a_{12} \rvert, (\phi_{12} - \phi_{22})$	$\phi_{22} - \phi_{11} = (\phi_{21} - \phi_{11}) - (\phi_{12} - \phi_{22})$
$a_{12} = a_{21} = 0$			
45° Linear	Parallel and perpendicular	$\lvert a_{11} \rvert, \lvert a_{22} \rvert, (\phi_{22} - \phi_{11})$	None

Table 11-3. Matrix Determination by Measurement of Amplitudes Only—General Target, Monostatic Case

Transmitted Polarization	Received Polarization	Measured Parameters	Calculated Parameters
Parallel	Parallel	$\mid a_{11} \mid$	—
Parallel	Perpendicular	$\mid a_{21} \mid$	$\mid a_{12} \mid$ †
Perpendicular	Perpendicular	$\mid a_{22} \mid$	—
Parallel	45° Linear	$\mid a_{45\parallel} \mid$	$\cos(\phi_{21} - \phi_{11})$ ‡
Parallel	Right circular	$\mid a_{RC\parallel} \mid$	$\sin(\phi_{21} - \phi_{11})$ #, $(\phi_{12} - \phi_{11})$ §
Perpendicular	45° Linear	$\mid a_{45\perp} \mid$	$\cos(\phi_{22} - \phi_{12})$ §§
Perpendicular	Right circular	$\mid a_{RC\perp} \mid$	$\sin(\phi_{22} - \phi_{12})$ ††, $(\phi_{22} - \phi_{11})$ ‡‡

† $\mid a_{12} \mid = \mid a_{21} \mid$.

‡ $$\cos(\phi_{21} - \phi_{11}) = \frac{2 \mid a_{45\parallel} \mid^2 - \mid a_{11} \mid^2 - \mid a_{21} \mid^2}{2 \mid a_{11} \mid \mid a_{21} \mid}.$$

$$\sin(\phi_{21} - \phi_{11}) = \frac{2 \mid a_{RC\parallel} \mid^2 - \mid a_{11} \mid^2 - \mid a_{21} \mid^2}{2 \mid a_{11} \mid \mid a_{21} \mid}.$$

§ $(\phi_{12} - \phi_{11}) = (\phi_{21} - \phi_{11})$.

§§ $$\cos(\phi_{22} - \phi_{12}) = \frac{2 \mid a_{45\parallel} \mid^2 - \mid a_{22} \mid^2 - \mid a_{12} \mid^2}{2 \mid a_{22} \mid \mid a_{12} \mid}.$$

†† $$\sin(\phi_{22} - \phi_{12}) = \frac{2 \mid a_{RC\parallel} \mid^2 - \mid a_{22} \mid^2 - \mid a_{12} \mid^2}{2 \mid a_{22} \mid \mid a_{12} \mid}.$$

‡‡ $(\phi_{22} - \phi_{11}) = (\phi_{22} - \phi_{12}) + (\phi_{21} - \phi_{11})$.

cannot be neglected, the amplitude-only expressions are sufficiently complex that high cumulative error results, unless the experimental error is very small. Procedures for scattering-matrix determination using amplitude measurements only are contained in Table 11-3 for the general target and in Table 11-4 for a symmetrical target.

Table 11-4. Matrix Determination by Measurement of Amplitudes Only—Symmetrical Target, Monostatic Case

Transmitted Polarization	Received Polarization	Measured Parameters	Calculated Parameters
Parallel	Parallel	$\mid a_{11} \mid$	—
Perpendicular	Perpendicular	$\mid a_{22} \mid$	—
45° Linear	45° Linear	$\mid a_{45,45} \mid$	$\cos(\phi_{22} - \phi_{11})$ †
Right circular	45° Linear	$\mid a_{45RC} \mid$	$\sin(\phi_{22} - \phi_{11})$ ‡

† $$\cos(\phi_{22} - \phi_{11}) = \frac{4 \mid a_{45,45} \mid^2 - \mid a_{11} \mid^2 - \mid a_{22} \mid^2}{2 \mid a_{11} \mid \mid a_{22} \mid}.$$

‡ $$\sin(\phi_{22} - \phi_{11}) = \frac{4 \mid a_{45RC} \mid^2 - \mid a_{11} \mid^2 - \mid a_{22} \mid^2}{2 \mid a_{11} \mid \mid a_{22} \mid}.$$

11.4. ELECTROMAGNETIC MODELS

The theory for modeling electromagnetic systems has been stated by Stratton[4] and Sinclair.[5] Table 11-5 is a tabulation of parameters and scale factors for an absolute model system and the full-scale system which is represented.[5] The four fundamental scale factors are: p, the scale factor for length; γ, the scale factor for time; α, the scale factor for electric field; β, the scale factor for magnetic field. It is theoretically possible to model an electromagnetic system using these four fundamental scale factors, to make absolute measurements of parameters in the model system, and to relate the measured quantities to the corresponding parameters in the full scale system. In practice, however, scale factors are limited by the values of conductivity, permittivity, and permeability obtainable in model materials. For most measurements, the permeability and permittivity of the model

Table 11-5. Conditions for an Absolute Model

Name of Quantity	Full-Scale System	Model System
Length	l	$l' = l/p$
Time	t	$t' = t/\gamma$
Electric field	$\mathbf{E}$	$\mathbf{E}' = \mathbf{E}/\alpha$
Magnetic field	$\mathbf{H}$	$\mathbf{H}' = \mathbf{H}/\beta$
Conductivity	σ	$\sigma' = (p\alpha/\beta)\sigma$
Inductive capacity	ϵ	$\epsilon' = (p\alpha/\beta\gamma)\epsilon$
Permeability	μ	$\mu' = (p\beta/\alpha\gamma)\mu$
Voltage	V	$V' = V/\alpha p$
Current density	$\mathbf{J}$	$\mathbf{J}' = p\mathbf{J}/\beta$
Current	$\mathbf{I}$	$\mathbf{I}' = \mathbf{I}/\beta p$
Power per unit area	W	$W' = W/\alpha\beta$
Total power	P	$P' = P/\alpha\beta p^2$
Charge density	ρ	$\rho' = (p^2/\beta\gamma)\rho$
Charge	Q	$Q' = (1/\beta\gamma p)Q$
Frequency	f	$f' = \gamma f$
Angular frequency	ω	$\omega' = \gamma\omega$
Wavelength	λ	$\lambda' = \lambda/p$
Phase velocity	ν	$\nu' = \gamma\nu/p$
Propagation constant	$\mathbf{k}$	$\mathbf{k}' = p\mathbf{k}$
Resistance	R	$R' = (\beta/\alpha)R$
Reactance	X	$X' = (\beta/\alpha)X$
Impedance	Z	$Z' = (\beta/\alpha)Z$
Electric flux density	$\mathbf{D}$	$\mathbf{D}' = (p/\beta\gamma)\mathbf{D}$
Capacitance	C	$C' = (\alpha/\gamma\beta)C$
Inductance	L	$L' = (\beta/\alpha\gamma)L$
Magnetic flux density	$\mathbf{B}$	$\mathbf{B}' = (p/\alpha\gamma)\mathbf{B}$

Table 11-6. Conditions for a Practical Geometrical Model

Name of Quantity	Full-Scale System	Model System†
Length	l	$l' = l/p$
Time	t	$t' = t/p$
Conductivity	σ	$\sigma' = p\sigma$
Inductive capacity	ϵ	$\epsilon' = \epsilon$
Permeability	μ	$\mu' = \mu$
Frequency	f	$f' = pf$
Wavelength	λ	$\lambda' = \lambda/p$
Phase velocity	v	$v' = v$
Propagation constant	k	$k' = pk$
Resistance	R	$R' = R$
Reactance	X	$X' = X$
Impedance	Z	$Z' = Z$
Capacitance	C	$C' = C/p$
Inductance	L	$L' = L/p$
Radar cross section	σ	$\sigma' = \sigma/p^2$
Antenna gain	g	$g' = g$

† p is the ratio of any full-scale length to the corresponding model length.

medium will necessarily be those of the full-scale system, so that $p\alpha/\beta\gamma = 1$ and $p\beta/\alpha\gamma = 1$. For these conditions, $\alpha = \beta$ and $p = \gamma$. Consequently, the conductivity in the model system must be p times the conductivity in the full-scale system.† Only two scale factors can be chosen, p and either α or β, to model a system absolutely. If only p is assigned a value, the absolute powers in the two systems cannot be measured, and the model system is called a geometrical model. Relative power can be measured in the model system and converted to absolute value by calibration procedures. The relationships for a geometrical model system are summarized in Table 11-6.[5]

11.5. MEASUREMENT METHODS

Almost all schemes for measuring radar cross section involve three basic elements: (1) a source or radiator of electromagnetic energy, (2) an obstacle or scatterer of energy, and (3) a receiving antenna or probe which measures the properties of an electromagnetic field at points in space. The electromagnetic field resulting from source currents in the absence of the scatterer is termed the incident field. The electromagnetic field resulting

† It is generally difficult to realize permeabilities at scale frequencies equal to full-scale values, as well as to scale the conductivity properly, although nonexact modeling of conductivity is often of little consequence.

from the same source currents but with the scatterer present is termed the total field. The vector difference at each point between the total field and the incident field is termed the scattered field.

Radar cross section is defined in the radar equation for large distances between radar and target. Consequently, the incident field is a spherical wave approximately plane over the target aperture, and the scattered field is a spherical wave approximately plane over the aperture of the receiving antenna. These conditions cannot be met in a laboratory measurement, but can be approximated. Errors due to imperfect illumination are discussed in Section 11.6.1.

Cross section is most often determined by comparison with the energy scattered by a calibration target with known cross section. The scattered power measured for the unknown target is, from Eq. (11.3-1),

$$P_r = P_t G_t G_r \lambda_0^2 \sigma_{\text{tar}} / (4\pi)^3 r_t^2 r_r^2 \tag{11.5-1}$$

where P_r is the received power; P_t the transmitted power; G_t the gain of the transmitting antenna; G_r the gain of the receiving antenna; λ_0 the wavelength; σ_{tar} the target cross section; r_t the distance between target and transmitting antennas; and r_r the distance between target and receiving antennas. If a calibration target is substituted for the unknown target with transmitted power and geometry unchanged, the ratio of the two received powers is

$$P_{\text{tar}}/P_{\text{cal}} = \sigma_{\text{tar}}/\sigma_{\text{cal}} \tag{11.5-2}$$

The value of σ_{cal} is generally obtained from theoretical solutions; a metal sphere is most often used (see Figure 3-3).

Measurement methods can be broadly classified according to how the transmitted and received fields are separated. The incident field and scattered field may be separated by a difference in propagation direction (magic-T, two-antenna, SWR systems), by a frequency difference (Doppler and FM systems), or by a time delay (pulsed systems).

11.5.1. Differences in Propagation Direction

11.5.1.1. *Hybrid Junction Cancellation*

The CW cancellation method requires a single antenna hybrid junction (magic-T) system. A magic-T is a waveguide junction which can be represented as a four-terminal network. If the four terminals are terminated in matched or balanced impedances, a signal entering any terminal will divide equally between the two adjacent terminals and be effectively isolated from the fourth. A typical arrangement is shown in Figure 11-2.

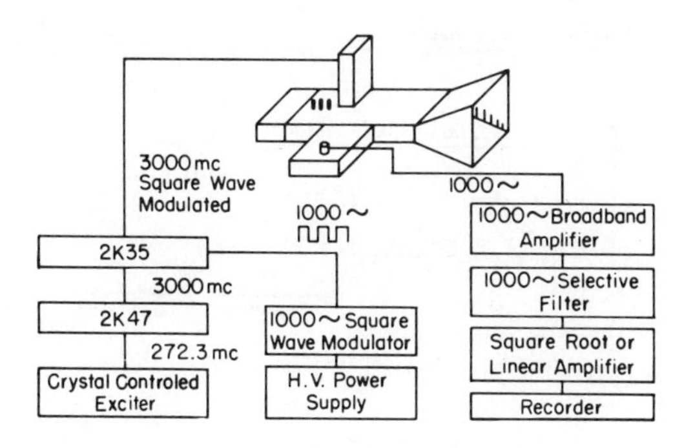

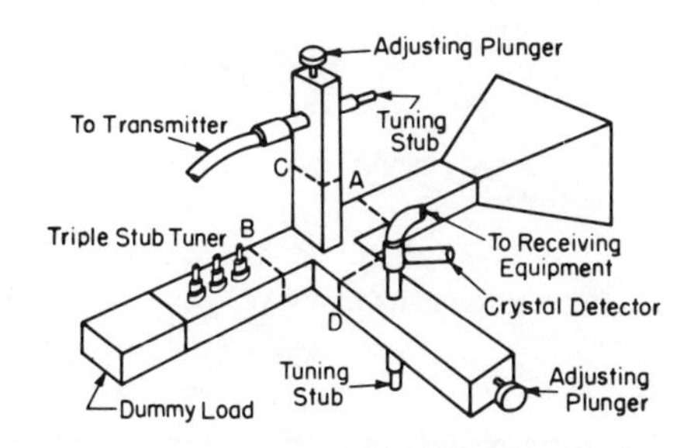

Fig. 11-2. Magic-T method [after Kouyoumjian[6]].

A transmitter is connected to arm C, a matched load to arm B, a receiver to arm D, and an antenna to arm A. Under balanced conditions, a continuous wave originating at the transmitter will divide equally between the transmitting antenna and the matched load. A received signal (which will be much smaller in amplitude than the transmitted signal) appearing at the antenna terminals will divide between the receiver and transmitter. Utilization of only half of the transmitted and received powers is not a problem, because target ranges are small compared to those of a field radar. The matched load can be adjusted to a small degree of mismatch, resulting in reflection of a small amount of transmitter signal back into the receiver arm. This signal can be adjusted in amplitude and phase to effectively cancel out any received signal. If this nulling is accomplished in the absence of a scatterer, the background return is effectively eliminated at the receiver. Upon introduction of a scatterer, the receiver output is then a measure of the target cross section. For this to be an accurate measurement, frequency stability of the source must be accurately maintained, and the introduction of the scatterer must not substantially change the magnitude and phase of the background return. Isolation between transmitter and receiver must be of the order of 100 db[6] for accurate measurement of target nulls as shown

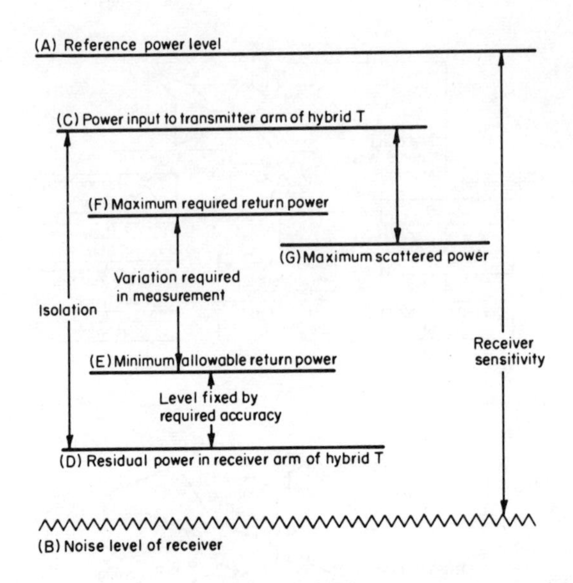

Fig. 11-3. Cancellation requirements in a CW system [after Hines and Tice[7]].

in Figure 11-3. The required isolation is the difference in power level between the power input to the transmitter arm of the magic-T and the residual power in the receiver arm after the background is cancelled. This is determined by the difference between transmitter power and maximum scattered power; by the difference between maximum scattered power and minimum scattered power (at nulls of the scattering pattern); and by the difference between the minimum scattered power and residual uncancelled power, the last difference being a function of required measurement accuracy. Figure 11-4 shows the variation in isolation with changing transmitter frequency for the system of Figure 11-2. It is seen that the transmitter must be highly monochromatic and maintain frequency stability for the duration of the measurement.

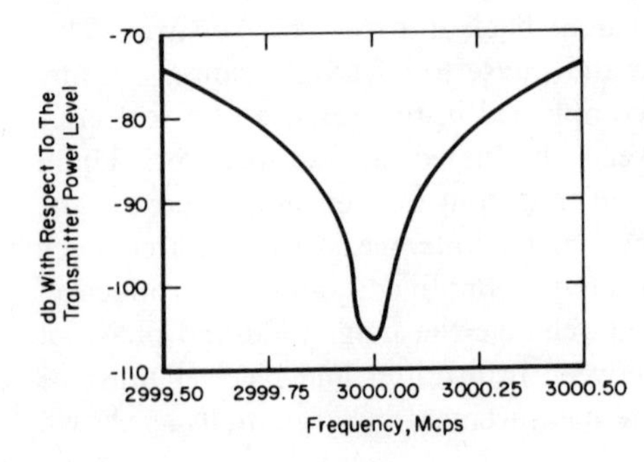

Fig. 11-4. Variation in isolation with changing frequency [after Kouyoumjian[6]].

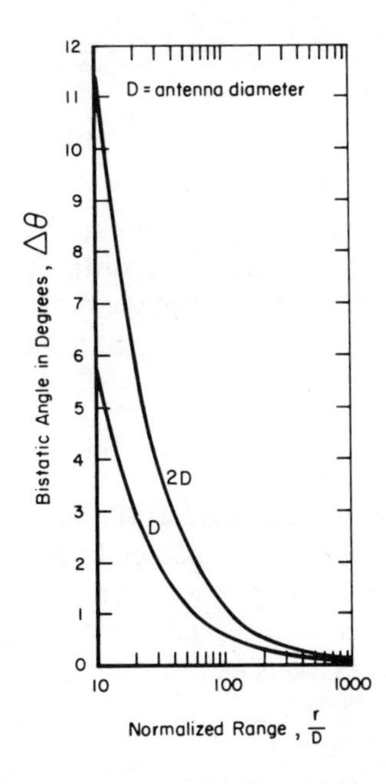

Fig. 11-5. Bistatic angle as a function of range and antenna separation for separate transmitting and receiving antennas [from Kouyoumjian[6]].

11.5.1.2. *Two-Antenna Method*

The need for a magic-T waveguide junction can be eliminated by using separate antennas for transmitting and receiving; bistatic and quasimonostatic measurements can be made with a two-antenna system. Figure 11-5 shows the magnitude of the bistatic angle as a function of range when the bore sights of transmitting and receiving antennas are separated by one and by two antenna diameters. For quasimonostatic measurements, the bistatic angle should be kept smaller than the smallest increment in aspect angle which gives significant variation in cross section. In many cases, this requires greater range separation between scatterer and antennas than is needed to achieve approximate plane-wave illumination (see Section 11.6.1). An important consideration with a two-antenna system is the isolation between transmitting and receiving antenna channels, especially when making cross-polarized measurements.

11.5.1.3. *SWR Method*

By using field probes as illustrated in Figure 11-6, direct measurements of the scattered and incident fields can be made, eliminating the need for calibration targets.[8] Small monopoles are usually used with an image plane which shields the probe transmission line and other necessary equipment from direct illumination and eliminates the need for model supports.

The monopole is moved along a line between the source and the target to measure the amplitudes of standing waves caused by interference between the incident field and scattered field. The reflection coefficient of the target is

$$b = lw_2 \frac{[\rho/(l - w_2)] - [1/(l - w_1)]}{\rho + (w_2/w_1)} \tag{11.5-3}$$

where w_2 is the distance from the scatterer to a null in the total field; w_1 the distance from the scatterer to a maximum in the total field; l the distance from source to scatterer; ρ the measured voltage standing-wave-ratio, $\rho = (E_i + E_s)/(E_i - E_s)$, with E_i the incident field and E_s the scattered field.

The backscattering cross section is given by

$$\sigma = 4\pi b^2 \tag{11.5-4}$$

Use of an image plane restricts the types of targets to those with planar symmetry, and the field polarization to a linear polarization perpendicular to the image plane. A single image plane simulates a free-space geometry, consisting of the geometry above the plane together with the mirror image of this geometry. Scattering from infinite cylindrical targets can be simulated

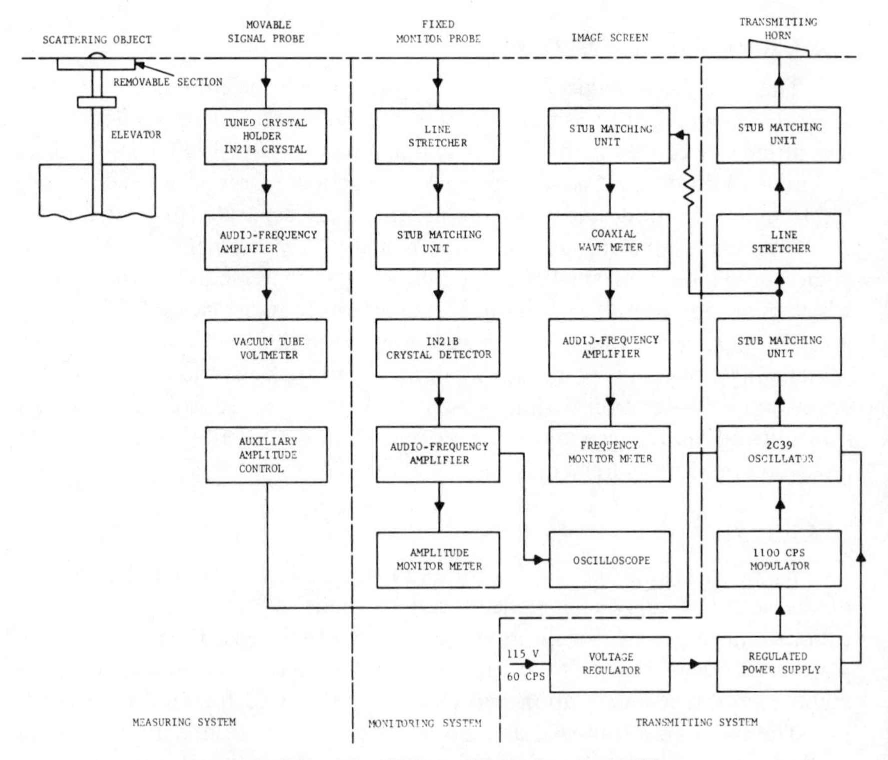

Fig. 11-6. The SWR method [after Iizrika[8]].

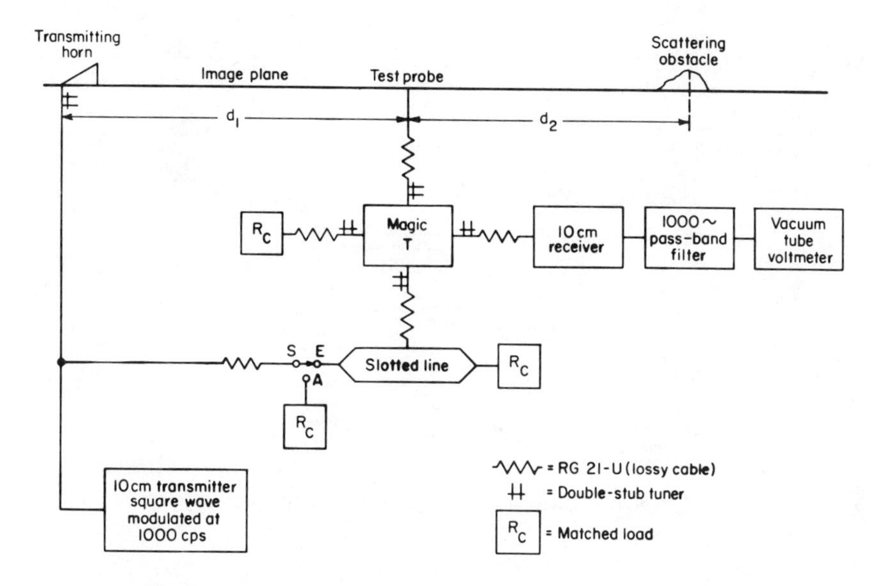

Fig. 11-7. The fixed-probe cancellation method [after Iizrika[8]].

for polarization parallel to the cylinder axis by using two parallel image planes.[9]

SWR determinations assume: (1) The probe is in the far field of both source and obstacle, so that energy propagates approximately as a plane wave. (2) The incident field polarization and scattered field polarization are parallel so that interference will occur. (3) The probe does not perturb the field in the vicinity of either the source or the obstacle.

11.5.1.4. *Fixed-Probe Cancellation Method*

The fixed-probe cancellation method, illustrated in Figure 11-7, can be used in the same geometries as the SWR method; however, the probe is fixed, and cancellation techniques are used to null the probe output voltage when the obstacle is not present.[8] When the obstacle is added, the probe voltage is proportional to the scattered field.

11.5.2. Modulation Methods (Frequency Differences)

11.5.2.1. *Doppler Method*

Separation of scattered field and incident field may also be accomplished by modulating either field at the transmitter or at the obstacle. If the obstacle is given a component of velocity along the direction of illumination, the scattered field will be shifted in frequency and may be separated on this basis. Figure 11-8 shows a Doppler system used with a "free-space" range, and Figure 11-9 shows a similar technique applied to a parallel-plane geometry.[8,9] In both cases, the obstacle and calibrating object are mounted on a rotating disc.

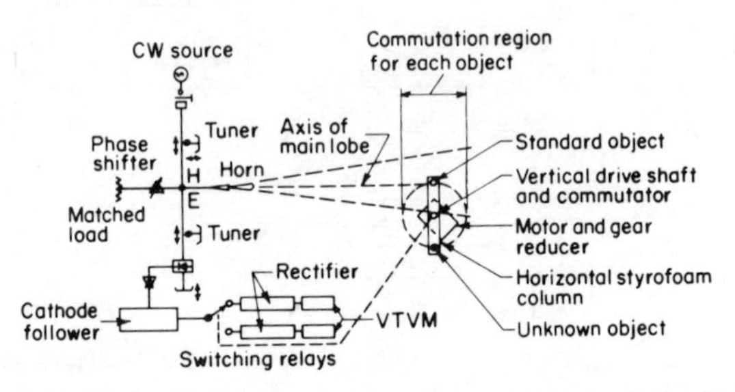

Fig. 11-8. A free-space doppler-measurement system [after Tang[9]].

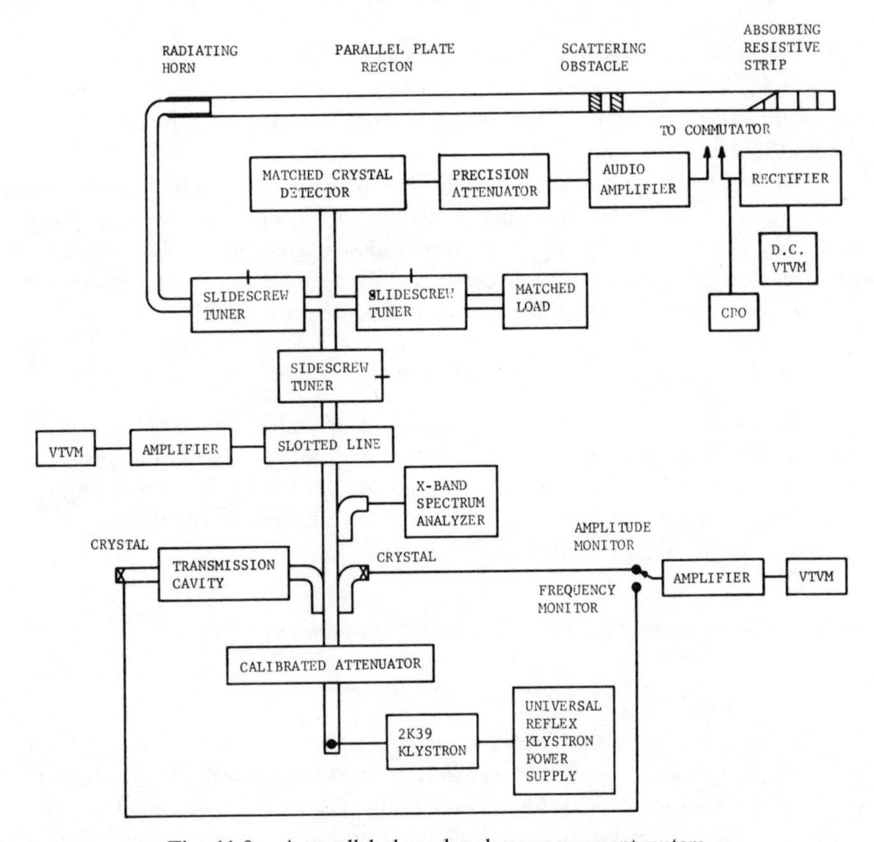

Fig. 11-9. A parallel-plane doppler-measurement system.

11.5.2.2. FM-CW *Method*

Separation of the scattered field from the incident and undesired background fields can also be accomplished by frequency modulating the incident field. An incident field of frequency f_0 at time t_0 will, after reflection from the scatterer, return to the receiver at time $t = t_0 + (2r/c)$, where r is the distance between the transmitter and the scatterer and c is the velocity of light. If the transmitted frequency is modulated at a rate of df/dt, then the frequency of the received signal will be

$$f = f_0 + (2r/c)\, df/dt \tag{11.5-5}$$

If this is mixed with a sample of the transmitted signal, then the difference frequency of $(2r/c)\, df/dt$ can be recovered, and the amplitude of this difference-frequency signal will be proportional to the cross section of the scatterer.

Discrimination against background echoes from sources at ranges other than that of the target can be accomplished by filtering so that only the difference frequency corresponding to the target range is retained. This procedure is made simpler if linear frequency modulation is used, thus giving a difference frequency which is constant with time. Linear, sawtooth frequency modulation is most often used. If the carrier frequency is swept linearly through a maximum deviation Δf in t' seconds, and repeats this $f_m = 1/t'$ times per second, then the difference frequency from a target at range r will be

$$f_d = (2r/c) f_m\, \Delta f \tag{11.5-6}$$

Since the carrier frequency is not constant, the cross section measured will not be a true CW cross section but some type of average over a portion of the deviation frequency. In order to approximate the CW cross section as closely as possible, the modulation rate should be kept low in order that the frequency change during the time required for the wave to pass over the scatterer is small.

Unfortunately, a spectral analysis of the difference frequency generated by such a system shows that energy occurs not at a single frequency f_d, but at f_d and all harmonics of f_d, and has a finite spread around these points. Thus complete rejection of undesired background signals is not possible. Particular difficulties may be encountered with large background sources close to the transmitter, which can produce significant energy in the spectral region where the desired scattered signal is located.

Similarly, any amplitude modulation of the transmitted signal can produce energy in that portion of the spectrum where the desired signal lies. Thus any residual amplitude modulation of the transmitted signal must be minimized.

11.5.3. Pulse Systems (Time Differences)

The scattered fields can be separated from the incident and background fields on the basis of time differences. This is accomplished by using a pulse system to measure the target cross section. A pulse measuring system is essentially a simplified radar adapted for measurements at short ranges. Such systems usually use pulse widths ranging from 0.1 to 1 μsec, and repetition rates of 500–25,000 cps.

In order that the cross section measured by a pulse system closely approximate the plane-wave cross section as measured by a CW system, the spatial extent of the pulse must be considerably larger than the maximum extent of the target.

11.6. MEASUREMENT ERRORS

11.6.1. Effects of Nonplanar Illumination

Several conflicting criteria concerning the required distance between target and illuminating antenna during reflectivity measurements have been developed.[10] The definition of radar cross section in Section 2.1.1 assumes that the incident field at the scatterer is a plane wave, and that the scattered field is determined at points sufficiently removed from the scatterer so that the scattered field varies as $1/r$, where r is the radial distance from the scatterer. Consequently, cross-section measurements can be made, even in the near field of a transmitting antenna, provided (1) the incident field approximates a plane wave over the volume of the scatterer sufficiently well, and (2) the scattered field is measured at distances far enough from the scatterer so that the scattered field has a $1/r$ dependence. If one considers the scatterer to be a radiating antenna, containing currents induced by the incident field which produce the scattered field, the second condition is equivalent to requiring the receiving antenna to be in the far zone of the scatterer. This would require a spacing of at least L^2/λ_0, where L is the maximum dimension of the target as projected in a plane perpendicular to the direction of observation.

In practice, the illuminating field will have a spherical wave front rather than a plane one. The various distance criteria are derived from limits on the phase and amplitude variations of the field within a volume of space containing the scatterer. If all aspects are to be seen, this volume is taken as a cube as large as the longest body dimension of the scatterer. The target aperture is defined as a square area transverse to the incident radiation with a side as long as the longest body dimension of the scatterer.

The field variations which have been used to develop distance criteria are: (1) the transverse phase variation, (2) the transverse amplitude variation, and (3) the longitudinal amplitude variation. A criterion applicable to all scatterers and all types of source radiators is not available.

Those criteria which have been discussed are each based on one type of field variation and an idealized source. Considering a point source and using the transverse variation of phase as the limiting parameter, the minimum separation r_m between source and scatterer is[10]

$$r_m = \pi L^2/4\phi\lambda_0 \tag{11.6-1}$$

where L is the longest dimension of the scatterer, λ_0 is the wavelength, and ϕ is defined by

$$\frac{E_i(L/2)}{E_i(0)} = Ae^{-i\phi} \tag{11.6-2}$$

where $E_i(L/2)$ equals the incident field amplitude at the edge of the target aperture, and $E_i(0)$ equals the incident field amplitude at the center of the target aperture. A value of ϕ commonly used is $(\pi/8)$ radians $(22\frac{1}{2}°)$, for which

$$r_m = 2L^2/\lambda_0 \tag{11.6-3}$$

Consideration of an antenna aperture other than a point source has resulted in several conflicting distance criteria, ranging in magnitude from aL^2/λ_0 to $b(L + l)^2/\lambda_0$, where a and b are constants and l is the antenna aperture dimension.[10] For scatterers many wavelengths long, these formulas yield distances too large to be obtained on a reflectivity range. Reduced distances must be used, depending on the measurement accuracy required at aspects where scattering nulls occur.

The transverse variations of field amplitude and phase over the target aperture must in general be determined by probing the field of the illuminating antenna. Figure 11-10 gives minimum range requirements in terms of transverse field variations of amplitude, and phase as calculated by Kouyoumjian and Peters[10] for an idealized source. The source used for this calculation was a square aperture with constant amplitude distribution. The dashed lines are lines of constant field amplitude variation at the extremities of the target aperture. The solid lines are lines of constant phase variation. The minimum range is determined by the phase variation for antenna apertures smaller than target apertures. For antenna apertures larger than target apertures, the amplitude variation dominates. Figure 11-11 is a plot of the minimum range versus the longitudinal amplitude dependence of the incident field. The amount of longitudinal field variation must in general be checked, particularly for axial illumination of very long scatterers.

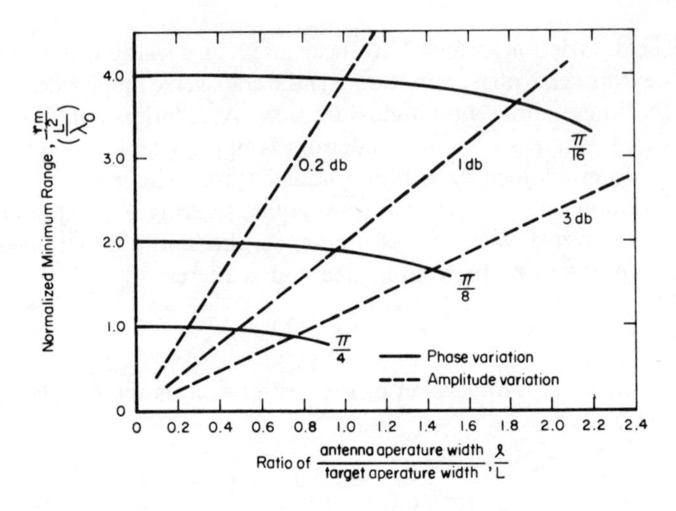

Fig. 11-10. Minimum range as a function of transverse amplitude and phase variations across the target aperture [after Kouyoumjian and Peters[10]].

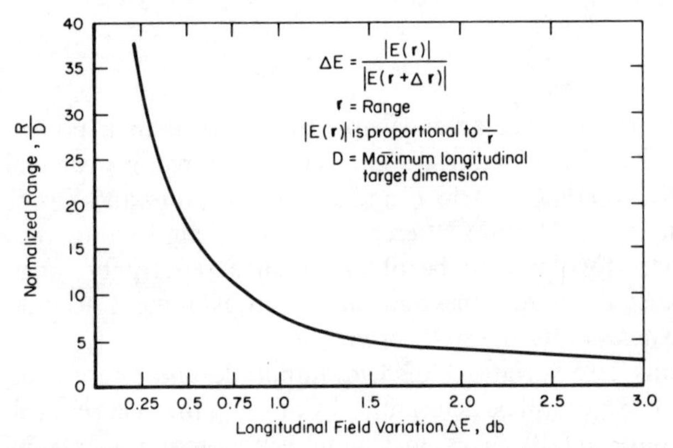

Fig. 11-11. Minimum range as a function of longitudinal $1/r$ field variation [after Kouyoumjian and Peters[10]].

11.6.2. Use of Microwave Lenses

Microwave lenses are sometimes used to reduce the transverse phase variation of the incident field across the target aperture. Dielectric lenses used for reflectivity measurements are usually designed with a hyperbolic contour surface, as shown in Figure 11-12.[11–13] The equation of the hyperbola is given by

$$r(\theta) = (m - 1)(r_0 - T)/[(m \cos \theta) - 1] \tag{11.6-4}$$

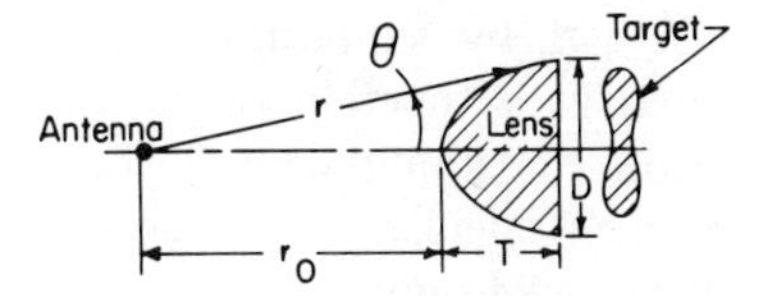

Fig. 11-12. Dielectric lens geometry.

where m is the index of refraction of the dielectric. Assuming the curved portion of the lense is illuminated by a spherical wave, only energy between the asymptotes can be focused. This angular region is given by

$$\theta \leqslant \cos^{-1}(1/m) \tag{11.6-5}$$

The maximum angular extent of the lens is

$$\theta_0 = \tan^{-1}(D/2R_0) \tag{11.6-6}$$

The lens thickness can be expressed by

$$\frac{T}{r_0 - T} = \frac{1 - (r_0/D)[(r_0/D)^2 + 0.25]^{-1/2}}{m(r_0/D)[(r_0/D)^2 + 0.25]^{-1/2} - 1} \tag{11.6-7}$$

Two dielectric materials which have been used for microwave lenses are styrofoam ($m \approx 1.015$)[11] and polystyrene ($m \approx 1.58$).[12]

11.6.3. Background Errors

Together with target supports, background is probably the most serious source of error in cross-section measurements. The background cross section is defined to be the cross-section measured in the absence of the scatterer. This signal includes energy scattered from other objects in the vicinity of the measurement range, i.e., anechoic chamber walls and target support devices, or trees, buildings, etc., in the vicinity of an outdoor range. In the case of an image-plane range, the principal contributions are due to reflections from the edges of the image plane. The background field is the vector sum of a number of fields resulting from interactions between scatterers in the vicinity of the measurement. If the presence of a target does not greatly change the background fields, their effects may be removed from the measurement by analog cancellation or by digital vector-field subtraction. Analog cancellation involves the introduction into the measuring device of an RF field or voltage of proper amplitude and phase to cancel the background signal. This is usually done with a magic-T or hybrid-T waveguide junction.

If analog-to-digital conversion and phase measurement capabilities are available, the background amplitude and phase with the supporting device present and the target absent can be measured as a function of rotator aspect angle, and can be digitally subtracted from the amplitude and phase measured with the target present. Samples of such calculations are shown in Figures 11-13 and 11-14. However, if the background signal has a substantial magnitude, this probably indicates a large amount of indirect illumination in the vicinity of the support device. A target with a substantial physical aperture and/or large forward-scattering characteristics will change the background sufficiently to introduce errors into the cancellation or subtraction measurement. Insertion of the target after balance results in loss of balance and a nonzero background contribution. In this event, the only means by which these errors can be reduced is to reduce the magnitude of the undesired fields scattered from background objects.

Considering the cumulative effect of the background to be represented by a cross-section amplitude, the total measured cross section is given by

$$\sigma = \sigma_B + \sigma_T + 2\sqrt{\sigma_B \sigma_T} \cos \psi \tag{11.6-8}$$

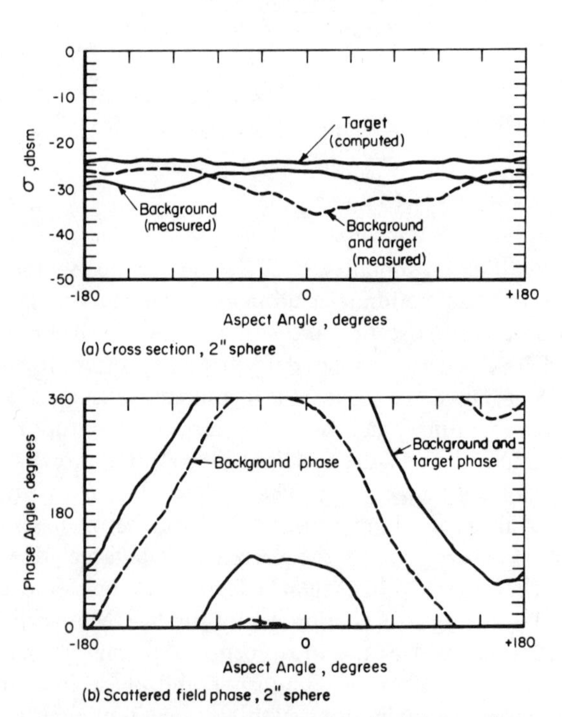

Fig. 11-13. Vector background subtraction—sphere. (a) Cross section, 2-inch sphere. (b) Scattered field phase, 2-inch sphere [after Falk[14]].

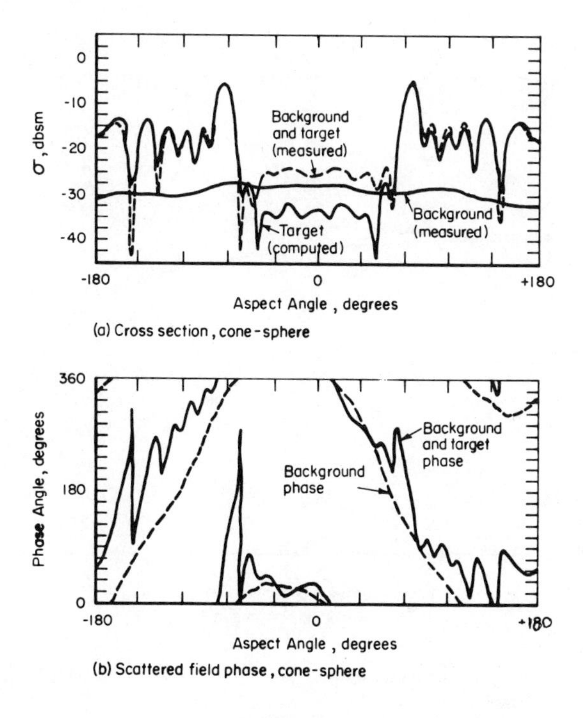

Fig. 11-14. Vector background subtraction—cone-sphere. (a) Cross section, cone-sphere. (b) Scattered field phase, cone-sphere [after Falk[14]].

where σ_B is the background cross section (cross section in the absence of the target), σ_T is the target cross section (cross section of the target when illuminated by both incident and background fields), and σ is the measured cross section. Equation (11.6-8) assumes the receiver noise voltages to be negligible compared to the voltage due to background return. Figure 11-15 is a plot of the upper and lower limits of the ratio σ/σ_T:

$$\sigma/\sigma_T = 1 + (\sigma_B/\sigma_T) \pm 2\sqrt{\sigma_B/\sigma_T} \tag{11.6-9}$$

Similarly, the limits of phase-measurement error due to background effects, plotted in Figure 11-16, is given by

$$\phi - \phi_T = \pm \tan^{-1}\sqrt{\sigma_B/\sigma_T} = \alpha \tag{11.6-10}$$

11.6.4. Target-Support Errors

Structures used to support targets for measurement purposes are usually cellular plastic columns or dielectric suspension lines. Factors which determine the choice of supporting system include target weight, target cross

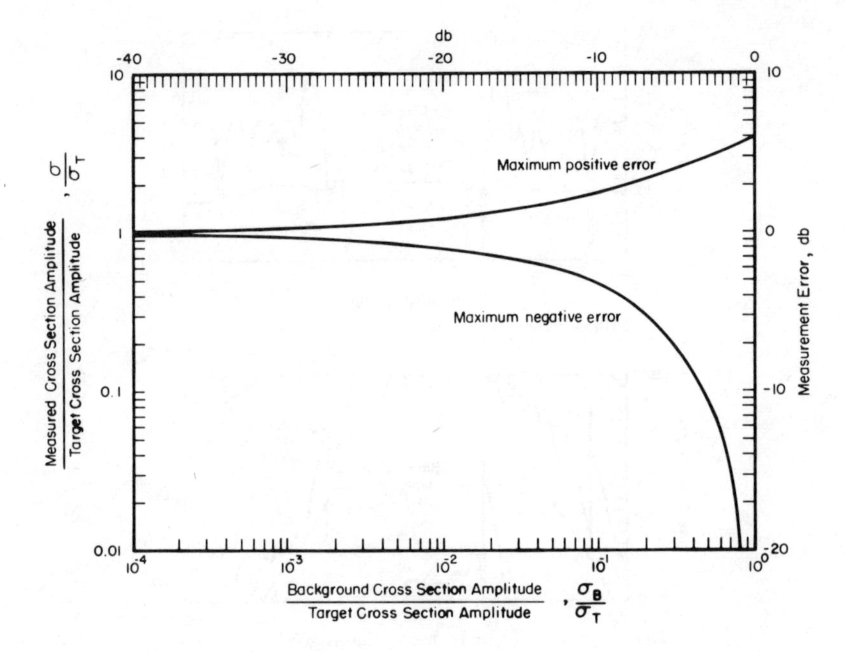

Fig. 11-15. Amplitude-measurement error as a function of background amplitude.

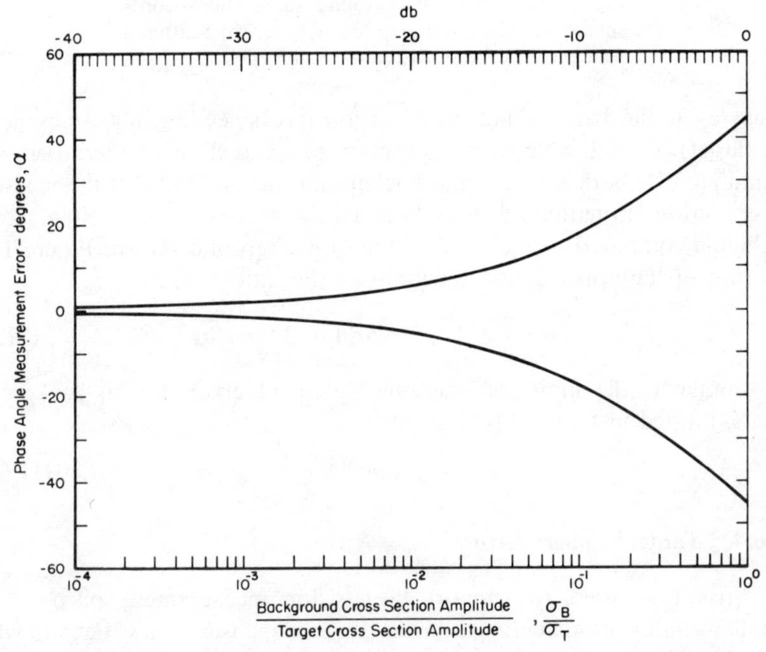

Fig. 11-16. Phase-measurement error as a function of background amplitude.

section, required cross-section measurement accuracy, frequency, required control of target aspect, range geometry, and target-support interaction.

11.6.4.1. *Plastic Columns*

Three factors determine the size of a plastic column: (1) problems of fabrication place an upper limit on diameter and length; (2) the weight of the target places a requirement on the number and size of columns; (3) the column diameter can be "tuned" and "shaped" to present a minimum backscattering cross section.

An estimate of the required column diameter needed to support the weight of a target is given by[15–17]

$$D = \left[\frac{256WL^2}{\pi^3 E}\right]^{1/4} \qquad (11.6\text{-}11)$$

where D is the column diameter, in inches; W the weight of the target, in pounds; L the column length, in inches; E the modulus of elasticity, in psi. This equation assumes failure by buckling with one end fixed. A safety factor of two is commonly used with Eq. (11.6-11); for very heavy targets, multiple columns are used. Table 11-7 gives moduli of elasticity and other parameters for several typical foamed plastic materials.[16]

Phase-measurement error due to column deflection caused by wind can be estimated from[18]

$$Y_m = (F + L^3)/8EI \qquad (11.6\text{-}12)$$

where Y_m is the maximum column deflection; $F = 2PRL$; $P = 0.00256V^2$; F is the force acting uniformly along the column length; P the wind pressure, in lb/foot^2; R the column radius, in feet; L the column length, in feet; V the wind velocity, in mph; and $I = \pi R^4/4$, the moment of inertia of a circular column.

The broadside cross section of a long, uniform, homogeneous, lossless dielectric column is[17,19]

$$\sigma = \pi(\epsilon_r - 1)^2 (k_0 D)^2 L^2 J_1^2(k_0 D)/16 \qquad (11.6\text{-}13)$$

where ϵ_r is the relative dielectric constant of the column; $k_0 = 2\pi/\lambda_0$, with λ_0 the free-space wavelength; D the column diameter; and L the column length. The cross-section pattern contains an interference pattern with nulls occurring approximately at

$$D/\lambda_0 = 0.61 + 0.5N, \qquad N = 0, 1, 2, 3,... \qquad (11.6\text{-}14)$$

Table 11-7. Properties of Foamed Plastics†

		Base Plastic			Foamed Plastic		
Material	Manufacturer	Type	Density, ρ (lb/ft^3)	Dielectric Constant, ϵ‡	Density, ρ	Elastic Modulus, E (lb/inch^2)	Cell Diameter $2a$ (inches)
Tyrifoam	Dow Chemical Co. Midland, Mich.	Styrene-acrylonitrile	66.8	2.56	0.70	207	0.461
Pelaspan	Dow Chemical Co.						
Eccofoam	Emerson & Cuming Canton, Mass.	Polystyrene	66.5	2.55	1.15	733	0.125
Styrofoam FB	Dow Chemical Co.	Polystyrene	66.5	2.55	1.76	2061	0.025
Styrofoam DB	Dow Chemical Co.	Polystyrene	66.5	2.55	1.80	1692	0.057
Styrofoam FR	Dow Chemical Co.	Polystyrene	66.5	2.55	1.97	3000	0.011
Thurane	Wyandotte Chem. Co., Wyandotte, Mich.	Polyurethane	70	2.06	2.04	710	0.019

† After Knott *et al.*[16]
‡ At 1 Gcps.

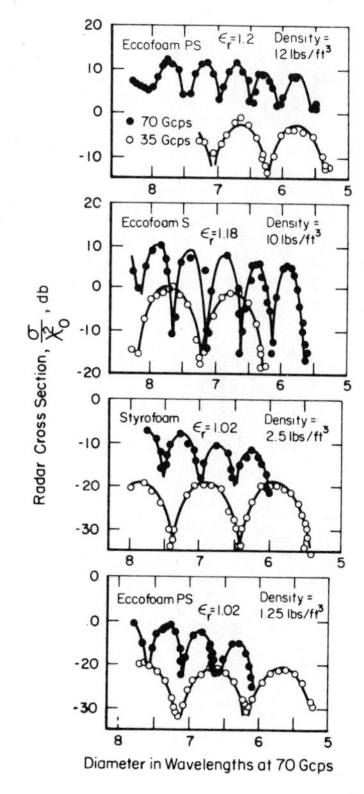

Fig. 11-17. Experimentally measured cross sections of polyfoam columns as a function of column diameter and dielectric constant [after Blore[19]].

Figure 11-17 shows good agreement between Eq. (11.6-14) and experimental data for columns constructed of four different plastic foams.[19]

An overall estimate of plastic foam column performance as a cylindrical target support is shown in Figure 11-18. This figure uses styrofoam parameters, and assumes the target cross section is at least 10 db greater than the column cross section (measurement errors limited to 3 db). Three frequency regions are distinguishable in Figure 11-18: (1) a region in which it is possible to "tune" the column for minimum return at a given frequency; (2) the Rayleigh region in which no tuning is possible; and (3) a transition region. Cylindrical columns are assumed, with a maximum obtainable diameter of 20 inches.

11.6.4.2. *Dielectric Suspension Lines*

A typical target arrangement using dielectric suspension lines is shown in Figure 11-19. The backscatter for the towers can be reduced by shaping; by using RAM; by placing the towers in null regions of the antenna patterns; and by placing the towers at different ranges from that of the target, so

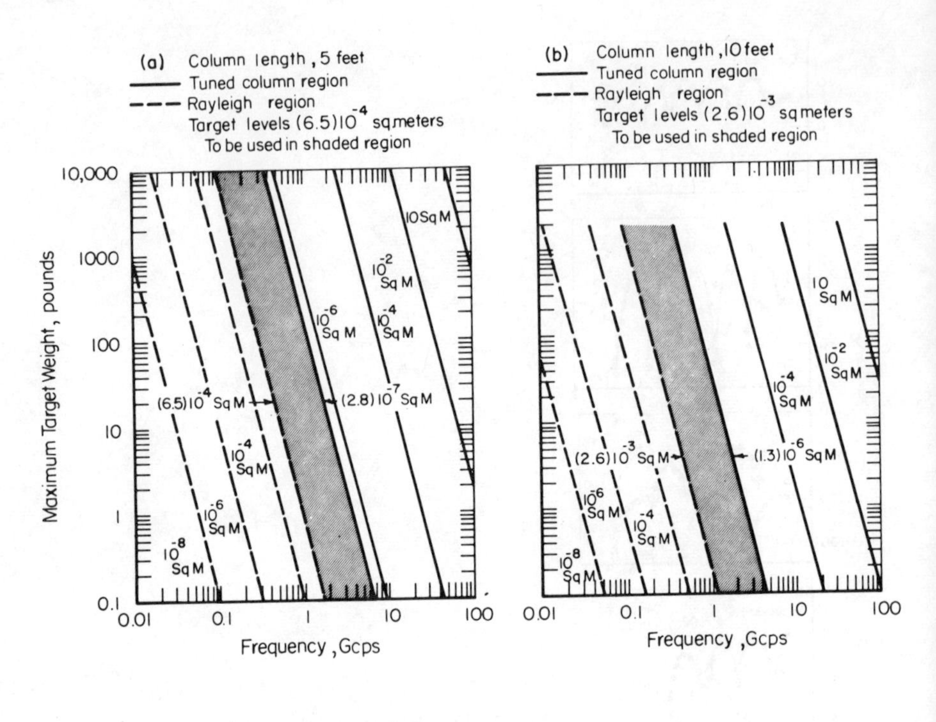

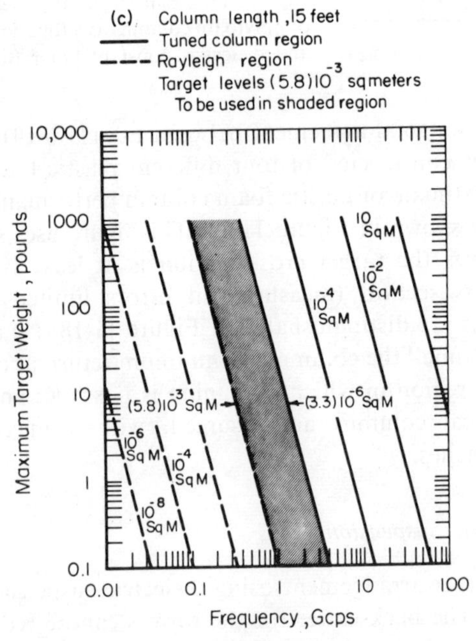

Fig. 11-18. Performance estimates for cellular plastic columns [after Freeny[17]].

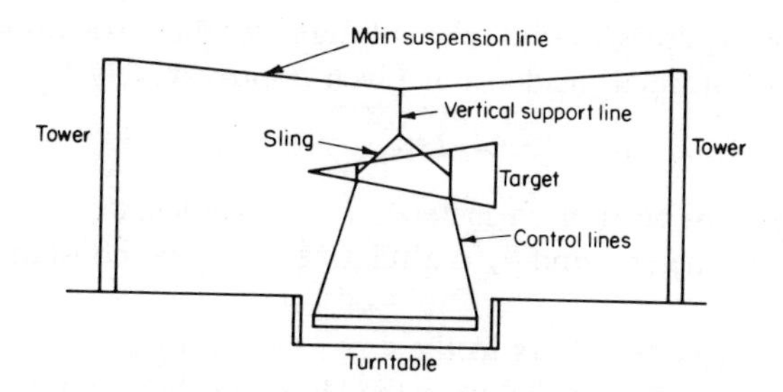

Fig. 11-19. Conventional suspension support.

that range gating can be used to discriminate against the tower return. The main horizontal suspension line is usually placed above the main antenna beam and at an oblique aspect angle to the direction of illumination. This suspension method is generally unsuitable for bistatic measurements.

Figure 11-20 shows breaking strengths for lines of several materials. Lines of dacron and polypropylene have also been used as target supports.[20]

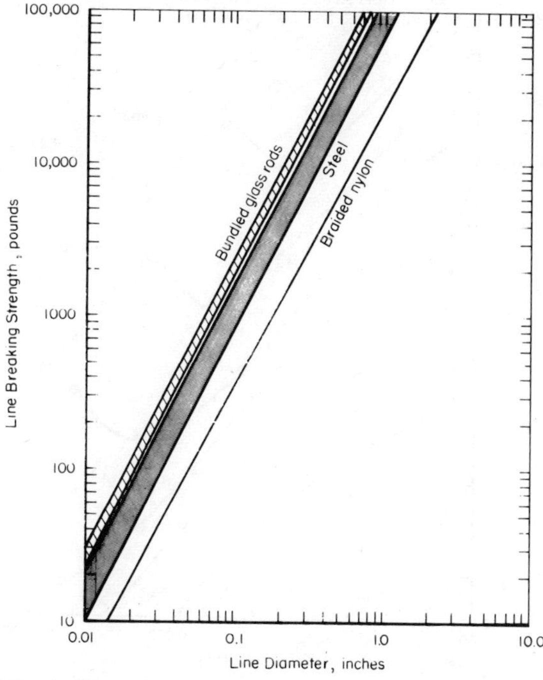

Fig. 11-20. Support-line breaking strength as a function of diameter.

The backscattering cross section of small-diameter dielectric lines for parallel polarization and normal incidence is given approximately by[17]

$$\sigma = (\pi/64)\, L^2(\epsilon_r - 1)^2\, (k_0 D)^4 \tag{11.6-15}$$

where σ is the cross section, in meters^2; L the line length, in meters; D the line diameter, in meters; and ϵ_r the relative dielectric constant of the line material.

The principle contributors to the cross section of a suspension support system are the vertical line (between the sling and horizontal line) and the sling. Figure 11-21 presents experimental estimates of the nylon vertical-line breaking strengths required to support targets of various cross sections as a function of frequency.[20] The ratio of target cross section to line cross section was assumed to be 10. These data were obtained with the incident field polarized along the support line. Figure 11-22 presents experimental estimates of conventional sling breaking strengths required to support targets of various cross sections as a function of frequency. The ratio of target cross section to sling cross section was assumed to be 10. Data for

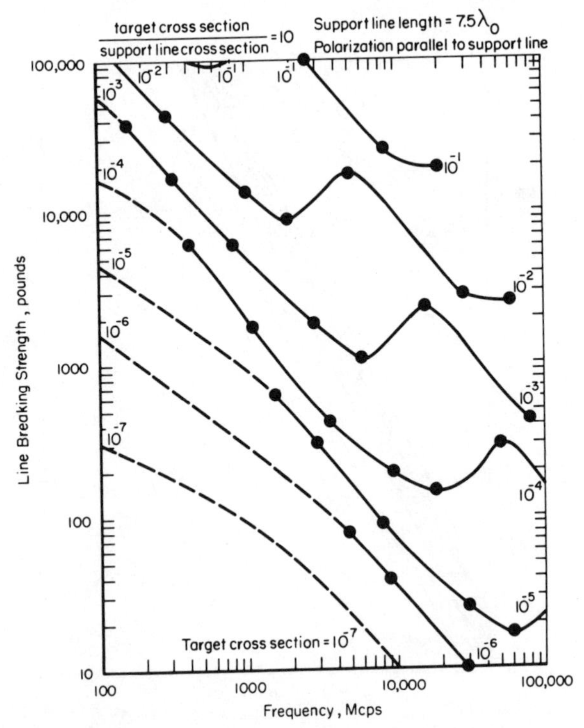

Fig. 11-21. Contours of constant cross section for Nylon vertical suspension lines.

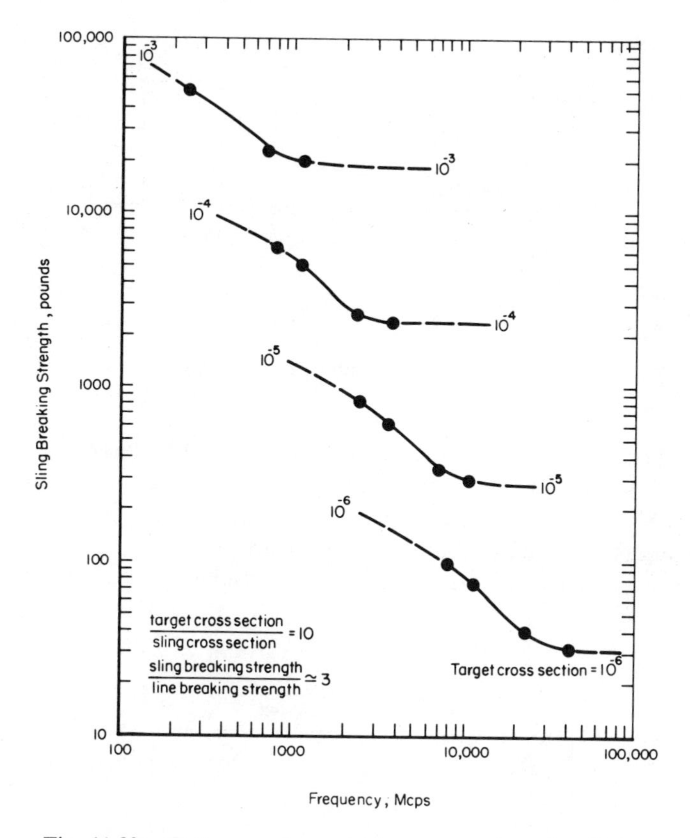

Fig. 11-22. Contours of constant cross section for Nylon support slings.

"worst-case" polarization was used for these results. A ratio of sling breaking strength to individual line breaking strength of about three can be used with the data of Figure 11-22.

11.6.5. Range-Geometry Effects

Measures used to reduce the level of the background cross section depend largely upon range geometry. With indoor ranges, as much of the extraneous scattered energy as possible is dissipated in absorbing materials. With outdoor ranges, the geometry generally is arranged so that the extraneous energy diverges or is reflected away from the region of measurement. If the outdoor range is situated over a flat region devoid of buildings and trees, the principal source of scattered energy, other than the target and its supporting structures, is the ground. If the surface of the ground is flat and

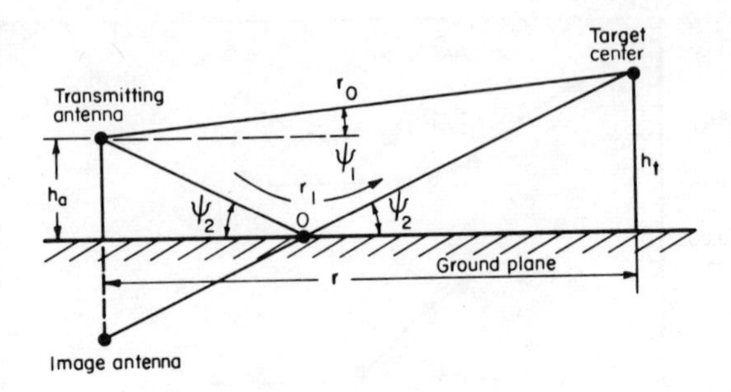

Fig. 11-23. Ground plane geometry.

uniform between the illuminating antennas and the target, and if the frequency is sufficiently high, the incident field at the target can be considered the sum of a direct wave and a specularly-reflected wave from the ground.

Distances involved in reflectivity measurements are short enough so that planar geometry (Figure 11-23) may be assumed. Figure 11-24 gives a

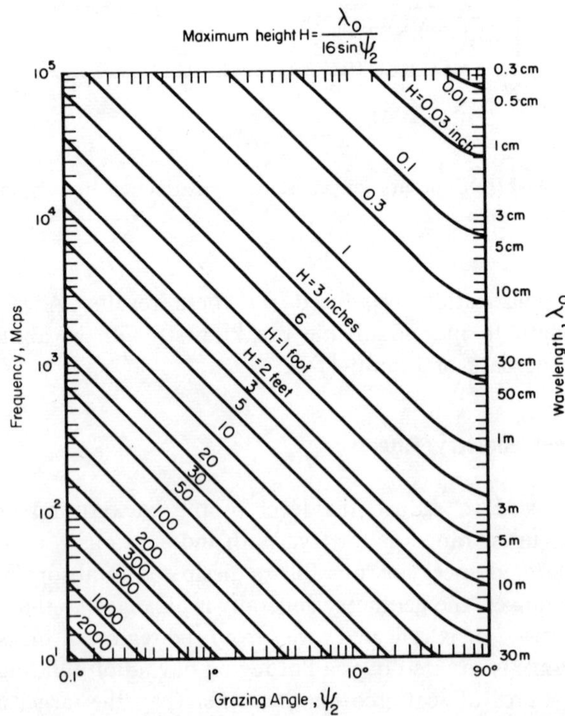

Fig. 11-24. Criterion for roughness of a reflecting surface [after Reed and Russell[21]].

smoothness criterion for describing ground reflection by a single ray. Using ray theory, the reflected wave can be described by

$$\begin{bmatrix} E_\parallel^R \\ E_\perp^R \end{bmatrix} = \begin{bmatrix} R_\parallel & 0 \\ 0 & R_\perp \end{bmatrix} \begin{bmatrix} E_\parallel^i \\ E_\perp^i \end{bmatrix} \tag{11.6-16}$$

where $E_\parallel^R$ is the parallel-polarized component of reflected field; $E_\perp^R$ the perpendicular-polarized component of reflected field; $E_\parallel^i$ the parallel-polarized component of incident field; $E_\perp^i$ the perpendicular-polarized component of incident field; $R_\parallel$ the reflection coefficient for parallel polarization; and $R_\perp$ the reflection coefficient for perpendicular polarization. Both $R_\parallel$ and $R_\perp$ are functions of the grazing angle ψ and the electrical properties of the reflecting medium:

$$R_\parallel = \frac{(\sin \psi) - \sqrt{m^2 - \cos^2\psi}}{(\sin \psi) + \sqrt{m^2 - \cos^2\psi}} \tag{11.6-17}$$

and

$$R_\perp = \frac{m^2(\sin \psi) - \sqrt{m^2 - \cos^2\psi}}{m^2(\sin \psi) + \sqrt{m^2 - \cos^2\psi}} \tag{11.6-18}$$

where the index of refraction m is defined by

$$m^2 = \frac{\epsilon + i(\sigma/\omega)}{\epsilon_0} = \epsilon_r + i\frac{\sigma}{\omega\epsilon_0} \tag{11.6-19}$$

with ϵ_0 the permittivity of free space; ϵ the permittivity of the reflecting medium; ϵ_r the relative permittivity of the reflecting medium; σ the conductivity of the reflecting medium; $\omega = 2\pi \times$ carrier frequency. Typical values for UHF frequencies (300–3000 Mcps) are shown in Table 11-8. Figures 11-25 to 11-27 show curves of the magnitude R and phase ϕ of $R_\perp$ and $R_\parallel$ evaluated for reflection from three types of ground at various frequencies. It is seen that $R_\perp$ and $R_\parallel$ both approach the value 1, $\angle 180°$ as the grazing angle goes to zero.[21]

For frequencies sufficiently low, due to the earth's dielectric properties, the possibility also exists for a surface wave component of the incident field.

Table 11-8

Material	ϵ_r	σ (mho-meters/meter^2)
Wet earth	5–30	10^{-1}–10^{-3}
Dry earth	2–5	10^{-4}–10^{-5}

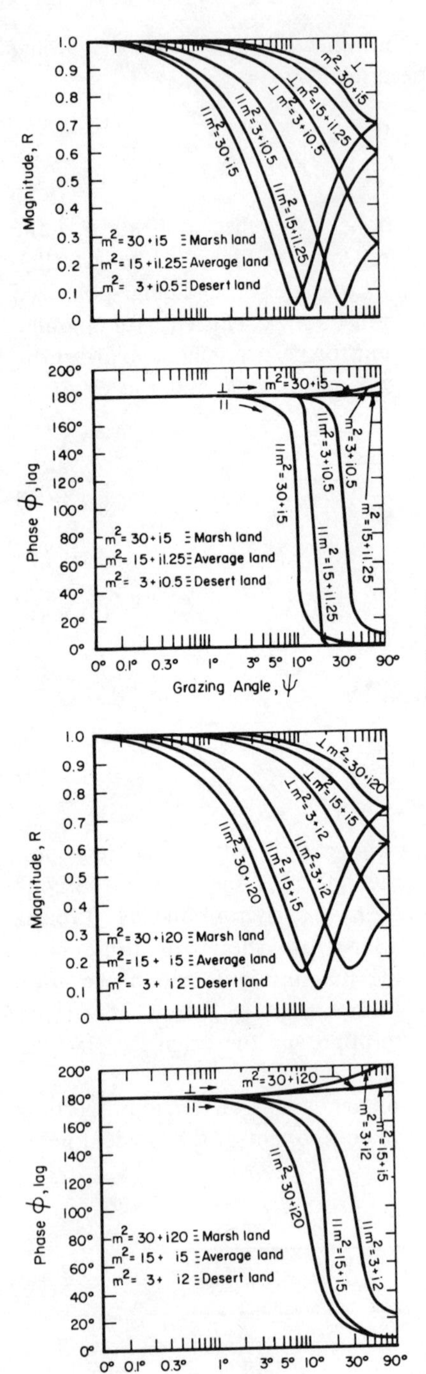

Fig. 11-25. Magnitude and phase angle of reflection coefficient as a function of grazing angle for smooth land. Frequency > 5000Mcps [after Reed and Russell[21]].

Fig. 11-26. Magnitude and phase of reflection coefficient as a function of grazing angle for smooth land. Frequency: 400 Mcps [after Reed and Russell[21]].

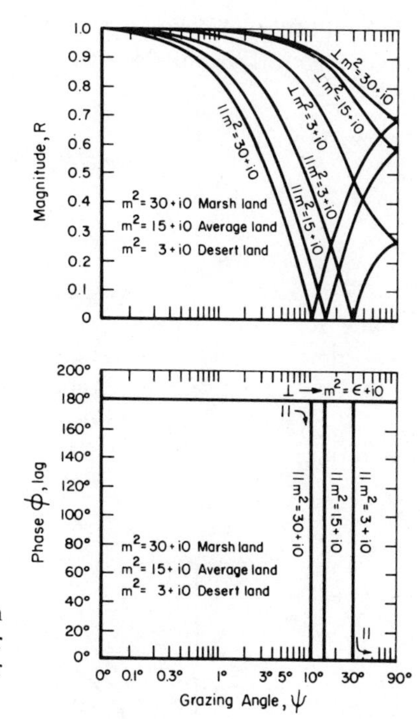

Fig. 11-27. Magnitude and phase of reflection coefficient as a function of grazing angle for smooth land. Frequency: 100 Mcps [after Reed and Russell[21]].

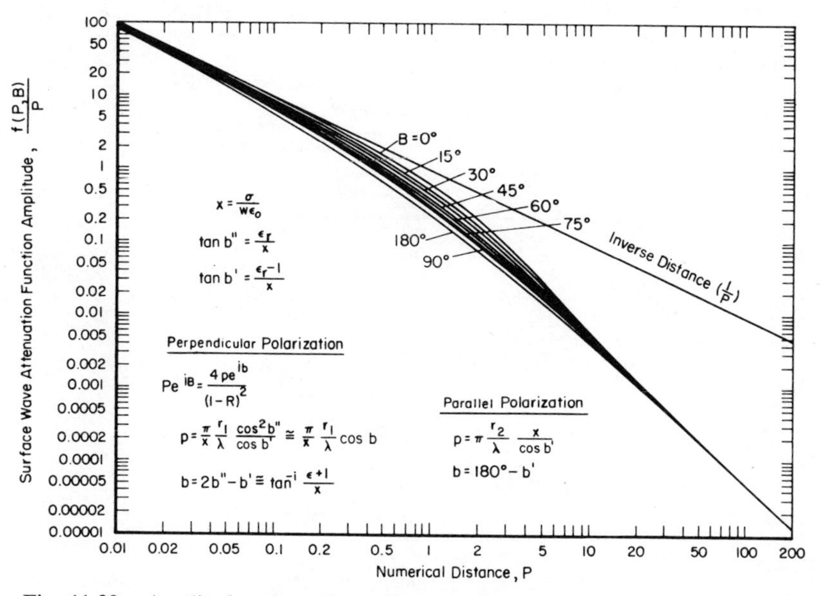

Fig. 11-28. Amplitude of surface-wave attenuation function versus numerical distance over a plane earth [after Norton[22]].

The amplitudes of field components due to the direct wave, ground-reflected wave, and surface wave for an isotropic radiator are related by (see Figure 11-23)

$$E = \frac{E_0}{r}\{\cos\psi_1 \exp(ik_0r_0) + R_i \cos\psi_2 \exp(ik_0r_1) + (1 - R_i) f(P, B) \cos\psi_2 \exp[-i(k_0R_1 + \phi)]\} \quad (11.6\text{-}20)$$

with R_i is the plane-wave reflection coefficient ($i = \parallel$ or $\perp$); E is the field amplitude for either parallel or perpendicular polarization; $\cos\psi_1 \exp(ik_0r_0)$ gives the relative amplitude of the direct wave; $R_i \cos\psi_2 \exp(ik_0r_1)$ gives the relative amplitude of the ground-reflected wave; and $(1 - R_i) f(P, B) \cos\psi_2 \exp[i(k_0r_1 + \phi)]$ gives the relative amplitude of the surface wave, with $f(P, B)\, e^{-i\phi}$, the surface-wave attenuation function,

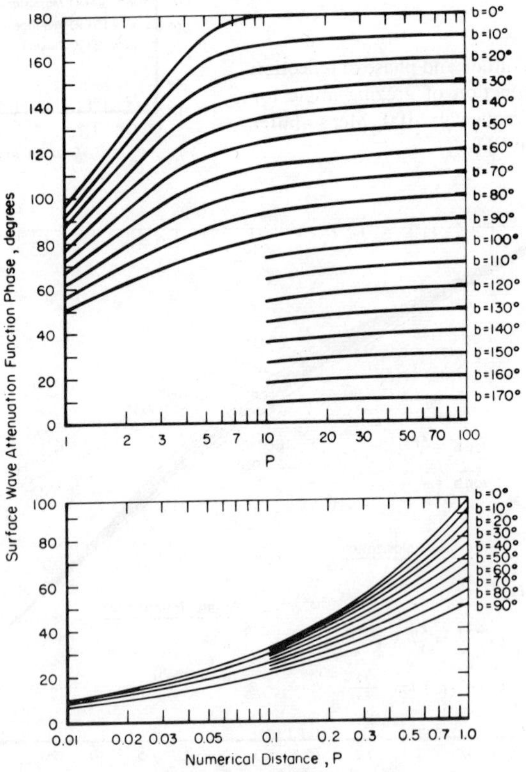

Fig. 11-29. Phase of surface-wave attenuation function versus numerical distance over a plane earth [after Norton[22]].

Table 11-9. Distance in Statute Miles at Which the Surface-Wave Field Strength is 40 db below the Free-Space Field Strength

Surface	ϵ_r	σ	Perpendicular Polarization (Mcps)			Parallel Polarization (Mcps)		
			30	100	1000	30	100	1000
Average land	15	0.03	4.8	1.0	0.10	0.0087	0.0039	0.0004
Desert land	3	0.01	1.4	0.3	0.03	0.032	0.021	0.0030

plotted in Figures 11-28 and 11-29. The parameters P and B are defined by

$$Pe^{iB} = 4pe^{ib}/(1 - R_i)^2 \tag{11.6-21}$$

where

$$p = \frac{k_0 r_1}{2} \frac{\cos^2 b''}{x \cos b'} \approx \frac{kr_1}{2} \frac{\cos b}{x}$$
$$b = 2b'' - b' \tag{11.6-22}$$

for perpendicular polarization;

$$p = \frac{k_0 r_1}{2} \frac{x}{\cos b'}, \qquad b = 180° - b' \tag{11.6-23}$$

for parallel polarization; and

$$x = \sigma/\omega\epsilon_0 = 60\sigma\lambda_0 \tag{11.6-24}$$

$$\tan b' = (\epsilon_r - \cos^2 \psi_2)/x \approx (\epsilon_r - 1)/x \tag{11.6-25}$$

$$\tan b'' = \epsilon_r/x \tag{11.6-26}$$

As an example of the distances at which the surface wave is negligible, Table 11-9 gives the distances at which the surface-wave amplitude is 40 db

Table 11-10. Transmitting- or Receiving-Antenna Height (in Feet) Which Yields Surface-Wave Field Strength 40 db below Free-Space Field Strength, at about 0.73 of the Distances Given in Table 11-9

Surface	Perpendicular Polarization (Mcps)				Parallel Polarization (Mcps)			
	30	100	300	1000	30	100	300	1000
Average land	257	64.6	31.0	6.3	10.94	4.05	1.39	0.42
Desert land	140	33.4	11.1	3.3	20.8	9.58	3.62	1.11

below the free-space wave amplitude for antennas located on the earth's surface.[21] Table 11-10 shows heights at which the surface-wave amplitude is 40 db below the free-space wave value. Table 11-10 assumes h_a and h_t are both at the same height, and that the range is about 0.73 of those given in Table 11-9.[21]

Neglecting the surface wave and using the approximations

$$r_1 = [r^2 + (h_t + h_a)^2]^{1/2} \approx r + [(h_t + h_a)^2/2r] \tag{11.6-27}$$

and

$$r_0 = [r^2 + (h_t - h_a)^2]^{1/2} \approx r + [(h_t - h_a)^2/2r] \tag{11.6-28}$$

then

$$r_1 - r_0 \approx 2h_t h_a/r \tag{11.6-29}$$

and at the target center,

$$E_i \approx \frac{E_0}{r_0} \exp(ik_0 r_0) + \frac{R_i E_0}{r_1} \exp(-ik_0 r_1) \tag{11.6-30}$$

where $i = \parallel$ or $\perp$. Assuming $r_0 \approx r_1 \approx r$ in amplitude terms, and $r_1 - r_0 \approx 2h_t h_a/r$ in phase terms, then

$$E_i \approx \frac{E_0}{r} \exp(ik_0 r_0) \left[1 + r_i \exp\left(ik_0 \frac{h_t h_a}{r}\right)\right] \tag{11.6-31}$$

If $R_i \approx -1$, then

$$E_i \approx i\frac{2E_0}{r} \exp\left[ik_0\left(r_0 + \frac{h_t h_a}{r}\right)\right] \sin\left(\frac{k_0 h_t h_a}{r}\right) \tag{11.6-32}$$

so that

$$| E_i | \approx (2E_0/r) \sin(k_0 h_t h_a/r) \tag{11.6-33}$$

and

$$\arg E_i \approx \frac{\pi}{2} - k_0 r_0 - \frac{k_0 h_t h_a}{r} = \frac{\pi}{2} - k_0 r - k_0 \frac{h_t^2 + h_a^2}{2r} \tag{11.6-34}$$

Lobes in the field amplitude occur for

$$h_t h_a = n\lambda_0 r/4, \qquad n = 1, 3, 5,... \tag{11.6-35}$$

Figure 11-30 shows the relationship between antenna height, range, and height of the first lobe above the ground plane. Figure 11-31 shows grazing angles corresponding to this relationship. Figure 11-32 compares the results of Eq. (11.6-32) with field-probe data obtained on the RATSCAT† ground-plane range at 3274 Mcps.[23] Similar agreement was obtained for frequencies in L, C, and X bands.

† Radar Target Scatter site located at Holloman Air Force Base, New Mexico.

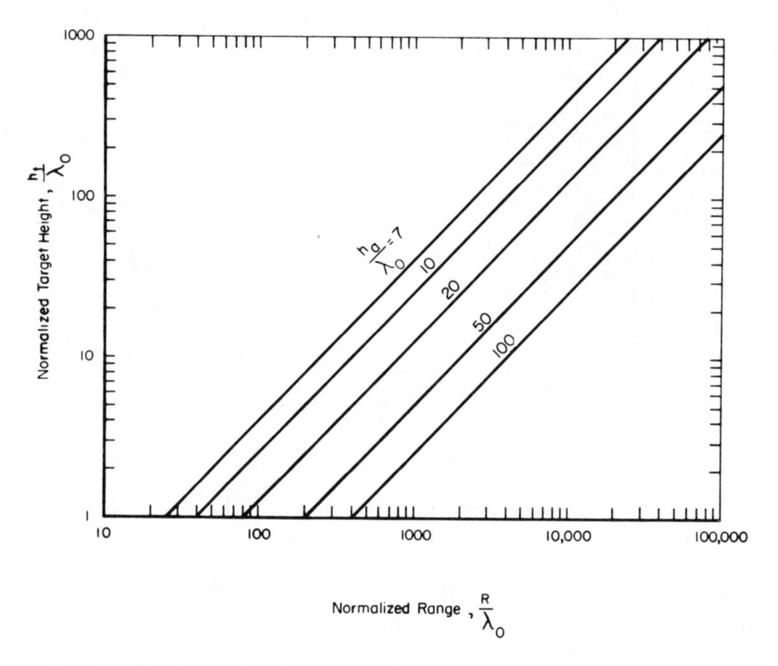

Fig. 11-30. Target height as a function of range and antenna height for a ground-plane range.

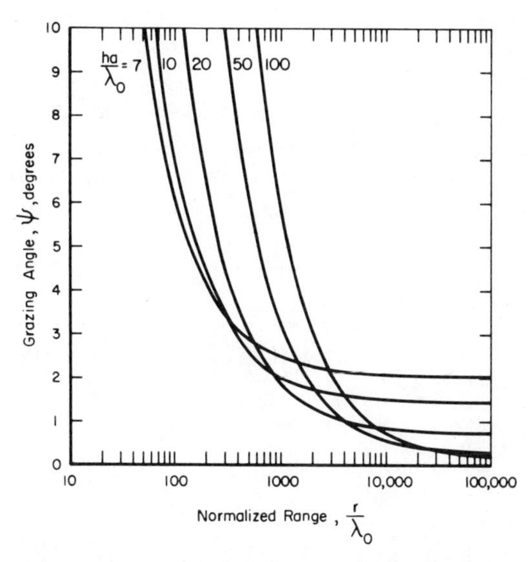

Fig. 11-31. Grazing angle as a function of target height and antenna height for a ground-plane range.

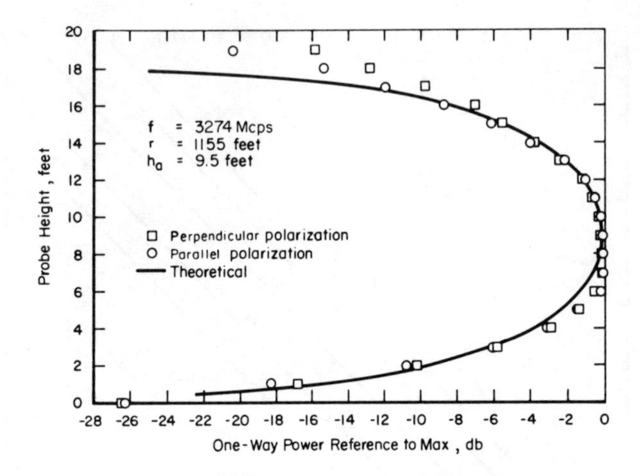

Fig 11-32. Field-power pattern above a ground-plane range.

When both principle plane polarizations are transmitted simultaneously, the ground-plane effects can be described by the matrix equation:

$$\begin{bmatrix} E_{\parallel}{}^{i} \\ E_{\perp}{}^{i} \end{bmatrix} \approx \frac{e^{ikr}}{r} \begin{bmatrix} 1 + R_{\parallel} \exp\left(ik_0 \dfrac{2h_t h_a}{r}\right) & 0 \\ 0 & 1 + R_{\perp} \exp\left(ik_0 \dfrac{2h_t h_a}{r}\right) \end{bmatrix} \begin{bmatrix} E_{\parallel}{}^{0} \\ E_{\perp}{}^{0} \end{bmatrix} \tag{11.6-36}$$

where $E_{\perp}{}^{i}$, $E_{\parallel}{}^{i}$ are components of the incident field at the target, and $E_{\perp}{}^{0}$, $E_{\parallel}{}^{0}$ are components of the transmitted field. Since $R_{\perp} \neq R_{\parallel}$, the polarization of the incident field will be different from the polarization of the transmitted field, except for the two cases when $E_{\parallel}{}^{0} = 0$, $E_{\perp}{}^{0} \neq 0$, and

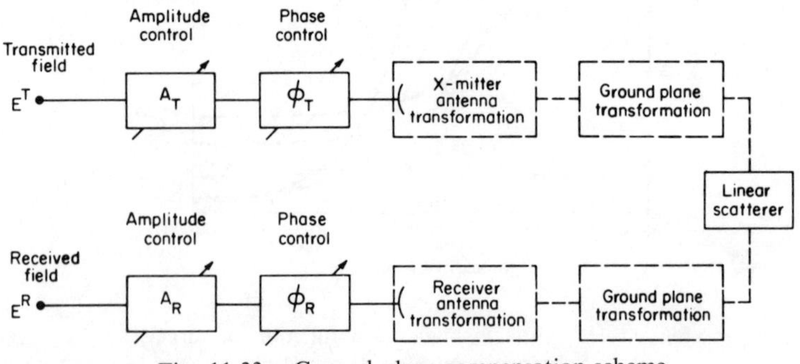

Fig. 11-33. Ground-plane compensation scheme.

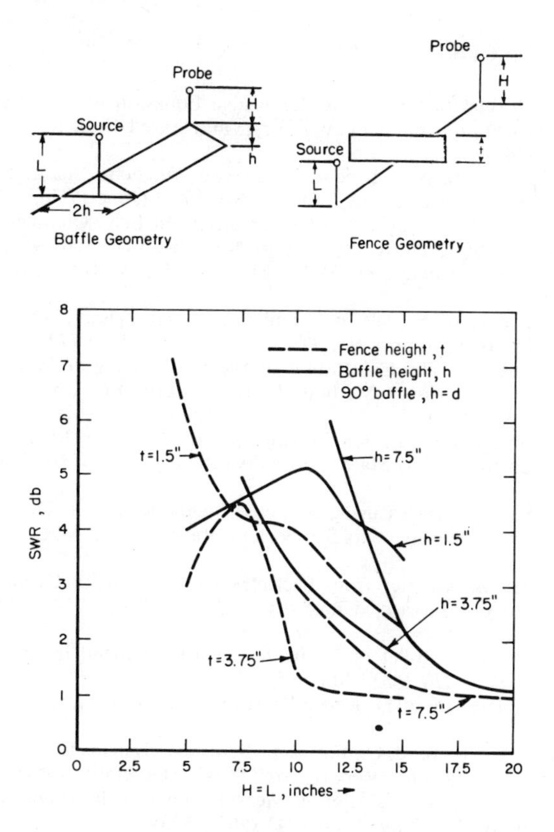

Fig. 11-34. Field-probe data for a transverse radar fence and a longitudinal baffle ($\lambda = 0.305$ inch, Range $= 62.2$ inch) [after Kay[25]].

$E_{\perp}{}^{0} = 0$, $E_{\parallel}{}^{0} \neq 0$. Consequently, on a ground-plane range, to illuminate the target with a circularly-polarized field or a linearly-polarized field other than parallel or perpendicular, an elliptically-polarized field must be transmitted to compensate for the depolarizing effects of the ground plane. Figure 11-33 shows a block diagram of the compensation scheme used at RATSCAT.[24]

On certain ranges, the ground-reflected energy is suppressed by placing transverse radar fences or longitudinal baffles between the source and the target. Figure 11-34 shows measured field-probe data comparing the effects of a longitudinal baffle and a single transverse fence.[25] The SWR was determined by measuring the interference pattern of the field as a function of height at the target region.†

† SWR $= 20 \log_{10} (|E_d| + |E_r|/|E_d| - |E_r|$, with E_d the direct field, and E_r the reflected field.

11.7. REFERENCES

1. Sinclair, G., Modification of the Radar Range Equation for Arbitrary Targets and Arbitrary Polarization, Antenna Laboratory, Ohio State University, Report No. 302-19 (September 1948).
2. Kennaugh, E. M., Effects of Type of Polarization on Echo Characteristics, Antenna Laboratory, Ohio State University, Report No. 389-4 (June 1950), ATI-88348.
3. Kennaugh, E. M., Effects of Type of Polarization on Echo Characteristics, Antenna Laboratory, Ohio State University, Report No. 389-13 (March 1952), AD 2493.
4. Stratton, J. A., *Electromagnetic Theory*, McGraw-Hill Book Co., New York (1940), p. 488.
5. Sinclair, G., Theory of Models for Simulating Radar Echoes, Antenna Laboratory, Ohio State University, Report No. 302-3 (July 1947), ATI-30770.
6. Kouyoumjian, R. G., The Calculation of the Echo Areas of Perfectly Conducting Objects by the Variational Method, Ph.D. Dissertation, Ohio State University, Columbus, Ohio (1953).
7. Hines, J. N., and Tice, T. E., On Investigation of Reflection Measuring Equipment, Antenna Laboratory, Ohio State University, Report No. 478-13 (June 1953), AD 12901.
8. Iizrika, K., Methods of Measuring Scattered Fields, Rome Air Development Center, Radar Reflectivity Measurements Symposium, Report No. RADC-TDR-64-25, Vol. I (1964), p. 2.
9. Tang, C. C. H., Backscatter from Dielectric-Coated Infinite Cylindrical Obstacles, Cruft Laboratory, Harvard University, AFCRC-TN-56-754 (September 1956), AD 98809.
10. Kouyoumjian, R. G., and Peters, L., Jr., Range Requirements in Radar Cross Section Measurements, *Proc. IEEE* **53**:920 (1965).
11. Mentzer, J. R., The Use of Dielectric Lenses in Reflection Measurements, *Proc. IRE* **41**:252 (1953).
12. Carswell, A. I., and Richard, C., Focused Microwave Systems for Plasma Diagnostics, RCA Victor Company, Ltd., Report No. 7-801-32 (December 1964).
13. Carswell, A. I., Microwave Scattering Measurements in the Rayleigh Region Using a Focused-Beam System, *Can. J. Phys.* **43**:1962 (1965).
14. Falk, B., A Method of Phase Measurement for Pulsed Radar Systems, Rome Air Development Center, Radar Reflectivity Measurements Symposium, Report No. RADC-TDR-64-25 (April 1964).
15. Radar Cross-Section Target Supports—Plastic Materials, Rome Air Development Center, Report No. RADC-TDR-64-381 (June 1964), AD 608252.
16. Knott, E., Plonus, M., and Senior, T., Design of Foamed-Plastic Target Supports, *Microwaves* **3**:38 (1964).
17. Freeny, C. C., Target Support Parameters Associated with Radar Reflectivity Measurements, *Proc. IEEE* **53**:929 (1965).
18. Theoretical and Experimental Investigation of a Technique for Reducing Extraneous Signals in Radar Scattering Measurements, Rome Air Development Center, Report No. RADC-TDR-64-418 (July 1964), AD 607136.
19. Blore, W. E., The Radar Cross Section of Polyfoam Towers, *IEEE Trans. Antennas Propagation* **AP-12**:237 (1964).
20. Radar Cross Section Target Supports—Metal Columns and Suspension Devices, Rome Air Development Center, Report No. RADC-TDR-64-382 (June 1964), AD 606122.
21. Reed, H. R., and Russell, C. M., *Ultra High Frequency Propagation*, Boston Technical Publishers, Boston (1964), p. 237.
22. Norton, K. A., The Calculation of Ground-Wave Field Intensity over a Finitely Conducting Spherical Earth, *Proc. IRE* **29**:623 (1941).
23. Investigation of Measurement Errors of the RAT SCAT Cross Section Facility, Rome Air Development Center, Report No. RADC-TDR-64-397 (July 1964), AD 608429.

24. Freeny, C. C., Experimental and Analytical Investigation of Target Scattering Matrices, Rome Air Development Center, Report No. RADC-TR-65-298 (December 1965), AD 477070.
25. Kay, A. F., A Comparison between Longitudinal Baffling and Transverse Fence for Reducing Range Ground Scattering, Rome Air Development Center, Radar Reflectivity Measurements Symposium, Report No. RADC-TDR-64-25, Vol. I (1964), p. 110.

INDEX